Springer Series in Solid and Structural Mechanics

Volume 3

Series Editors

Michel Frémond, Rome, Italy
Franco Maceri, Rome, Italy

For further volumes:
http://www.springer.com/series/10616

Vincenzo Vullo

Circular Cylinders and Pressure Vessels

Stress Analysis and Design

Springer

Vincenzo Vullo
Department of Mechanical Engineering
University of Rome Tor Vergata
Rome
Italy

ISSN 2195-3511 ISSN 2195-352X (electronic)
ISBN 978-3-319-00689-5 ISBN 978-3-319-00690-1 (eBook)
DOI 10.1007/978-3-319-00690-1
Springer Cham Heidelberg New York Dordrecht London

Library of Congress Control Number: 2013946471

Printed on acid-free paper

Springer is part of Springer Science+Business Media (www.springer.com)

Vir fuit, hic, ortu Samius …

… isque, licet caeli regione remotos,
mente deos adiit et, quae natura negabat
visibus humanis, oculis ea pectoris hausit.
Cumque animo et vigili perspexerat omnia
cura,
in medium discenda dabat, coetusque
silentum
dictaque mirantum magni primordia mundi
et rerum causas et, quid natura, docebat.

Ovidius P. Naso, Metamorphoses, XV, 60–68

To my wife Maria Giovanna
and my sons Luca and Alberto

Preface

Sizing circular cylindrical bodies of annular cross-section subjected to surface forces and thermal loads is one of the classic topics of Machine Design Theory and Methodology. And it is far from being straightforward, as a knowledge of elasticity theory is not sufficient for thorough analysis. It is often necessary to broaden the design horizons, to investigate what happens at pressures that stress the material beyond its yield point and at thermal loads that give rise to creep.

Many mechanical design textbooks and treatises devote ample space to calculating the strength of these structural components. However, the focus is often limited to stresses in the linear elastic field, and is seldom extended to load conditions that cause plastic flow as a result of stressing the material beyond the yield point, and/or viscous flow associated with creep. In current undergraduate and post-graduate programs, moreover, the subject is split up, covering the basic, specialized, and most advanced aspects in the first-cycle, second-cycle and third-cycle, and doctoral programs respectively.

This book has the two-fold aim of meeting the needs of university education and those of the engineering profession. In the first area, the goal is to provide a link between the material covered in the textbooks on the strength of materials intended for students in three-year first-cycle degree programs, and that addressed in the more advanced monographs on elasticity theory, plasticity theory, and creep theory used in second-cycle and third-cycle or doctoral programs. In the second area, the objective is provide practitioners and professional engineers working in research, industry, and services with an advanced knowledge of the subject that can serve as a basis for designing products that are safe and technologically sophisticated, or for developing innovative applications for these structural components.

Anyone who has ever worked with analysis and design of these components is well aware that in many actual engineering applications, cylindrical bodies are part of complex structures which must not only be analyzed as a whole, but must also be considered in terms of the interactions between their component parts and the resulting stress and strain states, including those that arise in dynamic conditions such as impact loading, vibration, high- and low-cycle fatigue, and so forth. Very often, the theoretical cylindrical geometry with annular cross-section is disregarded, both at the level of the overall structure, where it is obtained from roll formed tubular segments welded together (with longitudinal weld beads along the length of each segment, and circumferential beads joining the segments) and at the local level (e.g., where holes are provided for inserting pipe segments, here again with welded joints). In such cases, the use of numerical methods such as FEA is inescapable.

All of the problems presented in this text are approached and solved using theoretical methods, focusing chiefly on the analytical-methodological aspects. Dealing with problems characterized by generalized axis-symmetry, it was possible to obtain accurate closed-form solutions, and general analytical expressions for stresses, strains, and deformations (displacements and rotations). In fact, the analytical solutions allow any desired parametric study, and quick calculations, and provide the basis to evaluate the error of the numerical solutions obtained with approximate numerical methods.

The analytical solutions proposed here are formulated so as to be of interest not only to academics, as was once almost exclusively the case, but also to the designer who deals with real engineering problems. This is because such solutions, though sophisticated and complex, have become immediately useable by practitioners in the field thanks to today's computers, which can readily solve demanding equations. For the reader, moreover, these solutions provide the grounding needed to achieve a full understanding of the requirements laid out in the standards applying in this area. In the majority of cases, the analytic relations developed for use in the design analysis and/or in the response analysis (or verification analysis) are also represented in graphical form. This gives the reader an immediate grasp of the underlying physical phenomena that these formulas explain, and clarifies exactly which major magnitudes must be borne in mind by the designer.

In dealing with certain topics, including, for example, the interaction between the substructures making up a complex structure, we have included fully developed exercises to draw the reader's attention to the aspects that are of greatest interest to the designer, as well as to clarify the calculation procedure. We have not considered it necessary to provide such exercises in cases where they could be solved immediately with the relations presented in each chapter, as it was felt that they would add nothing to an understanding of the text.

Each topic is addressed from a theoretical standpoint, but in such a way as not to lose sight of the physical phenomena at the basis of the strain mechanics of circular cylinders bodies under various types of loads. Analysis proceeds in steps, with particular attention to each stress and strain state. The material is thus organized so that the knowledge gained in the beginning chapters provides the grounding needed to understand the topics covered in the chapters that follow. Consequently, the problems that are normally encountered in studying the elasto-plastic problem and the viscous problem associated with creep are significantly reduced, once a complete mastery of the elastic problem is achieved.

This text, which is also intended for the students in my course on Machine Design Theory and Methodology (*Progettazione Meccanica e Costruzione di Macchine*) at the Università di Roma Tor Vergata, consists of an introduction which describes and discusses the general hypotheses and assumptions underlying the proposed theoretical approaches and solution methods, plus 11 chapters.

The first three chapters deal with thin-walled circular cylinders under internal and/or external pressure and stressed in the linear elastic field, and their design analysis and response analysis on the basis of different strength theories. The instability affecting these components as a result of external pressure alone and external pressure with initial geometric imperfections is also addressed, as are other types of instability such as global and local instability resulting from axial load, instability due to torsional load and flexural ovalization instability during bending.

The Chap. 4 covers thick-walled circular cylinders under internal and/or external pressure, again stressed in the linear elastic field, giving relations for design analysis and response analysis on the basis of several strength theories; the results obtained are then compared to provide information that can be used in component design. The Chap. 5, which also deals with the linear elastic field, addresses the problem of circular cylinders assembled with a radial interference fit and of multilayer structures, as well these components' optimization on the basis of their constituent materials and strength properties. The influence of centrifugal load on the force-fit assembly is also analyzed.

The Chap. 6 covers thick-walled circular cylinders subjected in the elastic field to a temperature gradient as a result of radial heat flow according to Fourier's law of heat conduction. The relations for thin-walled bodies are determined as a limiting case of those for thick-walled bodies. The chapter also discusses the edge effects occurring at the free ends of a thin-walled circular cylindrical body of finite length, and of transient thermal load in a solid cylindrical structure.

The Chaps. 7–9 are devoted to the nonlinear problems of pressure loading beyond initial yielding. Specifically, Chaps. 7 and 8 address the problem of thick-walled cylindrical bodies made of a material showing elastic-perfectly plastic behavior, determining the stress and strain states starting from the pressure leading to the onset of the plastic state at the inner radius, up to the pressure at which the entire wall thickness is in the fully plastic state, as well as the residual stress states after the autofrettage pressure is released. These chapters also cover prestressing in terms of overstressing and overstrain, both with and without reverse yielding, as well as with cyclic repressurizing during a more complex autofrettage process, extending these concepts to composite multilayer structures.

In the Chap. 9, the restrictive assumption of an elastic-perfectly plastic material is relaxed and the problem is generalized in order to deal with real materials that undergo strain hardening when stressed beyond yield, or that exhibit viscous behavior associated with creep during service at high temperatures in steady state conditions. Here again, discussion is extended to multilayer structures. All three of these chapters provide relations whereby the designer will be able to develop structures capable of achieving high performance after an optimized preliminary prestressing process.

The Chaps. 10 and 11 address the bending problems arising from edge effects, or in other words resulting from the discontinuities in geometry and form that cause the substructures of a complex structure to have different responses in terms of stiffness. The Chap. 10 discusses the bending theory for cylindrical shells of annular cross-section under axisymmetric loads, presenting the fundamental relations. The chapter deals with both long and short cylindrical shells, i.e., those having axial lengths that are greater or lesser than the so-called extinction length or decay distance, under the concentrated and distributed axisymmetric loads that are of interest for design purposes.

The Chap. 11 covers cylindrical pressure vessels with flat ends and with dished ends whose generating meridians are hemispherical and semi-elliptical. To clarify its scope, both the membrane theory and the bending theory for double-curvature shells of revolution under axisymmetric loads are used. The stresses arising from discontinuities, which are superposed on the membrane stresses, are analyzed using an approximate bending theory which introduces well-known simplifying assumptions and hypotheses. This makes it possible to arrive at closed-form relations which are known to provide results of unquestioned utility for design.

The author would like to express his warmest thanks to Dr. Daniele Negretti, his highly promising student, for the helpfulness, enthusiasm, and meticulous care he showed in preparing the drawings and graphics for this book, and to Dr. Alberto Maria Vullo, his son, for the same reasons and for his valuable assistance in solving some of the systems of equations correlated with edge effects in circular cylindrical pressure vessels.

Vincenzo Vullo

Contents

Introduction

General

This volume addresses the stress and strain state in machine components and mechanical structures featuring axial symmetry in the broad sense (*generalized axisymmetry*), i.e., referring not only to the structure's geometrical shape (this is the so-called geometrical cylindrical axisymmetry, and the generic cross-section is thus circular or annular), but also to the distribution of surface forces (internal and external pressure), the thermal loads acting on the structure, the boundary conditions, the material's elastic properties, and if necessary the body forces. The textbook also describes methods for evaluating the strength of such structural components and thus ensure that they are correctly designed and meet the criteria specified for compliance checks in national and international standards. Here, analysis is restricted to structures consisting of homogeneous and isotropic material; for structures consisting of plastic or composite materials, the reader is referred to specialized texts.

The structures considered herein include, for example, pressure vessels, nuclear vessels, reactor liners, solid fuel elements, control rods, the reactors used in the chemical and petrochemical industries, the tubing and, more generally, the piping in manufacturing plants, the casings, tubing and drill pipes used in oil and gas fields, the piping in offshore drilling operations, the pipelines that carry oil and natural gas, heat exchanger tubes, the shrink-fit shaft/hub assemblies employed in the mechanical industry and the components of many other machines used in sectors spanning the gamut from aerospace to domestic appliances. Special topics that will be addressed in this area include composite material vessels, elastomer tubes, sleeves, box couplings, the petrochemical industry's storage tanks, coining dies, multilayer tanks and their optimization, the loosening of a shrink-fit shaft/hub assembly as a result of centrifugal load, limit analysis design, autofrettage, or the creation of a residual stress in an internally pressurized cylinder when the internal pressure is increased from the value causing initial yielding to the value causing yielding of the entire cross section (full yield), and the thermal stresses in long cylindrical bodies.

Today, the development of chemical and, in particular, nuclear installations and the growth of space propulsion and submarine exploration call for a more thorough understanding of the problems involved in designing the structures employed in these fields to ensure that safety requirements will be met under all operating conditions. We will never find ourselves dealing with simple circular cylinders, but with structures that are a part of far more complex systems, and thus feature closures (which may be flat, formed, or virtual), flanges for connection to piping systems, and other irregularities, with the resulting geometrical discontinuities at which stress concentrations occur.

In these cases, we must go beyond the standards, though compliance must always be guaranteed through engineering specifications. Even for simple geometries, in fact, the most advanced standards have certain limitations; for example, one of the most complete and long-established standards in existence today, the ASME code for designing ambient temperature pressure vessels, is based on assuming a linear elastic stress state. Consequently, a comparison of the working pressure and the collapse pressure, i.e., that pressure—a steady-state pressure, moreover—at which the vessel continues to yield due to plastic flow, will demonstrate that the code is much more conservative for thick-walled vessels than it is for those with thin-walled structures.

It is thus necessary to extend the analysis of stress and strain states with more elaborate theoretical models (for steady-state viscous flow, plastic flow, creep, etc.) and, where the latter prove to be too limited, we must go on to numerical models such as the finite difference, finite element and boundary element methods. All of the various analytical and numerical methods that can be used complement each other, and can coexist throughout the *design analysis* and *response analysis* procedures.

The problems that these procedures entail, as well as those involved in checking the compliance (compliance analysis) of circular cylinders under internal and external pressure and subjected to temperature gradients through the wall thickness are for the most part highly specific, not least because it is often necessary to strike a balance between contrasting needs. Very frequently, in fact, maximum performance must be guaranteed with minimum weight, while allowing for ever-higher operating pressures (which in some cases may vary rapidly over time) and thermal loads involving high gradients and elevated temperatures.

Because of the dangers that their structural collapse would pose, the stress and strain states of the circular cylinders considered in this volume must be analyzed as rigorously as possible to ensure that they are correctly sized. The design approach must also extend to a careful selection of materials, strict controls over production processes (welding in particular, as these structures rarely consist of a single piece), proof testing to failure, programmed maintenance and appropriate nondestructive tests for structural soundness at regular intervals. All of this results from the fact that these structures can be extremely hazardous, especially if they contain substances that are harmful to health or the environment or gases at high pressure, particularly if they are flammable.

Purpose and Basic Assumptions

In the following pages, we will address the issues involved in structural calculation and conceptual design of circular cylinders subjected to loads arising from pressures acting on the inner and outer radii as well as to non-uniform temperature distributions along the radius. For the sake of convenience, the circular cylinders considered here will be divided into the two families of thin-walled structures and thick-walled structures on the basis of the generally adopted convention which will be discussed in Chap. 1. However large or small the wall thickness of the circular cylinder may be, however, its stress and strain states will always be triaxial.

Given our concern with conceptual design, we will use analytical models of general validity that, by introducing simplifying assumptions, ensure that the real-world problem can be more readily addressed but in any case guarantee that the results thus obtained are meaningful in actual applications. Essentially, two simplifying assumptions will be introduced: axial symmetry in the broad sense (generalized axisymmetry), i.e., referring not only to the cylindrical body's geometrical shape, but also to the surface forces (pressures) and the thermal load acting on it, its boundary conditions and the material's elastic properties, and the assumption regarding the stress state and strain state (Sokolnikoff 1956, Timoshenko and Goodier 1970, Boresi and Chong 2000, Haupt 2000).

From the geometrical standpoint, however, assuming axisymmetry is not always entirely justified. Take, for example, the case of circular cylinders featuring weld beads, attachment holes for pipe sections and other non-uniformities, all of which make axisymmetry a rather rough approximation. If the irregularities are minor compared to the problem's significant magnitudes (the significant magnitudes or significant quantities are those magnitudes or quantities which characterize the mechanical behavior of a structure), it can first be assumed that the influence on the circular cylinder's stress state is negligible, introducing it later as a localized effect in the areas affected by the irregularities (Peterson 1974). This is the procedure followed for weld beads, for which the notch effect is considered. As indicated above, the assumption of generalized axisymmetry also includes that of the material's mechanical properties, which must thus be axisymmetrically orthotropic or isotropic. Only homogeneous and isotropic materials will be considered here.

The assumption concerning the stress state and the strain state will differ according to the individual problem to be addressed. More specifically, the plane stress state assumption ($\sigma_z = 0$) , is used for circular cylinders that can be regarded as rings, or in other words are of small axial thickness, whereas a plane strain state ($\varepsilon_z = 0$) is assumed for circular cylinders constrained at the ends or infinitely extended along the axis, and a strain state with $\varepsilon_z = const$ (the so-called generalized plane strain state) for circular cylinders of finite length and free at the ends.

If the axial dimensions of the circular cylinder are not too large (the structure can thus be regarded as a ring) and the loads acting on it are contained within the plane bisecting its longitudinal thickness (it is thus further assumed that the ring's geometry, including the boundary conditions, the mechanical properties of the ring material and the loads on it are symmetrical with respect to this mid-plane, at axial coordinate $z = 0$), it can be assumed that $\sigma_z = \tau_{xz} = \tau_{yz} = 0$ at every point of the ring, meaning that the stress state is plane. Strictly speaking, this is true only where the axial thickness is constant and infinitesimal, but the errors that result from using this assumption, even for sizable axial thicknesses, are sufficiently small that they can be considered negligible provided that the ring is subjected only to surface forces acting on the inner and outer radii and there is thus no thermal load.

With this assumption, all stress components are independent of the axial coordinate z, and two coordinates in the plane of the generic cross-section (x, y, as in Fig. 1.1b, or r, θ, as in Fig. 4.1a are sufficient to describe the phenomenon completely. In other words, we are dealing with a plane, i.e., two-dimensional, problem. Where the assumptions of a plane stress state and generalized axisymmetry coexist, the problem becomes one-dimensional, and the stress and strain components can be described by means of coordinate r alone.

The cases in which the material works in the linear elastic field (Filonenko-Borodich 1968, Timoshenko and Goodier 1970, Franciosi 1981, Barber 1999) and in which it is loaded beyond yielding (Hill 1950, Dill 2007) will be handled separately. In the latter case, it will be first be assumed that the material is homogeneous and isotropic, and shows non-hardening linear elastic-perfectly plastic behavior, or in other words, that behavior is perfectly elastic up to the yield stress (i.e., in this event, the proportional limit, the elastic limit and the yield point coincide), and perfectly plastic thenceforth up to failure. With this assumption of non-hardening linear elastic-perfectly plastic behavior of the material, the stress state in the circular cylinder stressed beyond yield can be solved in closed form by means of the mono-dimensional theory.

In a circular cylinder stressed symmetrically around its axis and loaded uniformly along the latter (here, assuming plane stress, we will have $\sigma_z = 0$), the radial, circumferential and axial directions will be the principal directions of both stresses and strains. It should be noted that the circumferential direction is that tangent to the generic circumference in the equally generic cross-section: circumferential direction and tangential direction are thus synonyms. However, we will use only the terms hoop or circumferential normal stress and, more briefly, hoop or circumferential stress to avoid confusion with shearing stress, which is the stress correlated with shear or torsion forces. As regards strains, there is no reason for confusion, as the normal strains, or unit elongations, are distinct from unit shear strains both in substance and terminologically.

From the design standpoint, most interest focuses on cases where strains are infinitesimal (the small strains or small deformations assumption) and those where strains have a non-negligible finite value (the large strains or large deformations assumption). Below, it will be assumed that the strains are small, and that the associated strain state is thus described by conventional strains (or nominal, engineering or Lagrangian strains) ε_r, ε_t and ε_z. In cases involving large strains, it will clearly be necessary to introduce the true strains (or natural or logarithmic strains) $\overline{\varepsilon}_r = \ln(1 + \varepsilon_r)$, $\overline{\varepsilon}_t = \ln(1 + \varepsilon_t)$ and $\overline{\varepsilon}_z = \ln(1 + \varepsilon_z)$.

The assumption that strains are small is fully verified for materials that show well-defined yielding in the $\sigma = \sigma(\varepsilon)$ curve. In such cases, there is a gradual passage from an elastic strain state to the plastic strain state, with a plastic-elastic interface that, from the permanent residual strain that first appears, shifts to mark the boundary between two regions, one plastic and one elastic. Because of the two regions' continuity and contiguity, strains in the plastic region must be of comparable magnitude with those in the elastic region, or in other words must be small because those of the elastic region are small.

With a single exception, this textbook deals with problems in which strains can be considered to be small. The exception concerns only the topics addressed in Sects. 3.5 (thin-walled circular cylinders under internal pressure, loaded beyond the elastic range) and 3.6 (thin-walled circular cylinders under internal pressure and axial tension, and stressed beyond yielding), which are characterized by large strains. In dealing with these topics, it is clearly necessary to use the natural or logarithmic strains. In this specific connection, however, we have preferred to express them in the form $\varepsilon_r = \ln\left(1 + \varepsilon_{r,0}\right)$, $\varepsilon_t = \ln\left(1 + \varepsilon_{t,0}\right)$ and $\varepsilon_z = \ln\left(1 + \varepsilon_{z,0}\right)$, where $\varepsilon_{r,0}$, $\varepsilon_{t,0}$ and $\varepsilon_{z,0}$ are the conventional or nominal strains.

This choice was dictated by the need to avoid over-complicating the relations involved with quantities having multiple subscripts, indices, and diacritics. However, as this leaves some room for confusion, the reader must bear in mind that for the problems involving small strains, ε_r, ε_t and ε_z are the conventional or nominal strains, whereas for the problems involving large strains, the same symbols ε_r, ε_t and ε_z represent the natural or logarithmic strains.

It should also be recognized that the stress state in a circular cylinder having large axial dimensions (the structure is thus tubular) and subjected simultaneously to surface forces at the inner and outer radii and to thermal load is not plane. In each elementary ring of axial thickness dz, in fact, thermal load generates a stress state which can in the first approximation be considered as plane. However, because of the material's lateral contraction, an axial strain ε_z arises which in general varies along radius r, and the sections tend to undergo distortion. This distortion is free to occur in rings of infinitesimal axial thickness, but is prevented in structures of large longitudinal thickness, where a state of coaction is consequently produced.

In analyzing tubular circular cylinders, it is thus more realistic to assume a plane strain state ($\varepsilon_z = 0$) if the structure is constrained at its ends or infinitely extended along the axis, or uniform axial translation of a generic cross-section ($\varepsilon_z = const$) if the structure has finite axial length and is free at the ends. Both assumptions of a plane strain state ($\varepsilon_z = 0$) and of uniform axial translation of a generic cross-section ($\varepsilon_z = const$) coincide with the plane stress state assumption ($\sigma_z = 0$) only where axial strain ε_z is independent of variable r.

This is the case of circular cylinders satisfying the above constraint and generalized axisymmetry conditions, consisting of homogenous isotropic material and subjected only to uniform pressures, internal and external if the structure is hollow and only internal if it is non-hollow, all providing that Saint-Venant's principle (Zaslavsky 1982) is respected (in this connection, it should be borne in mind that in a tubular circular cylinder with free and non-loaded ends, there is a gradual transition from a strain state near the ends characterized by $\varepsilon_z = const$ to a plane stress state, given that we must necessarily have $\sigma_z = 0$ on the unloaded edges). In other words, thick-walled tubes, circular cylindrical roads and thin-walled rings all feature the same stress state if they are subjected only to surface force distributions. If thermal gradients are also at work (and the same is true if body forces are involved), the assumptions of plane stress state, plane strain state and uniform translation of a generic cross-section will lead to different results.

A circular cylinder working in the elastic field has limited resistance to loads arising from internal pressure. Common methods for increasing this strength include autofrettage or self-hooping and the use of composite multilayer structures assembled with an interference fit. In the autofrettage procedure, which is a prestressing procedure, a plastic flow is generated in the circular cylinder by subjecting it at a particular point of the production cycle to an internal pressure between that which brings the inner radius to the onset of yield and that at which the entire wall thickness of the circular cylinder is in the fully plastic state. This gives rise to a state of coaction in the wall of the circular cylinder: as load increases, the permanent residual strain that the pressure acting at the inner radius produces in the inner plastic region prevents the material in the outer elastic region from contracting, thus exerting a self-hooping force on the inner plastic region. This plastic flow, which proceeds from the inner radius to the outer radius as load increases, is such that a residual stress state remains in the wall after the autofrettage pressure is released. This residual stress state is characterized around the bore by a compressive hoop stress distribution.

Over time, our understanding of the autofrettage procedure has grown, and now encompasses reverse yielding (and, more in general, successive yield cycles, first in tension, then in compression, then again in tension, and so on) of thick circular cylinders subjected to autofrettage until the entire wall thickness is in the plastic state in order to produce strong, high performance structures such as those used in hypervelocity launchers for ordnance of warships. These interesting issues will be addressed by assuming that, regardless of the material's condition (virgin or otherwise), the part of the stress–strain curve $\sigma = \sigma(\varepsilon)$ obtained for a specimen subjected to a compression test is the mirror image of the origin of the part of the same curve for a specimen subjected to a tensile test. In the entire test, in other words, we will assume the material is linear elastic-perfectly plastic with identical tensile and compressive behavior; accordingly we will ignore the Bauschinger effect, whereby there will be a reduction in the compressive yield point (or tensile yield point) of a specimen whose material has a non-zero plastic tensile strain history (or compressive strain history).

The assumption that the homogeneous and isotropic material exhibits linear elastic-perfectly plastic behavior, or in other words is non-hardening, is unrealistic. The steels that are currently used in the construction of the structures of interest to us here mostly show strain hardening behavior. Their stress–strain curve $\sigma = \sigma(\varepsilon)$ determined by means of a uniaxial tensile test on a standard specimen is thus nonlinear starting from the proportional limit, which thus differs from both the elastic limit and the yield stress (yield point). When the structure is stressed beyond the elastic range, the results obtained by assuming that the material exhibits linear elastic-perfectly plastic behavior consist of curves which are continuous for the stresses of interest (autofrettage stresses, residual stresses after releasing the autofrettage pressure, and effective stresses under actual operating conditions), but discontinuous as regards their gradient at the interface between the inner region where the material's behavior is plastic and the outer region where the material shows elastic behavior. In other words, the distributions of these stresses along the radius are characterized by cusps at this interface.

The fact that the distribution curves for the stresses of interest here are cusped is clearly a physical absurdity: in accordance with Linnaeus' dictum that *"natura non facit saltus"*, it has been demonstrated experimentally that the transition between these two regions is gradual, and takes place in a cylindrical band of annular cross-section between the two regions.

Though the purpose of the treatment based on the assumption that the material's behavior is linear elastic-perfectly plastic is merely to permit an initial approach to the problem, it can provide information that is useful for design purposes, without, however, claiming to interpret the actual behavior of the material. To overcome the limitations of this treatment and the physical absurdity deriving from it, and to reflect experimental findings more accurately, this text will tackle the problem in much more general terms by introducing a constitutive relation that can describe the behavior of the cylinder's material as faithfully as possible. To this end, the constitutive relation that will be considered here is capable of simulating not only the elasto-plastic behavior of a strain-hardening material, but also the viscous behavior in steady state conditions associated with creep at high temperatures.

In a section of the Chap. 1, thin-walled circular cylinders are treated as circular cylindrical shells. This concept is then generalized in the last two chapters of this text, with an extension to shells having a geometry of revolution. In the theory of elasticity, the term shell indicates three-dimensional structural bodies or components which, for the purpose of stress analysis, can be considered as the materialization of a curved surface, i.e., as a curved physical surface made of a material capable of transmitting loads and of undergoing deformations as a result. A shell is bounded by two curved surfaces, the faces, and its thickness, which is small compared to its other dimensions, may be the same everywhere, or may vary from point to point. We consider the shell's middle surface as the surface which passes midway between the two faces. The shell is geometrically fully described when we know the shape of the middle surface and the shell thickness for every one of its points. Here we consider shells having constant thickness, made of homogeneous and isotropic material which obeys Hooke's law, and we assume that the displacements at a point are small in comparison with the thickness of the shell (small displacement hypothesis). If the small displacements limitation is not imposed, the governing differential equations of the shells become nonlinear and their solution involves significant difficulties.

Though the stress and strain analysis in shell structures, which are three-dimensional bodies, can be analyzed using the methods provided by the three-dimensional theory of elasticity, an analysis based on these methods presents unusual difficulties, even for a simple structure under ordinary loads. To overcome these difficulties, the fundamentals of three-dimensional shell theory must be reduced to an approximate two-dimensional shell theory. In this connection, simplified methods have been developed which introduce approximate assumptions, and reduce the three-dimensional problem to a two-dimensional problem (the analysis is reduced to the shell middle surface), as was done in the development of the elastic theory of thin plates.

With this approach, various shell theories have been developed gradually over time, based on the theory of plates. In the classical theory of plates, two fundamental solution methods have been proposed, the first by Cauchy (1828) and Poisson (1829), and the second by Kirchhoff (1850). The first method is based on the expansion of displacements and stresses in the plate in power series of the distance of a given point from the plate's middle plane; the results differ according to the number of terms used in these series, the solution tending to the exact one as the number of terms becomes infinite. The second method, developed to overcome the limitations of the first, which relate to convergence problems (in this respect, St. Venant 1883, was the first to contest the validity of the method; see Novozhilov 1970) and to the difficulty of imposing the boundary conditions, introduces physical quantities, such as internal forces and moments, into the theory of plates, as well as assumptions analogous to those used in the theory of beams. This second method, even if approximate, simplifies the treatment of the problem and leads to very reliable results. For these reasons, it has remained in use until the present date.

In the development of a theory of shells, both the above methods can be effectively used, introducing the appropriate assumptions and generalizations. In fact, several authors have developed shell theories similar to that of plates formulated by Cauchy and Poisson (Krauss 1929; Kil'chevskii 1939). Others made fundamental contributions using different methods, all related to that of plates proposed by Kirchhoff (Aron 1874; Love 1888; Donnell 1933; Galerkin 1934; Lur'e 1940; Gol'denweizer 1940; Reissner 1912; Meissner 1915; Mushtari 1938; Vlasov 1944; Rabotnov 1945; Feinberg, 1939; Bassett 1890; Flügge 1932; Geckeler 1930; etc.).

In brief, we can say that the derivation of two-dimensional shell theories may follow, essentially, two different paths depending on whether it starts: (1) from a priori assumptions, generally regarding the kinematics of deformation, or (2) from expansions of the variables (displacement and stress components) in power series of a thickness parameter. Conventionally, the first approach is called the *aprioristic approach*, and the second the *asymptotic approach*. As aptly described by Cicala (1978):

> The asymptotic approach operates on the three-dimensional theory like a prism acting on composite light: it separates the spectrum of the stress states into monochromatic classes of solutions with different lengths of variation, governed by different parametric expansion. On the contrary, aprioristic formulations are capable of covering bands of the spectrum, but successive approximation steps cannot be attained on this basis.

Since only problems of thin shells will be considered here, we will use the first-order approximation shell aprioristic theory, first developed by Love on the basis of classical linear elasticity, and applying the hypotheses introduced by Kirchhoff for the plate bending theory (the straight fibers of a plate or shell which are perpendicular to the middle surface before deformation remain so after deformation and do not change their length, and the normal stresses acting on planes parallel to the middle surface may be neglected in comparison with the other stresses) together with the assumptions of small displacements and thinness of the shell. These assumptions are called the *Kirchhoff–Love assumptions* (see also Budiansky and Sanders 1963).

This first-order approximation shell theory, which we define the technical theory of shells, is not free from deficiencies and inconsistencies. In particular, the latter concern the small terms, some of which are retained and others rejected, although they are of the same order. The recognition of these limitations led to the development of higher-order approximation shell theories, one of which is the second-order approximation theory of shells independently developed by Lur'e (1940), Flügge (1960), Byrne (1944), Biezeno and Grammel (1953) and Novozhilov (1964). Nonlinear theories have also been developed, which take into account both geometrical nonlinearity (large deformations) and physical nonlinearity, with respect to the stress–strain relations (Reissner 1952; Nagdi and Nordgren 1963; Sanders 1963; Koiter 1970; Vlasov 1964; Mushtari and Galimov 1961; Simmond and Danielson (1972), etc.).

In this book, we will use the simplified shell theory, known as the Donnell-Mushtari-Vlasov shell theory, as it has proved extremely convenient for solving many engineering shell problems. This theory is based on the following Kirchhoff–Love assumptions: (1) straight fibers normal to the non-deformed middle surface are displaced into straight fibers normal to the deformed middle surface, and do not change their length; (2) stresses in the direction normal to the middle surface and shear strains are negligible; (3) the ratio between the thickness at any point of the shell and either one of the two principal radii of curvature is much smaller than unity; (4) deformations (displacements and rotations) are small; (5) the material is homogeneous and isotropic; and (6) Hooke's law applies. As we indicate in Chap. 11, this simplified theory can be extended to other more complex problems by introducing other hypotheses.

With this linear shell theory, which has been used by many investigators to solve various problems, and in particular to find analytic solutions, only a few closed-form and asymptotic solutions have been obtained for shells of simple geometrical shape (i.e., shells having zero Gaussian curvature, and shells of uniform thickness such as spheres and ellipsoids). Most of these solutions are expressed in terms of infinite series converging quite slowly, and their numerical values are rather difficult to obtain. For shells of arbitrary thickness and curvature, numerical solutions are the only practicable approach. In this respect, in addition to the universally known methods, such as Finite Element Method (FEM), Boundary Element Method (BEM) and Finite Difference Method (FDM), are also to remember the method of successive approximations, the small-parameter method, and the method of successive loadings (Petrov 1975). For numerical methods applied to the shell structures, see: Soare 1962, Cicala 1965 and 1977, Zienkiewitcz 1969, Rutten 1971, Fenner 1975, Cook 1981, Benarjee and Butterfield 1981, Reddy 1985, Moaveni 1999.

References

Aron, H. (1874). Das Gleichgewicht und die Bewegung einer unendlich dünnen, beliebig gekrümmten, elastischen Schale. *Journ. Für reine und ang. Math*, *78*, 136–174.

Banerjee, P. K., & Butterfield, R. (1981). *Boundary element methods in engineering science*. London: McGraw-Hill.

Barber, J. R. (1999). *Elasticity*. Dordrecht: Kluwer Academic Publishers.

Bassett, A. (1890) On the extension and flexure of cylindrical and spherical thin elastic shells. *Philosophical Transactions of the Royal Society, series A*, *181*(6), 433–480.

Bassett, A. (1890) On the extension and flexure of thin elastic shells. *Philosophical Transactions of the Royal Society, series A*, *179*.

Biezeno, C. B., & Grammel, R. (1953). *Technische Dynamik*. Erster Band, Grundlagen und Einzelne Maschinenteile, Berlin: Springer.

Boresi, A. P., & Chong, K. P. (2000). *Elasticity in engineering mechanics*. New York: Wiley.

Budiansky, B., & Sanders, J. L. (1963). *On the "best" first-order linear shell theory*. Progress in applied mechanics (pp. 129–140). New York: McMillan.

Byrne, R. (1944). Theory of small deformations in a thin elastic shell, Seminar report in mathematics, University of California. *Publications in Mathematics*, *2*(1), 103–152.

Cauchy, A. L. (1828). Sur l'equilibre et le mouvement d'une plaque solide. *Exercise de Mathématique*, *3*, 328.

Cicala, P. (1977). Asymptotic approach to linear shell theory. *AIMETA Research Report*, *6*, 129.

Cicala, P. (1978). Linear Shell, Theories, An Asymptotic Approach, Libreria Editrice Universitaria Levrotto & Bella, Torino.

Cicala, P. (1965). *Systematic approximation approach to linear shell theory*. Torino: Libreria Editrice Universitaria Levrotto & Bella.

Cook, R. D. (1981). *Concepts and applications of finite element analysis* (2nd ed.). New York: Wiley.
Dill, E. H. (2007). *Continuum mechanics: Elasticity, plasticity, viscoelasticity*. Boca Raton: CRC Press, Taylor & Francis Group.
Donnell, L.H. (1933). Stability of Thin Walled Tubes under Torsion. NACA, Report no. 479, Washington.
Feinberg, S. (1939). On the problem of the construction of an approximate theory of thin shells of arbitrary shape, In Studies in the Theory of Structures, Gosstroiizdat.
Fenner, R. T. (1975). *Finite element methods for engineers*. London: The Macmillan Press Ltd.
Filonenko-Borodich, M. (1968). *Theory of elasticity*. Moscow: MIR Publishers.
Flügge, W. (1932). Die stabilität der Kreiszylinderschale. Ing-Arch, 3, 463–506.
Flügge, W. (1960). Stresses in shells (2nd ed.). Springer-verlag, Berlin, reprint of 1973.
Franciosi, V. (1981). *Fondamenti di Scienza delle Costruzioni*, Teoria dell'Elasticità (Vol. I). Napoli: Liguori Editore.
Galerkin, B. G. (1934). On the theory of elastic cylindrical shells. D.A.N., S.S.S.R., 4, 5–6.
Geckeler, J. (1930). Die festigkeit achsensymmetrischer schalen. Ing-Arch, Bd. 1, 22–71.
Gol'denweizer, A. L. (1940). The equations of the theory of shells. *Prikl. Mat. Mekh., Akademiya Nauk. S.S.S.R*, *4*(2).
Haupt, P. (2000). *Continuum mechanics and theory of materials*. Berlino: Springer.
Hill, R. (1950). *Mathematical theory of plasticity*. Oxford: Clarendon Press.
Kil'chevskii, N. A. (1939). A generalization of contemporary shell theory. *Prikl. Mat. Mekh, Akademiya Nauk. S.S.S.R*, *2*(4).
Kirchhoff, G. R. (1850). Uber das gleichgewichi und die bewegung einer elastishem scheibe. *J Fuer die Reine und Angewandte Mathematik*, *40*, 51–88.
Koiter, W. T. (1970). The stability of elastic equilibrium (Chapter 5: Shell theory for finite deflections), Diss., Tech. Rept., AFFDL-TR-70-25, Air Force System Command, Wright-Patterson AFB.
Krauss, F.(1929). Über die Grundgleichungen der Elastizitätstheorie schwachdeformierter Schalen. *Mathematische Annalen*, 101(1), 61–92.
Lur'e, A. I. (1940). The general theory of thin elastic shells. *Prikl. Mat. Mekh., Akademiya Nauk. S.S.S.R*, *4*(2).
Love, A. E. H. (1888). On the small free vibrations and deformations in the elastic shells. *Philosophical Transactions of the Royal Society A*, *179* , 491–549.
Meissner, E. (1915). Über elastizität und festigkeit dünner schalen, Vierteljahrsschr d. Naturf., Gesellsch., Bd. 60, Zürich.
Moaveni, S. (1999). *Finite element analysis. Theory and application with ANSYS*. Upper Saddle River: Prentice-Hall.
Mushtari K. M. (1938). Certain Generalizations of the Theory oh Thin Shells. *Izv. Kaz. fiz-mat. o-va., pri Kaz Un-te*, *11*(8), (in Russian).
Mushtari, K. M., & Galimov, K. Z. (1961). Nonlinear Theory of Thin Shells, The Israel Program for Scientific Translations, Jerusalem.
Naghdi, P. M., Nordgren, R. P. (1963). On the linear theory of shells under the Kirchhoff hypothesis. *Quarterly of Applied Mathematics*, *21*, 49–60.
Novozhilov, V. V. (1964). *The Theory of Thin Elastic Shells*. Groningen: P. Noordhoff LTS.
Novozhilov, V. V. (1970). *Thin Shell Theory* (2nd ed.). Groningen, The Netherlands: Wolters-Noordhoff Publishing.
Peterson, R. E. (1974). *Stress concentration factor*. New York: Wiley.
Petrov, V. V. (1975). The method of successive loadings in the nonlinear theory of plates and shells. Saratov: S.S.S.R, Izd-vo Saratov University.
Poisson, S. D. (1829). Mémoire sur l'équilibre et le mouvement des corps solides. *Mém. de l'Acad. Sci*, *8*, 357.
Rabotnov, Y. N. (1945). *The fundamental equations of the theory of shells* (Vol. XLVII, No. 2). D.A.N., S.S.S.R.

Reddy, J. N. (1985). *An introduction to the finite element method.* New York: McGraw-Hill Book Company.

Reissner, E. (1952). On axisymmetrical deformation of thin shells of revolution. *Proceedings of Symposia in Applied Mathematics*, *3*, 27–52.

Reissner, H. (1912). Spannungen in Kugelschalen, Müller, Breslau-Festscher, Leipzig.

Rutten, H. S. (1971). *Asymptotic approximation in the three-dimensional theory of thin and thick elastic shells.* Hertogenbosch: Nederl. Boekdruk Ind., 's.

Sanders, J. L. (1963). Nonlinear theories for thin shells. *Quarterly of Applied Mathematics*, 21, 21–36.

Simmond, J. G., & Danielson, D. A. (1972). Nonlinear shell theory with finite rotation and stress-function vectors. *Journal of Applied Mechanics*, 39, 1085–1090.

Soare, M. (1962). *Application des equations aux differences Finis au Calcul des Coques*. Paris: Eyrolles.

Sokolnikoff, I. S. (1956). *Mathematical theory of elasticity*. New York: McGraw-Hill Book Company.

Timoshenks, S. P., & Goodier, J. N. (1970). *Theory of elasticity* (3rd ed.). New York: McGraw-Hill.

Vlasov, V. Z. (1944). The fundamental differential equations of the general theory of elastic shells. *Prikl. Mat. Mekh., Akademiya Nauk. S.S.S.R*, 8(2) (also NACA T.M., No. 1241).

Vlasov, V. Z. (1964). *General Theory of Shells and its Applications in Engineering*. NASA Technical Translation TTF-99 (translated from Russian), Washington, DC.

Zienkiewitcz, D. C. (1969). *The finite element method in engineering science*. New York: McGraw-Hill.

Chapter 1
Thin-Walled Circular Cylinders Under Internal and/or External Pressure and Stressed in the Linear Elastic Range

1.1 Foreword

Circular cylinders are generally divided into two families, according to the equations governing their stress state. If the wall thickness is small compared to the inside diameter, within the limits indicated below, it is assumed that the stresses are uniformly distributed through it, which simplifies the treatment to a considerable extent (Bickell and Ruiz 1967, Iurzolla 1981, Burr 1982, Ventsel and Krauthammer 2001, Ugural and Fenster 2003). In this case, we are dealing with a cylindrical membrane shell, i.e., a thin-walled circular cylinder, where the axial dimension is large, and a thin-walled circular ring where the axial dimension is small.

In this chapter, we will discuss thin-walled circular cylinders with large axial dimension, or in other words, tubular structures having an annular cross-section of small radial thickness stressed in the linear elastic field. Assuming that the stresses are uniformly distributed through the wall thickness makes it possible to address the problem using only the equilibrium equations. It is this not necessary to employ the compatibility equations or advance hypotheses regarding the stress state or strain state.

Geometrical axisymmetry and the uniform distribution of stresses through the wall thickness means that the radial, circumferential (or tangential) and axial directions are the principal directions of both stresses and strains. The stress and strain states of thin-walled circular cylinders under internal and external pressure are analyzed here by assuming that displacements are small and strains are infinitesimal. Consequently, the geometrical magnitudes characterizing deformation mechanics remain essentially unchanged after the loads associated with pressure are applied. It will also be assumed that the material is homogeneous and isotropic.

If the wall is thick, it is no longer possible to assume that stresses are uniformly distributed through it. In this case, the approach to the problem is more complex, as it involves simultaneously employing the equilibrium equations and the compatibility equations, both of which must be satisfied, as well as the boundary conditions. Here we are dealing with vessels or, more generally, thick-walled circular cylinders and rings, depending on whether their axial dimensions are large or small.

V. Vullo, *Circular Cylinders and Pressure Vessels*, Springer Series in Solid and Structural Mechanics 3, DOI: 10.1007/978-3-319-00690-1_1,

It should be specified that there is no difference between the two families of thin-walled and thick-walled circular cylinders from the standpoint of continuum mechanics. The difference is purely conventional. Here, we will follow the more common international convention, whereby the ratio of wall thickness s and inside diameter d_i for the family of thin-walled circular cylinders is less than or at most equal to 0.05, i.e.:

$$\frac{s}{d_i} \leq \frac{1}{20}. \tag{1.1}$$

Noting that $s = (d_e - d_i)/2$, where d_e is the outside diameter of the circular cylinder, Eq. (1.1) is usually expressed in equivalent form by introducing the dimensionless diameter ratio $\delta = (d_e/d_i)$ or its reciprocal $\beta = 1/\delta = (d_i/d_e)$. We thus have:

$$\delta \leq 1.10 \text{ or } \beta \geq 0.91(\sim 0.90). \tag{1.2}$$

As we will have occasion to demonstrate below, this conventional demarcation between thin-walled circular cylinders and thick-walled circular cylinders has a theoretical foundation, based as it is on the fact that the curve representing the maximum normal stress strength theory, σ_{max} (the so-called Hamburg formula), intersects the curve for the maximum distortion energy strength theory, ED_{max}, in the diagram expressing $p_i/\sigma_{e,i} = f(\delta)$, where p_i is the internal pressure and $\sigma_{e,i}$ is the equivalent or ideal stress calculated at the inner radius r_i (at this radius, the equivalent stress is maximum, and the material is at greatest risk). The two curves, in fact, intersect for $\delta = 1.101$, i.e., in round numbers, for approximately $\delta = 1.10$.

There are also other conventional criteria based on the intersection between the various curves representing the ratio $p_i/\sigma_{e,i} = f(\delta)$, and correlated with the different strength theories used in this specific area of mechanical design. One example will suffice: that based on the intersection of the curve representing the Hamburg formula and that for the ε_{max} maximum normal strain theory: as the two curves intersect for $\delta = 1.16$, thin-walled and thick-walled circular cylinders are defined as those satisfying the inequalities $\delta \leq 1.16\,(\beta \geq 0.86)$ and $\delta > 1.16$ $(\beta < 0.86)$ respectively.

It should be recognized that the international, national and European standards covering pressure vessel design analysis and response analysis do not always agree on the value of ratio δ to be used as the demarcation between thin-walled and thick-walled circular cylinders. At times, these discrepancies are rooted in empirical rather than theoretical considerations; as a result, the simpler relations deriving from the thin-wall assumption are extended well beyond these conventional limits to include a sizable portion of the values of ratio δ (the lower values) that, at least from the standpoint of the above convention, should be considered as being in the category of thick-walled circular cylinders. It should also be noted that a few standards introduce an intermediate range midway between thin-walled and thick-walled circular cylinders.

For $\delta > 1.10$, the assumption that stresses are uniformly distributed through the wall thickness, in itself a rough approximation, becomes even rougher the larger the thickness is by comparison with the diameter, or in other words, the larger δ becomes. We will return to this point in Sect. 1.5. For this reason, all references in the text to the conventional upper limit of the ratio δ for a circular cylinder to be considered as thin-walled will mean $\delta = 1.10$.

1.2 Stress State in Thin-Walled Circular Cylinders Under Internal and External Pressure

We will now consider a thin-walled circular cylinder (circular cylindrical shell) indefinitely extended in the direction of its axis of symmetry (the z axis) and subjected in the linear elastic field to both internal pressure p_i and external pressure p_e; here and below, all pressures are to be interpreted as relative to atmospheric pressure.

The following demonstration applies both when external pressure is less than internal pressure ($p_e < p_i$), and in the opposite case. In the latter circumstance, however, the strength analysis (design analysis and response analysis) is not exhaustive and must be completed by a structural stability analysis. In this case attention must be directed towards those problems arising mainly from situations where applied loads cause the structure to wrinkle, buckle, or otherwise collapse. As a matter of fact the various instability phenomena (elastic Euler's instability of columns, buckling of circular rings and tubes under uniform external pressure, symmetrical buckling of a cylindrical shell under the action of uniform axial compression, inextensional forms of bending or concertina modes of cylindrical shells due to local instability, torsional buckling, cross-section ovalization or ellipticity due to bending moments, etc.) can lead to failures even before stresses reach the strength limit values. In other words, it is necessary to determine the critical stresses that cause the structure to collapse under combined compression and bending, and which are aggravated by the deterioration of the material's mechanical properties when it works beyond the elastic limit, and by local instability.

If we assume generalized axisymmetry (axisymmetry in shape, loading, constraints and in the mechanical behavior of the material, which thus must be considered isotropic or axisymmetrically orthotropic, as well as homogeneous) and that magnitudes are independent of the axial coordinate z and depend only on radius r, the problem becomes mono-dimensional.

The pressures exerted on the wall of the circular cylinder generate three types of stress: radial σ_r, hoop or circumferential σ_t and axial or longitudinal σ_z. All three are principal stresses, given that:

- Axial or longitudinal strain ε_z is zero or constant, depending on the constraint and load conditions at the end of the body (clamped ends, free ends with no load, free ends under load).
- Unit shearing strains γ_{rz} and γ_{tz} are zero, as the magnitudes on which they depend are independent of the axial coordinate z.
- Shear stress τ_{rt} and unit shearing strain γ_{rt} are zero, as generalized axisymmetry is assumed.

We are thus dealing here with a generalized plane strain state, while the three radial, tangential and axial directions are the principal directions of both stresses and strains.

Hoop stress σ_t is generated by the pressures exerted on the inner and outer cylindrical surfaces, which tend to divide or compress the structure along axial planes (or diameter planes), forming two cylindrical half shells. Longitudinal stress σ_z arises from the pressures exerted on the closures, whether real or virtual (for example, an elbow pipe fitting or a curved pipe fitting would be a virtual closure), and may not be present, as is the case in open vertical tanks, where the pressure is that resulting from hydrostatic head, or in containers where the closures are not positively connected to the walls (hydraulic and pneumatic cylinders with mobile pistons and the like). Radial stress σ_r, which is always negative because it is compressive, arises as a result of pressures p_i and p_e which at all points act normal to the inner and outer surfaces delimiting the circular cylinder.

Under the simplifying assumption that the three principal stresses are uniformly distributed through the wall thickness, which though not strictly true, is a sufficiently close approximation to provide results that are of design significance for thin-walled circular cylinders, these stresses, which entail a triaxial stress state, can be readily obtained by simply setting equilibrium conditions, i.e., without making use of compatibility conditions.

1.2.1 Hoop Stress

We will consider a thin-walled circular cylinder of thickness s, with inside and outside diameters d_i and d_e and indefinite axial length, subjected to internal pressure p_i and external pressure p_e. By means of two planes perpendicular to the axis, we will then isolate a small element of length dz located between the cross-section of abscissa z and that of abscissa $z + dz$ (Fig. 1.1a) which is sufficiently far from the ends as not to be affected by the perturbations at the edges contemplated by Saint–Venant's principle (Timoshenko and Goodier 1970, Belluzzi 1971, Zaslavsky 1982). This principle states that a self-equilibrating distribution of force, applied on a small part of an elastic body, produces only local stress and local strain.

We then section this small element along axial plane $y - y$ (Fig. 1.1b) and consider equilibrium of one of the two half shells thus obtained in the direction

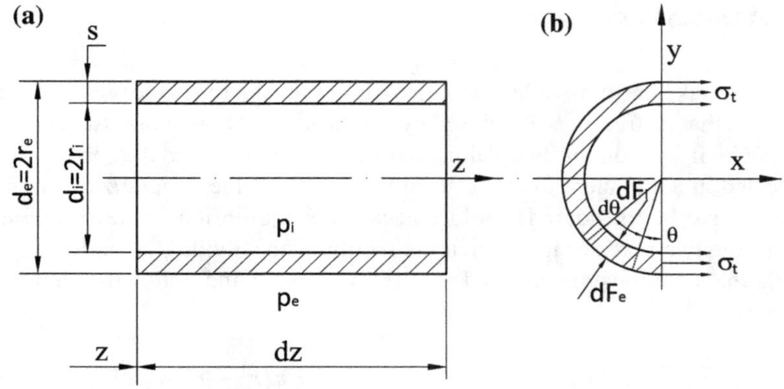

Fig. 1.1 **a** Small element of a cylindrical shell of length dz. **b** Half shell obtained by sectioning the small element along a axial plane

perpendicular to the section plane, i.e., along the x axis. Using the notations shown in Fig. 1.1 and introducing the inner radius r_i, we obtain:

$$2 \cdot \sigma_t \cdot s \cdot dz = \int_o^{\pi} (dF_i - dF_e) \cdot sin\vartheta \tag{1.3}$$

where

$$dF_i = p_i\, r_i\, d\vartheta\, dz \text{ and } dF_e = p_e (r_i + s)\, d\vartheta\, dz. \tag{1.4}$$

Substituting relations (1.4) in relation (1.3) and integrating gives the following expression for hoop stress, written in terms of inside diameter d_i:

$$\sigma_t = \frac{d_i(p_i - p_e)}{2s} - p_e. \tag{1.5}$$

In reality, hoop stress σ_t is a function of radius r, but the differences between the values it assumes at the inner and outer radii are negligible if the body can be regarded as thin-walled. Consequently, Eq. (1.5) represents the average effective hoop stress σ_{te}, viz.:

$$\sigma_{te} = (1/s) \int_{r_i}^{r_e} \sigma_t\, dr. \tag{1.6}$$

If, as is fairly frequent in design applications, external pressure is zero ($p_e = 0$) or internal pressure is zero ($p_i = 0$), Eq. (1.5) leads to the following respective relations:

$$\sigma_t = \frac{p_i\, d_i}{2s} \text{ and } \sigma_t = -\frac{p_e\, d_e}{2s}. \tag{1.7}$$

These are Mariotte's formulas for boilers.

1.2.2 Axial Stress

We will now take the thin-walled circular cylinder considered in the previous section, but assume that its free ends are closed by two heads, either real or virtual as defined above. We will section this body along a plane perpendicular to the z axis (Fig. 1.2) and located at a sufficient distance from the ends that the generalized strain state assumption can be considered to hold, and consider equilibrium in the axial direction of one of the two portions into which the section plane divides the body.

With the notations shown in Fig. 1.2, we obtain the following equilibrium equation:

$$2\pi \int_{r_i}^{r_e} \sigma_z r dr = \pi\left(p_i r_i^2 - p_e r_e^2\right) = \pi\left(r_e^2 - r_i^2\right)\sigma_z \tag{1.8}$$

which, after carrying out a few further passages and remembering that $\left(r_e^2 - r_i^2\right) = (r_e + r_i)(r_e - r_i)$, $(r_e + r_i) = 2r_i + s$ and $(r_e - r_i) = s$, gives the following expression for axial stress σ_z in terms of inside diameter d_i:

$$\sigma_z = \frac{d_i(p_i - p_e) - 4p_e s(1 + s/d_i)}{4s(1 + s/d_i)}. \tag{1.9}$$

Assuming a thin-walled structure, the ratio s/d_i can be regarded as negligible with respect to unity. Consequently, relation (1.9) can be simplified as follows:

$$\sigma_z \cong \frac{d_i(p_i - p_e) - 4p_e s}{4s}. \tag{1.10}$$

Again starting from relation (1.8), but writing it in terms of outside diameter d_e rather than inside diameter d_i, we obtain the following expressions for σ_z, which are equivalent to relations (1.9) and (1.10) respectively:

$$\sigma_z = \frac{d_e(p_i - p_e) - 4p_i s(1 - s/d_e)}{4s(1 - s/d_e)}; \tag{1.11}$$

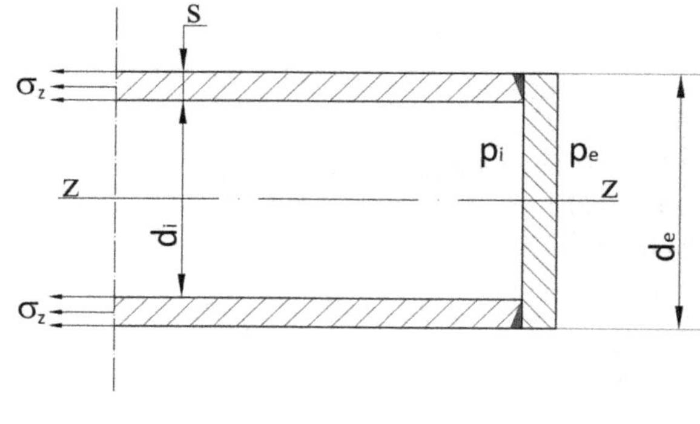

Fig. 1.2 One of the two portions of the circular cylinder obtained by sectioning it along a plane perpendicular to the axis

$$\sigma_z \cong \frac{d_e(p_i - p_e) - 4p_i s}{4s}. \tag{1.12}$$

Finally, in the fairly frequently encountered cases with zero external pressure ($p_e = 0$) and zero internal pressure ($p_i = 0$), relations (1.10) and (1.12) lead respectively to the following:

$$\sigma_z \cong \frac{p_i\, d_i}{4s} \quad \text{and} \quad \sigma_z \cong -\frac{p_e\, d_e}{4s}. \tag{1.13}$$

As a comparison of relations (1.13) and (1.7) shows, the hoop stress σ_t in thin-walled circular cylinders under internal and external pressure with closed ends is approximately twice the axial stress σ_z; consequently, from the strength standpoint alone, failure will occur along the cylindrical body's generators.

In conclusion, it should be noted that the assumption of uniform distribution of axial stress σ_z through the thickness does not apply only to the thin-walled circular cylinder, but also holds true for the thick-walled circular cylinder providing that it is not subject to stresses due to thermal loads.

1.2.3 Radial Stress

Internal pressure p_i and external pressure p_e compress the shell radially, generating radial compression stresses. In accordance with the convention commonly adopted in continuum mechanics, these stresses are obviously negative (Timoshenko 1956, Cicala 1963, Faupel 1964).

The radial stresses at the inner radius r_i and the outer radius r_e are respectively:

$$\sigma_{ri} = -p_i \text{ and } \sigma_{re} = -p_e. \tag{1.14}$$

When the three principal stresses are assumed to be uniformly distributed through the thickness of the structure, the arithmetic mean of stresses σ_{ri} and σ_{re} is conventionally taken as the radial stress σ_r, i.e.:

$$\sigma_r = \frac{\sigma_{ri} + \sigma_{re}}{2} = -\frac{p_i + p_e}{2}. \tag{1.15}$$

In the particular cases where $p_e = 0$ or $p_i = 0$, relation (1.15) yields the following:

$$\sigma_r = -p_i/2 \text{ and } \sigma_r = -p_e/2. \tag{1.16}$$

All other conditions remaining equal, it should be noted that radial stress with current working pressures is an order of magnitude lower than the hoop and longitudinal stresses. Assuming a thin-walled circular cylinder ($d_i/s \geq 20$) under, for example, internal pressure alone ($p_i \neq 0$ and $p_e = 0$), a comparison of the first relations given by expressions (1.7), (1.13) and (1.16) shows that the hoop stress is approximately twice the longitudinal stress, and that the absolute value of radial stress is less than or at most equal to 1/20 and l/10 of the hoop stress and longitudinal stress respectively.

Accordingly, as the radial stress is an order of magnitude lower than the other two stresses, the stress state can be considered as biaxial when sizing thin-walled circular cylinders under pressure.

1.3 Design of Thin-Walled Circular Cylinders Under Internal and External Pressure

As was demonstrated in the foregoing sections, the stress state in thin-walled circular cylinders under internal and external pressure is triaxial, as it is characterized by the three principal stresses $\sigma_1 = \sigma_t$, $\sigma_2 = \sigma_z$ and $\sigma_3 = \sigma_r$, with $\sigma_t > \sigma_z > \sigma_r$. Where radial stress σ_r can be regarded as negligible, and thus zero, the stress state becomes biaxial. According to whether the stress state is considered to be triaxial ($\sigma_t \neq 0$; $\sigma_z \neq 0$; $\sigma_r \neq 0$) or biaxial ($\sigma_t \neq 0$; $\sigma_z \neq 0$; $\sigma_r = 0$), the diameter of the principal Mohr's circle (Mohr 1882, Chen 1976), on which the material's risk of failure depends, will be $(\sigma_t - \sigma_r)$, with $|\sigma_t - \sigma_r| > \sigma_t$, and σ_t respectively.

In sizing circular cylinders, it is in any case necessary to calculate the equivalent (or ideal) stress $\sigma_{e.}$ This presupposes that an appropriate strength theory has been selected (strength theory, strength criterion, and theory of failure are synonyms). The summary given in Table 1.1 shows:

- The four strength theories currently used in the design analysis and in the response (or verification) analysis of this type of structure in the first column,
- The general relations that express the equivalent stress for a triaxial stress state for each of these theories in the second column, and
- The corresponding specific relations for equivalent stresses obtained by substituting relations (1.5), (1.10) and (1.15) in the general relations (as indicated earlier, these substituted relations respectively give hoop stress, axial stress—under the assumption that ratio s/d_i is negligible with respect to unity—and radial stress) in the third column. Here, the circular cylinders mean diameter $d_m = (d_i + s) = (d_e - s)$ is introduced for two strength theories (maximum shear stress theory, τ_{max}, and maximum distortion energy, which will henceforth also be indicated as the ED_{max} theory).

In connection with the summary shown in Table 1.1, it should be noted that:

- With the maximum stress theory, σ_{max} (Lamé or Rankine theory, see Juvinall 1967; but also Galileo or Navier theory, see Cicala 1963), the equivalent stress depends on a single principal stress, hoop stress σ_t.
- With the maximum shear stress theory, τ_{max} (Coulomb 1773, Tresca 1864, Guest 1900 theory, see Juvinall 1967), the equivalent stress depends on two principle stresses, the maximum and the minimum, or in other words on stresses σ_t and σ_r. If radial stress σ_r can be regarded as negligible, the τ_{max} theory coincides with the σ_{max} theory.

Table 1.1 Design of thin-walled circular cylinders under internal and external pressure: summary of strength theories, the associated general relations for equivalent stress and the specific relations for free closed ends ($\sigma_z \neq 0$)

Strength theory	σ_e (General relation)	σ_e (Specific relation)
Maximum normal stress σ_{max} (Galileo, Lamé, Navier or Rankine theory)	$\sigma_e = \sigma_{max} = \sigma_t$	$\sigma_e = \frac{d_i}{2s}(p_i - p_e) - p_e$
Maximum principal strain ε_{max} (St. Venant's theory)	$\sigma_e = E\varepsilon_{max} = \sigma_t - \nu(\sigma_r + \sigma_z)$	$\sigma_e = \frac{d_i}{2s}(p_i - p_e)\left(1 - \frac{\nu}{2}\right) + \frac{\nu p_i - p_e(2 - 3\nu)}{2}$
Maximum shear stress τ_{max} (Tresca, Guest or, more appropriately, Coulomb theory)	$\sigma_e = \sigma_{max} - \sigma_{min} = \sigma_t - \sigma_r$	$\sigma_e = \frac{d_i + s}{2s}(p_i - p_e) = \frac{d_m}{2s}(p_i - p_e)$
ED_{max} (Huber, von Mises, Henckey or, more appropriately, Maxwell theory)	$\sigma_e^2 = \sigma_t^2 + \sigma_z^2 + \sigma_r^2 - (\sigma_t\sigma_z + \sigma_t\sigma_r + \sigma_z\sigma_r)$	$\sigma_e = \frac{(p_i - p_e)}{2s}\sqrt{3\left(\frac{d_i}{2}\right)^2 + 3s\frac{d_i}{2} + s^2} = \frac{(p_i - p_e)}{4s}\sqrt{3d_m^2 + s^2}$

- With the maximum normal strain theory or maximum principal strain theory, ε_{max} (Saint-Venant theory 1870), and maximum distortion energy theory, ED_{max} (Maxwell, Huber, Mises or Hencky theory, see Timoshenko 1953 and Juvinall 1967), the equivalent stress depends on all three principal stresses.

In the contest of mechanics of materials, we could be interested in both response analysis and in design analysis. In response (or verification) analysis, the structure's properties are given, and the response is to be determined in terms of stresses, strains, and deformations produced by loads, load-carrying capacity, allowable load, etc. The design analysis is the inverse process with respect to response analysis, where we must determine the structure's properties to ensure that the structure will support the required loads and perform its intended function (Gere and Timoshenko 1997).

Response analysis for the circular cylinder is carried out by establishing that the equivalent stress correlated with the selected strength theory is less than or at most equal to the material's allowable stress σ_a; this yields the relations given in the second column of Table 1.2.

In design analysis, on the other hand, the only unknown is thickness, as the inside diameter is already known from other design data. To determine thickness, it is necessary to establish that the equivalent stress is equal to the allowable stress for the material, and then solve the equality thus obtained for s. Performing these calculations yields the relations shown in the third column of Table 1.2. It should be pointed out that the design relation deriving from the maximum distortion energy criterion applies only if $(p_i – p_e)/\sigma_a < 4$.

The relations shown in Tables 1.1 and 1.2 were formulated for one of the two cases of practical interest, viz., that of the thin-walled circular cylinder under internal and external pressure with free closed ends (in this case, the ends are loaded, and $\sigma_z \neq 0$), which, as stated earlier, is characterized by a triaxial stress state.

The other case of practical interest is that of the thin-walled circular cylinder under internal and external pressure, but with free open (i.e., unloaded) ends,

Table 1.2 Design of thin-walled circular cylinders under internal and external pressure: summary of response and design analysis relations for the four strength theories in the case of bodies with free closed ends ($\sigma_z \neq 0$)

Strength theory	Response analysis relations	Design analysis relations
σ_{max}	$\frac{d_i}{2s}(p_i - p_e) - p_e \leq \sigma_a$	$s = \frac{d_i(p_i - p_e)}{2(\sigma_a + p_e)}$
ε_{max}	$\frac{d_i}{2s}(p_i - p_e)\left(1 - \frac{v}{2}\right) + \frac{vp_i - p_e(2 - 3v)}{2} \leq \sigma_a$	$s = \frac{d_i(p_i - p_e)\left(1 - \frac{v}{2}\right)}{2\sigma_a - [vp_i - p_e(2 - 3v)]}$
τ_{max}	$\frac{d_i + s}{2s}(p_i - p_e) = \frac{d_m}{2s}(p_i - p_e) \leq \sigma_a$	$s = \frac{d_i(p_i - p_e)}{2\sigma_a - (p_i - p_e)}$
ED_{max}	$\frac{(p_i - p_e)}{2s}\sqrt{3\left(\frac{d_i}{2}\right)^2 + 3s\frac{d_i}{2} + s^2} = \frac{(p_i - p_e)}{4s}\sqrt{3d_m^2 + s^2} \leq \sigma_a$	$s = \frac{d_i(p_i - p_e)}{\sqrt{\frac{16\sigma_a^2 - (p_i - p_e)^2}{3}} - (p_i - p_e)}$

which by contrast is characterized by a biaxial stress state ($\sigma_z = 0$). The relations for this case can be readily obtained from general equations given in the second column of Table 1.1. It is obvious that nothing will change if we use the σ_{max} and τ_{max} strength theories, as they do not contain stress σ_z. If, however, we use the ε_{max} and ED_{max} theories, the relations deriving from them will differ, as σ_z must be omitted from the general expressions.

Lastly, it should be observed that the third possible case, that of the thin-walled circular cylinder under internal and external pressure with clamped ends, is of little practical interest. In piping systems, which are almost always subject to thermal loads, constraining the ends would generate stresses high enough to lead to breakage or Euler instability, even for modest changes in temperature. This is why the piping systems in chemical and nuclear plants are supported by constraints that allow axial displacement, and sometimes also employ special expansion joints (expansion loops, bellows joints, etc.).

1.3.1 Thin-Walled Circular Cylinders Under Internal Pressure

For thin-walled circular cylinders under internal pressure with free closed ends, the relations in the last column of Table 1.1 expressing the equivalent stress in accordance with the four strength theories considered here are simplified, as p_e is omitted. Consequently, starting from these relations written for $p_e = 0$ leads to the results shown in Table 1.3.

To compare the results of using the strength theories which yield these results in the design of thin-walled circular cylinders under internal pressure only and assess their pros and cons quantitatively, it is advisable to introduce the dimensionless ratio p_i/σ_e, which is readily obtained from the equality relations (i.e., the relations preceding the inequality sign) given in the second column of Table 1.3.

Table 1.3 Design of thin-walled circular cylinders under internal pressure only: summary of expressions for equivalent stress and the response and design analysis relations for the four strength theories

Strength theory	Equivalent stresses and response analysis relations	Design analysis relations
σ_{max}	$\sigma_e = \frac{p_i\, d_i}{2s} \le \sigma_a$	$s = \frac{p_i\, d_i}{2\sigma_a}$
ε_{max}	$\sigma_e = \frac{p_i\, d_i}{2s}\left(1 - \frac{\nu}{2}\right) + \frac{\nu p_i}{2} \le \sigma_a$	$s = \frac{p_i\, d_i\left(1 - \frac{\nu}{2}\right)}{2\sigma_a - \nu p_i}$
τ_{max}	$\sigma_e = \frac{p_i\, d_m}{2s} \le \sigma_a$	$s = \frac{p_i d_i}{2\sigma_a - p_i}$
ED_{max}	$\sigma_e = \frac{p_i}{2s}\sqrt{3\left(\frac{d_i}{2}\right)^2 + 3s\frac{d_i}{2} + s^2} = \frac{p_i}{4s}\sqrt{3d_m^2 + s^2} \le \sigma_a$	$s = \frac{p_i\, d_i}{\sqrt{\frac{16\sigma_a^2 - p_i^2}{3}} - p_i}$

Table 1.4 Thin-walled circular cylinders under internal pressure only: dimensionless ratios $p_i/\sigma_e = f(\delta)$ and $p_i/\sigma_e = f(\beta)$

Strength theory	$p_i/\sigma_e = f(\delta)$	$p_i/\sigma_e = f(\beta)$
σ_{max}	$\frac{p_i}{\sigma_e} = \delta - 1$	$\frac{p_i}{\sigma_e} = \frac{(1-\beta)}{\beta}$
ε_{max}	$\frac{p_i}{\sigma_e} = \frac{2(\delta-1)}{(2-\nu)+\nu(\delta-1)}$	$\frac{p_i}{\sigma_e} = \frac{2(1-\beta)}{2\beta(1-\nu)+\nu}$
τ_{max}	$\frac{p_i}{\sigma_e} = \frac{2(\delta-1)}{(\delta+1)}$	$\frac{p_i}{\sigma_e} = \frac{2(1-\beta)}{(1+\beta)}$
ED_{max}	$\frac{p_i}{\sigma_e} = \frac{2(\delta-1)}{\sqrt{\delta^2+\delta+1}}$	$\frac{p_i}{\sigma_e} = \frac{2(1-\beta)}{\sqrt{1+\beta+\beta^2}}$

Expressing this ratio as a function of δ or $\beta = 1/\delta$ gives the results shown in Table 1.4. It should be noted that the p_i/σ_e ratios deriving from the σ_{max} strength theory, written either as a function of δ or as a function of β (see the third row in Table 1.4) are the well known Hamburg formula.

The different strength theories can be more readily compared by plotting the p_i/σ_e ratios they yield versus δ and/or β in a single diagram (Fig. 1.3). For the sake of completeness, Fig. 1.3 shows the abscissas in terms of δ, β and the ratio s/d_i. The curve for the ε_{max} theory is plotted for $\nu = 0.3$.

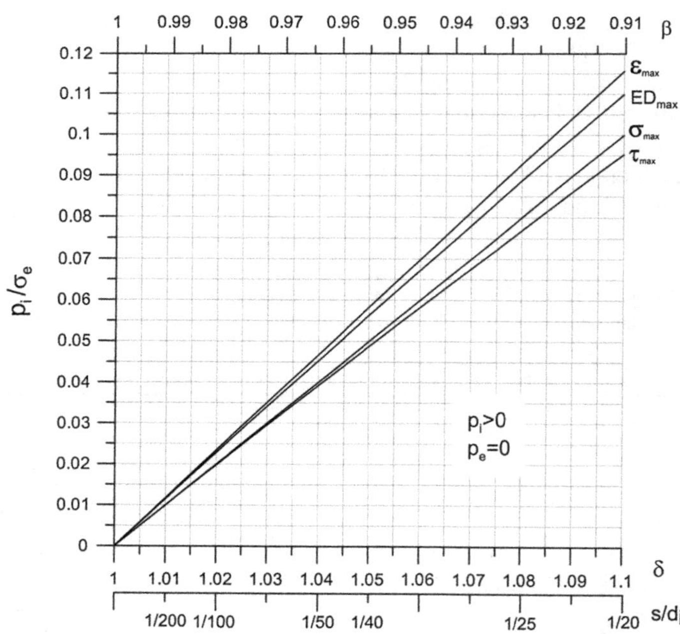

Fig. 1.3 Ratios p_i/σ_e versus δ, β and the ratio s/d_i for a thin-walled circular cylinder under internal pressure

As can be seen from the curves shown in Fig. 1.3, the τ_{max} strength theory is definitely the most conservative, as it results in the highest wall thickness for the structure for any given ratio p_i/σ_e. The σ_{max} strength theory leads to similar values for thickness, though they are somewhat less conservative than those from the τ_{max} strength theory. For any given ratio p_i/σ_e, the ED_{max} strength theory and, to an even greater extent, the ε_{max} strength theory result in thicknesses that are significantly lower than the other two strength theories.

As they provide a higher degree of safety, because of the greater formal simplicity of the relations stemming from them, the σ_{max} and τ_{max} strength theories are thus to be preferred to the others.

In addition, the τ_{max} strength theory, which establishes the most severe conditions for sizing thin-walled circular cylinders under internal pressure, is not only the basis for the best-established national standards for pressure vessels (ASME Section VIII, Division 1 is an exception, as it uses the σ_{max} strength theory) and pipes, but has also had its range of applications extended well beyond the conventional limit for thin-walled bodies, i.e., for $\delta > 1.10$ (see Sect. 1.1) and has at times been used as the strength theory for determining the relations for thick-walled circular cylinders.

In conclusion, as the curves in Fig. 1.3 show, thin-walled circular cylinders can withstand internal pressures equal at most to one tenth of the material's allowable or working stress.

1.3.2 *Thin-Walled Circular Cylinders Under External Pressure*

As was indicated earlier, the strength analysis (in terms both of design analysis and response analysis) for this type of structure is not exhaustive, and must be completed by checking elastic instability, a topic covered in Chap. 2.

Here again, it will be assumed that the thin-walled circular cylinder under external pressure alone has free closed ends. As was the case for the body under internal pressure, the relations in the last column of Table 1.1 expressing the equivalent stress in accordance with the four strength theories considered here are also simplified in this case, as p_i is omitted. Starting from these relations written for $p_i = 0$ leads to the results shown in Table 1.5.

For a clearer comparison of the results of using the strength theories which yield these results in the design of thin-walled circular cylinders under external pressure alone and to assess their advantages and disadvantages quantitatively, it is once again advisable to introduce the dimensionless ratio $-p_e/\sigma_e$ which is readily obtained from the equality relations (i.e., the relations preceding the inequality sign) given in the second column of Table 1.5. Expressing this ratio as a function of δ or $\beta = 1/\delta$ gives the results shown in Table 1.6.

Table 1.5 Design of thin-walled circular cylinders under external pressure only: summary of expressions for equivalent stress and the response and design analysis relations for the four strength theories

Strength theory	Equivalent stresses and response analysis relations	Design analysis relations
σ_{max}	$\sigma_e = -\dfrac{p_e\,d_e}{2s} \le -\sigma_a$	$s = \dfrac{p_e\,d_e}{2\sigma_a}$
ε_{max}	$\sigma_e = -\dfrac{p_e\,d_e}{2s}\left(1-\dfrac{\nu}{2}\right) + \nu p_e \le -\sigma_a$	$s = \dfrac{p_e\,d_e\left(1-\dfrac{\nu}{2}\right)}{2(\sigma_a + \nu p_e)}$
τ_{max}	$\sigma_e = -\dfrac{p_e\,d_m}{2s} \le -\sigma_a$	$s = \dfrac{p_e d_{i,e}}{2\sigma_a + p_e}$
ED_{max}	$\sigma_e = -\dfrac{p_e}{2s}\sqrt{3\left(\dfrac{d_i}{2}\right)^2 + 3s\dfrac{d_i}{2} + s^2}$ $= -\dfrac{p_e}{4s}\sqrt{3d_m^2 + s^2} \le -\sigma_a$	$s = \dfrac{p_e\,d_e}{\sqrt{\dfrac{16\sigma_a^2 - p_e^2}{3}} + p_e}$

Table 1.6 Thin-walled circular cylinders under external pressure only: dimensionless ratios $-p_e/\sigma_e = f(\delta)$ and $-p_e/\sigma_e = f(\beta)$

Strength theory	$-p_e/\sigma_e = f(\delta)$	$-p_e/\sigma_e = f(\beta)$
σ_{max}	$-\dfrac{p_e}{\sigma_e} = \dfrac{(\delta-1)}{\delta}$	$-\dfrac{p_e}{\sigma_e} = 1-\beta$
ε_{max}	$-\dfrac{p_e}{\sigma_e} = \dfrac{2(\delta-1)}{\delta(2+\nu)-2\nu}$	$-\dfrac{p_e}{\sigma_e} = \dfrac{2(1-\beta)}{2+\nu(1-2\beta)}$
τ_{max}	$-\dfrac{p_e}{\sigma_e} = \dfrac{2(\delta-1)}{(\delta+1)}$	$-\dfrac{p_e}{\sigma_e} = \dfrac{2(1-\beta)}{(1+\beta)}$
ED_{max}	$-\dfrac{p_e}{\sigma_e} = \dfrac{2(\delta-1)}{\sqrt{\delta^2+\delta+1}}$	$-\dfrac{p_e}{\sigma_e} = \dfrac{2(1-\beta)}{\sqrt{1+\beta+\beta^2}}$

In this as in the earlier case, different strength theories can be more readily compared by plotting the $-p_e/\sigma_e$ ratios they yield versus δ and/or β in a single diagram (Fig. 1.4). For the sake of completeness, Fig. 1.4 shows the abscissas in terms of δ, β and the ratio s/d_i (the latter can also be given as a percentage). The curve for the ε_{max} strength theory is plotted for $\nu = 0.3$. As can be seen from the curves in Fig. 1.4, the most prudential criterion for sizing thin-walled circular cylinders under external pressure is the σ_{max} strength theory which, by comparison with the other strength theories, has the major advantage of yielding a very simple expression for thickness (see Table 1.5).

1.4 Strain State in Thin-Walled Circular Cylinders

The strain state in thin-walled circular cylinders under internal and external pressure, as defined by the principal strains (radial strain ε_r, tangential or circumferential strain ε_t and axial or longitudinal strain ε_z), is found by using the

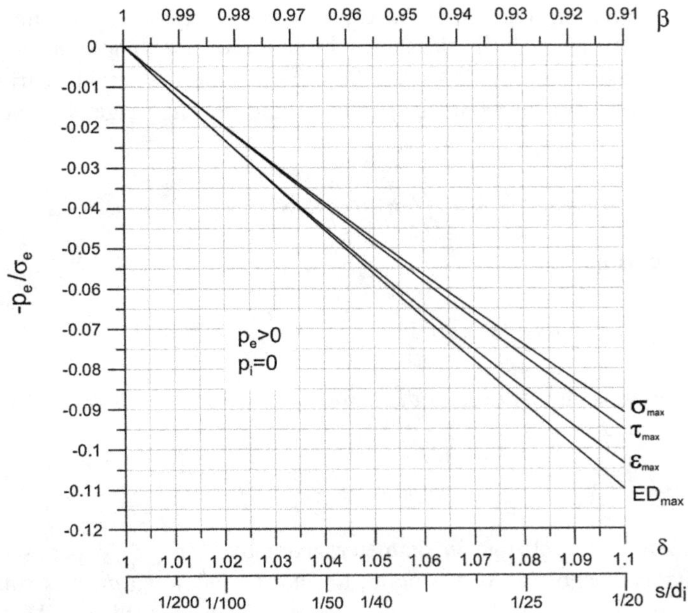

Fig. 1.4 Ratios $-p_e/\sigma_e$ versus δ, β and the ratio s/d_i for thin-walled circular cylinders under external pressure

following thermoelastic stress–strain relations, also called generalized Hooke's relations or constitutive laws (Desai and Sirilwardane 1984, Belloni and Bernasconi 1984, Shames and Cozzarelli 1997):

$$\begin{aligned}\varepsilon_r &= \frac{1}{E}[\sigma_r - \nu(\sigma_t + \sigma_z)] + \alpha T\\ \varepsilon_t &= \frac{1}{E}[\sigma_t - \nu(\sigma_z + \sigma_r)] + \alpha T\\ \varepsilon_z &= \frac{1}{E}[\sigma_z - \nu(\sigma_r + \sigma_t)] + \alpha T,\end{aligned} \tag{1.17}$$

from which temperature terms have been omitted, after substituting relations (1.5), (1.9) or (1.10) and (1.15), which respectively express hoop stress σ_t, axial stress σ_z (in exact or approximate terms) and radial stress σ_r.

It must be recognized, however, that quantifying the principal strains ε_r, ε_t and ε_z through this procedure is of no great engineering significance. From the design standpoint, it is important only to evaluate the diameter's expansion (or contraction) as a result of the pressures acting on the inner and outer radii. This evaluation can be performed using the following procedure which, though simplified, leads to results that can be used in design procedures and practical applications.

Let d_i' be the inside diameter of the circular cylinder after deformation, and let $L = \pi(d_i + s)$ and $L' = \pi(d_i' + s)$ be the length of the mean circumference before and after deformation. Obviously, L' can also be expressed as the sum $L + \Delta L$, where ΔL is the change in the mean circumference's length as a result of the diameter's expansion. As is known, however, the normal strain ε deriving from a uniaxial equivalent stress σ_e is given, in the elastic range, by the relation $\varepsilon = \Delta L/L = \sigma_e/E$, so that:

$$\Delta L = L\frac{\sigma_e}{E} = \frac{\sigma_e}{E}\pi(d_i + s). \tag{1.18}$$

We will thus have:

$$L' = L + \Delta L = \pi d_i\left(1 + \frac{s}{d_i}\right)\left(1 + \frac{\sigma_e}{E}\right) = \pi d_i'\left(1 + \frac{s}{d_i'}\right). \tag{1.19}$$

Assuming as usual that ratios $s/d_i \cong s/d_i'$ are negligible with respect to unity, it can be concluded that

$$d_i' = d_i\left(1 + \frac{\sigma_e}{E}\right). \tag{1.20}$$

Consequently, the change in diameter given by $d_i(\sigma_e/E)$ depends on which strength theory is chosen. At this point, to find the unknown d_i', it is sufficient to substitute the expressions for equivalent stress given in Tables (1.1), (1.3) and (1.5) which apply, in that order, to the general case ($p_i \neq 0$ and $p_e \neq 0$) and to the particular cases ($p_i \neq 0$ and $p_e = 0$) and ($p_i = 0$ and $p_e \neq 0$), in relation (1.20). This task will be left to the reader.

For a more rigorous approach, radial displacement u can be calculated by recalling that $\varepsilon_t = u/r$. Consequently, using the second relation (1.17) without the temperature term and ignoring the contribution of radial stress σ_r, inasmuch as it is an order of magnitude lower than σ_t and σ_z, and substituting the first relation (1.7) and the first relation (1.13) in it gives the following relation for the radial displacement u_i undergone by the generator of radius r_i as a result of internal pressure p_i alone:

$$u_i = \frac{p_i r_i^2}{Es}\left(1 - \frac{\nu}{2}\right). \tag{1.21}$$

It can then be concluded that $d_i' = d_i + 2u_i$. Proceeding in the same way, we can determine the radial displacement of the same generator when the circular cylinder is under external pressure p_e alone.

1.5 Effective Hoop Stress Distribution Through the Wall Thickness

The assumption that the principal stresses in a thin-walled circular cylinder under internal and/or external pressure are uniformly distributed through the wall thickness is a rough approximation, and the larger the thickness is by comparison

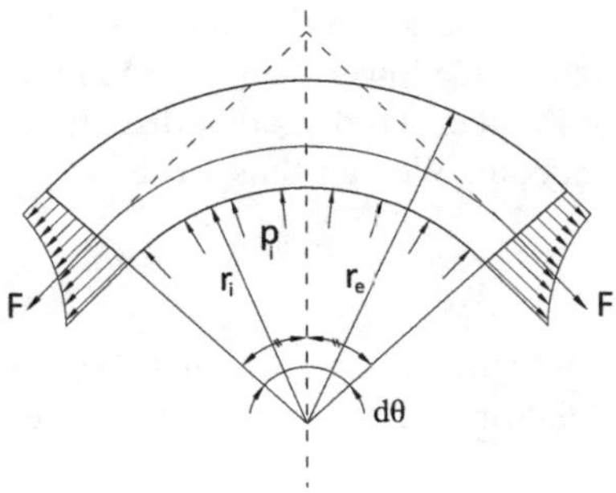

Fig. 1.5 Circumferential small element of a circular cylinder under internal pressure and effective hoop stress distribution through its wall thickness

with the diameter, the rougher this approximation will be. With the exception of the zones near the closed ends, assuming that the longitudinal stress is constant is a sufficiently close approximation if there is no thermal load. The same cannot be said for radial and hoop stress even if there is no thermal load. As was pointed out earlier, radial stress is an order of magnitude lower than hoop stress. We will thus concentrate our attention on the latter.

From the conceptual standpoint, there is no reason to distinguish between the thin-walled and thick-walled circular cylinder. Even in the thin-walled circular cylinder, the hoop stress distribution through the thickness cannot be constant as shown in Fig. 1.1b, but varies along a cubic hyperbola, as will be demonstrated in Chap. 4. For a circular cylinder under internal pressure alone, the hoop stress will assume its maximum value at the inner radius and its minimum value at the outer radius, as shown in Fig. 1.5.

We will now consider radial equilibrium condition for the small element shown in Fig. 1.5, which is assumed to be of unitary axial length, under internal pressure $\boldsymbol{p_i}$ and open at the ends, so that we will have $\sigma_z = 0$. Following the same procedure used to arrive at Mariotte's formulas for boilers (1.7), but considering a circumferential small element of a circular cylinder having angular width $d\theta$ rather than π as in Fig. 1.1b, and designating the resultant of stresses σ_t distributed along the radius as

$$F = \int_{r_i}^{r_e} \sigma_t \, dr, \tag{1.22}$$

we obtain the relation

$$p_i r_i d\theta = 2F sin\frac{d\theta}{2} \tag{1.23}$$

which, given $sin(d\theta/2) = d\theta/2$, leads to:

$$p_i r_i = F. \tag{1.24}$$

If we assume that both radial stress σ_r (which as we have said is an order of magnitude lower than σ_t), and radial strain $\varepsilon_r = du/dr$, are zero, so that $u = const$, or in other words, radial displacement is constant, it follows that the stress state is uniaxial. We can thus write:

$$\sigma_t = E\varepsilon_t = E\frac{u}{r} \tag{1.25}$$

where $\sigma_t r = \sigma_{t,i} r_i = const$, and where $\sigma_{t,i}$ represents hoop stress at radius r_i. Relation (1.22) can thus be written in the following form:

$$F = \sigma_{t,i} r_i \int_{r_i}^{r_e} \frac{dr}{r}. \tag{1.26}$$

Recalling relations (1.24) and (1.26) and considering that in a uniaxial stress state the equivalent stress is equal to the maximum hoop stress at the inner radius, i.e., $\sigma_{e,i} = \sigma_{t,i}$, we will have:

$$\sigma_{e,i} = \frac{p_i}{\ln(r_e/r_i)} = \frac{2p_i}{\ln(r_e/r_i)^2}. \tag{1.27}$$

If we now perform a power series expansion of the logarithm appearing in this relation and set $(r_e/r_i)^2 = x$, with x obviously greater than 1, we obtain:

$$\ln x = 2\left[\frac{x-1}{x+1} + \frac{(x-1)^3}{3(x+1)^3} + \cdots\right]. \tag{1.28}$$

By stopping the expansion at the first term in the series, and thus using a type of approximation that is entirely different from the usual one employed earlier (uniform distribution of stresses through the wall thickness), we will have:

$$\ln\frac{r_e^2}{r_i^2} = 2\frac{r_e^2 - r_i^2}{r_e^2 + r_i^2} \tag{1.29}$$

which, substituted in relation (1.27), enables us to write:

$$\sigma_{e,i} = p_i \frac{r_e^2 + r_i^2}{r_e^2 - r_i^2} = p_i \frac{\delta^2 + 1}{\delta^2 - 1}. \tag{1.30}$$

This relation coincides with that shown at the top left of Table 4.3 for the σ_{max} strength theory applied to thick-walled circular cylinders under internal pressure. Paradoxically, however, the design importance of the latter is less than that of relation (1.27), which can also be expressed in the following form:

$$\frac{r_e}{r_i} = e^{\frac{p_i}{\sigma_{e,i}}}. \tag{1.31}$$

As a number of researchers have demonstrated, this form reflects the experimental findings more accurately.

It should also be emphasized that, as we will demonstrate in Chap. 4 on the basis of the theory applying to thick-walled circular cylinders under internal pressure and stressed in the elastic range, relation (1.30) is an exact formula. It was obtained here using a procedure as singular as it is interesting, given that it is based on different kinds of approximation whose effects in fact compensate for each other.

1.6 Thin-Walled Circular Cylinders and Membrane Theory of Shells

The stress state in a thin-walled circular cylinder under internal pressure p_i and stressed in the linear elastic field can be determined from the relations obtained with the membrane theory of shells of revolution (Timoshenko and Woinowsky-Krieger 1959, Flügge 1960). As is known, the in-plane resultant forces N_φ and N_θ, to which a membrane of revolution loaded and constrained symmetrically with respect to the rotation axis is subjected, are linked by the following relations:

$$\frac{N_\varphi}{r_I} + \frac{N_\theta}{r_{II}} = Z \tag{1.32}$$

and

$$N_\varphi = \frac{Q}{2\pi r \sin\varphi} \tag{1.33}$$

where:

- N_φ and N_θ are respectively the meridian and circumferential normal in-plane forces (stress resultants) per unit of length acting at a generic point of the membrane having spherical coordinates φ and θ (these stress resultants act, in order, along the directions tangent to the meridian and to the parallel passing through that point).
- φ is the meridian angle (or colatitude angle), i.e., the angle formed by the extended normal in the generic point of the shell and the axis of rotation, and thus defines the position of the parallel.
- θ is the circumferential angle, i.e., the angle formed by the meridian passing through the point in question relative to a meridian taken as a reference, and thus defines the position of the meridian.
- r_I and r_{II} are the principal radii of curvature for this generic point, respectively in the meridian plane (the principal radius of the meridian) and in the plane perpendicular to the later (the principal radius of the parallel circle) containing the extended normal. These two planes are thus the principal planes of curvature, and the straight line obtained intersecting these planes is inclined in the

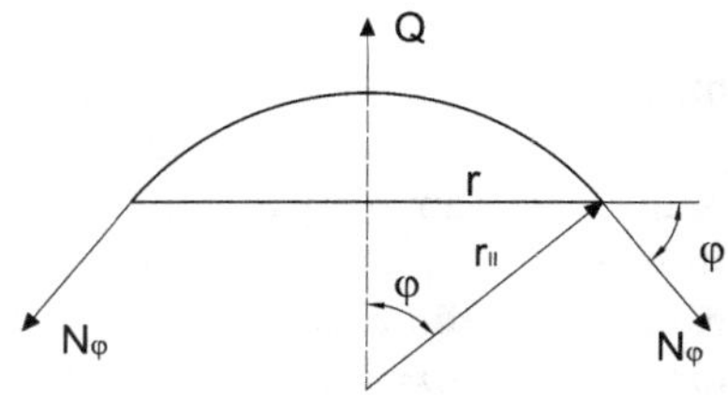

Fig. 1.6 Portion of the shell symmetrical to the axis above the parallel of angular coordinate φ and loads acting on it

meridian plane by an angle φ relative to the axis of rotation; thus r_{II} is the distance along the extended normal to the meridian curve drawn from the point in question to the axis of revolution of the surface.

- $r = r_{II}\sin\varphi$ is the radius of the parallel circle, i.e., the distance of the point in question from the axis of rotation.
- Z is the component in the direction of the normal of the distributed load per unit of surface acting on the shell (here, $Z = p_i$).
- Q is the resultant in the direction of the rotational axis of the external total load acting on a portion of the membrane symmetrical to the axis and bounded by the parallel of angular coordinate φ (Fig. 1.6): this resultant is here assumed to be positive if it is directed towards the exterior of the shell.

For a thin-walled cylindrical shell of annular cross-section, we have: $r_I = \infty$, $r = r_{II}$ and $\varphi = \pi/2$. Consequently, relations (1.32) and (1.33) will become respectively:

$$N_\theta = r_{II}Z \tag{1.34}$$

and

$$N_\varphi = \frac{Q}{2\pi r_{II}}. \tag{1.35}$$

These relations can also be obtained directly by imposing equilibrium on the radial and axial directions of a small element of the cylindrical shell bounded by two planes perpendicular to the rotation axis, of abscissa φ and $\varphi + d\varphi$ (in this case, spherical coordinate φ coincides with axial coordinate z, so that these two planes are those having coordinates z and $z + dz$), and by two meridian planes forming the elementary angle $d\theta$.

In our case, we have $Z = p_i$ and $r_{II} = d_i/2$; consequently, relation (1.34) yields:

$$N_\theta = \frac{d_i p_i}{2} = N_t; \tag{1.36}$$

but since $N_\theta = N_t = \sigma_t s$, relation (1.36) coincides with the first relation (1.7).

For the same shell assumed to be closed at the ends, we have $Q = \pi d_i^2 p_i/4$. Relation (1.35) thus gives:

$$N_\varphi = \frac{d_i p_i}{4} = N_z. \tag{1.37}$$

Here again, however, $N_\varphi = N_z = \sigma_z s$, so relation (1.37) coincides with the first relation (1.13).

According to the membrane theory of shells, the third principal stress is assumed to be zero, i.e., $\sigma_r = 0$.

For other assumptions, the membrane theory thus confirms the results given in Sect. 1.2, which were obtained by imposing simple boundary conditions on non-small elements of the structure in question.

However, the strain state of the cylindrical shell can also be obtained as a particular case of that for the shell of revolution under membrane stress state (stress couples and thus bending and twisting moments are both zero). For this membrane stress state, in fact we have:

$$\begin{aligned}\varepsilon_\varphi &= \frac{1}{E}\left(\sigma_\varphi - \nu\sigma_\theta\right) = \frac{1}{Es}\left(N_\varphi - \nu N_\theta\right)\\ \varepsilon_\theta &= \frac{1}{E}\left(\sigma_\theta - \nu\sigma_\varphi\right) = \frac{1}{Es}\left(N_\theta - \nu N_\varphi\right).\end{aligned} \tag{1.38}$$

Specializing these relations for the thin-walled annular cylindrical body considered here, relations (1.36) and (1.37) will yield:

$$\begin{aligned}\varepsilon_t = \varepsilon_\theta &= \frac{p_i d_i}{2Es}\left(1 - \frac{\nu}{2}\right)\\ \varepsilon_z = \varepsilon_\varphi &= \frac{p_i d_i}{2Es}\left(\frac{1}{2} - \nu\right).\end{aligned} \tag{1.39}$$

Finally, let u, v and w be the displacements of a generic point of the cross-section in the directions of the radius, the tangent to the circumference passing through this point and the z axis respectively. Because of axial symmetry, we will obviously have $v = 0$. Recalling that $\varepsilon_t = u/r$, the first of these relations gives the following expression for the generator's radial displacement at cylinder inner radius r_i:

$$u_i = \frac{p_i d_i^2}{4Es}\left(1 - \frac{\nu}{2}\right). \tag{1.40}$$

This displacement coincides with that given by relation (1.21).

Remembering then that $\varepsilon_z = dw/dz$, the second relation (1.39) yields:

$$\frac{dw}{dz} = \frac{p_i d_i}{2Es}\left(\frac{1}{2} - \nu\right) \tag{1.41}$$

which, integrated, gives:

$$w = \frac{p_i d_i}{2Es}\left(\frac{1}{2} - \nu\right)z + C, \tag{1.42}$$

where C is an integration constant to be calculated by setting boundary conditions. Placing the origin of the z axis in the cylindrical shell's end plane and imposing $w = 0$ for $z = 0$ yields $C = 0$, whereby:

$$w = \frac{p_i d_i}{2Es}\left(\frac{1}{2} - \nu\right)z. \tag{1.43}$$

Lastly, we remember that there are two different classes of shells: thin shells and thick shells. A shell is called thin if the maximum value of the ratio $s/r_{I,II,min}$ ($r_{I,II,min}$ is the minimum principal radius of curvature of the shell's middle surface) can be neglected in comparison with unity. As Novozhilov (1964) indicates, for design applications, a shell may be regarded as thin if the following condition is satisfied:

$$max\left(\frac{s}{r_{I,II}}\right) \leq \frac{1}{20}. \tag{1.44}$$

Hence, shells for which this inequality is not satisfied are referred to as thick shells. The thickness of shells utilized for a large number of practical applications lies in the range of thin shells, because generally we have:

$$\frac{1}{1000} \leq max\left(\frac{s}{r_{I,II}}\right) \leq \frac{1}{20}. \tag{1.45}$$

Consequently the theory of thin shells is of great engineering importance. However the boundary between thin and thick shells is very roughly defined by the inequality (1.44), because it depends also upon other parameters (geometry of the shells, type of boundary conditions, the variation of external surface forces and its smoothness, etc.).

Chapter 2
Instability of Thin-Walled Circular Cylinders Under External Pressure

2.1 General

A wide variety of chemical and petrochemical processes involve using equipment operating under partial vacuum. Examples include vacuum condensers for evaporators and distillation columns, the vacuum columns used in manufacturing lubricating oils, vacuum crystallizers, and so on. In other industrial processes, vessels are sometimes jacketed and heated (or cooled) by means of fluids circulating under pressure between the inner and outer shells of the jacket. These vessels are generally cylindrical and feature formed heads at the ends as closures (Brownell and Young 1968).

Innumerable examples of cylindrical structures under external pressure are also found in other technological sectors; here, mention should be made of the jacketed vessels used in the wine, and more generally, food processing industries, boiler and heat exchanger tubes, internal and external circumferential stiffening rings (stiffeners) of reinforced vessels and hulls of submarines and bathyscaphes, and many more. These latter stiffening rings increase the collapsing strength of such structures.

In the previous sections, it was demonstrated that the circumferential compressive stress σ_t in a thin-walled circular cylinder under external pressure is equal to twice the longitudinal stress σ_z, which is likewise compressive. But rings or cylindrical bodies of annular cross-sections are liable to collapse as a result of combined uniform compression and bending due to an external pressure alone if their flexural rigidity (or bending stiffness) is insufficient. In addition, the structure's collapse may take place at stresses well below the yield point of the material. The pressure at which the annular shape becomes unstable and begins to buckle, and the deflections increase indefinitely, is called the *critical pressure* (Cicala 1940 and 1940, Batdorf 1947, Timoshenko and Gere 1961, Belluzzi 1971, Bruhn 1973).

The collapsing strength of cylindrical structures of annular cross-section, however, can be increased by the use of internal or external stiffening rings uniformly spaced in the direction of the axis. From the standpoint of elastic stability, such stiffening rings have the effect of subdividing the structure into substructures whose length is equal to the center-to-center spacing of each stiffener.

V. Vullo, *Circular Cylinders and Pressure Vessels*, Springer Series in Solid and Structural Mechanics 3, DOI: 10.1007/978-3-319-00690-1_2,

Long, thin circular cylinders without stiffeners or with stiffeners spaced beyond the so-called *critical length* will buckle at stresses below the yield point of the material. The corresponding critical pressure at which buckling occurs is a function only of the modulus of elasticity, E, of the material and of the s/d_i ratio. If the length of the shell with closures, l, or the distance between uniformly spaced stiffeners, l, are less than the critical length, the critical pressure at which collapse occurs is a function of the modulus of elasticity, E, and of the s/d_i ratio as well as of the l/d_i ratio.

2.2 Instability of a Thin-Walled Ring

Consider a thin curved beam with a circular centerline, that is a thin-walled ring with a circular axis, having the geometry shown in Fig. 2.1 and subject to a pressure $p = (p_e - p_i)$ which is greater than zero.

As a result of this pressure, a hoop stress σ_t is generated which is given by the second relation (1.7), writing p instead of p_e. As p rises, the absolute value of the stress σ_t increases in accordance with this relation and, at the same time, there is a contraction in the diameter which can be calculated in approximate terms using relation (1.20), bearing in mind that the equivalent stress is negative. Again in approximate terms and also because of the small thickness, we can substitute the outside diameter d_e in the second relation (1.7) with the initial middle diameter d_m or the diameter $d_0 = 2r_0$ of the middle fiber immediately before buckling begins. This latter diameter, which differs from the initial middle diameter by a small second-order quantity, is that corresponding to the elastically deformed shape,

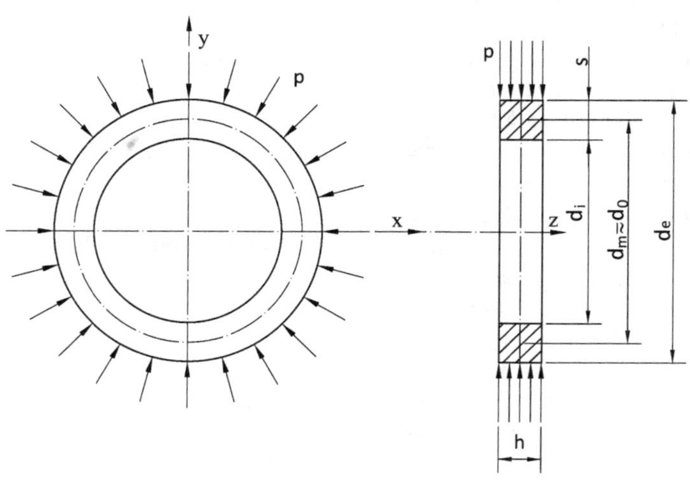

Fig. 2.1 Thin-walled annular ring; reference system and geometry

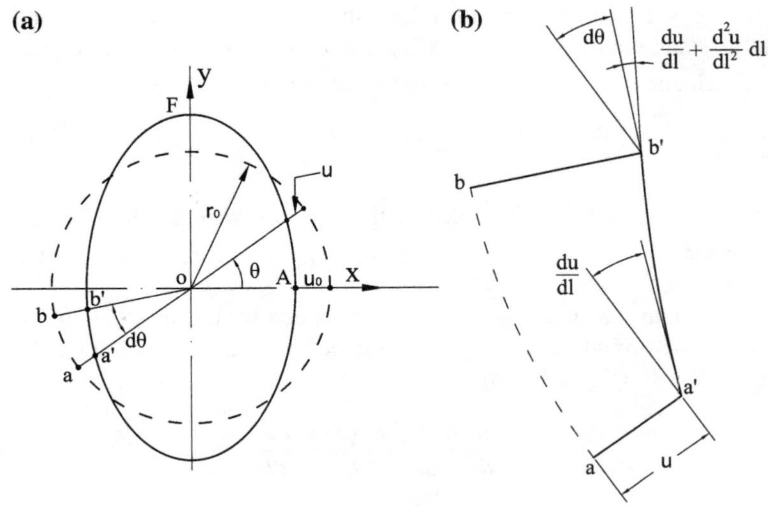

Fig. 2.2 **a** Axisymmetrically deformed shape (*dashed line*) and buckled shape (*solid line*); **b** small elements, with displacements and rotations

which is still circular, at incipient buckling, and is represented with a dashed line in Fig. 2.2a. In this figure, the dashed line also indicates the initial circular shape of the ring centerline, because it is assumed that the strains and deformations are small.

We will consider the buckled shape of the ring's mean fiber, as represented with a solid line in Fig. 2.2a (this solid line represents the slightly deflected centerline of the ring on which a uniformly distributed pressure is acting) and assume its two axes of symmetry x and y as the Cartesian reference system. The figure also shows the polar coordinate reference system, with angular coordinate θ between the x axis and the generic radial line from the center O of the annular beam, and regarded as positive when measured counterclockwise. Let u be the radial displacements resulting from bending, which are assumed to be very small compared to the ring's radius r_0 (the initial radius of curvature of the centerline of the ring) and positive if directed towards the center O of the annular beam.

To formulate the differential equation of the deflection curve, we must first express the middle fiber's change in curvature $(1/r_1 - 1/r_0)$ resulting from bending; $r_1 = r_1(\theta)$ is the radius of curvature after deformation at any point of the centerline, defined by the angle θ. We will consider small elements $\widehat{ab}$ and $\widehat{a'b'}$ of the ring (shown on a normal scale in Fig. 2.2a, and on a larger scale in Fig. 2.2b), bounded by the same radii starting from the center O and forming angle $d\theta$, and belonging respectively to the axisymmetrically deformed shape and to the buckled shape. The initial length and the initial curvature of small element $\widehat{ab}$ are given by:

$$dl = r_0 \cdot d\theta \quad \text{and} \quad 1/r_0 = d\theta/dl \tag{2.1}$$

Assuming small displacements (Tomoshenko 1956, Timoshenko and Gere 1961), this small element's post-buckling curvature can be supposed equal to that of small element $\widehat{a'b'}$ which can be written in the following form:

$$\frac{1}{r_1} = \frac{d\vartheta + \Delta d\vartheta}{dl + \Delta dl}, \tag{2.2}$$

where $(d\vartheta + \Delta d\vartheta)$ and $(dl + \Delta dl)$ denote the angle between the normal sections of the deformed ring at a' and b', and the length of the small element $\widehat{a'b'}$ respectively.

Under the same assumption, the angles bounded by tangent lines in points a' and b' to the deformed middle fiber and the normal lines to radii aO and bO are given respectively (Fig. 2.2b) by:

$$\frac{du}{dl} \quad \text{and} \quad \frac{du}{dl} + \frac{d}{dl}\left(\frac{du}{dl}\right)dl = \frac{du}{dl} + \frac{d^2u}{dl^2}dl. \tag{2.3}$$

The change in angle $\Delta d\vartheta$ is then given by:

$$\Delta d\vartheta = \frac{d^2u}{dl^2}dl. \tag{2.4}$$

The effect of angle *du/dl* can be ignored in calculating the length of small element $\widehat{a'b'}$, which can thus be assumed to be equal to $(r_0 - u)d\vartheta$. Thus, as $(dl + \Delta dl) = (r_0 - u)d\vartheta$, we have:

$$\Delta dl = -ud\vartheta = -udl/r_0. \tag{2.5}$$

Substituting expressions (2.4) and (2.5) in relations (2.2) gives:

$$\frac{1}{r_1} = \frac{d\vartheta + \frac{d^2u}{dl^2}dl}{dl\left(1 - \frac{u}{r_0}\right)}; \tag{2.6}$$

from which, omitting small higher-order quantities, we obtain the following relation:

$$\frac{1}{r_1} = \frac{d\vartheta}{dl}\left(1 + \frac{u}{r_0}\right) + \frac{d^2u}{dl^2} = \frac{1}{r_0}\left(1 + \frac{u}{r_0}\right) + \frac{d^2u}{dl^2}. \tag{2.7}$$

This then leads to the following expression for the change in curvature:

$$\frac{1}{r_1} - \frac{1}{r_0} = \frac{u}{r_0^2} + \frac{d^2u}{dl^2}. \tag{2.8}$$

On the other hand, the approximate theory of curved beams formulated by Résal and by Winkler, further developed by Grashof and Pearson, and perfected by Golovin (Timoshenko 1953, Timoshenko 1956), states that the neutral axis (which in the more general case does not pass through the center of gravity) will be at the centroid when the cross-sectional height, which here is the same as thickness s, is small by comparison with the radius of curvature. Consequently, we have:

$$\Delta d\vartheta = \frac{M r_0 d\vartheta}{EI} = \frac{M dl}{EI}, \tag{2.9}$$

where M is the bending moment, E the Young's modulus of the material, and $I = h\, s^3/12$ is the moment of inertia of the effective cross-sectional area, of surface $A = sh$ (Fig. 2.1). However, since in this case $(1/r_1 - 1/r_0) = \Delta d\vartheta/dl$, we have:

$$\frac{1}{r_1} - \frac{1}{r_0} = -\frac{M}{EI} \tag{2.10}$$

where the minus sign is due to the fact that the bending moment is assumed to be positive when it causes a reduction in the annular beam's initial curvature.

Finally, relations (2.8) and (2.10) yield:

$$\frac{d^2u}{dl^2} + \frac{u}{r_0^2} = -\frac{M}{EI}, \tag{2.11}$$

which is the differential equation for the deflection curve of a thin beam with a circular centerline. It should be noted that, where r_0 is infinitely large, this equation is reduced to that for straight beams.

To solve our problem, bearing the first relation (2.1) in mind, it is advisable to write relation (2.11) in the following form:

$$\frac{d^2u}{d\theta^2} + u = -r_0^2 \frac{M}{EI}. \tag{2.12}$$

To calculate the bending moment M in any cross-section B identified by angle ϑ relative to the reference section A (the x axis), we will consider the deformed portion $\widehat{AB}$ of the annular beam bounded by said section B and the x axis, which, coinciding with one of the two axes of symmetry, locally intersects the tangent to the deflection curve portion under a right angle (Fig. 2.3). By imposing rotational equilibrium around point B for all loads acting on this portion, we obtain:

$$M = M_0 + N_0 \cdot \overline{AD} - q\frac{\overline{AB}^2}{2} \tag{2.13}$$

where $q = ph$ is the uniform external load (or uniform normal pressure) per unit of length of the ring centerline and $N_0 = q(r_0 - u_0)$ is the normal compressive force in section A. Consequently, relation (2.13) can be written in the following form:

$$M = M_0 + q \cdot \overline{AO} \cdot \overline{AD} - q\frac{\overline{AB}^2}{2} = M_0 - q\left(\frac{\overline{AB}^2}{2} - \overline{AO} \cdot \overline{AD}\right) \tag{2.14}$$

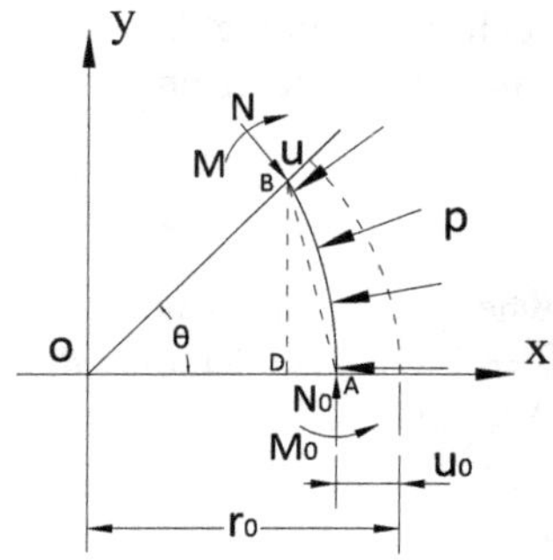

Fig. 2.3 Stress resultants and displacements of ends *A* and *B* of the deformed portion

We can now apply Carnot's theorem to the triangle $\widehat{ABO}$ or the Pythagorean theorem to triangles $\widehat{ABD}$ and $\widehat{OBD}$ to obtain the following in successive steps:

$$\frac{\overline{AB}^2}{2} - \overline{AO}\cdot\overline{AD} = \frac{1}{2}\left(\overline{OB}^2 - \overline{OA}^2\right) = \frac{1}{2}\left[(r_0 - u)^2 - (r_0 - u_0)^2\right] \tag{2.15}$$

where u_0 is the radial displacement at *A* (Fig. 2.3).

Expanding this latter expression and ignoring the small second-order terms in u_0^2 and u^2 then yields:

$$\frac{\overline{AB}^2}{2} - \overline{AO}\cdot\overline{AD} = r_0(u_0 - u). \tag{2.16}$$

Finally, relation (2.14) gives the following expression for the bending moment:

$$M = M_0 - qr_0(u_0 - u). \tag{2.17}$$

Recalling that $q = ph$ and substituting relation (2.17) in relation (2.12) yields the following non-homogeneous second order differential equation in terms of u:

$$\frac{d^2u}{d\vartheta^2} + u = -\frac{r_0^2}{EI}\left[M_0 - phr_0(u_0 - u)\right]. \tag{2.18}$$

This can be rewritten in extended form:

$$\frac{d^2u}{d\vartheta^2} + u\left(1 + \frac{phr_0^3}{EI}\right) = \frac{-M_0r_0^2 + phr_0^3u_0}{EI} \tag{2.19}$$

or in the more compact form:

$$\frac{d^2u}{d\vartheta^2} + k^2u = k' \tag{2.20}$$

where k^2 and k' are given respectively by:

$$k^2 = 1 + \frac{phr_0^3}{EI} \quad \text{and} \quad k' = \frac{-M_0r_0^2 + phr_0^3u_0}{EI}. \tag{2.21}$$

The general solution of differential equation (2.20) is as follows:

$$u = C_1 sink\vartheta + C_2 cosk\vartheta + k'' \tag{2.22}$$

where the first two terms in the second member are the solution of the associated homogeneous equation, with C_1 and C_2 being integration constants to be determined by imposing boundary conditions at sections A and F of the deformed ring (Fig. 2.2a), while k'' is the particular solution given by the relation:

$$k'' = \frac{k'}{k^2} = \frac{-M_0 r_0^2 + phr_0^3 u_0}{EI + phr_0^3}. \tag{2.23}$$

The boundary conditions to be imposed, which are governed by the symmetry of the deflection curve as indicated above (Fig. 2.2a), are as follows:

$$\left(\frac{du}{d\vartheta}\right)_{\vartheta=0} = 0 \quad \text{and} \quad \left(\frac{du}{d\vartheta}\right)_{\vartheta=\pi/2} = 0. \tag{2.24}$$

After deriving relation (2.22) for ϑ, we can conclude from the first of these conditions that $C_1 = 0$, while the second gives:

$$sin\frac{k\pi}{2} = 0. \tag{2.25}$$

The subsequent non-zero roots of relation (2.25) are found for $k\,\pi/2 = n\pi$, or in other words for $k = 2n$ (with $n = 1, 2, 3, \ldots$). For $n = 1$, i.e., for $k = 2$, we have the smallest root of Eq. (2.25), which corresponds to two-wave deflection curve (Fig. 2.4, upper left). Substituting this value of k in the first relation (2.21) and remembering that $I = h\,s^3/12$ gives the following expression for critical pressure p_{cr}:

$$p_{cr} = \frac{3EI}{h\,r_0^3} = \frac{E}{4}\left(\frac{s}{r_0}\right)^3. \tag{2.26}$$

This relation, first obtained by Bresse, 1866 (see Timoshenko and Gere 1961), defines the theoretical value of external pressure, which is assumed to be uniformly distributed, required to maintain the ring in neutral equilibrium regardless of the slightly deformed configuration we have considered. When this critical value is exceeded, the structure will collapse instantaneously.

If the external pressure is relatively small, the action of the restoring moment will predominate over the action of the external pressure and the annular beam will return to its initial position defined by the circular centerline. Under these conditions, the structure is said to be stable (in stable equilibrium). However, if the external pressure is large, the radial displacements will increase until the structure collapses. Under these conditions, the structure is unstable (in unstable equilibrium) and fails by buckling.

The radial displacements (or radial deflections) of the deflected ring, with $k = 2$, are determined from relation (2.22), first calculating constant C_2 with the

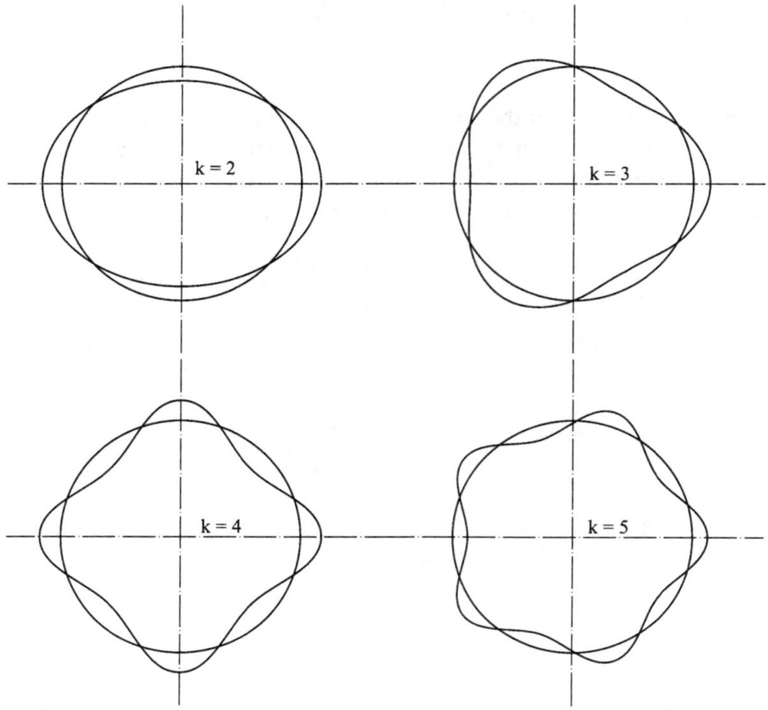

Fig. 2.4 Two-, three-, four- and five-wave buckling deflection curves

condition $u = u_0$ for $\vartheta = 0$ and recalling the first relation (2.21) written for $k = 2$. We thus arrive at the relation:

$$u = \frac{1}{4}\left(\frac{M_0 r_0^2}{EI} + u_0\right) cos2\vartheta - \frac{M_0 r_0^2}{4EI} + \frac{3}{4} u_0. \tag{2.27}$$

The tangential displacements v, on the other hand, are calculated on the basis of the inextensional condition of the ring's middle centerline. The unit elongation, ε, of this middle centerline as a result of deflection is in general the sum of two contributions: that of tangential displacement v, which is equal to $dv/dl = dv/(r_0 d\vartheta)$, and that of radial displacement u, equal to $-u/r_0$. We thus have:

$$\varepsilon = \frac{1}{r_0}\left(\frac{dv}{d\vartheta} - u\right). \tag{2.28}$$

Consequently, the inextensional condition is given by the equation

$$\left(\frac{dv}{d\vartheta} - u\right) = 0 \tag{2.29}$$

which yields $v = \int u d\vartheta + C$, where C is an integration constant to be determined by imposing boundary conditions. By substituting relation (2.27) in (2.29), integrating and determining the integration constant by setting $v = 0$ for $\vartheta = 0$, which gives $C = 0$, we obtain:

$$v = \frac{1}{8}\left(\frac{M_0 r_0^2}{EI} + u_0\right) sin2\vartheta + \left(\frac{3}{4}u_0 - \frac{M_0 r_0^2}{4EI}\right)\vartheta. \tag{2.30}$$

For $\vartheta = 0$ and $\vartheta = \pi/2$, the symmetry conditions require that $v = 0$, so that relation (2.30) yields:

$$\frac{3}{4}u_0 - \frac{M_0 r_0^2}{4EI} = 0 \tag{2.31}$$

and hence:

$$M_0 = \frac{3u_0 EI}{r_0^2} = p_{cr} h u_0 r_0. \tag{2.32}$$

Substituting the latter result in relations (2.27) and (2.30) gives:

$$u = u_0 cos2\vartheta \quad \text{and} \quad v = \frac{1}{2}u_0 sin2\vartheta. \tag{2.33}$$

Relation (2.32) indicates that bending moment M_0 can be obtained by applying the compressive force $p_{cr}hr_0$ with an eccentricity u_0 in the sections of the ring cut by the x axis (Fig. 2.2a). In this case, the circumference of radius r_0, represented with a dashed line in Fig. 2.2a, can be regarded as the funicular curve of the uniform external pressure, while the area between this curve and the deflected ring's middle centerline is the diagram for the ring's bending moment. The same result is reached by substituting relation (2.32) in Eq. (2.17). For $\vartheta = \pm\pi/4$ and $\vartheta = \pm 3\pi/4$, radial displacement and bending moment are both zero.

For $n = 2, 3, \ldots$, i.e., for $k = 4, 6, \ldots$, we can obtain a series of possible shapes of the buckled ring's centerline with a larger and larger number of waves in them (the deflected shapes will have 4, 6,... waves). Here, the deflected shapes with an even number of waves are obtained assuming that the deflection curve is characterized by two axes of symmetry. If we assume a single axis of symmetry (only the x axis, for instance) and that the bending moments are zero at the ends of the vertical axis, we will obtain solutions where k is odd, i.e., with $k = 3, 5, \ldots$

The case $k = 1$ is not relevant in elastic stability problems, as it represents a rigid body translation of the annular beam. From the practical and design standpoint, in fact, only the solution with $k = 2$ is of interest, as deflection curves with more than two waves can be obtained only if there are certain additional constraints that prevent a two-wave buckling deflection curve. Without such additional constraints, buckling will always be of the type shown in Fig. 2.4, upper left. The higher-order buckling deflection curves (the first three are shown in

Fig. 2.4) are of a certain practical interest for short thin-walled circular cylindrical shells with clamped ends.

Returning to p_{cr}, it should be noted that it is not necessary to calculate the constants C_1 and C_2 appearing in Eq. (2.22) in order to determine the critical pressure. It is sufficient to bear in mind that since the deflection curve must be a closed curve, u must assume the same value if ϑ increases by 2π; it follows that k must be an integer. Accordingly, the first relation (2.21) yields:

$$p_{cr} = \left(k^2 - 1\right)\frac{EI}{hr_0^3}. \tag{2.34}$$

Obviously, it is not possible to have $k = 1$, as this would result in $p_{cr} = 0$. Consequently, we must have $k = 2, 3, 4\ldots$ For $k = 2$ we obtain relation (2.26) and, for $k > 2$, we obtain all of the multi-wave buckling deflection curves indicated earlier.

The critical stress is determined from the second relation (1.7) written in terms of outer radius r_e, substituting p with the p_{cr} given by relation (2.26) and r_e with the radius r_0 at incipient buckling. This gives:

$$\sigma_{cr} = \frac{3EI}{shr_0^2} = \frac{E}{4}\left(\frac{s}{r_0}\right)^2. \tag{2.35}$$

It should be recalled that since the entire treatment is based on the assumption of linear elasticity, relation (2.35) holds true only as long as the critical stress does not exceed the proportional limit σ_p.

Assuming that the material exhibits linear elastic-perfectly plastic behavior, and that the proportional limit, the elastic limit and the yield point thus coincide, relation (2.35) will apply for $\sigma_{cr} \leq \sigma_s$, where σ_s is the yield stress.

It should also be recalled that as it is advisable to select a safety factor of 4 in elastic instability response analysis, the allowable external pressure and the allowable stress will be given respectively by:

$$p_{cr,a} = \frac{E}{16}\left(\frac{s}{r_0}\right)^3 \quad \text{and} \quad \sigma_{cr,a} = \frac{E}{16}\left(\frac{s}{r_0}\right)^2. \tag{2.36}$$

2.3 Instability of a Thin-Walled Cylinder of Annular Cross-Section

The treatment presented above for the instability of a thin-walled circular ring can be readily extended to long thin-walled cylinders of annular cross-section (thin-walled circular cylindrical shells or thin-walled circular cylindrical tubes). In fact, using the last two constitutive equations (1.17) without temperature terms and written in the assumption that radial stress is negligible compared to the hoop and axial stresses, and solving them for stresses σ leads to the following two relations:

$$\begin{aligned}\sigma_t &= \frac{E}{1-\nu^2}(\varepsilon_t + \nu\varepsilon_z).\\ \sigma_z &= \frac{E}{1-\nu^2}(\varepsilon_z + \nu\varepsilon_t).\end{aligned} \tag{2.37}$$

For an indefinitely long circular cylinder, axial strain ε_z is zero (plane strain state). It can thus be seen from the first of these relations that the effect of biaxiality can be considered simply by substituting $E/(1-\nu^2)$ for the modulus of elasticity, E, as the shape of the cross-section does not change during deflection.

In the case of interest to us here, relations (2.26) and (2.35) respectively become:

$$p_{cr} = \frac{E}{4(1-\nu^2)}\left(\frac{s}{r_0}\right)^3 \tag{2.38}$$

$$\sigma_{cr} = \frac{E}{4(1-\nu^2)}\left(\frac{s}{r_0}\right)^2. \tag{2.39}$$

We thus arrive at the conclusion that the effect of biaxiality is to increase the critical pressure and critical stress by a ratio of $1/(1-\nu^2)$ compared to the uniaxial structure of the thin-walled annular ring in Fig. 2.1. This is due to the fact that biaxiality is a constraint on longitudinal strain. Relation (2.38), known as Bresse's formula (1859), leads to results that overestimate the critical pressure. In tests on commercial tubing and pipe, Stewart (see Timoshenko and Gere 1961, Brownell and Young 1968) found that collapse occurred at a critical pressure of 27 % less than the theoretically predicted pressure given by Bresse's formula. This is why, here as elsewhere, a safety factor of 4 is recommended in design analysis and response analysis of thin-walled long circular cylinders operating under external pressure. Accordingly, the allowable external pressure and the allowable stress are given respectively by:

$$p_{cr,a} = \frac{E}{16(1-\nu^2)}\left(\frac{s}{r_0}\right)^3 \quad \text{and} \quad \sigma_{cr,a} = \frac{E}{16(1-\nu^2)}\left(\frac{s}{r_0}\right)^2. \tag{2.40}$$

Once again, relation (2.39) is valid providing that the critical stress does not exceed the material's proportional limit, σ_p. Considering a steel having $E = 210$ GPa and $\nu = 0.3$, and plotting critical stress σ_{cr} versus ratio $s/2r_0$ gives the curve shown in Fig. 2.5 and representing relation (2.39), which applies as long as $\sigma_{cr} \le \sigma_p$. Beyond this limit, the curve gives exaggerated and unrealistic values for the critical stress. To find the real value of σ_{cr} corresponding to a predetermined ratio s/d_i (note that, given the small thickness, $s/2r_0 = s/d_0 \cong s/d_i$), we can proceed in the same way as for ideal columns subjected to an axial compressive force and stressed beyond the proportional limit and use the relation (2.39) after

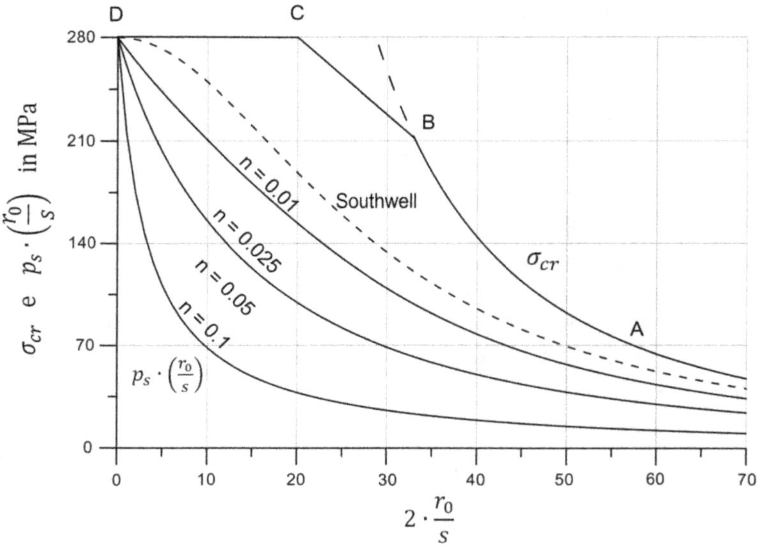

Fig. 2.5 Mixed curve *ABCD*, Southwell's curve and $p_s(r_0/s)$ curves for various initial imperfection ratios $n = u_{i,0}/r_0$ (steel with $\sigma_p = 210$ MPa, $\sigma_s = 280$ MPa, $E = 210$ GPa, $\nu = 0.3$)

substituting the tangent modulus E_t for Young's modulus. Relation (2.39) thus becomes:

$$\sigma_{cr} = \frac{E_t}{4(1 - \nu^2)} \left(\frac{s}{r_0}\right)^2. \tag{2.41}$$

If the material's $\sigma = \sigma(\epsilon)$curve determined from a uniaxial compression test on a standard specimen is available, it can be used to find the value of the tangent modulus $E_t = d\sigma/d\varepsilon$ corresponding to each stress value over σ_p. Relation (2.41) can then be used to find the ratio $2r_0/s$ deriving from it. The $\sigma_{cr} = \sigma_{cr}(2r_0/s)$ curve can thus be completed in the range $\sigma_{cr} > \sigma_p$. For reasons of practical utility, this latter portion of the curve is usually approximated by two straight lines.

For materials exhibiting a pronounced yield point and for thick-walled circular cylinders ($d_i/s < 20$), the yield-point stress (line $\overline{CD}$ in Fig. 2.5) can be assumed as the critical stress. In the intermediate portion, a straight line segment joins point *C*, the limit of validity for a thick-walled body, to point *B*, the limit of validity for relation (2.39).

If we take, for example, a steel having $\sigma_p = 210$ MPa and $\sigma_s = 280$ MPa, it can be seen that the curve representing relation (2.39) applies for ratios $2r_0/s \cong d_i/s \geq 33$ (the limit of validity *B* for this curve is characterized by a ratio $2r_0/s$ of approximately 33) and that, in the section where $2r_0/s \leq 20, \sigma_{cr} = \sigma_s$; in the intermediate portion where $20 < 2r_0/s < 33$ it is assumed that critical stress σ_{cr} is represented by the points of segment joining points B and C. Consequently, the mixed curve $ABCD$ in Fig. 2.5 represents the critical stress for any d_i/s ratio used in designing the circular cylinder. As it is clear that higher safety factors must be used for higher d_i/s ratios and vice versa, each portion of the mixed curve corresponds to a different safety factor.

Instead of the mixed curve $ABCD$ in Fig. 2.5, it is sometimes useful to consider a continuous curve such as that below, which was proposed by Southwell (1913) and is represented with a dashed line in Fig. 2.5:

$$\sigma_{cr} = \frac{\sigma_s}{1 + \frac{\sigma_s(1-\nu^2)}{E}\left(\frac{2r_0}{s}\right)^2}. \quad (2.42)$$

Southwell's formula is analogous to Rankine's formula for columns. For thick-walled circular cylinders, this relation gives a critical stress which is close to the yield stress, while for thin-walled circular cylinders it leads to critical stress values that are a bit lower than those given by relation (2.39). For d_i/s ratios included among those usually employed in actual design, Southwell's relation (see also Fig. 2.5) provides values of σ_{cr} that are much lower than those given by the mixed curve $ABCD$. This is a further safety measure that compensates for the effect of any initial deviations from the circular cylinder's annular shape, which are always to be expected in practice. This leads inevitably to the question we will take up in the next section, viz., the effect of initial geometrical imperfections on the elastic instability of thin-walled circular cylinders under external radial pressure.

2.4 Effect of Geometrical Imperfections

The collapse of thin-walled circular cylinders under uniform external pressure depends to a large extent in their geometrical imperfections, the most important of which undoubtedly is initial ovalization (or initial ellipticity), whose maximum values and tolerance ranges, which are codified by national and international standards (see, for example, the ASME standards), can be determined by means of numerous inspection measurements. From the design standpoint, it is thus extremely useful to have an expression in which this initial out-of-roundness (initial ellipticity) from a perfect circular form, which reduces the critical collapse pressure, appears explicitly (Donnell and Wan 1950, Timoshenko and Gere 1961).

To arrive at such an expression, we will assume that the initial radial deviation of the circular cylinder's middle line (the oval represented by a solid line in Fig. 2.6) from its theoretical circular shape (shown with a dashed line in Fig. 2.6) is given by the equation:

$$u_i = u_{i,o} cos 2\vartheta \quad (2.43)$$

where $u_{i,o}$ is the maximum initial radial deviation from the perfect circular form (maximum initial ovalization), regarded as small compared to r_0, and ϑ is the

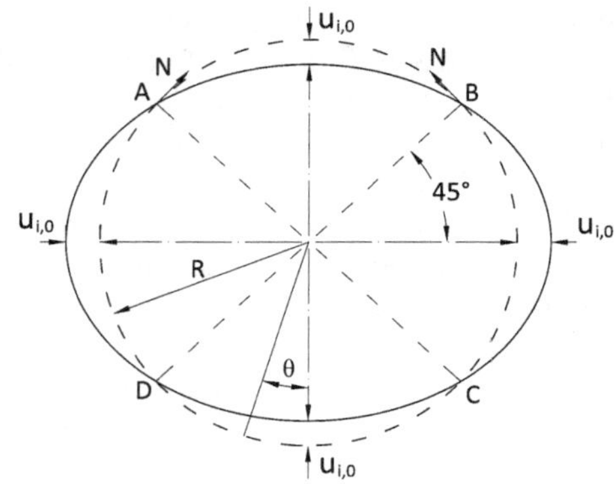

Fig. 2.6 Theoretical middle line (*dashed*) and initial deformated shape (*solid line*) from any initial ellipticity

central angle measured starting from one of the two axes of symmetry of the deformated shape, where $u_i = u_{i,o}$ (the vertical line in Fig. 2.6).

If a uniform external pressure p is applied to the circular cylinder whose middle line has undergone an initial ellipticity, and is thus no longer circular, the structure will be further flattened, with additional radial displacements $u = u(\vartheta)$. These additional radial displacements are determined using differential equation (2.12), but substituting the annular beam's flexural rigidity, EI, with the cylindrical shell's flexural rigidity, $D = Es^3/12(1 - \nu^2)$, so that the equation thus becomes:

$$\frac{d^2u}{d\vartheta^2} + u = -r_0^2 \frac{M}{D}. \tag{2.44}$$

As can be seen from Fig. 2.6, a uniform external pressure produces a decrease in the initial curvature in portions $\widehat{AB}$ and $\widehat{CD}$ of the deflection curve, while the initial curvature increases in portions $\widehat{AD}$ and $\widehat{BC}$; in accordance with the convention adopted for the bending moment, M will thus be positive in the first two portions and negative in the other two.

At points A, B, C and D, the bending moment is zero, and the mutual actions between the portions of the deflection curve take place through forces N, which are tangent to the circumference (represented by the dashed line in Fig. 2.6) that define the ideal theoretical shape of the structure's middle centerline. This circumference can be regarded as the funicular curve of the external pressure p; the compressive force along it, on the other hand, remains constant at $N = qr_0 = pr_0$, given that the approach to the problem is based on considering a small element of a circular cylinder having unitary axial length ($h = 1$).

Consequently, the bending moment in the generic cross-section identified by central angle ϑ is obtained by multiplying force N by the total radial displacement $(u + u_i)$ that occurs in this cross-section, so that:

$$M = pr_0\left(u + u_{i,o}cos2\vartheta\right). \tag{2.45}$$

Equation (2.44) then becomes:

$$\frac{d^2u}{d\vartheta^2} + u = -\frac{pr_0^3}{D}(u + u_{i,o}cos2\vartheta) \tag{2.46}$$

and can be written as:

$$\frac{d^2u}{d\vartheta^2} + u\left(1 + p\frac{r_0^3}{D}\right) = -\frac{1}{D}pr_0^3u_{i,o}cos2\vartheta. \tag{2.47}$$

The solution to this equation, which satisfies the continuity conditions at points *A, B, C* and *D*, is as follows:

$$u = \frac{pu_{i,o}}{p_{cr} - p}cos2\vartheta \tag{2.48}$$

where p_{cr} is given by Eq. (2.38). As can be seen, radial displacement u and its second derivative $d^2u/d\vartheta^2$ drop to zero at points *A, B, C* and *D* at 45° from the axes of symmetry. Consequently, the bending moments at these two points are also zero, as was already indicated. The maximum bending moment occurs at cross-sections $\vartheta = 0$ and $\vartheta = \pi$, and is:

$$M_{max} = pr_0\left(u_{i,0} + \frac{u_{i,0}p}{p_{cr} - p}\right) = \frac{pu_{i,o}r_0}{1 - (p/p_{cr})}. \tag{2.49}$$

As is immediately clear from Eq. (2.49), the change in the circular cylinder's ellipticity as a result of an uniform external pressure p can be ignored where ratio p/p_{cr} is small, and the maximum bending moment is obtained by multiplying the compressive force $N = pr_0$ by initial displacement $u_{i,0}$. Where ratio p/p_{cr} is not small, the effects of the change in shape are significant and must be taken into consideration, while Eq. (2.49) must be used to calculate the maximum bending moment.

The maximum compressive stress σ_{max} is the sum of the stress produced by the compressive force N and the compressive stress due to the maximum bending moment M_{max}. We thus have:

$$\sigma_{max} = \frac{pr_0}{s} + \frac{6pu_{i,o}r_0}{s^2(1 - p/p_{cr})} \tag{2.50}$$

where $s^2/6$ is the circular cylinder's section modulus in bending per unit of length. The dangerous value of pressure p is that at which stress σ_{max} equals yield stress σ_s. Assuming that relation (2.50) can be used with sufficient accuracy up to the yield-point stress of the material, we have:

$$\sigma_s = \frac{p_sr_0}{s} + 6p_s\left(\frac{r_0}{s}\right)^2\frac{u_{i,o}}{r_0(1 - p_s/p_{cr})}. \tag{2.51}$$

This relation makes it possible to calculate the constant value of pressure p_s at which yielding begins in the outermost fibers, when σ_s and u_{io} are known. Using the notations $r_0/s = m$ and $u_{i,o}/r_0 = n$, relation (2.51) can be transformed, after a few manipulations, into:

$$p_s^2 - \left[\frac{\sigma_s}{m} + (1 + 6\,mn)p_{cr}\right]p_s + \frac{\sigma_s p_{cr}}{m} = 0. \tag{2.52}$$

This relation makes it possible to calculate p_s.

It should be noted that the pressure p_s determined with relation (2.52) is lower than the critical collapse pressure p_{cr} and becomes equal to the latter only for circular cylinders with no initial imperfections. Using pressure p_s for design purposes thus swings the balance in favor of greater safety.

Figure 2.5 also shows several curves calculated with relation (2.52) which give the values of mean compressive hoop stress $\mathrm{p_s}(r_{0/s})$ at which yield begins versus ratio $2r_0/s = 2m$ for $n = 0.1$, 0.05, 0.025 and 0.01 and for $\sigma_s = 280$ MPa. The necessary wall thickness can be calculated quickly by using these curves and an appropriate safety factor.

2.5 Instability of a Thin-Walled Cylinder of Annular Cross-Section with Circumferential Stiffening Rings

In the section covering the elastic instability of a thin-walled cylinder of annular cross-section under uniform external pressure, it was assumed that the structure is indefinitely extended in the direction of its axis. The relations obtained in this context apply to circular cylinders with large ratios l/d of axial length to diameter (here again, given the small thickness, it is assumed that diameters d_0, d_i, d_e, d_m and d are practically equal).

For shorter circular cylinders, whose ends have flat closures or formed heads or are constrained at their ends (built-in edges, simply supported edges and so forth), and for long circular cylinders featuring circumferential stiffeners uniformly or non-uniformly spaced in the direction of the axis at less than the critical length, the value of p_{cr} is greater than that given by relation (2.38) and depends on the ratios l/d and d/s. Investigating the elastic instability of these structures is more complex, as they will exhibit waves along the middle circumference during deflection. The number of these waves will depend on the ratio l/d. With stiffened structures of this kind, however, it is possible to achieve significant savings in weight and material.

Here, we will restrict ourselves to presenting the results for thin-walled circular cylinders under external pressure featuring circumferential stiffening rings uniformly spaced in the longitudinal direction at a distance of less than the *critical length*, defined as the particular axial distance beyond which these rings will no longer have a stiffening effect. For further details on the subject see Chap. 10.

This critical length, determined by Southwell (1913) with a complicated procedure which will not be addressed here, is:

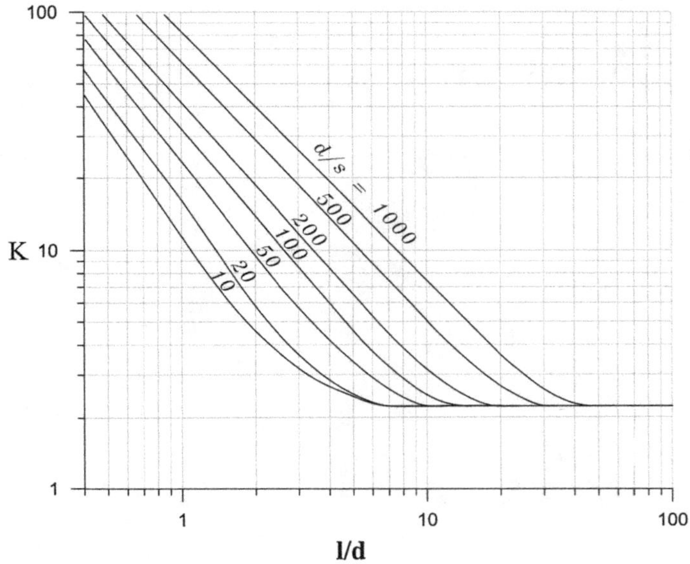

Fig. 2.7 Factor $K = K(l/d,\ d/s)$ for determining critical pressure in circular cylinders with circumferential stiffeners

$$l_c = \frac{4\pi\sqrt{6}}{27}\sqrt[4]{(1-\nu^2)}d\sqrt{d/s}. \tag{2.53}$$

For $\nu = 0.3$, this relation becomes:

$$l_c \cong 1.11\,d\sqrt{d/s}. \tag{2.54}$$

For circular cylinders with circumferential stiffeners spaced at a length $l < l_c$, the critical pressure is given by the following relation:

$$p_{cr} = KE(s/d)^3, \tag{2.55}$$

where the factor K is a function of ratios l/d and d/s; relation (2.55) is formally equivalent to relations (2.38). For steel structures ($\nu = 0.3$), K can be found from the plot in Fig. 2.7 (Brownell and Young 1968). It should be noted that where ratio l/d is very large ($l/d > 40$), regardless of ratio d/s, K tends to the asymptotic value of $8/[4(1 - \nu^2)] \cong 2.2$, which characterizes relation (2.38) applying to circular cylinders of unlimited length.

The circumferential stiffening rings, on the other hand, must be calculated as annular beams subjected to an external radial load equal to the resultant of the pressure forces acting on a length $l_c/2$ from both sides of its axis line. In this connection, the reader is referred to the previous section.

Chapter 3
Other Types of Instability of Thin-Walled Circular Cylinders

3.1 Foreword

Buckling is one of the major causes of failures in thin-walled structures, and therefore the possibility of buckling should always be considered in design analysis and response analysis (Timoshenko 1956, Timoshenko and Gere 1961, Belluzzi 1971, Bruhn 1973, Cicala 1974, Chajes 1974, Brush and Almroth 1975). In many practical applications, the thin-walled circular cylindrical structures analyzed in the previous chapter are also subjected to axial loads as well as torsional and bending loads, that can lead to other types of instability (Donnell 1934, Cicala 1940 and 1940, Hoff and Rehfield 1964, Obodan, Lebedeyev and Gromov 2013). In other circumstances, it is necessary to determine how a thin-walled circular cylinder behaves under internal pressures that stress it beyond the yield-point stress of the material, up to the burst pressure. In these cases, we are dealing with another kind of instability, caused by plastic flow.

In this chapter, a review of the concepts underlying the Euler instability (or Euler buckling) to which slender columns, with edges variously constrained and loaded by longitudinal compressive forces applied through the centroid of the end cross-sections, are subject will be followed by a brief overview of the other kinds of instability that occur most frequently, and that the designer must necessarily bear in mind when performing design and response analyses for thin-walled tubular structures under various types of load and constraint. The types of instability addressed in this overview are those that entail a change in the geometrical shape of the structure in question, without substantial variations in its effective cross-sectional area.

In the last two sections of the chapter, we will discuss plastic instability caused by overloading. Though this leads to dimensional changes in the structure, the latter retains the initial shape it had with no load, i.e., before internal pressure was applied. This type of instability includes that associated with the plastic expansion of a thin-walled circular cylinder under internal pressure and, in some cases, subjected to an external axial load large enough to stress the structure beyond the yield-point stress of the material (Hill 1950, Nadai 1950, Drucker and Shield 1958, Prager 1959, Nadai 1963, Faupel 1964).

V. Vullo, *Circular Cylinders and Pressure Vessels*, Springer Series in Solid and Structural Mechanics 3, DOI: 10.1007/978-3-319-00690-1_3,

3.2 Instability of Thin-Walled Circular Cylinders Subjected to Axial Load

Long slender structural members (also called columns) loaded axially in compression can fail by bending or deflecting laterally rather than by direct compression of the material. Under an increasing longitudinal load, the lateral deflections will also increase, the equilibrium will become unstable, and eventually the column will collapse completely, even if the normal compressive stresses in the structure are lower than those at which failure would occur as a result of direct compression. This is the phenomenon referred to as the *overall instability* (or *Euler instability or Euler buckling*) of axially compressed columns.

Where the slender structure under axial compression has thin walls, another type of instability is possible which can lead to the structure's collapse, and begins with an alteration in the shape of the middle generator that does not occur in Euler buckling. The middle generators undergo lateral deflections that are symmetrical with respect to the axis of the cylinder, showing a typical collapse shape known as *concertina mode* or *concertina buckling*. This type of instability thus shows itself with the formation of accordion-like waves or pleats in the wall whose wavelength is much less than the length of the cylindrical structure. To distinguish it from overall or Euler instability, this type is called *local instability*, as it is not influenced by the length of the slender cylinder.

3.2.1 Compendia on Elementary Column Theory (Overall Instability or Euler Buckling)

Very generally, we can say that a long slender column under axial compressive load fails through overall instability if, supposing that the structure is subjected to a lateral displacement of any kind and intensity, the work performed by force P (Fig. 3.1) for the axial displacement Δl that the load application point undergoes as a result of this lateral displacement is greater than the elastic potential energy accumulated during the bending of the column. This criterion, however, is difficult to apply, both because the lateral deflection to be considered is indeterminate, and because, with the usual assumption of very small displacements, that is very small deflections due to bending, the axial displacements of a long slender deflected column are very much smaller, i.e., are second-order small.

Under particular conditions, however, the phenomenon can be investigated independently of energy considerations. This is the case of a constant cross-section column under compressive axial load. As this topic is extensively covered in other courses (*Scienza delle Costruzioni, Strength of Materials,* etc.), we will limit ourselves here to presenting the results that the designer will use in structural analysis of columns (Timoshenko 1956, Cicala 1963, Faupel 1964, Almroth 1966, Belluzzi 1971).

Fig. 3.1 Column with hinged ends subjected to compressive axial load

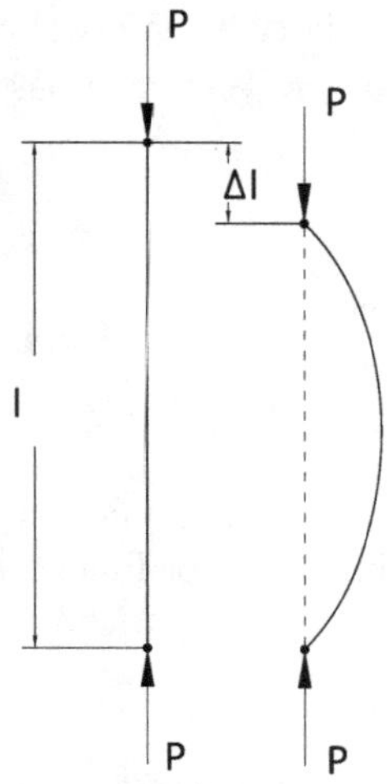

With reference to Fig. 3.1, the critical load P_{cr} (called *Euler's critical buckling load* or *Euler critical load*, as Euler was the first to determine it in the appendix *De Curvis Elasticis* to his celebrated work of 1744, *Methodus Inveniendi Lineas Curvas Maximi Minimive Proprietate Gaudentes*) at which unstable elastic equilibrium occurs is given by the following relation, known as *Euler's formula*:

$$P_{cr} = \frac{\pi^2 EI}{L^2}, \tag{3.1}$$

where E is the Young's modulus of the material used for the column, I is the minimum moment of inertia of the column's cross-sectional area, and L is the equivalent column length (also called free length), i.e., the distance between two successive points of inflection of the deflection curve. This length will be $2l$, l, $l/\sqrt{2}$ and $l/2$ respectively, with l as the actual length of the column, depending on whether the column is built-in at one end and free at the other, hinged at both ends, hinged at one end and built-in at the other, or built-in at both ends. The equivalent length of a column is that used in Euler's formula for a column as in Fig. 3.1, hinged at both ends (pinned-end column) to give the buckling load for the actual column.

Introducing the minimum radius of gyration of the cross section in the plane of bending $\rho_{min} = \sqrt{I/A}$, where A is the column's cross-sectional area, and slenderness ratio $\lambda = L/\rho_{min}$, i.e., the dimensionless ratio of equivalent column length to the minimum radius of gyration, relation (3.1) becomes:

$$P_{cr} = \frac{\pi^2 EA}{\lambda^2}. \tag{3.2}$$

For a thin-walled circular cylinder, where $I = \pi r_m^3 s$, $A = 2\pi r_m s$ and $\rho = r_m/\sqrt{2}$, relations (3.1) and (3.2) respectively become:

$$P_{cr} = \frac{\pi^3 E r_m^3 s}{L^2} \tag{3.3}$$

and

$$\sigma_{cr} = \frac{\pi^2 E r_m^2}{2L^2}. \tag{3.4}$$

It can be inferred from relation (3.2) that the critical stress at which overall instability or Euler buckling occurs is given by the following expression:

$$\sigma_{cr} = \frac{\pi^2 E}{\lambda^2}. \tag{3.5}$$

This expression applies as long as the critical stress remains below, or at most equal to, the proportional limit σ_p, i.e., as long as $\sigma_{cr} \leq \sigma_p$. Euler's theory does not hold beyond this stress value, because the buckling equation of compressed slender columns due to Euler was derived using Hooke's law. In addition, care must be taken in selecting the value for I and, consequently, λ. In fact, I is the minimum moment of inertia of area (or minimum second moment of area) only if the column is free to deflect in any of the infinite number of planes containing the axis in its pre-buckling configuration, and if the constraints are such as to react in the same way to all possible bending deflections.

If the column is not free to deflect in all directions, I is the smallest of the moments of inertia of area calculated for the axis around which the cross-sectional area can rotate during bending. Conversely, if the constraints are such that they do not show the same behavior in all axial planes (e.g., a cylindrical hinge that behaves as a hinge in the plane of its axis, and as a clamp in the plane perpendicular to that of the axis, as is generally the case of the connecting rod in any connecting rod-crank mechanism), it will be necessary to evaluate the equivalent column lengths and, consequently, the slenderness ratios in the various directions, and then select the maximum slenderness ratio to be used in calculations.

Except for the case where the column is fixed at one end and hinged at the other, its buckled deflection curve upon reaching the critical load is an arc of a sinusoid, while the equivalent column length L is the distance between two successive points of inflection of the deflection curve.

As can be seen from Euler's relations (3.3) and (3.4), to have higher values for the critical stress σ_{cr} in a thin-walled tubular structure of annular cross-section with assigned constraint conditions, equivalent column length and effective cross-sectional area, it is necessary to increase the circular cylinder's diameter and consequently reduce its thickness, so that the cross-sectional area, and hence the weight, remain constant. It would even be conceivable to increase the diameter to a point that would prevent overall instability from occurring for any value of equivalent column length L. If this were done, however, the structure would quickly be affected by the local instability discussed in the following section. There is thus a diameter at which the two kinds of instability—overall and local—arise simultaneously.

As was indicated above, relation (3.5) applies as long as $\sigma_{cr} \leq \sigma_p$. This condition, which is obviously to be extended with the necessary changes to relations (3.1)–(3.4), is self-evident when we consider that modulus of elasticity E, which remains constant only up to the proportional limit stress, appears in all five relations (3.1)–(3.5).

The instability of slender axially compressed columns which, as they have sufficiently small slenderness ratio λ and thus a critical stress such that $\sigma_{cr} > \sigma_p$, are subject to stress beyond the proportional limit (elasto-plastic or plastic instability), is addressed using approximate methods and empirical formulas.

A first and very simple method consists of continuing to use Euler's relation (3.5) for the elasto-plastic and plastic field, but substituting the tangent modulus E_t (Fig. 3.2) for Young's modulus E. As Fig. 3.2 shows, the tangent modulus E_t, which is equal to the derivative $d\sigma/d\varepsilon$, is a point function which varies along with slenderness ratio λ. To plot the $\sigma_{cr} = \sigma_{cr}(\lambda)$ curve to be used in the range where the Euler cubic hyperbola given by relation (3.5) can no longer be employed, we must proceed pointwise, once the material's curve (true stress—true strain, as shown in Fig. 3.2) has been determined experimentally on a standard specimen.

To this end, several values of stress σ beyond σ_p are established on the curve $\sigma = \sigma(\varepsilon)$, and the values of the tangent modulus for each of these points are determined. Relation (3.5) is then used after substituting E_t for E in it to find the corresponding values of λ. This procedure, which enables us to complete the $\sigma_{cr} = \sigma_{cr}(\lambda)$ curve beyond that for the Euler hyperbola as shown in Fig. 3.3, is clearly the converse of the one followed for slender axially compressed columns. Rather than finding the value of σ_{cr} at which overall instability arises for an assigned column whose slenderness ratio λ is known, we find the column, or in other words its slenderness ratio λ, which is subject to overall instability at a given value of $\sigma_{cr} > \sigma_p$.

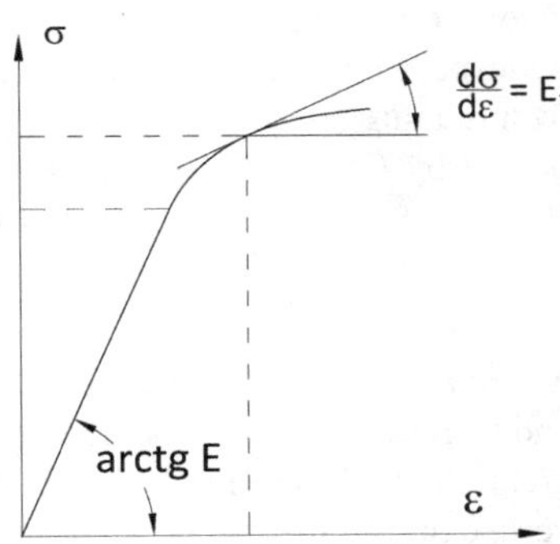

Fig. 3.2 True stress–true strain curve $\sigma = \sigma(\varepsilon)$ showing the modulus of elasticity E (or Young's modulus) and tangent modulus E_t

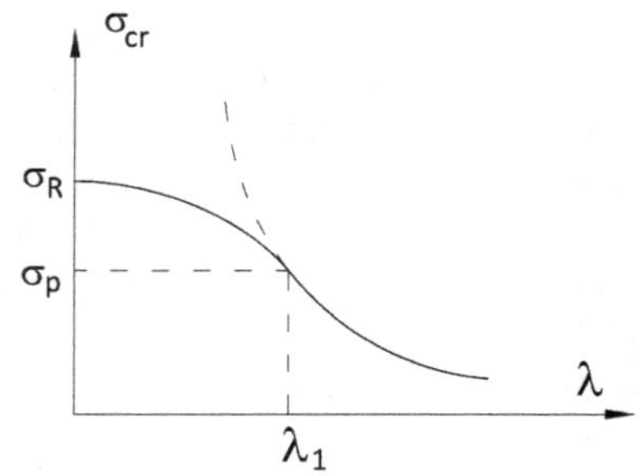

Fig. 3.3 $\sigma_{cr} = \sigma_{cr}(\lambda)$ curve in the elasto-plastic field ($\lambda \le \lambda_1$; $\sigma_{cr} \ge \sigma_p$) and the proportional elastic field ($\lambda \ge \lambda_1$; $\sigma_{cr} \le \sigma_p$)

With this procedure, the $\sigma_{cr} = \sigma_{cr}(\lambda)$ curve for a given material is obtained. The portion of the curve to the right of the vertical line of abscissa λ_1 is simply the plot of the Euler cubic hyperbola (Euler's curve) given by relation (3.5), while the portion to the left of this vertical line is the curve plotted as we have just described. Clearly, for $\lambda = 0$, we must have $\sigma_{cr} = \sigma_R$, with σ_R being the material's ultimate stress, as instability phenomena cannot occur for infinitely short columns ($\lambda \to 0$). However, we rarely need go as far as the material's ultimate stress σ_R. In most design problems of practical interest, it is necessary and sufficient to plot the $\sigma_{cr} = \sigma_{cr}(\lambda)$ up to the offset yields strength, defined as the stress σ_s corresponding to the intersection of the stress–strain curve with a parallel to the straight-line portion, with origin at some finite strain value (the strain offset or permanent set is usually 0.01, 0.02 or 0.20 % and the corresponding stress is called offset yield strength or proof stress).

Fortunately, given the difficulties of theoretical analysis in the elasto-plastic field, experience has been of invaluable aid in providing a number of empirical formulas for calculating critical stress. Of the many such formulas in common use, mention should be made of the linear formulas, such as:

$$\sigma_{cr} = a - b\lambda, \tag{3.6}$$

and of the quadratic formulas, of the following type

$$\sigma_{cr} = a - b\lambda - c\lambda^2, \tag{3.7}$$

where a, b and c are experimental constants that depend on the material. The most widespread quadratic formula that should be mentioned here is the parabolic formula suggested by Johnson (Johnson's formula, Johnson J.B. 1894), and supported by Ostenfeld, which can be written in the following form:

$$\sigma_{cr,J} = \sigma_s - \frac{\sigma_s^2}{4\sigma_E}, \tag{3.8}$$

where $\sigma_{cr,J}$ and $\sigma_E = \sigma_{cr}$ are respectively the critical stresses according to Johnson and to Euler [the latter being expressed by relation (3.5)], while σ_s is the material's yield stress. Relation (3.8), which represents a concave down parabola having its vertex on the ordinate, of coordinates $\lambda = 0$ and $\sigma_{cr} = \sigma_s$, is tangent to the Euler cubic hyperbola at the point where $\sigma_E = \sigma_s/2$. This means that Euler's theory is to be adopted when $\sigma_{cr} \leq \sigma_s/2$ and Johnson's relation is to be used when $\sigma_{cr} \geq \sigma_s/2$. The proportional limit σ_p is assumed to be half of σ_s, or in other words $\sigma_p = \sigma_s/2$.

In reality, for the materials used in the structures of interest to us here, proportional limit σ_p can be very different from $\sigma_s/2$. However, any inaccuracy in locating point *M* (Fig. 3.4) leads to negligible errors in designing these structures, as the two curves—hyperbola and parabola—are tangent at this point.

Using Johnson's relation to calculate the critical stress in the elasto-plastic field is very simple, providing that the material's yield stress σ_s is known. In fact, once the Euler's stress σ_E has been calculated with relation (3.5), it can be assumed as the critical stress if it is less than or at most equal to $\sigma_s/2$; otherwise, if $\sigma_E > \sigma_s/2$, this value is substituted in relation (3.8) and the $\sigma_{cr,J}$ value to be assumed as the critical stress is calculated. Thus, even if $\sigma_{cr} = \sigma_{cr}(\lambda)$ curves such as that in Fig. 3.3 are not available, reliable results of undoubted practical utility are obtained.

Similarly, critical stresses in the elasto-plastic field can be calculated using linear formulas such as relation (3.6), the best-known and most widely used being Tetmajer's formula. The coefficients appearing in relation (3.6) can be readily determined if we bear in mind that, in Tetmajer's original proposal, the critical stress line joins point *M*, which is the limit of validity for Euler's theory, with point *N* lying on the ordinate and having coordinates $\lambda = 0$ and $\sigma_{cr} = \sigma_s$ (Fig. 3.4).

In reality, as this approach underestimates the critical stresses for short columns (small λ values), the various codes based on Tetmajer's formula have replaced the straight line *MN* (not shown in Fig. 3.4) with the broken line *MPN* shown in Fig. 3.4. In addition, as Tetmajer suggested using a single safety factor for the different slenderness ratios λ, whereas it is clearly necessary to have higher safety factors for higher λ values and vice versa, the codes have opted for safety factors that vary with λ. Johnson's formula and the codes based on it also contemplate safety factors varying with λ, though giving rise to critical stresses that are a better approximation.

According to the empirical formulas we have discussed so far, the yield stress σ_s is the critical stress that can be reached in the limiting case where $\lambda \to 0$.

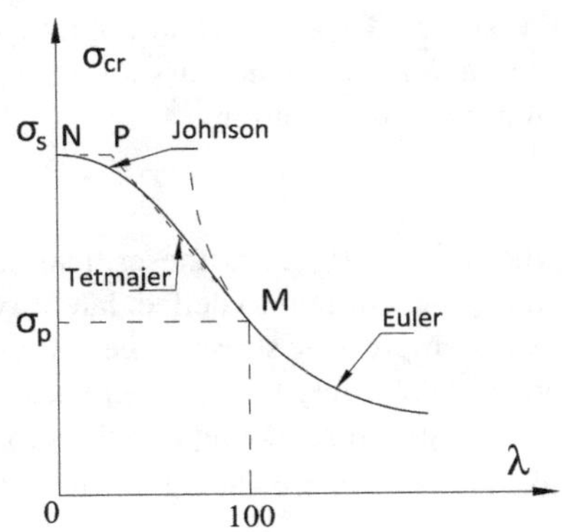

Fig. 3.4 Euler's cubic hyperbola, Johnson's parabola, Tetmajer's straight line *MN*, not shown, and broken line based on Tetmajer (modified)

However, when very short structural components are subjected to simple compression, the absolute value of the allowable compressive stress can exceed σ_s. But as cases where stresses are higher than yield-point stress are of very limited importance from the design standpoint, the critical stresses [those calculated with relation (3.8), for instance], though slightly below those which very short columns can in fact withstand, can be assumed in designing such columns, practically without committing errors of any kind.

It should also be borne in mind that the dimensions to be assigned to a compressed column are generally selected by trial and error. For this purpose, once the working load P acting on the column has been given, certain rough dimensions of the effective cross-sectional area are tentatively assumed and the radius of gyration ρ and slenderness ratio $\lambda = l/\rho$ are then calculated. Subsequently, the corresponding critical stress σ_{cr} is calculated, either using curves such as those shown in Figs. 3.3 and 3.4, if available, or analytically by means of relations (3.5) and (3.8) or the like. Multiplying the stress σ_{cr} thus found by the cross-sectional area that was assumed gives the critical load P_{cr}, which, after introducing the safety factor, gives the allowable load for the column. If the load obtained through this procedure is neither less than the working load P nor too much above it, the cross-section that was assumed can be regarded as correct. Otherwise, it will be necessary to make further assumptions and repeat the calculation however many times are needed.

Response analysis of the column's instability is simpler. In this case, the column has already been sized on the basis of other design considerations. Relations (3.5) or (3.8) will be used, verifying that the ratio of allowable stress to the critical stress thus calculated is higher than the predetermined safety factor. Response analysis for thin-walled tubular structures subject to overall instability is covered by existing national and international codes. Here we will summarize the recommendations of two standards, the first national and the second international.

Standard 182/1997 of the CNR (Italian National Research Council), which substantially conform to the German standard DIN 4114, requires that the slenderness ratio of the main and secondary members of an axially compressed structure be $\lambda \leq 200$ and $\lambda \leq 250$ respectively, and that these limits drop to $\lambda \leq 150$ and $\lambda \leq 200$ in cases where significant dynamic actions are involved. According, the standard specifies that column safety in the working conditions must be verified against buckling, assuming that its cross-section is under uniform compression, with the relation

$$\sigma_{cr}/\sigma \geq k \tag{3.9}$$

where $\sigma_{cr} = P_{cr}/A$ is the critical stress corresponding to critical load P_{cr} which causes the column to deflect laterally in the plane considered, $\sigma = P/A$ is the mean axial compressive stress in the column's cross-section A due to axial working load P, and k is the safety factor (k here has a very different meaning than in Sect. 2.2). For a tubular structure having annular cross-section and thickness $s \leq 40$ mm, the values for ratio σ_{cr}/σ_s are shown in Fig. 3.5 versus ratio λ/λ_c, where $\lambda_c = \pi\sqrt{E/\sigma_s}$ is the

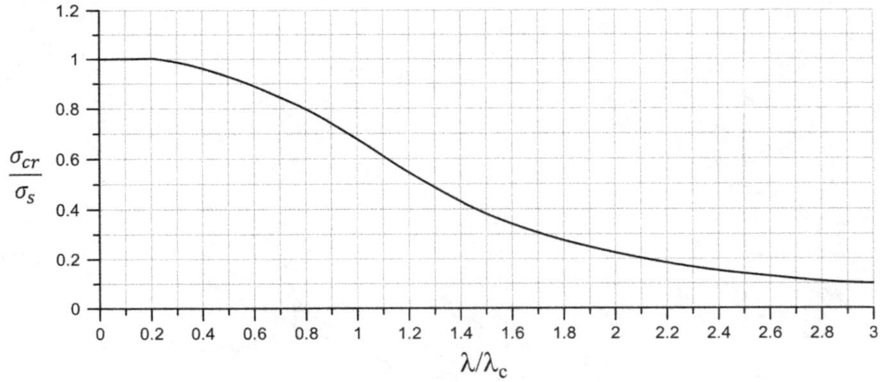

Fig. 3.5 Ratio σ_{cr}/σ_s versus λ/λ_c, for tubular structures of circular cross-section, with $s \leq 40$ mm

slenderness ratio at the limit beyond which the column no longer exhibits purely linear elastic behavior (the standard thus clearly assumes that the proportional limit σ_p, elastic limit σ_{el} and yield-point stress σ_s coincide). The diagram in Fig. 3.5 also applies to welded and rolled square or rectangular hollow sections (and thus not only circular sections) provided that $s \leq 40$ mm.

This standard also permits safety verifications on columns under compression to be carried out using the so-called *omega method* (ω method). With this simple method, verifications are performed assuming that the column's cross-section is compressed by the axial working force P increased by the ω coefficient, which is thus the factor by which the compressive working load must be multiplied in order to take lateral deflection into account. We must thus have:

$$\frac{\omega P}{A} \leq \sigma_a. \tag{3.10}$$

Coefficient ω varies with the slenderness ratio λ as a function of the material used for the column. Figure 3.6 shows the distribution curves of the function $\omega = \omega(\lambda)$ for welded or rolled circular, square or rectangular hollow sections with $s \leq 40$ mm, consisting of Fe360, Fe430 and Fe510 grade steel.

European Standard EN 1993-1.1 requires that the overall instability of a column structure of annular cross-section assumed to be under uniform compression be verified against buckling by imposing

$$\frac{P_{cal}}{P_{b,res}} \leq 1 \tag{3.11}$$

where P_{cal} and $P_{b,res}$ are respectively the design value of the compressive force (axial design load introduced for the calculation), and the design buckling resistance of the compressed structural member in question. The latter is given by the relation:

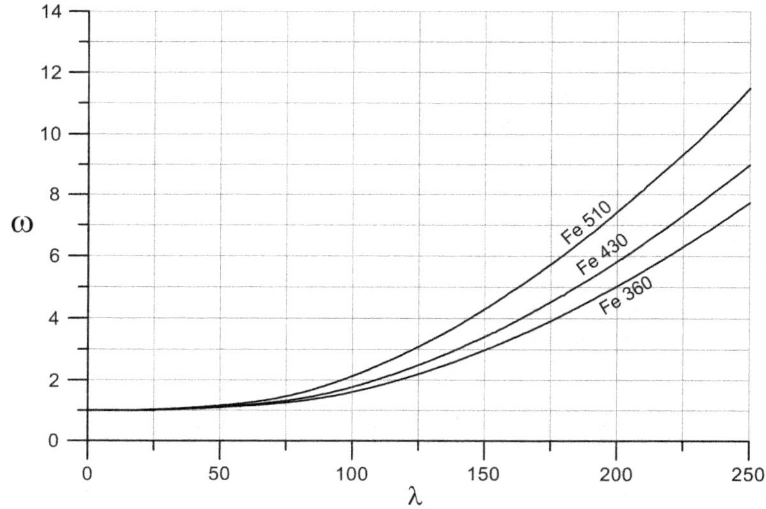

Fig. 3.6 Distribution curves of the function $\omega = \omega(\lambda)$ for Fe360, Fe430 and Fe510 tubes

$$P_{b,res} = \frac{\chi A \sigma_s}{\gamma_{M1}}, \tag{3.12}$$

where σ_s is the material's yield stress, A is the gross cross-sectional area or the effective cross-sectional area subjected to uniform compression for Class 1, 2 and 3 cross-sections and for Class 4 cross-sections respectively (in both cases, holes for fasteners at the column ends, which weaken the cross-section, need not be taken into account), γ_{M1} is the partial factor for resistance of members to instability assessed by member checks, and χ is a reduction factor for the relevant buckling mode, which depends on the type of cross-section and the steel used. This factor, which may not exceed a maximum of 1, is given by the relation:

$$\chi = \frac{1}{\phi + \sqrt{\phi^2 - \bar{\lambda}^2}} \leq 1 \tag{3.13}$$

where $\phi = 0.5\left[1 + \alpha\left(\bar{\lambda} - 0.2\right) + \bar{\lambda}^2\right]$, $\bar{\lambda} = \sqrt{A\sigma_s/P_{cr}}$ is the non-dimensional slenderness (and A is as specified above), P_{cr} is the elastic critical load for the relevant buckling mode based on the gross cross-sectional properties and the structure's equivalent column length L, and α is an imperfection factor obtained from specific tables in the standard.

To calculate reduction factor $\chi = \chi(\bar{\lambda})$, the standard provides five curves whereby the factor can be immediately determined for many effective cross-sectional areas and for the various steel classes considered (from S 235 to S 460). The five buckling curves, which are designated a_0, a, b, c and d and correspond, in that order, to imperfection factors α equal to 0.13, 0.21, 0.34, 0.49 and 0.46, are similar to that given in Fig. 3.5 and all start from the point having coordinates ($\bar{\lambda} = 0.2$; $\chi = 1.0$). The hollow sections of interest to us here are divided into two separate families, according to whether they are hot-finished or cold-formed. For cold-formed structures, buckling curve c is to be used, regardless of steel grade. For hot- finished structures, curve a_0 is to be used for S 460 grade steel, and curve a for steel in any of grades S 235, S 275, S 255 and S 420.

It should be noted that curve a coincides with the curve in Fig. 3.5, where the abscissa and ordinate are $\bar{\lambda}$ and χ respectively. It should also be noted that the standard specifies that buckling effects for beam structures under compression may be ignored for $\bar{\lambda} \leq 0.2$ or $P_{cal} \leq 0.04\,P_{cr}$.

3.2.2 Overview of Local Instability: Linear Buckling of Circular Cylinders

Even where it is aggravated by the deterioration in mechanical properties that occurs beyond the material's proportional limit, the overall instability or Euler buckling discussed in the foregoing section depends on the length of the slender structural component in question. For thin-walled cylindrical columns under axial load, however, another type of instability is possible which entails deflection curves of the structure's generators which are symmetrical with respect to the axis. This deflection mode transforms the generators into sinusoids, all of which are equal, so that the structure assumes an accordion-like shape; this type of axisymmetric lateral deflection of the shell middle surface and the corresponding collapse shape is known as *concertina mode* or *concertina buckling* (Von Kármán and Tsien 1941, Timoshenko and Gere 1961, Salvini et al. 1994). The wavelength of these sinusoids and the resulting wrinkling or corrugations in the wall are very short by comparison with column's actual length l (Fig. 3.7). Hence the term local elastic instability, as the column's length has no influence on it.

Consider a circular cylindrical shell of mean radius r_m, length l and very small thickness s, compressed uniformly at the hinged ends in the axial direction by forces of intensity q per unit of circumferential length (Fig. 3.7). The case of loading analyzed here is of great practical importance (tubular structures used in the automotive industry as impact absorbers to improve crashworthiness, aircraft fuselages subjected to compressive loads transmitted from the engines at the acceleration path, etc.). This kind of local instability can be investigated either by formulating and solving the fourth order differential equation for the deformed shape of a generator (equilibrium method), or by using the energy method. More simply, we can arrive at the solution of the problem by recalling that each elementary longitudinal strip cut off from the cylindrical shell by two neighboring longitudinal sections ϑ and $\vartheta + d\vartheta$, and thus having circumferential dimension $r_m d\theta$, is in the same conditions as a strip-beam resting on an elastic Winkler foundation and subjected to longitudinal compressive load (Winkler 1867, Belluzzi 1971); in this case the elastic medium consists of the annular strips that hoop the elementary strip-beams.

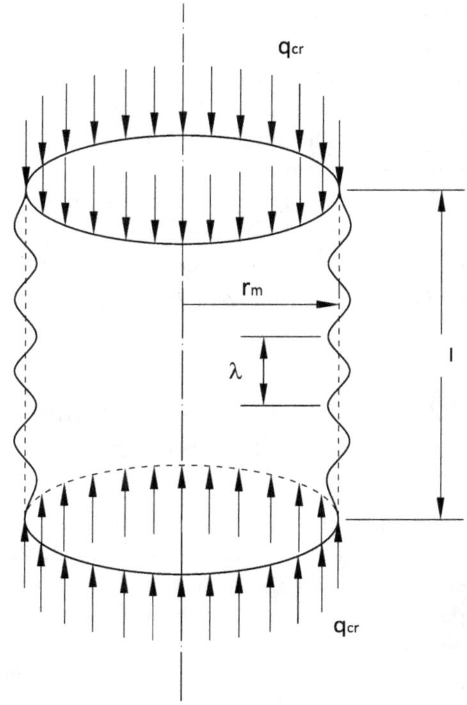

Fig. 3.7 Thin-walled cylindrical shell subjected to local buckling resulting from uniformly distributed axial load q_{cr}

By considering any elementary longitudinal strip-beam of length l and unitary circumferential dimension ($r_m d\theta = 1$), we arrive at the following relations:

$$q_{cr} = 2\sqrt{\xi D} = \frac{Es^2}{r_m\sqrt{3(1-\nu^2)}} \tag{3.14}$$

and

$$\sigma_{cr} = \frac{E}{\sqrt{3(1-\nu^2)}}\left(\frac{s}{r_m}\right). \tag{3.15}$$

Respectively, these relations give the critical unit load, q_{cr} (critical load per unit of circumferential length), and the critical stress, σ_{cr}. In relation (3.14), $D = Es^3/12(1-\nu^2)$ is the cylindrical shell's flexural rigidity and $\xi = Es/r_m^2$ is a parameter depending on material and geometry. The total critical load will obviously thus be given by: $P_{cr} = 2\pi r_m q_{cr}$. For steels ($\nu = 0.3$), relations (3.14) and (3.15) respectively become:

$$q_{cr} = 0.605\frac{Es^2}{r_m} \tag{3.16}$$

and

$$\sigma_{cr} = 0.605 \frac{Es}{r_m}. \tag{3.17}$$

The length λ_0 of the natural half-wave (half-wave length λ is not to be confused with column slenderness) is given by the relation:

$$\lambda_0 = \pi \sqrt[4]{D/\xi} = \frac{\pi}{\sqrt[4]{12(1-\nu^2)}} \sqrt{r_m s}, \tag{3.18}$$

which for steels ($\nu = 0.3$) becomes:

$$\lambda_0 = 1.73\sqrt{r_m s}. \tag{3.19}$$

The critical unit load given by relation (3.14) is the minimum possible ($q_{cr,min}$) and occurs when length l is a multiple of the natural half-wave length λ_0 and the edges can be considered as hinged; in this case $q_{cr,min}$ is independent of l. If l is not a multiple of λ_0, we have a half-wave length $\lambda \neq \lambda_0$ and q_{cr} is slightly above the minimum value given by relation (3.14). If $l < \sqrt{2}\lambda_0 = 2.44\sqrt{r_m s}$, there will be a single half-wave with $\lambda = l$, whereas if $l = 2.44\sqrt{r_m s}$, there may be one or two half-waves and q_{cr} reaches the maximum value $q_{cr,max} = 1.25 q_{cr,min}$. If l is many times higher than λ_0, the value of q_{cr} given by relation (3.14) can be accepted, even if l is not a multiple of λ_0. Finally, if the two edges, rather than being hinged, are free to move in the lateral direction, the critical unit load will be approximately half of that calculated with relation (3.14).

Here again, relation (3.15) is valid providing that $\sigma_{cr} \leq \sigma_p$. This is true for very thin cylindrical shells, i.e., those for which the ratio r_m/s is greater than or at minimum equal to $E/[\sigma_p\sqrt{3(1-\nu^2)}]$ (for steels $r_m/s \geq 0.605(E/\sigma_p)$).

It can be demonstrated that when the circular cylindrical shell does not have very thin walls and the critical stress exceeds the proportional limit, the critical load can be determined with the same fourth order differential equation for the deflection curve of the generator, providing that the tangent modulus E_t is substituted for Young's modulus E in the expression for the shell's flexural rigidity D appearing in this equation. We thus arrive at the following relation:

$$\sigma_{cr} = \frac{\sqrt{EE_t}}{\sqrt{3(1-\nu^2)}} \left(\frac{s}{r_m}\right). \tag{3.20}$$

Once again, then, after setting $\sigma_{cr} = \sigma_p$ in relation (3.15) and using the latter to determine the limiting value of ratio r_m/s for which relation (3.15) is still valid, we must proceed point-wise to plot the $\sigma_{cr} = \sigma_{cr}(r_m/s)$ curve to be used in the range where said relation can no longer be applied. The procedure to be used for this purpose is that same as that which enabled us to obtain function $\sigma_{cr} = \sigma_{cr}(\lambda)$ in Fig. 3.3. Thus, several values of critical stress σ_{cr} beyond σ_p are established on the material's experimental $\sigma = \sigma(\varepsilon)$ curve and the values of the tangent modulus E_t

are determined for each of these points. Once σ_{cr} and E_t are known, the values of ratio r_m/s for them can be calculated using relation (3.20).

For instability beyond the proportional limit, the relation whereby the natural half-wave length can be calculated is as follows:

$$\lambda_0 = \pi \frac{\sqrt{r_m s}}{\sqrt[4]{12(1-\nu^2)}} \left(\sqrt[4]{\frac{E_t}{E}} \right). \tag{3.21}$$

For steels ($\nu = 0.3$), this becomes:

$$\lambda_0 \cong 1,73\sqrt{r_m s} \left(\sqrt[4]{\frac{E_t}{E}} \right). \tag{3.22}$$

Consequently, in the case of local instability with critical stress beyond the proportional limit, the wavelength of the accordion-like deflection curve is shorter than in the case where $\sigma_{cr} \leq \sigma_p$.

If l is very large compared to r_m, overall buckling of the entire structure may occur first, with the circular cylindrical shell behaving like a column subjected to longitudinal compressive load. Assuming that the structure is hinged at the ends, recalling relation (3.5) and given $\rho^2 = r^2/2$, this will take place if:

$$\pi^2 \frac{E r_m^2}{2l^2} < \frac{1}{\sqrt{3(1-\nu^2)}} \left(\frac{Es}{r_m} \right) \tag{3.23}$$

that is if

$$\frac{l^2}{r_m^2} > \frac{\pi^2 \sqrt{3(1-\nu^2)}}{2} \left(\frac{r_m}{s} \right) \tag{3.24}$$

and, for steels, if

$$\frac{l^2}{r_m^2} > 8.15 \left(\frac{r_m}{s} \right). \tag{3.25}$$

In this case, we have:

$$\sigma_{cr} = \pi^2 \frac{E r_m^2}{2l^2} \tag{3.26}$$

and

$$P_{cr} = 2\pi r_m s \sigma_{cr} = \pi^3 \frac{E s r_m^3}{l^2}. \tag{3.27}$$

Conversely, if the circular cylindrical shell is very short compared to the radius (in this case there will be no stiffening ring effect), or if length l, which is equal to the half-wave length λ, is significantly less than the length of the natural half-wave

λ_0, q_{cr} will tend to the value corresponding to a strip-beam of unitary width considered in isolation. In this case, we have:

$$q_{cr} \cong \pi^2 \frac{D}{l^2} \tag{3.28}$$

and

$$\sigma_{cr} \cong \frac{\pi^2}{12(1-\nu^2)} \frac{Es^2}{l^2} \tag{3.29}$$

and, for steels,

$$\sigma_{cr} \cong 0.9 \frac{Es^2}{l^2}. \tag{3.30}$$

Finally, if length l and mean radius r_m are such as to satisfy the inequality

$$\frac{l^2}{r_m^2} < \frac{\pi^2\sqrt{3(1-\nu^2)}}{2} \left(\frac{r_m}{s}\right) \tag{3.31}$$

and, for steels, if

$$\frac{l^2}{r_m^2} < 8.15 \left(\frac{r_m}{s}\right), \tag{3.32}$$

the circular cylindrical shell, still assumed to be hinged at the ends and under uniform axial load, will fail through local instability. In this case, the critical stress and the critical load per unit of length can be found from relations (3.15) and (3.14) respectively, while $P_{cr} = 2\pi r_m q_{cr}$.

In the more general case, however, another type of local instability may occur, with a deformation of the structure which is not symmetrical with respect to the axis of the circular cylindrical shell under axial compression (Baker et al. 1968, Belluzzi 1971, Como 1973, Bruhn 1973). To investigate this type of instability, we start from the general partial differential equations of equilibrium and continuity for cylindrical shell structures and formulate a three-equation system whose unknowns are the displacements u, v and w in the radial, circumferential and axial directions respectively. These equations and the boundary conditions (the usual hinged or simply supported edges) are satisfied by expressions of the following kind

$$u(z,\theta) = C \sin\left(\frac{m\pi z}{l}\right) \sin(n\theta), \tag{3.33}$$

where z is the axial coordinate measured from one end of the circular cylindrical shell, and θ is the angular coordinate of the longitudinal cross-section measured from a longitudinal reference plane, while m and n are two integers and C is a constant to be determined by applying boundary conditions.

Deformation in the circular cylindrical shell takes place with m half-waves along the structure's axis and with n half-waves along the generic circumference in the transversal cross-section, or in other words, in a diamond pattern of dimples. The buckling equation obtained by equating a determinant to zero gives $q_{cr} = q_{cr}(r_m, l, s, m, n)$. The integers m and n for which q_{cr} is minimum will vary according to the relationships between r_m, l and s.

This approach, which we have necessarily described in barest outline, leads to the particular cases indicated earlier. Specifically, for $n = 0$, we have the symmetric accordion-like deformation discussed above; with very large ratios l/r_m and $n = 1$, we have the typical deformation associated with Euler buckling [relations (3.26) and (3.27)]; for very small ratios l/r_m and $n = 0$, we obtain relations (3.28)–(3.30). In the general case, or in other words when the mn dimples appear, we obtain an expression for q_{cr} like relation (3.14) applying to the symmetric case. From the theoretical standpoint, this demonstrates that the formation of the n half-waves in the circumferential direction takes place when l is not an exact multiple of λ_0; it follows that q_{cr} is approximately equal to the critical value given by relation (3.14), without undergoing the small increases that are found in symmetric buckling.

The foregoing considerations would suggest that relation (3.14) should be taken as the basis for the calculations performed for non-symmetric buckling. In reality, because of the initial imperfections that are inevitably found in thin-walled circular cylindrical shells, deflection modes can also occur with m half-waves that are not all of the same length λ. In this case, there will be a lower critical unit load given by the relation:

$$q_{cr} = \frac{3Es^2}{5r_m\sqrt{3(1-\nu^2)}}. \tag{3.34}$$

Lastly, it should be borne in mind that the experimental campaigns carried out to verify the theoretical results demonstrate that measured values of the critical unit load are much lower than those calculated with this relation (i.e., are 20–60 % of the calculated loads), and that the gap widens as ratio s/r_m drops.

3.3 Overview of Torsional Buckling of Circular Cylinders

All structures whose cross-sections show low torsional rigidity are prone to buckling under torsion. Though this is chiefly true of thin-walled structures with open cross-section, it should be recognized that closed section structural components such as the thin-walled cylindrical shells with annular cross-section (circular cylindrical shells) of interest to us here may also be subject to this type of instability, particularly around structural discontinuities such as openings, manholes, nozzles and the like (Timoshenko and Gere 1961, Belluzzi 1971).

Torsional buckling of a thin-walled circular cylindrical shell subjected to twisting couple M_t produces helical and diagonal wrinkles in the structure. This phenomenon is also investigated on the basis of the general partial differential equations of equilibrium and continuity for cylindrical shell structures, from which a three-equation system can be formulated whose unknowns are the displacements u, v and w in the radial, circumferential and axial directions respectively. These equations and the boundary conditions (the usual hinged or simply supported edges) are satisfied by expressions of the following kind:

$$u(z,\theta) = C\sin\left(\frac{m\pi z}{l} - n\theta\right). \tag{3.35}$$

This relation represents n helices with geometrical pitch $\lambda = 2l/m$. In this relation, z is the axial coordinate measured from one end of the circular cylindrical shell, and θ is the angular coordinate of the longitudinal cross-section measured from a longitudinal reference plane, while m and n are two integers and C is a constant to be determined by applying boundary conditions.

The critical value of the twisting moment (critical twisting moment or critical torque) is given by the relation:

$$M_{t,cr} = \frac{\pi\sqrt{2}}{3(1-\nu^2)^{3/4}} Es^2\sqrt{r_m s} = 1.481\frac{Es^2\sqrt{r_m s}}{(1-\nu^2)^{3/4}}, \tag{3.36}$$

which corresponds to the maximum critical shear stress

$$\tau_{cr,max} = \frac{M_{t,cr}}{2\pi r_m^2 s} = \frac{E}{3\sqrt{2}(1-\nu^2)^{3/4}}\left(\frac{s}{r_m}\right)^{3/2}. \tag{3.37}$$

For steels ($\nu = 0.3$), these relations respectively become:

$$M_{t,cr} = 1.590Es^2\sqrt{r_m s}, \tag{3.38}$$

and

$$\tau_{cr,max} = 0.253E\left(\frac{s}{r_m}\right)^{3/2}. \tag{3.39}$$

Torsional buckling occurs in the elastic field if $\tau_{cr,max} \leq \tau_p$ (τ_p is the shear stress at the proportional limit), i.e., if:

$$\frac{r_m}{s} \geq \frac{1}{\sqrt[3]{18}\sqrt{1-\nu^2}}\left(\frac{E}{\tau_p}\right)^{2/3}. \tag{3.40}$$

and, for steels ($\nu = 0.3$), if

$$\frac{r_m}{s} \geq 0.40\left(\frac{E}{\tau_p}\right)^{2/3}. \tag{3.41}$$

3.4 Overview of Flexural Ovalization Buckling of Circular Cylinders

The flexural ovalization buckling of circular cylindrical shells under a bending moment acting in a longitudinal plane (plane of bending) is a somewhat different phenomenon than those we have addressed so far, as the main deformation resulting from pure bending is accompanied by a secondary deformation which causes a progressive reduction in the structure's flexural rigidity. If a constant bending moment M_f is applied in a longitudinal plane of the thin-walled circular cylinder (circular cylindrical shell), that can be treated as an annular cross-section beam (Fig. 3.8a), the cross-section will ovalize progressively as M_f increases. Consequently, the elementary arcs $r_m d\theta$ will approach the neutral axis and the moment of inertia of area (or second moment of area) with respect to diameter, initially equal to I_0, will assume smaller and smaller values I. In addition, the unit angle φ around which two cross-sections at a unit distance from each other rotate as the structure bends will increase not only because of the increase in M_f, but also because of the decrease in I. Essentially, then, φ is no longer a linear function of M_f, but increases much more rapidly (Belluzzi 1971).

It is thus necessary to determine the relation which correlates the bending moment to the cross-sections' rotation angle, as well as that which links stress to the bending moment, which will no longer be linear, given the synergistic influence of the reduction in the section modulus of the cross-sectional area resulting from the reduction in I. This latter reduction, moreover, can be so considerable as to lead to a spontaneous increase in angle φ with no further rise in bending moment M_f, so that equilibrium becomes unstable.

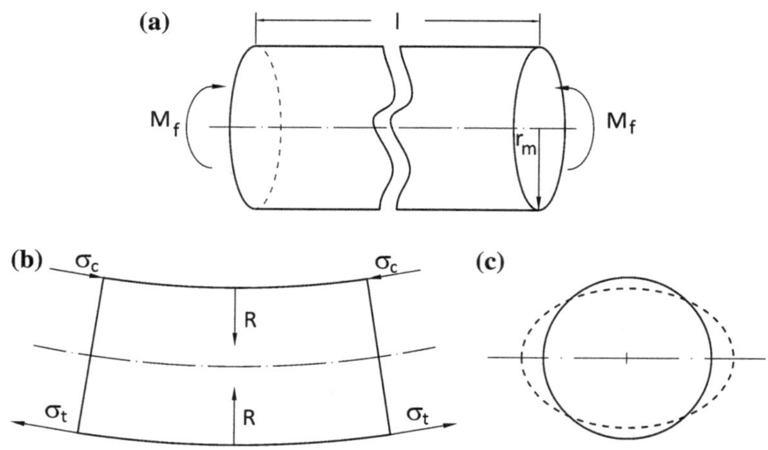

Fig. 3.8 **a** Annular cross-section cylindrical shell subject to constant bending moment M_f; **b** compressed elementary arc and stretched elementary arc of the upper and lower generators; **c**) cross-sectional ovalization due to the centripetal resultants of σ_c and σ_t

Figure 3.8 also illustrates the ovalization phenomenon. As a result of bending, two initially parallel generic cross-sections rotate until they form a given angle. The cylinder's generators located above and below the neutral axis will thus be compressed and stretched respectively. The compressive stress σ_c and tensile stress σ_t acting at the ends of the arcs of the upper and the lower generator (Fig. 3.8b), bounded by these two cross-sections, will be inclined relative to each other by the same angle formed by the cross-section, and will thus have resultants R that are both in the centripetal direction, or in other words towards the neutral axis, and thus tend to deform the cross-section, ovalizing it if it is thin. Though this phenomenon, which exists in any deflected beam, can be ignored for solid cross-sections, it is anything but negligible for the thin-walled circular cylindrical shells of interest to us here.

To determine the critical bending moment which triggers flexural ovalization buckling, it is first necessary to determine the intensity of secondary deformation as compared to primary deformation. This is accomplished by means of the minimum total energy principle and, specifically, the first theorem of minimum work. For the analytical developments involved, the reader is directed to specialized texts. Here, we will present only those results that are of interest in practical design applications.

The critical bending moment is given by the relation:

$$M_{f,cr} = 0.388\pi \frac{E r_m s^2}{\sqrt{1-\nu^2}} = 1.219 \frac{E r_m s^2}{\sqrt{1-\nu^2}}. \tag{3.42}$$

This moment corresponds to the following maximum critical stress:

$$\sigma_{cr,max} = 0.461 \frac{E s}{r_m \sqrt{1-\nu^2}}. \tag{3.43}$$

When bending moment M_f reaches the critical value given by relation (3.42) and exceeds it by however small an amount, equilibrium becomes impossible and the moment of inertia of area I continues to decrease, while φ increases (to arrest this self-perpetuating process, it would be necessary to reduce M_f very rapidly).

This type of buckling takes place in the elastic field if $\sigma_{cr,max} \leq \sigma_p$, i.e., if:

$$\frac{r_m}{s} \geq \frac{0.461 E}{\sigma_p \sqrt{1-\nu^2}}. \tag{3.44}$$

It should be borne in mind that relations (3.42) and (3.43) do not take any end constraints on the circular cylindrical shell into account and apply under the assumption that there is a constant bending moment at all of the structure's cross-sections. If such constraints are able to prevent ovalization, even if only in a localized area, these relations will apply only for structures that are very long by comparison with the radius. If the structure is not very long, the stiffening effect of the constraints will cause the critical bending moment to be higher that the theoretical value given by relation (3.42). If M_f is not constant as assumed (as is typically the case of horizontal tanks resting on saddle supports), instability starts at the cross-section where the bending moment is highest. The critical value $M_{f,cr}$, however, is slightly higher than that given by relation (3.42), as the portions of the circular cylindrical shell adjacent to the most highly stressed cross-section are subject to lower bending moments and thus exert a containment action.

3.5 Thin-Walled Circular Cylinders Under Internal Pressure, Loaded Beyond the Elastic Range

We will consider a thin-walled circular cylinder such as that shown in Fig. 1.1. In the initial condition, i.e., before internal pressure p_i is applied, this structure has inside diameter $d_{i,0}$ and thickness $s_{i,0}$. We will subject this thin-walled tube to an internal pressure p_i such as to stress it beyond yielding. Let d and s be the instantaneous values of the inside diameter and thickness of this circular cylinder, corresponding to the generic value of the internal pressure that stresses it beyond the elastic range. Once deformation has taken place, let d_p and s_p be respectively the inside diameter of the structure in the plastic state (the subscript p thus designates magnitudes referring to plastic conditions). If the cylinder is open at the ends ($\sigma_z = 0$), the two non-zero principal stresses, i.e., the true radial stress and the true hoop stress, are given respectively by the following relations [see the first relation (1.16) and the first relation (1.7)]:

$$\sigma_r = -\frac{p_i}{2} \text{ and } \sigma_t = \frac{p_i d}{2s}. \tag{3.45}$$

As usual, we assume here that the material is isotropic and incompressible, and also that the material's behavior is the same for loading in compression as for tension (then, there is no Bauschinger effect). In this case, the volume remains constant during deformation and volume expansion will thus be zero, i.e., the sum of the three principal true strains will be zero. As the incompressibility condition we will thus have:

$$\varepsilon_r + \varepsilon_t + \varepsilon_z = 0, \tag{3.46}$$

where ε_r, ε_t and ε_z denote the three principal natural strains (or true or logarithmic strains). Relation (3.46) is rigorously true if the plastic component of strain is predominant; it leads to the following relation which expresses the true radial strain as a function of the true hoop strain and of the true axial strain:

$$\varepsilon_r = -(\varepsilon_t + \varepsilon_z). \tag{3.47}$$

The natural radial strain, however, can also be expressed by the following relation:

$$\varepsilon_r = \int_{s_{i,0}}^{s} \frac{ds}{s} = \ln \frac{s}{s_{i,0}}. \tag{3.48}$$

By integrating this equation and recalling relation (3.47), we obtain:

$$s = s_{i,0}e^{\varepsilon_r} = s_{i,0}e^{-(\varepsilon_t+\varepsilon_z)} \tag{3.49}$$

where e is the base of the natural logarithms.

Likewise, the true hoop strain can be expressed as a function of the change in diameter, so that we will have

$$\varepsilon_t = \ln\frac{d}{d_{i,0}} \tag{3.50}$$

and thus, through integration

$$d = d_{i,0}e^{\varepsilon_t}. \tag{3.51}$$

Substituting relations (3.49) and (3.51) in the second relation (3.45) yields the following relation expressing the true hoop stress as a function of the initial dimensions of the circular cylinder in question:

$$\sigma_t = \frac{p_i d_{i,0}e^{\varepsilon_t}}{2s_{i,0}e^{-(\varepsilon_t+\varepsilon_z)}}, \tag{3.52}$$

from which we obtain

$$p_i = \frac{2\sigma_t s_{i,0}}{d_{i,0}}e^{-\varepsilon_t(2+\varepsilon_z/\varepsilon_t)}. \tag{3.53}$$

The following flow rules, independently due to Lévy 1870 and 1871, and Mises 1913 (thus called Lévy-Mises equations or plastic constitutive laws) for an ideal plastic body, where the elastic strains are negligible, are utilized:

$$\begin{aligned}
\varepsilon_r &= \frac{\varepsilon_e}{\sigma_e}\left[\sigma_r - \frac{1}{2}(\sigma_t + \sigma_z)\right] \\
\varepsilon_t &= \frac{\varepsilon_e}{\sigma_e}\left[\sigma_t - \frac{1}{2}(\sigma_z + \sigma_r)\right] \\
\varepsilon_z &= \frac{\varepsilon_e}{\sigma_e}\left[\sigma_z - \frac{1}{2}(\sigma_r + \sigma_t)\right].
\end{aligned} \tag{3.54}$$

In these relationships the ratio ε_e/σ_e, where ε_e and σ_e are the equivalent plastic strain (or effective or significant strain) and the equivalent stress (or effective or significant stress) respectively, changes throughout the course of the plastic deformation, and can be evaluated from an effective stress-effective strain curve. Bearing in mind that the radial stress is an order of magnitude lower than the hoop stress and the axial stress, the equations (3.54) yield:

$$\frac{\varepsilon_z}{\varepsilon_t} = \frac{2\sigma_z/\sigma_t - 1}{2 - \sigma_z/\sigma_t}. \tag{3.55}$$

For the circular cylinder with unloaded open ends (free open ends), where $\sigma_z = 0$, we obtain $\varepsilon_z/\varepsilon_t = -1/2$; accordingly, substituting this value of ratio $\varepsilon_z/\varepsilon_t$ in equation (3.53) gives:

$$\sigma_t = \frac{p_i}{2(s_{i,0}/d_{i,0})} e^{\frac{3}{2}\varepsilon_t}. \tag{3.56}$$

As in the elastic range hoop strain ε_t is very small, and indeed close to zero, relation (3.56) can be reduced to the second relation (3.45), where we set $d = d_{i,0}$ and $s = s_{i,0}$.

For the circular cylinder with closed ends, the axial stress is half of the hoop stress, i.e., $\sigma_z = \sigma_t/2$, and it can be concluded from relation (3.55) that $\varepsilon_z/\varepsilon_t = 0$. Consequently, relation (3.53) gives the following expression linking true hoop stress to the initial dimensions:

$$\sigma_t = \frac{p_i e^{2\varepsilon_t}}{2(s_{i,0}/d_{i,0})}. \tag{3.57}$$

Relations (3.56) and (3.57) indicate that true hoop stress increases exponentially with true hoop strain. For a circular cylinder with closed ends and a ductility of 50 %, for example (in this case, the instantaneous value of the diameter will be one and a half times the initial diameter, or $d = 1.5\, d_{i,0}$), we will have $\varepsilon_t = \ln(1.5\, d_{i,0}/d_{i,0}) = 0.405$, and relation (3.57) will thus yield $\sigma_t = 2.250(\sigma_t)_{nominal}$, while for ductility of 100 and 500 %, the same procedure will give $\sigma_t = 3.998(\sigma_t)_{nominal}$ and $\sigma_t = 25.00(\sigma_t)_{nominal}$ respectively. For a circular cylinder with unloaded open ends, all other conditions being equal, the $(\sigma_t)_{nominal}$ multiplication factors are reduced from 2.25 to 1.84, from 4.00 to 2.83, and from 25.00 to 11.18 for ductility of 50, 100 and 500 % respectively. Conditions permitting, it is clear that the circular cylinder with unloaded open ends, has advantages over its counterpart with closed ends as regards the ability to withstand pressures high enough to cause pronounced plastic flow.

We remember that ductility is a measure of the deformability of a material. Ductility is measured in tension tests by nominal strain or nominal unit elongation, $\varepsilon = (L - L_0)/L_0$, and reduction of area, $q = (A - A_0)/A_0$, where L and L_0 are the final and initial (or original) gage length respectively, and A and A_0 are the final and initial (or original) cross-sectional area respectively.

For steels, European Standard EN 1993 1.1 requires a minimum ductility that should be expressed in terms of limits for: ratio σ_u/σ_s of the specified minimum ultimate tensile strength σ_u to the specified minimum yield strength σ_s greater or at least equal to 1 ($\sigma_u/\sigma_s \geq 1$); elongation at failure on a gauge length of $5.65\sqrt{A_0}$ not less than 15 %; ratio $\varepsilon_u/\varepsilon_s$ of the ultimate strain ε_u, with ε_u corresponding to the ultimate strength σ_u, to the yield strain $\varepsilon_s = \sigma_s/E$ greater or at least equal to 1 ($\varepsilon_u/\varepsilon_s \geq 1$). However steel conforming with one of the steel grades specified by the same standard can be accepted as satisfying these requirements.

3.6 Thin-Walled Circular Cylinders Under Internal Pressure and Axial Tension, and Stressed Beyond Yielding

We will now consider a thin-walled circular cylinder, again like that shown in Fig. 1.1, but subjected simultaneously to internal pressure p_i and to axial tension as a result of an axial load F applied to the free ends. For this combined load, we will determine what the wall thickness will be when plastic flow instability occurs (Faupel 1964). There can be two cases, both of which will be analyzed below: that in which the hoop stress is greater than the axial stress ($\sigma_t > \sigma_z$), and that in which the axial stress is greater than the hoop stress ($\sigma_z > \sigma_t$).

3.6.1 Hoop Stress Greater than Axial Stress

In this case, there is a limiting value of internal pressure that triggers plastic flow instability. The objective is thus to determine this pressure, which is known as the *instability pressure*.

In the plastic range, the principal true stresses are given by the following relations:

$$\begin{aligned}\sigma_r &= -\frac{p_i}{2}\\ \sigma_t &= \frac{p_i d_p}{2s_p}\\ \sigma_z &= \frac{p_i d_p}{4s_p} + \frac{F}{\pi d_p s_p}.\end{aligned} \tag{3.58}$$

As the critical conditions in this case are determined by the hoop stress, the second relation (3.58) must be expressed as a function of true strains. Recalling relations (3.47) and (3.48), in which the upper and lower integration limits are substituted with s and s_p respectively, we obtain:

$$\varepsilon_r = -(\varepsilon_t + \varepsilon_z) = \ln\frac{s_p}{s}. \tag{3.59}$$

Integrating this relation gives

$$s_p = se^{-(\varepsilon_t+\varepsilon_z)}. \tag{3.60}$$

In the circumferential directions, given relations (3.50) and (3.51), we will have

$$\varepsilon_t = \ln\frac{d_p}{d} \tag{3.61}$$

and

$$d_p = de^{\varepsilon_t}. \tag{3.62}$$

Substituting relations (3.60) and (3.62) in the second relation (3.58) gives:

$$\sigma_t = \frac{pde^{\varepsilon_t}}{2se^{-(\varepsilon_t+\varepsilon_z)}} \tag{3.63}$$

and thus

$$p = 2\sigma_t\left(\frac{s}{d}\right)e^{-\varepsilon_t(2+\varepsilon_z/\varepsilon_t)}. \tag{3.64}$$

Recalling relation (3.55), relation (3.64) can be rewritten in the following form:

$$p = 2\sigma_t\left(\frac{s}{d}\right)e^{-\varepsilon_t[3/(2-\sigma_z/\sigma_t)]}. \tag{3.65}$$

Plastic flow instability will occur when pressure reaches a critical value that causes an increase in strain with no further rise in pressure, or in other words when the total pressure differential dp (the differential operator d is not to be confused with diameter d) drops to zero. As the pressure is a function of both σ_t and ε_t, the instability condition will be expressed by the following partial differential equation, which is also written in the form obtained by recalling relation (3.65):

$$\begin{aligned} dp = \frac{\partial p}{\partial \sigma_t}d\sigma_t + \frac{\partial p}{\partial \varepsilon_t}d\varepsilon_t = 0 = 2\left(\frac{s}{d}\right)e^{-\varepsilon_t[3/(2-\sigma_z/\sigma_t)]}d\sigma_t + \\ - 2\sigma_t\left(\frac{s}{d}\right)\frac{3}{2-\sigma_z/\sigma_t}e^{-\varepsilon_t[3/(2-\sigma_z/\sigma_t)]}d\varepsilon_t = 0. \end{aligned} \tag{3.66}$$

This yields:

$$\frac{d\sigma_t}{d\varepsilon_t} = \sigma_t\frac{3}{2-\sigma_z/\sigma_t}. \tag{3.67}$$

We will now assume that the experimental true stress–true strain curve $\sigma = \sigma(\varepsilon)$ obtained by means of a tensile test on a standard specimen can be expressed starting from the onset of plastic flow, i.e., from yield-point stress σ_s, up to the ultimate stress σ_u by means of the following Ramberg and Osgood formula (1943)

$$\sigma = K\varepsilon^n, \tag{3.68}$$

where K and n are constants for the material representing its strength coefficient and strain-hardening coefficient respectively.

In this connection, it should be noted that in a true stress–true strain diagram $\sigma = \sigma(\varepsilon)$ plotted using bi-logarithmic coordinates (true stress-true strain curve on log–log coordinates), the parabola given by relation (3.68) becomes a straight line of slope n relative to the axis of true strain ε on the abscissa, while K expresses the true stress σ, on the ordinate and corresponding to a unit true strain. It should also

be noted that relation (3.68) applies to the entire range of uniform plastic flow (plastic range), bounded on the left by strain ε_s corresponding to yield stress (yield strength) σ_s and on the right by strain ε_u corresponding to ultimate stress (tensile strength) σ_u, which is the maximum design stress that can be considered in design calculations for the ultimate limit state. The latter stress is reached when the effect of strain-hardening, which increases the material's strength, is exactly balanced by the reduction in the specimen's effective cross-sectional area due to necking down (or local instability of the experimental stress–strain curve). Unstable plastic flow begins from this moment, rapidly leading the material to the failure stress (fracture strength) σ_f, which corresponds to strain ε_f; this is the range of unstable flow.

Relation (3.68), however, represents the equivalent combined stress–strain curve $\sigma = \sigma(\varepsilon)$, i.e., the uniaxial tension stress–strain curve equivalent to a triaxial stress state. Accordingly, determining ε_e as a function of σ_e from relation (3.68) and substituting the expression thus found in the second relation (3.54) yields:

$$\varepsilon_t = \frac{\sigma_e^{(1-n)/n}}{K^{1/n}} \sigma_t \left[1 - \frac{\sigma_z}{2\sigma_t} - \frac{\sigma_r}{2\sigma_t}\right]. \tag{3.69}$$

The latter, after multiplying and dividing by $\sigma_t^{(1-n)/n}$ and recalling that on the basis of the maximum distortion energy strength theory, the equivalent stress σ_e is given by the relation shown in the second column, last row, of Table 1.1, yields:

$$\varepsilon_t = \left(\frac{\sigma_t}{K}\right)^{1/n} \left[\left(\frac{\sigma_r}{\sigma_t}\right)^2 + 1 + \left(\frac{\sigma_z}{\sigma_t}\right)^2 - \frac{\sigma_r}{\sigma_t} - \frac{\sigma_z}{\sigma_t} - \frac{\sigma_z \sigma_r}{\sigma_t^2}\right]^{(1-n)/2n} \left(1 - \frac{\sigma_z}{2\sigma_t} - \frac{\sigma_r}{2\sigma_t}\right). \tag{3.70}$$

In our case, as σ_r is an order of magnitude lower than σ_t and σ_z, and thus negligible (as a result, the stress state can be considered to be plane, with $\sigma_r = 0$), relation (3.70) becomes:

$$\varepsilon_t = \left(\frac{\sigma_t}{K}\right)^{1/n} \left[1 + \left(\frac{\sigma_z}{\sigma_t}\right)^2 - \frac{\sigma_z}{\sigma_t}\right]^{(1-n)/2n} \left(1 - \frac{\sigma_z}{2\sigma_t}\right). \tag{3.71}$$

From the latter, raising the first and second members to n and solving for σ_t, we obtain:

$$\sigma_t = f\left(\frac{\sigma_z}{\sigma_t}\right) \varepsilon_t^n \tag{3.72}$$

in which:

$$f\left(\frac{\sigma_z}{\sigma_t}\right) = \frac{K}{\left[1 + \left(\frac{\sigma_z}{\sigma_t}\right)^2 - \frac{\sigma_z}{\sigma_t}\right]^{(1-n)/2} \left(1 - \frac{\sigma_z}{2\sigma_t}\right)^n}. \tag{3.73}$$

Deriving relation (3.72) for ε_t gives:

$$\frac{d\sigma_t}{d\varepsilon_t} = nf\left(\frac{\sigma_z}{\sigma_t}\right)\varepsilon_t^{n-1}. \tag{3.74}$$

Finally, by equaling the second members of relations (3.67) and (3.74) and recalling relation (3.72), we obtain:

$$\varepsilon_t = n\frac{2 - \sigma_z/\sigma_t}{3}. \tag{3.75}$$

This is the equation which defines the instability condition in the case where hoop stress σ_t is predominant by comparison with axial stress σ_z.

To determine the burst pressure of the circular cylinder in question, we can use the relation for equivalent stress correlated with maximum distortion energy strength theory (shown, as mentioned above, in the second column, last row, of Table 1.1), substituting σ_u for σ_e. Remembering that σ_r is negligible compared to σ_t and σ_z, this equivalent stress relation can be written in the following form:

$$\sigma_u = \sigma_t\left[1 + \left(\frac{\sigma_z}{\sigma_t}\right)^2 - \frac{\sigma_z}{\sigma_t}\right]. \tag{3.76}$$

For a cylindrical body with closed ends subject only to internal pressure, for which $\sigma_z/\sigma_t = 1/2$ and axial load $F = 0$, and recalling (3.57) and (3.75), this relation gives the following burst pressure p_b:

$$p_b = \frac{4\sigma_u\left(s_{i,0}/d_{i,0}\right)}{\sqrt{3}e^n}. \tag{3.77}$$

The same procedure can be used in cases where there is an axial load $F \neq 0$ and the σ_z/σ_t ratio is thus in the interval $1/2 < \sigma_z/\sigma_t < 1$.

It is still necessary to calculate the wall thickness in the instability condition. For this purpose, we will equal the strain ε_t defining the instability condition given by relation (3.75) with the strain ε_t found from the plastic flow equation and given by relation (3.71). We will thus have:

$$n\frac{2 - \sigma_z/\sigma_t}{3} = \left(\frac{\sigma_t}{K}\right)^{1/n}\left[1 + \left(\frac{\sigma_z}{\sigma_t}\right)^2 - \frac{\sigma_z}{\sigma_t}\right]^{(1-n)/2n}\left(1 - \frac{\sigma_z}{2\sigma_t}\right). \tag{3.78}$$

By solving this equality for σ_t and recalling the second relation (3.58), we obtain:

$$\sigma_t = K\left(\frac{2n}{3}\right)^n \frac{1}{\left[1 + \left(\frac{\sigma_z}{\sigma_t}\right)^2 - \frac{\sigma_z}{\sigma_t}\right]^{(1-n)/2}} = \frac{pd_p}{2s_p}. \tag{3.79}$$

The values for diameter d_p and wall thickness s_p in the condition of plastic flow appear in this relation. To express σ_t as a function of the initial magnitudes, it must be recalled that

$$\begin{aligned}\varepsilon_r &= \ln(1+\varepsilon_{r,0}) = \ln(s_p/s)\\ \varepsilon_t &= \ln(1+\varepsilon_{t,0}) = \ln(d_p/d)\\ \varepsilon_z &= \ln(1+\varepsilon_{z,0}) = \ln(l/l_0)\end{aligned} \tag{3.80}$$

where $\varepsilon_{r,0}$, $\varepsilon_{t,0}$ and $\varepsilon_{z,0}$ are the conventional strains (or nominal, engineering or Langrangian strains), s_p, d_p and l are the circular cylinder's thickness, diameter and axial length in the plastic state, while s, d and l_0 are the same magnitudes in the initial conditions. As in this case it is the internal pressure that governs the phenomenon, the maximum principal strain is ε_t as given by the second relation (3.80). We will thus have:

$$\varepsilon_t = \ln\left(\frac{d_p}{d}\right) \text{ and } d_p = de^{\varepsilon_t} \tag{3.81}$$

where ε_t is given by relation (3.70) or (3.71), depending on whether the problem is triaxial or biaxial (in the latter case, σ_r is negligible). Relations (3.81) are obviously equal to relations (3.61) and (3.62). To calculate s_p, it is first necessary to assess which of the two remaining strains, ε_r and ε_z is the minimum principal strain. If the minimum principal strain is ε_r, we will have:

$$\varepsilon_r = \ln\left(\frac{s_p}{s}\right) \text{ and } s_p = se^{\varepsilon_r}, \tag{3.82}$$

whereas if ε_z is the minimum principal strain, we will have:

$$\varepsilon_z = \ln\left(\frac{s_p}{s}\right) \text{ and } s_p = se^{\varepsilon_z}. \tag{3.83}$$

The plastic flow laws (3.54) can be rewritten to introduce the correlation $\sigma = \sigma(\varepsilon)$ from the Ramberg and Osgood formula (3.68), expressing the equivalent stress σ_e on the basis of the maximum distortion energy strength theory, as well as indicating the maximum principal stress (σ_t in our case) and relating it to the other two principal stresses. In this way, relations (3.54) yield the following relations:

$$\begin{aligned}\varepsilon_r &= \left(\frac{\sigma_t}{K}\right)^{1/n}\left[\left(\frac{\sigma_r}{\sigma_t}\right)^2+1+\left(\frac{\sigma_z}{\sigma_t}\right)^2-\frac{\sigma_r}{\sigma_t}-\frac{\sigma_z}{\sigma_t}-\frac{\sigma_z\sigma_r}{\sigma_t^2}\right]^{(1-n)/2n}\left(\frac{\sigma_r}{\sigma_t}-\frac{1}{2}-\frac{\sigma_z}{2\sigma_t}\right)\\ \varepsilon_t &= \left(\frac{\sigma_t}{K}\right)^{1/n}\left[\left(\frac{\sigma_r}{\sigma_t}\right)^2+1+\left(\frac{\sigma_z}{\sigma_t}\right)^2-\frac{\sigma_r}{\sigma_t}-\frac{\sigma_z}{\sigma_t}-\frac{\sigma_z\sigma_r}{\sigma_t^2}\right]^{(1-n)/2n}\left(1-\frac{\sigma_z}{2\sigma_t}-\frac{\sigma_r}{2\sigma_t}\right)\\ \varepsilon_z &= \left(\frac{\sigma_t}{K}\right)^{1/n}\left[\left(\frac{\sigma_r}{\sigma_t}\right)^2+1+\left(\frac{\sigma_z}{\sigma_t}\right)^2-\frac{\sigma_r}{\sigma_t}-\frac{\sigma_z}{\sigma_t}-\frac{\sigma_z\sigma_r}{\sigma_t^2}\right]^{(1-n)/2n}\left(\frac{\sigma_z}{\sigma_t}-\frac{\sigma_r}{2\sigma_t}-\frac{1}{2}\right).\end{aligned} \tag{3.84}$$

Depending on whether the minimum principal strain is ε_r or ε_z, we will use relation (3.82), introducing the first relation (3.84) in it, or relation (3.83), introducing in it the third relation (3.84). Obviously, if the stress state is biaxial inasmuch as the radial stress is considered to be negligible, we will use relations (3.84) but omitting the terms in σ_r. Assuming, for example, that the minimum principal strain is ε_r, for the biaxial stress state ($\sigma_r = 0$) and recalling relations (3.81), (3.82) and the first two relations (3.84), we will have:

$$\sigma_t = \frac{pd_p}{2s_p} = \frac{pd}{2s}\frac{e^{\left\{(\sigma_t/K)^{1/n}\left[1+(\sigma_z/\sigma_t)^2-\sigma_z/\sigma_t\right]^{(1-n)/2n}(2-\sigma_z/\sigma_t)/2\right\}}}{e^{\left\{(\sigma_t/K)^{1/n}\left[1+(\sigma_z/\sigma_t)^2-\sigma_z/\sigma_t\right]^{(1-n)/2n}[-(1+\sigma_z/\sigma_t)]/2\right\}}}. \tag{3.85}$$

This yields:

$$s = \frac{pd}{2\sigma_t}e^{\frac{3}{2}\left\{(\sigma_t/K)^{1/n}\left[1+(\sigma_z/\sigma_t)^2-\sigma_z/\sigma_t\right]^{(1-n)/2n}\right\}} \tag{3.86}$$

where σ_t is given by the expression between the two equal signs in relation (3.79).

For a better understanding of how to use the fogoing relations, we will consider a cylindrical steel circular cylinder ($\sigma_s = 350$ MPa; $v = 0.3$; $E = 210$ GPa; $K = 400$ MPa; $n = 0.16$) of initial inside diameter $d_{i,0} = 0.5$ m and subjected to an internal pressure $p_i = 150$ bar $= 15$ MPa. We will determine both the wall thickness with purely elastic behavior, and the wall thickness in conditions of plastic flow instability, under the assumption that hoop stress is greater than axial stress.

Assuming that our circular cylinder has closed ends, recalling the first relations (1.7), (1.13) and (1.16), applying the maximum distortion energy strength theory with a biaxial stress state (and thus ignoring the radial stress as compared to the hoop and axial stresses), establishing that the material's behavior is elastic up to yield, i.e., that the proportional limit, the elastic limit and the yield-point stress coincide, and considering the limit of elastic behavior, with $\sigma_e = \sigma_s = (\sqrt{3}/2)\sigma_t$, gives the following value for thickness: $s = (\sqrt{3}pd)/4\sigma_s = 9.3\,10^{-3}$ m $= 9.3$ mm.

Setting $\sigma_z/\sigma_t = 1/2$ in relation (3.79) yields the following value for the hoop stress which, in the assigned conditions, leads to plastic flow instability: $\sigma_t =$ 315.5 MPa. With this value for σ_t, relation (3.86) gives the following value of thickness s in conditions of plastic flow instability: $s = 5.7\,10^{-3}$ m $= 5.7$ mm. Under the assigned conditions, in other words, a circular cylinder with a 5.7 mm thick wall will become unstable when subjected to an internal pressure of 15 MPa.

3.6.2 Axial Stress Greater than Hoop Stress

Here, by contrast with the previous case, there is a limiting value of axial load that triggers plastic flow instability. Consequently, the objective is to determine this value of axial load, known as the *axial instability load*. As the maximum principal stress is the axial stress $\sigma_z(\sigma_z > \sigma_t)$, the total axial force will be given by the relation:

$$F = \pi d_p s_p \sigma_z. \tag{3.87}$$

Substituting the expressions for d_p and s_p given by relations (3.62) and (3.60) respectively in relation (3.87) gives:

$$F = \pi d s \sigma_z e^{-\varepsilon_z}. \tag{3.88}$$

Here again, plastic flow instability will occur when axial force F reaches a critical value that causes an increase in strain with no further rise in F, or in other words when the total differential dF drops to zero. As F is a function of both σ_z and ε_z, the instability condition will be expressed by the following partial differential equation, which, as in the case of equation (3.66), is also written in the form obtained by recalling relation (3.88):

$$dF = \frac{\partial F}{\partial \sigma_z} d\sigma_z + \frac{\partial p}{\partial \varepsilon_z} d\varepsilon_z = 0 = \pi d s \sigma_z e^{-\varepsilon_z} d\sigma_z - \pi d s \sigma_z e^{-\varepsilon_z} d\varepsilon_z = 0. \tag{3.89}$$

This yields

$$\frac{d\sigma_z}{d\varepsilon_z} = \sigma_z. \tag{3.90}$$

Following the same procedure described for the previous case where $\sigma_t > \sigma_z$, we determine ε_e as a function of σ_e from relation (3.68) and substitute the expression thus found in the third relation (3.54). This gives:

$$\varepsilon_z = \frac{\sigma_e^{(1-n)/n}}{K^{1/n}} \sigma_z \left(1 - \frac{\sigma_r}{2\sigma_z} - \frac{\sigma_t}{2\sigma_z}\right). \tag{3.91}$$

The latter, after multiplying and dividing by $\sigma_z^{(1-n)/n}$ and using the maximum distortion energy strength theory to express the equivalent stress σ_e, yields:

$$\varepsilon_z = \left(\frac{\sigma_z}{K}\right)^{1/n} \left[\left(\frac{\sigma_r}{\sigma_z}\right)^2 + \left(\frac{\sigma_t}{\sigma_z}\right)^2 + 1 - \frac{\sigma_r \sigma_t}{\sigma_z^2} - \frac{\sigma_t}{\sigma_z} - \frac{\sigma_r}{\sigma_z}\right]^{(1-n)/2n} \left(1 - \frac{\sigma_r}{2\sigma_z} - \frac{\sigma_t}{2\sigma_z}\right). \tag{3.92}$$

where σ_r is negligible, this equation becomes:

$$\varepsilon_z = \left(\frac{\sigma_z}{K}\right)^{1/n} \left[\left(\frac{\sigma_t}{\sigma_z}\right)^2 + 1 - \frac{\sigma_t}{\sigma_z}\right]^{(1-n)/2n} \left(1 - \frac{\sigma_t}{2\sigma_z}\right). \tag{3.93}$$

From the latter, raising the first and second members to n and solving for σ_z, we obtain:

$$\sigma_z = f\left(\frac{\sigma_t}{\sigma_z}\right) \varepsilon_z^n \tag{3.94}$$

in which

$$f\left(\frac{\sigma_t}{\sigma_z}\right) = \frac{K}{\left[\left(\frac{\sigma_t}{\sigma_z}\right)^2 + 1 - \frac{\sigma_t}{\sigma_z}\right]^{(1-n)/2}\left(1 - \frac{\sigma_t}{2\sigma_z}\right)^n}. \tag{3.95}$$

Deriving relation (3.94) for ε_z and recalling relation (3.90) as well as (3.94) leads to the following relation which expresses the instability condition in the case where stress σ_z is predominant by comparison with σ_t:

$$\varepsilon_z = n. \tag{3.96}$$

Here as in the previous case, it is still necessary to calculate the wall thickness in the instability condition. As the phenomenon is governed here by the axial load, the maximum principal strain is ε_z as given by the last relation (3.79). This strain, recalling relations (3.80) as well as the constant-volume strain condition given by relation (3.46), can be expressed as follows:

$$\varepsilon_z = \ln(l/l_0) = \ln(s/s_p) + \ln(d/d_p). \tag{3.97}$$

Once again, it is assumed that the minimum principal strain is the radial strain, given by relation (3.82). Proceeding as in the previous case where $\sigma_t > \sigma_z$, but now introducing σ_z as the maximum principal stress, the plastic flow relations (3.54) can be rewritten in the following form:

$$\begin{aligned}
\varepsilon_r &= \left(\frac{\sigma_z}{K}\right)^{1/n}\left[\left(\frac{\sigma_r}{\sigma_z}\right)^2 + \left(\frac{\sigma_t}{\sigma_z}\right)^2 + 1 - \frac{\sigma_r\sigma_t}{\sigma_z^2} - \frac{\sigma_t}{\sigma_z} - \frac{\sigma_r}{\sigma_z}\right]^{(1-n)/2n}\left(\frac{\sigma_r}{\sigma_z} - \frac{\sigma_t}{2z} - \frac{1}{2}\right)\\
\varepsilon_t &= \left(\frac{\sigma_z}{K}\right)^{1/n}\left[\left(\frac{\sigma_r}{\sigma_z}\right)^2 + \left(\frac{\sigma_t}{\sigma_z}\right)^2 + 1 - \frac{\sigma_r\sigma_t}{\sigma_z^2} - \frac{\sigma_t}{\sigma_z} - \frac{\sigma_r}{\sigma_z}\right]^{(1-n)/2n}\left(\frac{\sigma_t}{\sigma_z} - \frac{1}{2} - \frac{\sigma_r}{2\sigma_z}\right)\\
\varepsilon_z &= \left(\frac{\sigma_z}{K}\right)^{1/n}\left[\left(\frac{\sigma_r}{\sigma_z}\right)^2 + \left(\frac{\sigma_t}{\sigma_z}\right)^2 + 1 - \frac{\sigma_r\sigma_t}{\sigma_z^2} - \frac{\sigma_t}{\sigma_z} - \frac{\sigma_r}{\sigma_z}\right]^{(1-n)/2n}\left(1 - \frac{\sigma_r}{2\sigma_z} - \frac{\sigma_t}{2\sigma_z}\right).
\end{aligned} \tag{3.98}$$

Obviously, as the minimum principal strain is ε_r, the first relation (3.98) must be substituted in (3.82). In these conditions for the case examined here, we will have:

$$\sigma_z = \frac{F}{\pi d_p s_p} = \frac{F}{\pi d e^{\varepsilon_t} s e^{-\varepsilon_r}}, \tag{3.99}$$

where ε_r and ε_t are given by the first and second relations (3.98). Substituting the latter relations in (3.99) for the biaxial stress state ($\sigma_r = 0$) yields:

$$\sigma_z = \frac{F}{\pi d s e^{\left\{(\sigma_z/K)^{1/n}\left[(\sigma_t/\sigma_z)^2 + 1 - \sigma_t/\sigma_z\right]^{(1-n)/2n}\left[(\sigma_t/2\sigma_z) - 1\right]\right\}}} \tag{3.100}$$

which gives:

$$s = \frac{F}{\pi d \sigma_z e^{\left\{ (\sigma_z/K)^{1/n} \left[(\sigma_t/\sigma_z)^2 + 1 - \sigma_t/\sigma_z \right]^{(1-n)/2n} \left[(\sigma_t/2\sigma_z) - 1 \right] \right\}}} \tag{3.101}$$

where σ_z is given by relation (3.103), which is obtained by recalling that in this case the maximum principal strain in the axial direction can be expressed by means of the third relation (3.98). Consequently, for a plane stress state, equating relation (3.96) with the third relation (3.98) will give:

$$n = \left(\frac{\sigma_z}{K}\right)^n \left[\left(\frac{\sigma_t}{\sigma_z}\right)^2 - \frac{\sigma_t}{\sigma_z} + 1 \right]^{(1-n)/2n} \left(1 - \frac{\sigma_t}{2\sigma_z}\right). \tag{3.102}$$

By solving the latter for σ_z and recalling relation (3.87), we obtain:

$$\sigma_z = K(2n)^n \frac{1}{\left[\left(\frac{\sigma_t}{\sigma_z}\right)^2 - \frac{\sigma_t}{\sigma_z} + 1 \right]^{(1-n)/2}} \frac{1}{\left(2 - \frac{\sigma_t}{\sigma_z}\right)^n}. \tag{3.103}$$

It should be pointed out that in the foregoing analysis, as well as in the previous Sect. 3.6.1, the condition of instability depends on a single load factor, which may be internal pressure or axial load. As long as one, and only one, of these factors dominates, the analysis is entirely valid. In the particular case where the axial stress is equal to the hoop stress ($\sigma_z/\sigma_t = 1$), the condition of instability will depend on both factors, i.e., on internal pressure and axial load together. Where $\sigma_z = \sigma_t$, or in other words $\sigma_z/\sigma_t = 1$, relations (3.79) and (3.103) give different results and are thus not valid.

Chapter 4
Thick-Walled Circular Cylinders Under Internal and/or External Pressure Stressed in the Linear Elastic Range

4.1 General

As indicated in Chap. 1, thick-walled circular cylinders are by convention considered to be cylinders which satisfy any one of the following three equivalent conditions (Bickell and Ruiz 1967, Iurzolla 1981, Burr 1982, Ventsel and Krauthammer 2001, Ugural and Fenster 2003):

$$\frac{s}{d_i} > \frac{1}{20}; \quad \delta = \frac{d_e}{d_i} = \frac{r_e}{r_i} > 1.10; \quad \beta = \frac{d_i}{d_e} = \frac{r_i}{r_e} = \frac{1}{\delta} < 0.91. \tag{4.1}$$

Analyzing the stress state is somewhat more complicated for a thick-walled cylinder than for thin-walled cylinders, as it is no longer possible to assume that radial and hoop stress distributions are uniform along the thickness. The axial stress distribution, however, can still be considered uniform along the thickness provided that there are no thermal stresses (Prosciutto 1952, Den Hartog 1952, Crossland, Jorgensen and Bones 1959, Faupel 1964, Favretti 1967, Caboni 1971, Belluzzi 1971, Feodosyev 1973, Iurzolla 1981, Manna 1983, Saada 1993).

The range of validity of this latter assumption is that based on the Bernoulli-Navier hypothesis (Navier 1864), viz., that during deformation as a result of pressure alone, cross sections normal to the axis do not distort, but remain plane and translate parallel to themselves. Obviously, this assumption no longer holds when thermal stresses are involved, given that if the material's coefficient of thermal expansion α is independent of the temperature and is thus maintained constant, the fibers at higher temperatures will undergo greater elongation and vice versa, and the sections that were initially plane will distort.

Chapter 1 demonstrated that the stress state and the strain state in a thick-walled cylindrical body under internal and external pressures are triaxial. Analyzing these states with both stresses and strains varying along the radius involves considering the equilibrium equations and the compatibility equations simultaneously.

V. Vullo, *Circular Cylinders and Pressure Vessels*, Springer Series in Solid and Structural Mechanics 3, DOI: 10.1007/978-3-319-00690-1_4,

4.2 Radial Stress and Hoop Stress

For a thick-walled axisymmetric cylindrical body indefinitely extended in the direction of its z axis, we will consider a small element of infinitesimal axial dimension dz located between two cross sections having coordinates z and $z + dz$ (Fig. 1.1, aside from thickness). The inner and outer boundaries of this structure of revolution are subject to surface forces, which are assumed to be positive if in tension, negative if in compression. Here, we will be concerned with pressures p_e and p_i acting at the outer and inner radius respectively, which are obviously to be considered as negative surface forces.

In these conditions, which entail symmetrical load, shape, constraints, displacements and, consequently, strains, and assuming that the material is homogeneous and isotropic and that the stresses do not exceed the material's elastic limit stress, it is clear that the stresses generated in the cylinder will depend only on radius r and not on the angle θ (Fig. 4.1).

To investigate the behavior of these stresses along the radius, we will isolate the small element shaded in Fig. 4.1a and delimited by two cylindrical surfaces of radius r and $r + dr$ and two axial cross sections having angular coordinates θ and $\theta + d\theta$. To write the equilibrium equation in the radial direction for the volume element thus isolated, we will examine the elementary forces acting on it. With the notation given in Fig. 4.1b, we have:

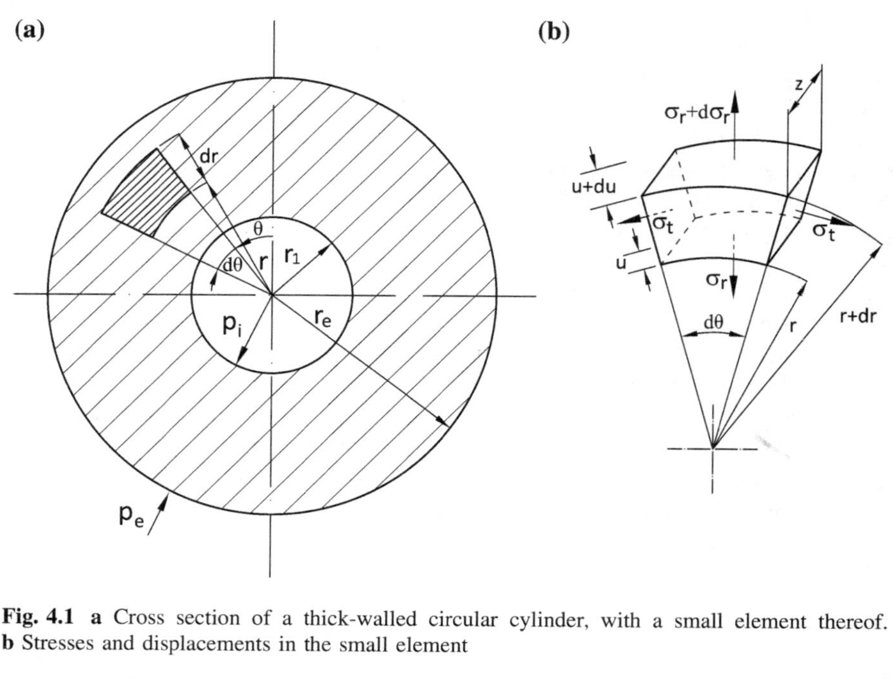

Fig. 4.1 **a** Cross section of a thick-walled circular cylinder, with a small element thereof. **b** Stresses and displacements in the small element

- Elementary radial force at radius r

$$dF_r = \sigma_r r \, d\theta \, dz. \tag{4.2}$$

- Elementary radial force at radius $r + dr$

$$dF_{r+dr} = dF_r + \frac{d}{dr}(dF_r)dr = \left[\sigma_r r + \frac{d}{dr}(\sigma_r r)dr\right] d\theta \, dz. \tag{4.3}$$

- Elementary tangential forces

$$dF_t = \sigma_t dr \, dz. \tag{4.4}$$

With reference to Fig. 4.2, which shows the orientation of the elementary forces acting on this volume element, the equilibrium equation in the radial direction is:

$$dF_{r+dr} - dF_r - 2dF_t \sin\frac{d\theta}{2} = 0. \tag{4.5}$$

Given that angle $d\theta$ is infinitesimal, $\sin(d\theta/2)$ may be replaced by $d\theta/2$ because $d\theta/2$ is an infinitesimal angle, and the above equilibrium equation becomes:

$$dF_{r+dr} - dF_r - dF_t d\theta = 0. \tag{4.6}$$

By substituting the expressions for elementary radial and tangential forces determined above in the latter relation, elaborating and omitting the higher-order infinitesimals, we arrive at the following relation expressing the equilibrium equation in the radial direction for the infinitesimal element in question:

$$\frac{d}{dr}(\sigma_r r) - \sigma_t = 0 \tag{4.7}$$

We thus obtain a first order differential equation containing the unknowns σ_r and σ_t. To obtain a second equation with the same unknowns and independent of this one, the displacement compatibility condition must be imposed. When axi-symmetry is assumed, the radial strain ε_r and tangential strain ε_t are linked to radial displacement u (with u, v and w designating displacements in the radial, tangential and axial directions) by the following geometric relations:

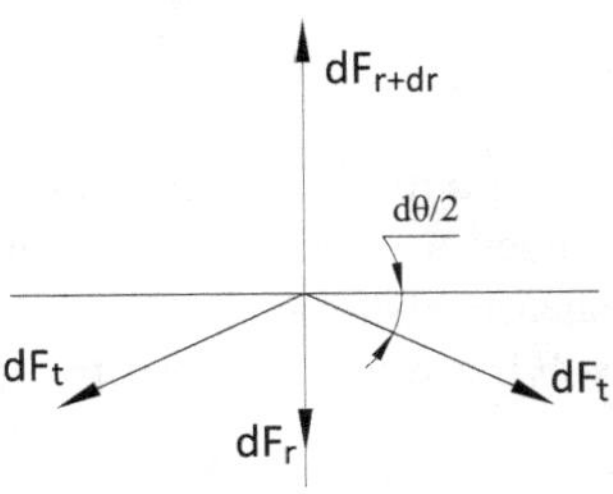

Fig. 4.2 Radial and tangential elementary forces acting on the small element shown in Fig. 4.1b

Fig. 4.3 Elementary radial and circumferential fiber portions, and displacements of their ends, assuming axisymmetry

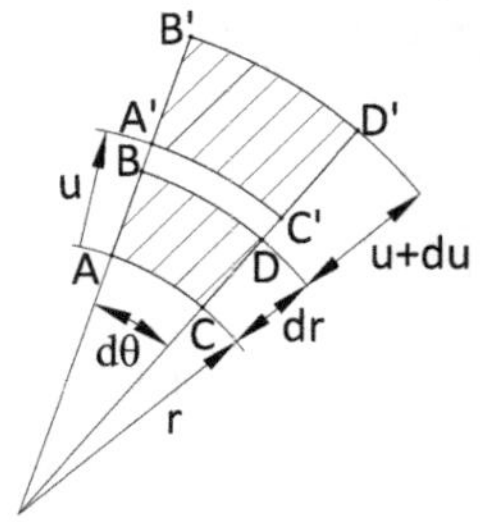

$$\varepsilon_r = \frac{du}{dr} \text{ and } \varepsilon_t = \frac{u}{r}. \tag{4.8}$$

These relations are readily obtained if we consider (Fig. 4.3) two infinitesimal portions of fiber, the first being radial and located between points A and B at distances r and $r + dr$ from the center and thus having length dr prior to deformation, and the second being circumferential, located between points A and C on the circumference of radius r and having length $rd\theta$ prior to deformation. After deformation, end A of the radial fiber portion moves to A', displaced by u, while end B of the same fiber moves to B', displaced by $u + du$ (Fig. 4.3). We thus have $\epsilon_r = (\overline{A'B'} - \overline{AB})/\overline{AB}$, from which we obtain the first relation (4.8). Again after displacement, the ends A and C of the circumferential fiber portion move to A' and C' respectively (Fig. 4.3). We thus have $\epsilon_r = (\overline{A'B'} - \overline{AB})/\overline{AB}$, from which we obtain the second relation (4.8).

The two relations (4.8), which apply when axisymmetry is assumed, give the following compatibility equation in terms of strain:

$$\frac{d}{dr}(\varepsilon_t r) - \varepsilon_r = 0. \tag{4.9}$$

For a homogeneous and isotropic material working in the linear elastic field, the constitutive stress–strain law strain law is given by relations (1.17). If there is no thermal load ($T = const$) and for a plane strain state, for which $\varepsilon_z = 0$ (this is the case of interest to us here, as the thick-walled cylindrical body is assumed to be infinitely extended in the direction of its axis and subjected only to surface forces acting at the inner and outer radii), this law takes the following matrix form:

$$\begin{Bmatrix} \varepsilon_r \\ \varepsilon_t \\ \gamma_{rt} \end{Bmatrix} = \begin{bmatrix} 1/E & -\nu/E & 0 \\ -\nu/E & 1/E & 0 \\ 0 & 0 & 1/G \end{bmatrix} \begin{Bmatrix} \sigma_r \\ \sigma_t \\ \tau_{rt} \end{Bmatrix}. \tag{4.10}$$

As a result of axisymmetry, the tangential displacement v in the direction of the tangent to the generic circumference of radius r is obviously nil. Accordingly, the shearing strain γ_{rt} and the shearing stress τ_{rt} are also nil. In these conditions, solving relation (4.10) for stresses σ gives the following two relations:

$$\begin{aligned}\sigma_r &= \frac{E}{1-\nu^2}(\varepsilon_r + \nu\varepsilon_t) = \frac{E}{1-\nu^2}\left(\frac{du}{dr} + \nu\frac{u}{r}\right)\\ \sigma_t &= \frac{E}{1-\nu^2}(\varepsilon_t + \nu\varepsilon_r) = \frac{E}{1-\nu^2}\left(\frac{u}{r} + \nu\frac{du}{dr}\right).\end{aligned} \tag{4.11}$$

These relations express normal stresses σ_r and σ_t as a function of the normal strains ε_r and ε_t (or of radial displacement u), as well as of the material's elastic magnitudes (modulus of elasticity in tension E, and Poisson's ratio ν, Poisson 1829). Solving relations (4.10) for ε_r and ε_t, deriving. ε_t with respect to r and substituting the expressions thus obtained for ε_r, ε_t and $d\varepsilon_t/dr$ in the compatibility equation in terms of ε given by (4.9), we obtain the following compatibility equation in terms of stress:

$$(\sigma_t - \sigma_r)(1+\nu) + r\frac{d\sigma_t}{dr} - \nu r\frac{d\sigma_r}{dr} = 0. \tag{4.12}$$

This equation applies assuming that elastic magnitudes E and ν are independent of r and thus constant throughout the cylinder. Were these magnitudes variable (a case that will not be examined below), the following more general compatibility equation would be obtained instead of Eq. (4.12):

$$(\sigma_t - \sigma_r)(1+\nu) + r\frac{d\sigma_t}{dr} - \nu r\frac{d\sigma_r}{dr} + Er\frac{d}{dr}\left(\frac{1}{E}\right) - Er\sigma_r\frac{d}{dr}\left(\frac{\nu}{E}\right) = 0. \tag{4.13}$$

From the two equilibrium and compatibility Eqs. (4.7) and (4.12), which are first order differential equations in two unknowns σ_r and σ_t, we can obtain a second order differential equation in one unknown. There are two alternatives in selecting this unknown: the first is to express stresses as a function of radial displacement u by means of the compatibility equation, and then introduce these expressions in the equilibrium equation, while the second alternative is to first obtain the expression of stress σ_t as a function of σ_r from the equilibrium equation and then introduce this expression in the compatibility equation. The two approaches are equivalent and lead to very similar solving equations. The first approach will be followed below, assuming that elastic magnitudes E and ν are constant, i.e., independent of variable r.

To this end, deriving the first relation (4.11) written in terms of u with respect to radius r gives:

$$\frac{d\sigma_r}{dr} = \frac{E}{1-\nu^2}\left(\frac{d^2u}{dr^2} + \frac{\nu}{r}\frac{du}{dr} - \frac{\nu u}{r^2}\right). \tag{4.14}$$

Developing Eq. (4.7) and introducing relations (4.11) written in terms of u in it together with (4.14) yields the following homogeneous second order differential equation in terms of displacement $u = u\ (r)$:

$$\frac{d^2u}{dr^2} + \frac{1}{r}\frac{du}{dr} - \frac{u}{r^2} = 0. \tag{4.15}$$

Integrating this equation is relatively simple, writing it in the following form:

$$\frac{d}{dr}\left[\frac{1}{r}\frac{d}{dr}(ur)\right] = 0. \tag{4.16}$$

Relation (4.15) and its compact form (4.16) are the differential equations governing the displacement field in the cylindrical body indefinitely extended in the direction of its axis and subjected to pressures p_i and p_e acting at the inner and outer radii respectively.

Integrating relation (4.16) twice and designating the two integration constants as C_1 and C_2 gives the following relation which expresses radial displacement u as a function of radius r:

$$u = C_1 r + \frac{C_2}{r}. \tag{4.17}$$

Considering this relation, the geometric relations (4.8) become:

$$\begin{aligned} \varepsilon_r &= \frac{du}{dr} = C_1 - \frac{C_2}{r^2} \\ \varepsilon_t &= \frac{u}{r} = C_1 + \frac{C_2}{r^2}. \end{aligned} \tag{4.18}$$

Substituting these relations in Eq. (4.11) written in terms of u gives the following expressions of stresses σ_r and σ_t as a function of radius r:

$$\begin{aligned} \sigma_r &= \frac{E\,C_1}{(1-\nu)} - \frac{E\,C_2}{r^2(1+\nu)} \\ \sigma_t &= \frac{E\,C_1}{(1-\nu)} + \frac{E\,C_2}{r^2(1+\nu)}. \end{aligned} \tag{4.19}$$

By combining constants, and then calling

$$A = \frac{E\,C_1}{(1-\nu)} \text{ and } B = \frac{E\,C_2}{(1+\nu)} \tag{4.20}$$

from relations (4.19) we obtain the following relations which express the normal stresses as a function of r, but in more compact form:

$$\begin{aligned} \sigma_r &= A - \frac{B}{r^2} \\ \sigma_t &= A + \frac{B}{r^2}. \end{aligned} \tag{4.21}$$

These are Lamé's equations (Lamé 1852); these relations indicate that normal stresses σ_r and σ_t vary according to cubic hyperbolas through the wall thickness of the cylindrical body examined. These relations can also be deduced by considering the equilibrium Eq. (4.7) and the third relation of the generalized of Hooke's law (the third relation (1.17) from which the temperature term has been omitted). In fact, assuming $\sigma_z = 0$ on a preliminary basis in the latter relation (this is the condition obtaining in a cylindrical body with open ends), if there is no longitudinal strain ε_z we must have $\sigma_r + \sigma_t = const$, and this constant can be indicated with 2*A*, so that we can write $\sigma_r + \sigma_t = 2A$. Substituting this sum in Eq. (4.7), we obtain:

$$r\frac{d\sigma_r}{dr} = 2(A - \sigma_r) \tag{4.22}$$

and thus:

$$\frac{d(A - \sigma_r)}{(A - \sigma_r)} = -2\frac{dr}{r}. \tag{4.23}$$

Integrating the latter relation and designating the integration constant as *B*, we have:

$$ln(A - \sigma_r) = -2lnr + lnB. \tag{4.24}$$

Consequently, passing from logarithms to functions, we obtain:

$$(A - \sigma_r) = \frac{B}{r^2}. \tag{4.25}$$

This gives the first relation (4.21). Since $\sigma_r + \sigma_t = 2A$, we can then obtain the second relation (4.21). In the two relations (4.21), *A* and *B* are the integration constants to be calculated by imposing boundary conditions. Accordingly, if the cylindrical body is subjected to pressures p_e and p_i, on the outer and inner surfaces, and noting that stresses are to be considered positive if they are tensile stresses and negative if they are compression stresses, we will have:

$$\sigma_{r,r=r_e} = -p_e \text{ and } \sigma_{r,r=r_i} = -p_i. \tag{4.26}$$

The following system is obtained from the first relation (4.21):

$$\begin{aligned} -p_e &= A - \frac{B}{r_e^2} \\ -p_i &= A - \frac{B}{r_i^2}. \end{aligned} \tag{4.27}$$

Solving this system for *A* and *B*, we obtain the following expressions of the two integration constants, as a function of pressures and of the inner and outer radii, i.e.:

$$A = \frac{p_i r_i^2 - p_e r_e^2}{r_e^2 - r_i^2} \text{ and } B = \frac{(p_i - p_e) r_i^2 r_e^2}{r_e^2 - r_i^2} \tag{4.28}$$

Finally, substituting relations (4.28) in relations (4.21) yields the following expressions for radial and hoop stresses:

$$\begin{aligned}\sigma_r &= \frac{p_i r_i^2 - p_e r_e^2}{r_e^2 - r_i^2} - \frac{(p_i - p_e) r_i^2 r_e^2}{r^2 (r_e^2 - r_i^2)} \\ \sigma_t &= \frac{p_i r_i^2 - p_e r_e^2}{r_e^2 - r_i^2} + \frac{(p_i - p_e) r_i^2 r_e^2}{r^2 (r_e^2 - r_i^2)}.\end{aligned} \tag{4.29}$$

On the basis of relations (4.21), and considering relations (4.28), it can be concluded that stresses σ_r and σ_t do not depend on the material's elastic magnitudes, but only on the boundary conditions. This is in agreement with Michell's theorem, which states that in a multiply-connected body subjected to a plane stress state, stressed in the linear elastic field with neither body forces nor thermal stresses and loaded by a system of external forces with a zero translation resultant on each boundary, separately considered, the stress state does not depend on the material's elastic constants.

The assumptions on which Michell's theorem is based are all fulfilled here, including that of loads with zero resultant along all boundaries. The latter assumption, in fact, is subsumed in that of generalized axisymmetry, which requires that only uniformly distributed loads act on each boundary. The assumption of a plane stress state is also fulfilled, as it coincides with the generalized plane stress state ($\varepsilon_z = 0$ or $\varepsilon_z = const$) in the case of a multiply-connected body subjected only to forces acting on the inner and outer radii. By contrast, as will be shown below, unit normal elongations ε_r and ε_t and radial displacement u are functions of the material's elastic magnitudes.

4.3 Axial Stress

When there are no thermal stresses, axial (or longitudinal) normal stress in thick-walled circular cylinders is calculated in the same way described for thin-walled circular cylinders, i.e., by imposing longitudinal equilibrium conditions. Basically, there are two different cases: cylinders with open ends and cylinders with closed ends. In both cases, the body is not subject to axial constraints.

It can be concluded from the axial equilibrium condition for one of the two portions into which a plane normal to the axis divides the cylinder in question (see Fig. 1.2) that the axial stress is given by the following two relations for the first and second case respectively:

$$\sigma_z = 0 \tag{4.30}$$

and

$$\sigma_z = \frac{p_i r_i^2 - p_e r_e^2}{r_e^2 - r_i^2} = A. \tag{4.31}$$

A comparison of relations (4.29) and (4.31) indicates that there is a relationship of interdependence between the three principal stresses σ_r, σ_t and σ_z for the circular cylinders with closed ends given by:

$$\sigma_z = \frac{1}{2}(\sigma_t + \sigma_r). \tag{4.32}$$

As σ_z is constant, this relation demonstrates that the sum $(\sigma_t + \sigma_r) = 2A$ is an invariant.

For a number of reasons which will be discussed below, it is advisable to avoid axial constraints on the circular cylinder, especially if it is subjected to thermal loads resulting from radial and/or axial heat flow in addition to surface forces acting at the outer and inner radii. Should such constraint conditions exist, however, we would be dealing with a third case, viz., that of the plane strain state defined by the relation $\varepsilon_z = 0$. In this third case, bearing Hooke's third relation for the triaxial stress state (see relation (1.17) without temperature terms) in mind, we obtain the following expression for axial stress:

$$\sigma_z = \nu(\sigma_t + \sigma_r) = 2\nu A. \tag{4.33}$$

4.4 Radial and Hoop Stress Distributions Through the Wall Thickness

When the dimensionless ratio $\delta = d_e/d_i = r_e/r_i$ (whose corresponding relations in ratio $\beta = 1/\delta$ need not be shown here) is introduced, relations (4.29) become:

$$\begin{aligned} \sigma_r &= \frac{p_i - p_e\delta^2}{\delta^2 - 1} - \frac{(p_i - p_e)r_e^2}{r^2(\delta^2 - 1)} \\ \sigma_t &= \frac{p_i - p_e\delta^2}{\delta^2 - 1} + \frac{(p_i - p_e)r_e^2}{r^2(\delta^2 - 1)}. \end{aligned} \tag{4.34}$$

If relations (4.34) are plotted versus r, we obtain two cubic hyperbolas which are symmetric about the common horizontal asymptote defined by the relation:

$$\sigma_{r,r=\infty} = \sigma_{t,r=\infty} = A \tag{4.35}$$

and whose vertical asymptote is the σ axis, given that:

$$\sigma_{r,r=0} = -\infty \text{ and } \sigma_{t,r=0} = +\infty. \tag{4.36}$$

The horizontal asymptote gives physical meaning to the integration constant A, as it represents axial normal stress σ_z for a cylinder with closed ends. The vertical asymptote poses a singularity problem around the axis, i.e., for $r_i \to 0$. As the inner radius is decreased and, at the limit, for $r_i \to 0$, radial stress cannot tend to a $-\infty$, as it must satisfy the boundary condition given by the second relation (4.26).

Consequently, we must put $r_e^2/r_i^2 = \delta^2$ also for $r_i \to 0$ in relations (4.34), resulting in finite values for both σ_r and σ_t. This singularity, however, is of little practical importance, as actual tubular structures almost invariably have an inner radius which is far from being infinitesimal.

If we look at the radial and hoop stress distributions as a function of r, we can see that the cylinder material is not put to effective use, as the innermost fibers of the wall are under much higher stresses than the outermost fibers. From the standpoint of structural strength, this poor use of the material's strength properties becomes more critical as ratio δ increases. We can conclude from relations (4.29) or the equivalent relations (4.34) that:

- Radial stress at the inner radius and at the outer radius assumes the following values:

$$\sigma_{r,r=r_i} = -p_i \text{ and } \sigma_{r,r=r_e} = -p_e \tag{4.37}$$

- Hoop stress at the inner radius and at the outer radius assumes the following values:

$$\sigma_{t,r=r_i} = \frac{p_i(\delta^2+1) - 2p_e\delta^2}{\delta^2 - 1} \text{ and } \sigma_{t,r=r_e} = \frac{2p_i - p_e(\delta^2-1)}{\delta^2-1}. \tag{4.38}$$

In absolute value, the highest principal stress is the hoop stress. As demonstrated in Sect. 4.4.3, for a circular cylinder under internal and external pressure simultaneously, this maximum stress can occur at either the inner radius or the outer radius, depending on the values of load ratio $\bar{p} = p_e/p_i$.

4.4.1 Stress Distributions in Circular Cylinders Under Internal Pressure

Where there is zero external pressure and the circular cylinder is subjected only to internal pressure ($p_e = 0$ and $p_i \neq 0$), relations (4.29) and (4.31) can be simplified as follows:

$$\begin{aligned}
\sigma_r &= \frac{p_i r_i^2}{r_e^2 - r_i^2}\left(1 - \frac{r_e^2}{r^2}\right) = \frac{p_i}{\delta^2 - 1}\left(1 - \frac{r_e^2}{r^2}\right) = K'\left(1 - \frac{r_e^2}{r^2}\right) \\
\sigma_t &= \frac{p_i r_i^2}{r_e^2 - r_i^2}\left(1 + \frac{r_e^2}{r^2}\right) = \frac{p_i}{\delta^2 - 1}\left(1 + \frac{r_e^2}{r^2}\right) = K'\left(1 + \frac{r_e^2}{r^2}\right) \\
\sigma_z &= \frac{p_i r_i^2}{r_e^2 - r_i^2} = \frac{p_i}{\delta^2 - 1} = K'.
\end{aligned} \tag{4.39}$$

As can be seen from these relations, the radial and hoop stresses are opposite in sign and their sum, as indicated above for the general case of a circular cylinder under internal and external pressure, is invariant and independent of the generic radius r. Given that $r_e^2/r^2 \geq 1$, radial stress σ_r is always negative (compressive stress) except at the outer radius where it is zero, whereas it reaches its absolute maximum value at the inner radius. For the same reason, hoop stress σ_t is always positive (tensile stress) and its absolute maximum value is also reached at the inner radius, while it is minimum at the outer radius.

The distribution curves for stresses σ_r and σ_t through the wall thickness (Fig. 4.4, where they are represented as dimensionless relative to pressure p_i) are cubic hyperbolas symmetric to the horizontal asymptote, at the ordinate $\sigma = K'$. Axial stress σ_z is zero for the circular cylinder with open ends, whereas for the cylinder with closed ends, we have $\sigma_z = K'$ if the structure is axially free and $\sigma_z = 2\nu K'$ if it has axial constraints.

As Fig. 4.4 shows, the elastic stress state at and around the inner radius exhibits high peaks in hoop stress σ_t and its gradient $d\sigma_t/dr$, indicating that the material's strength properties are not used to best advantage. As will be discussed in the following chapters, any measures capable of producing more favorable stress states in these areas are amply justified.

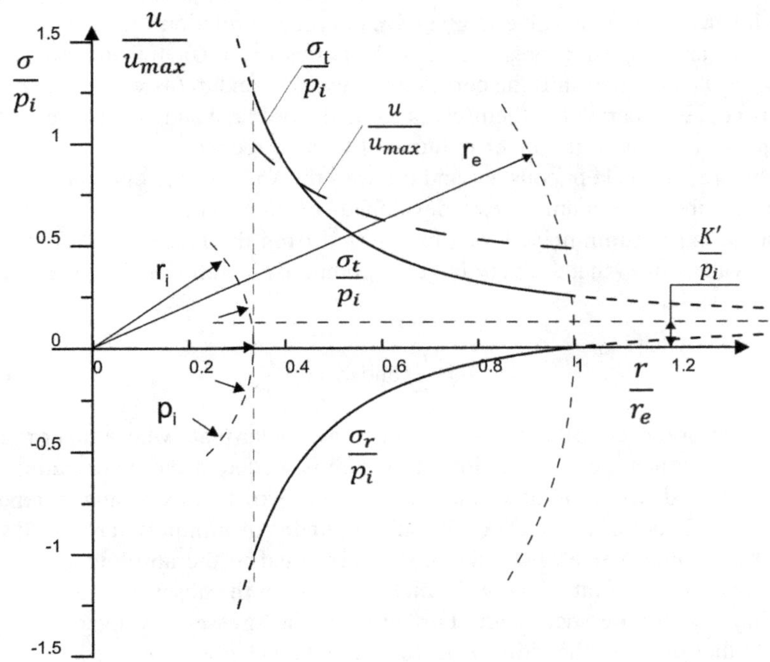

Fig. 4.4 Dimensionless stresses σ_r/p_i and σ_t/p_i and ratio u/u_{max} versus dimensionless radius in a thick-walled circular cylinder with $\delta = 3$ under internal pressure only ($p_e = 0$)

The radial and hoop stress limit values at inner and outer radii respectively are as follows:

- Inner radius

$$\sigma_{r,r=r_i} = -p_i \text{ and } \sigma_{t,r=r_i} = p_i \frac{\delta^2 + 1}{\delta^2 - 1} \tag{4.40}$$

- Outer radius

$$\sigma_{r,r=r_e} = 0 \text{ and } \sigma_{t,r=r_e} = \frac{2p_i}{\delta^2 - 1} = 2K'. \tag{4.41}$$

The relations expressing radial and hoop stresses at the mean wall radius r_m are:

$$\begin{aligned} \sigma_{r,r=r_m} &= \frac{p_i}{\delta^2 - 1}\left[1 - \frac{4\delta^2}{(\delta+1)^2}\right] \\ \sigma_{t,r=r_m} &= \frac{p_i}{\delta^2 - 1}\left[1 + \frac{4\delta^2}{(\delta+1)^2}\right]. \end{aligned} \tag{4.42}$$

The absolute maximum value of the stress field is that assumed by hoop stress at the inner radius. This value is given by the second relation (4.40). It should be noted that any opposing pressure at outer radius ($p_e \neq 0$), for any given $\Delta p = p_i - p_e > 0$ and δ, will shift the curves for stresses σ_r and σ_t (as well as for σ_z if the latter is not zero) parallel to themselves along the ordinate and downwards, and the hoop stress σ_t peak at the inner radius will thus be reduced.

If the pressure is kept constant and the wall thickness of the circular cylinder is increased, the maximum stress developed in the cylinder's wall thickness decreases, but not infinitely. It can be inferred from the first two relations (4.39) that for values of r_e that are very large by comparison with r_i and, at the limit, for $r_e \to \infty$, we have:

$$\sigma_r = -p_i \frac{r_i^2}{r^2} \text{ and } \sigma_t = p_i \frac{r_i^2}{r^2}. \tag{4.43}$$

These relations describe the stress state in a circular plate whose inner radius is much smaller than the outer radius and which is loaded at the inner radius by an axisymmetric distribution of surface forces. The two stresses σ_r and σ_t tend to 0 for $r \to \infty$. This means that for a circular cylinder of infinitely large wall thickness, the absolute value of the radial stress is equal to the absolute value of the hoop stress at any point of the wall thickness, and in the absence of axial stresses all points are in pure shear state. Furthermore, the stresses are inversely proportional to the square of the radius r. If, for example, we assume $r = 4r_i$, the stresses at points located at this distance from the axis are only 1/16 = 6.25 % of the maximum stresses. Consequently, if a calculation accuracy of 5–6 % is acceptable (greater accuracy is unattainable in practice because of elastic imperfections of the material), a circular cylinder having ratio $r_e/r_i = \delta > 4$ may be regarded as an infinitely large wall thickness cylinder.

It is important to note that the shape of the outer surface may be considered as immaterial. If all points of this outer immaterial surface are located at a distance of more than $4r_i$ from the axis of the inner surface, the shape of the outer surface may be arbitrary. Accordingly, elastic analysis around the holes of perforated plates, with holes spaced at a sufficient distance and loaded by an uniform distribution of surface forces can be performed with the same approach used for a circular cylinder of infinitely large wall thickness.

According to the maximum shearing stress theory (see Table 4.3), the equivalent stress when $r_e \rightarrow \infty$ is $\sigma_{e,i} = 2p_i$. Consequently, if for example the elastic limit of the material is 500 MPa, the strains in a circular cylinder of infinitely large thickness will be elastic for internal pressures not exceeding 250 MPa.

Lastly, it is important to determine how the radial and hoop stresses vary as the wall thickness of the circular cylinder decreases. As $r_e = (r_i + s)$, the second relation (4.39) yields:

$$\sigma_{t(r=r_i)} = \frac{p_i\left[(r_i + s)^2 + r_i^2\right]}{s(2r_i + s)} \text{ and } \sigma_{t(r=r_e)} = \frac{2p_i\, r_i^2}{s(2r_i + s)}. \tag{4.44}$$

For small values of s, these relations give:

$$\sigma_{t(r=r_i)} \cong \sigma_{t(r=r_e)} \cong \frac{p_i r_i}{s}. \tag{4.45}$$

Regardless of the cylinder's wall thickness, we have $\sigma_r = -p_i$ at the inner radius and $\sigma_r = 0$ at the outer radius. Thus we can see that for a circular cylinder of small wall thickness, the hoop stresses are distributed almost uniformly across the thickness, and the absolute value of the radial stress for $r = r_i$ has the same absolute value of the hoop stress when $r_i = s$.

4.4.2 *Stress Distributions in Circular Cylinders Under External Pressure*

Without going into questions of unstable elastic equilibrium, if internal pressure is zero and the circular cylinder is under external pressure only ($p_e \neq 0$ and $p_i = 0$), relations (4.29) and (4.31) can be simplified respectively as follows:

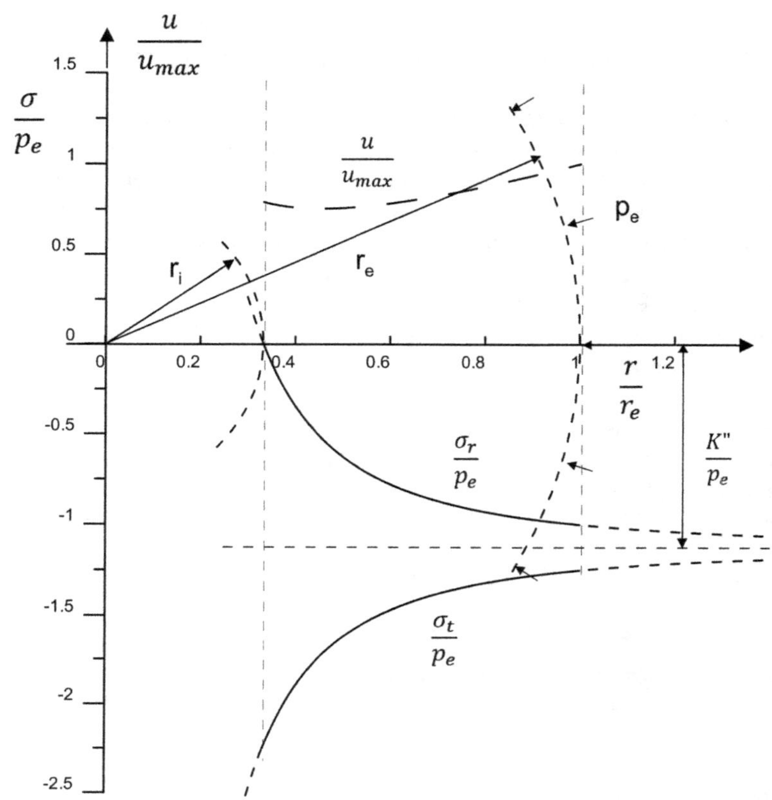

Fig. 4.5 Dimensionless stresses σ_r/p_e and σ_t/p_e and ratio u/u_{max} versus dimensionless radius r/r_e in a thick-walled circular cylinder with $\delta = 3$ under external pressure only ($p_i = 0$)

$$
\begin{aligned}
\sigma_r &= -\frac{p_e r_e^2}{r_e^2 - r_i^2}\left(1 - \frac{r_i^2}{r^2}\right) = K''\left(1 - \frac{r_i^2}{r^2}\right) \\
\sigma_t &= -\frac{p_e r_e^2}{r_e^2 - r_i^2}\left(1 + \frac{r_i^2}{r^2}\right) = K''\left(1 + \frac{r_i^2}{r^2}\right) \\
\sigma_z &= -\frac{p_e r_e^2}{r_e^2 - r_i^2} = K''.
\end{aligned}
\tag{4.46}
$$

As can be seen from the foregoing relations, radial and hoop stresses always have the same sign and their sum, as indicated for the general case, is invariant and independent of the generic radius r. The distribution curves for stresses σ_r and σ_t through the wall thickness (see Fig. 4.5, where they are represented as dimensionless relative to pressure p_e), are once again cubic hyperbolas symmetric to the horizontal asymptote, at the ordinate $\sigma_r = k''$.

The radial and hoop stress limit values at inner and outer radii respectively are as follows:

- Inner radius

$$\sigma_{r,r=r_i} = 0 \text{ and } \sigma_{t,r=r_i} = -p_e \frac{2\delta^2}{\delta^2 - 1} = 2K''. \tag{4.47}$$

- Outer radius

$$\sigma_{r,r=r_e} = -p_e \text{ and } \sigma_{t,r=r_e} = -p_e \frac{\delta^2 + 1}{\delta^2 - 1}. \tag{4.48}$$

The relations expressing radial and hoop stresses at the mean wall radius r_m are:

$$\begin{aligned}
\sigma_{r,r=r_m} &= -\frac{p_e\delta^2}{\delta^2 - 1}\left[1 + \frac{4}{(\delta+1)^2}\right] \\
\sigma_{t,r=r_m} &= -\frac{p_e\delta^2}{\delta^2 - 1}\left[1 - \frac{4}{(\delta+1)^2}\right].
\end{aligned} \tag{4.49}$$

The absolute maximum value of the stress field is that assumed by hoop stress at the inner radius. This value is given by the second relation (4.47). It should be noted that we have $|\sigma_t| > |\sigma_r|$ in all cases. It should also be noted that, here as in the previous case, any opposing pressure at inner radius ($p_i \neq 0$), for any given $\Delta p = p_e - p_i > 0$ and δ, will shift the curves for stresses σ_r and σ_t (as well as for σ_z if the latter is not zero) parallel to themselves along the ordinate and downwards. In this case, however, the absolute peak value of hoop stress σ_t at the inner radius will increase.

It can be deduced from the first two relations (4.46) that in a solid circular cylinder ($r_i = 0$) such as a drive shaft attached to a hub which stresses its outer radius as a result of a force-fit or shrink-fit pressure $p_f = p_e$, we have (for further details, see Sect. 5.2):

$$\sigma_r = \sigma_t = const = -p_f = -p_e. \tag{4.50}$$

Stresses σ_r and σ_t are thus constant, or in other words independent of the radius.

4.4.3 Remarks on Maximum Hoop Stress

As can be seen from Figs. 4.4 and 4.5, the maximum hoop stress for a circular cylinder under internal pressure alone or external pressure alone is found at the inner radius. This, however, cannot be generalized to cases where the cylinder is subjected to internal and external pressure simultaneously. There are, in fact, load

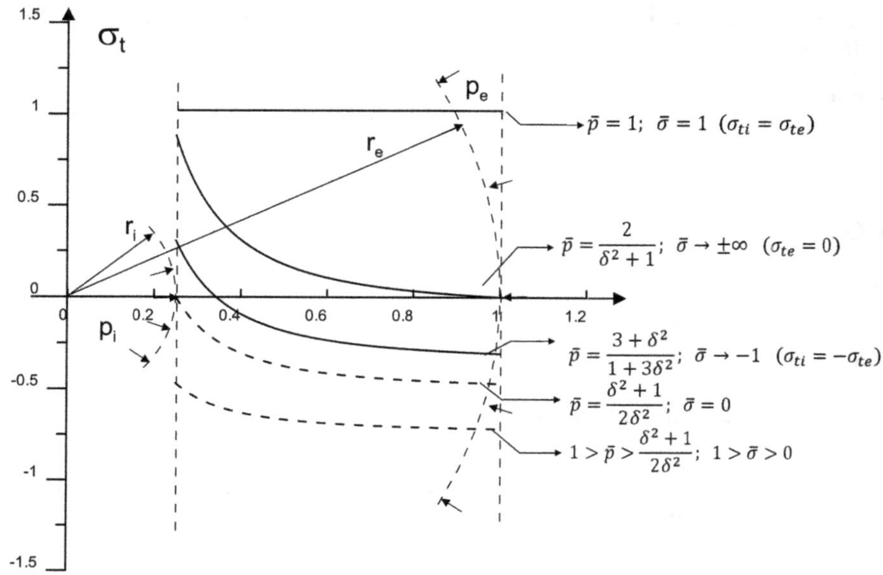

Fig. 4.6 Hoop stress σ_t through the wall thickness of a circular cylinder with $\delta = 4$ versus load ratio $\bar{p}$

conditions due to surface forces whereby maximum hoop stress occurs at the outer radius (see also Ugural and Fenster 2003).

To demonstrate this, we will consider a thick-walled circular cylinder under internal and external pressure, writing the second relation (4.29) or the second relation (4.34) in the following form:

$$\sigma_t = p_i\left[\frac{1 - \bar{p}\delta^2}{\delta^2 - 1} + r_e^2\frac{1 - \bar{p}}{r^2(\delta^2 - 1)}\right] \tag{4.51}$$

where $\bar{p} = p_e/p_i$ is the load ratio. We will then introduce the hoop stress ratio $\bar{\sigma} = \sigma_{ti}/\sigma_{te}$ given by:

$$\bar{\sigma} = \frac{\sigma_{ti}}{\sigma_{te}} = \frac{1 + \frac{(1-\bar{p})\delta^2}{(1-\bar{p}\delta^2)}}{1 + \frac{(1-\bar{p})}{(1-\bar{p}\delta^2)}}. \tag{4.52}$$

On the basis of this relation, it can be immediately demonstrated that:

- For $\bar{p} = 1$, we have $\bar{\sigma} = 1$, i.e., the hoop stress ratio is constant and unitary.
- For $\bar{p} = 2/(\delta^2 + 1)$, we have $\bar{\sigma} = \pm\infty$ and, consequently, $\sigma_{te} = 0$.

- For $\bar{p} = (3 + \delta^2)/(1 + 3\delta^2)$, we have $\bar{\sigma} = -1$, i.e., $\sigma_{ti} = -\sigma_{te}$ (the two maximum values at the inner radius and the outer radius are equal in modulus, but the first is positive and the second is negative).
- For $\bar{p} = (\delta^2 + 1)/2\delta^2$, we have $\bar{\sigma} = 0$, i.e., $\sigma_{ti} = 0$, while σ_{te} is maximum and negative at the outer radius.
- For $1 > \bar{p} > (\delta^2 + 1)/2\delta^2$, we have $0 < \bar{\sigma} < 1$, i.e., σ_t is again negative, with absolute maximum value at the outer radius.

Figure 4.6 shows hoop stress σ_t through the wall thickness of the circular cylinder for different load ratios $\bar{p}$ corresponding to different hoop stress ratios $\bar{\sigma}$. The lower, dashed-line curves apply to load conditions for which hoop stress is in fact negative. The load ratio given by relation $\bar{p} = (3 + \delta^2)/(1 + 3\delta^2)$ is the limit at which the maximum hoop stress still occurs at the inner radius (though absolute value is the same as at that at the outer radius).

4.5 Analysis of Strain State and Displacements

Analyzing the strain state of a thick-walled circular cylinder involves no difficulties. Analysis is carried out on the basis of the constitutive Eqs. (1.17) from which temperature terms have been omitted. In this connection, it is sufficient to substitute the general expressions of stresses given by relations (4.21) and (4.30) or (4.31) or (4.33) in these constitutive equations. Two possible cases are of interest from the standpoint of practical applications: that of the circular cylinder with closed ends, whose stress state is triaxial (all three principal stresses σ_r, σ_t and σ_z are other than zero) and that of the circular cylinder with open ends, for which the stress state is biaxial ($\sigma_z = 0$).

4.5.1 Circular Cylinders with Closed Ends

For the circular cylinder with closed ends ($\sigma_z \neq 0$), the three principal unit elongations (principal strains) are given by the following relations:

$$\begin{aligned} \varepsilon_r &= \frac{du}{dr} = \frac{1}{E}\left[A(1-2\nu) - \frac{B}{r^2}(1+\nu)\right] \\ \varepsilon_t &= \frac{u}{r} = \frac{1}{E}\left[A(1-2\nu) + \frac{B}{r^2}(1+\nu)\right] \\ \varepsilon_z &= \frac{dw}{dz} = const = \frac{A}{E}(1-2\nu). \end{aligned} \tag{4.53}$$

As the second geometric relation (4.8) gives $u = r\varepsilon_t$, the first of the relations shown below yields $u = u(r)$. Subsequently integrating the third relation (4.53), and calculating the integration constant by setting $w = 0$ for $z = 0$ (zero plane for the axial coordinate) gives the second of the following equations. It should also be noted that, as a result of axisymmetry, the tangential displacement v in the direction of the tangent to the generic circumference of radius r is nil.

$$\begin{aligned} u &= \frac{r}{E}\left[A(1-2v) + \frac{B}{r^2}(1+v)\right] \\ w &= \frac{A}{E}1(1-2v)z. \end{aligned} \tag{4.54}$$

We can conclude from relations (4.53) that the value of the first invariant of the strain tensor $\varepsilon_r + \varepsilon_t + \varepsilon_z$ is $3A(1-2v)/E$ and that the following relationship of mutual interdependence exists between the three principal strains:

$$\varepsilon_z = \frac{1}{2}(\varepsilon_t + \varepsilon_r). \tag{4.55}$$

This relationship is analogous to that between the three principal stresses expressed by (4.32).

In completely general terms, relations (4.53) and (4.54) express the strain state and displacements in a circular cylinder with closed ends under internal and external pressure. By substituting relations (4.32) which give constants A and B as a function of geometry and of pressures p_i and p_e in them, we obtain the complete relations. For radial displacement u, we thus have:

$$u = \frac{r}{E}\left[\frac{p_i r_i^2 - p_e r_e^2}{r_e^2 - r_i^2}(1-2v) + \frac{(1+v)}{r^2}\frac{(p_i - p_e)r_i^2 r_e^2}{r_e^2 - r_i^2}\right]. \tag{4.56}$$

In the two particular cases of a circular cylinder under internal pressure ($p_i \neq 0 \text{ and } p_e = 0$) or under external pressure ($p_i = 0 \text{ and } p_e \neq 0$), we will have respectively:

$$u = \frac{r}{E}\frac{p_i r_i^2}{\left(r_e^2 - r_i^2\right)}\left[(1-2v) + \frac{r_e^2}{r^2}(1+v)\right] \tag{4.57}$$

and

$$u = -\frac{r}{E}\frac{p_e r_e^2}{\left(r_e^2 - r_i^2\right)}\left[(1-2v) + \frac{r_i^2}{r^2}(1+v)\right]. \tag{4.58}$$

The following relations give the radial displacements at the inner radius and at the outer radius for the cylinder under internal pressure only or under external pressure only:

- Cylinder under internal pressure

$$
\begin{aligned}
u_{r=r_i} &= \frac{r_i}{E} \frac{p_i}{(\delta^2 - 1)} [(1 - 2\nu) + \delta^2 (1 + \nu)] \\
u_{r=r_e} &= \frac{r_e}{E} \frac{p_i}{(\delta^2 - 1)} (2 - \nu).
\end{aligned} \tag{4.59}
$$

- Cylinder under external pressure

$$
\begin{aligned}
u_{r=r_i} &= -\frac{r_i}{E} \frac{p_e \delta^2}{(\delta^2 - 1)} (2 - \nu) \\
u_{r=r_e} &= -\frac{r_e}{E} \frac{p_e}{(\delta^2 - 1)} [\delta^2 (1 - 2\nu) + (1 + \nu)].
\end{aligned} \tag{4.60}
$$

As indicated earlier, it should be noted that the strain state defined by principal unit elongations ε_r, ε_t and ε_z, and the displacement field defined by displacements u and w are in both cases functions of the material's elastic magnitudes, as can be seen from relations (4.53) and (4.54).

4.5.2 Circular Cylinders with Open Ends

For the circular cylinder with open ends ($\sigma_z = 0$), the three principal strains are given by the following relations:

$$
\begin{aligned}
\varepsilon_r &= \frac{du}{dr} = \frac{1}{E} \left[A(1 - \nu) - \frac{B}{r^2} (1 + \nu) \right] \\
\varepsilon_t &= \frac{u}{r} = \frac{1}{E} \left[A(1 - \nu) + \frac{B}{r^2} (1 + \nu) \right] \\
\varepsilon_z &= \frac{dw}{dz} = \text{const} = -\frac{2\nu A}{E}.
\end{aligned} \tag{4.61}
$$

The same procedure used for the cylinder with closed ends leads to the following expressions for the displacements in an open-ended cylinder under internal and external pressure:

$$
\begin{aligned}
u &= \frac{r}{E} \left[A(1 - \nu) + \frac{B}{r^2} (1 + \nu) \right] \\
w &= -\frac{2\nu A}{E} z.
\end{aligned} \tag{4.62}
$$

We can conclude from relations (4.61) that the value of the first invariant of the strain tensor $\varepsilon_r + \varepsilon_t + \varepsilon_z$ is $2A(1-2v)/E$ and that the following relationship of mutual interdependence exists between the three principal strains:

$$\varepsilon_r + \varepsilon_t = -\frac{1-v}{v}\varepsilon_z. \tag{4.63}$$

Here again, relations (4.61) and (4.62) express the strain state and displacements in a circular cylinder with open ends under internal and external pressure, doing so in completely general terms. By substituting relations (4.28) into Eqs. (4.61) and (4.62) we obtain the complete relations. For radial displacement u, we thus have:

$$u = \frac{r}{E}\left[\frac{p_i r_i^2 - p_e r_e^2}{r_e^2 - r_i^2}(1-v) + \frac{(1+v)}{r^2}\frac{(p_i - p_e) r_i^2 r_e^2}{r_e^2 - r_i^2}\right]. \tag{4.64}$$

In the two particular cases of a circular cylinder under internal pressure ($p_i \neq 0$ and $p_e = 0$) or under external pressure ($p_i = 0$ and $p_e \neq 0$), we will have respectively:

$$u = \frac{r}{E}\frac{p_i r_i^2}{\left(r_e^2 - r_i^2\right)}\left[(1-v) + \frac{r_e^2}{r^2}(1+v)\right]. \tag{4.65}$$

and

$$u = -\frac{r}{E}\frac{p_e r_e^2}{\left(r_e^2 - r_i^2\right)}\left[(1-v) + \frac{r_i^2}{r^2}(1+v)\right]. \tag{4.66}$$

The following relations give the radial displacements at the inner radius and at the outer radius for the cylinder under internal pressure only or under external pressure only:

- Cylinder under internal pressure

$$\begin{aligned} u_{r=r_i} &= \frac{r_i}{E}\frac{p_i}{\left(\delta^2 - 1\right)}\left[(1-v) + \delta^2(1+v)\right] \\ u_{r=r_e} &= \frac{2r_e}{E}\frac{p_i}{\left(\delta^2 - 1\right)}. \end{aligned} \tag{4.67}$$

- Cylinder under external pressure

$$\begin{aligned} u_{r=r_i} &= -\frac{2r_i}{E}\frac{p_e\delta^2}{\left(\delta^2 - 1\right)} \\ u_{r=r_e} &= -\frac{r_e}{E}\frac{p_e\delta^2}{\left(\delta^2 - 1\right)}\left[(1-v) + \frac{(1+v)}{\delta^2}\right]. \end{aligned} \tag{4.68}$$

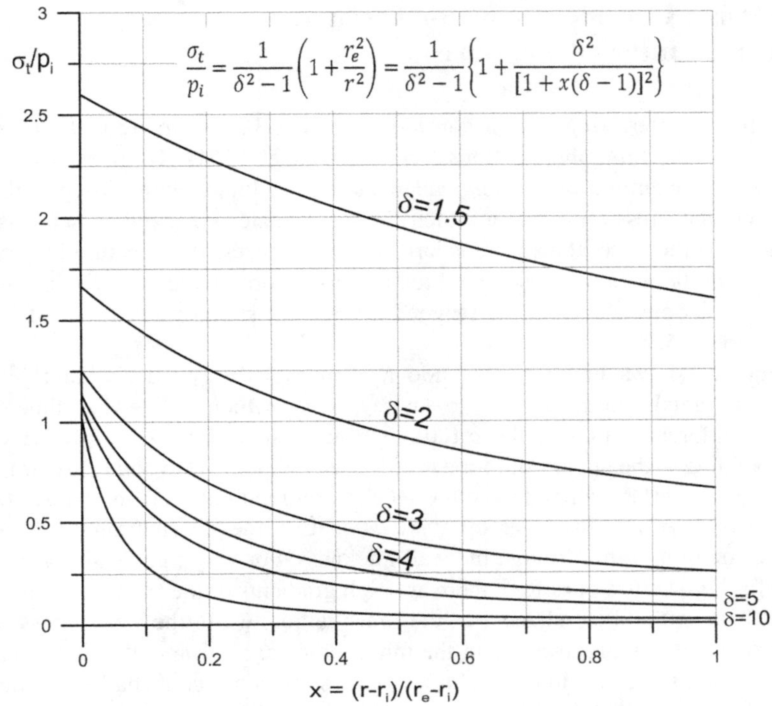

Fig. 4.7 Dimensionless ratio σ_t/p_i versus dimensionless radial coordinate $x = (r - r_i)/(r_e - r_i)$, for different values of ratio δ in a thick-walled circular cylinder under internal pressure

Here again, it should be noted that the strain state defined by principal unit elongations ε_r, ε_t and ε_z, and the displacement field defined by displacements u and w are functions of the material's elastic magnitudes, as can be seen from relations (4.61) and (4.62). For the cylindrical body subjected only to internal pressure, maximum displacement u_{max} is at the inner radius. Ratio u/u_{max} through the wall thickness of this cylinder, which is assumed to be steel ($\nu = 0.3$), is shown in Fig. 4.4. For the circular cylinder subjected only to external pressure, by contrast, the maximum displacement u_{max} is at the outer radius; for this cylinder, also assumed to be steel ($\nu = 0.3$), ratio u/u_{max} through the wall thickness is shown in Fig. 4.5.

4.6 Design Considerations for Circular Cylinders Under Internal Pressure

A number of interesting design considerations can be drawn from the relations presented in the foregoing sections (see also Faupel 1964). To give an idea, we will take the example of a thick-walled circular cylinder with closed ends and under internal pressure only, in which the stress state is triaxial. It will also be assumed that the internal pressure is large enough to stress the structure in question in the elastic field. In this case, the three principal normal stresses σ_r, σ_t and σ_z are given by relations (4.39), while tangential principal strain ε_t is expressed by the second Eq. (4.53).

Figure 4.7 shows dimensionless ratio σ_t/p_i through the cylinder's wall thickness versus the radial coordinate $x = (r - r_i)/(r_e - r_i)$, which is likewise dimensionless, for different values of the δ ratio starting from $\delta = 1.5$ (the $\delta = 1.1$ curve corresponding to the upper limit for a thin-walled cylinder is beyond the scale). This ratio σ_t/p_i is that determined from the second relation (4.39). As the figure shows, gradient $d(\sigma_t/p_i)/dx$ drops along with δ, so that the distribution of stress σ_t becomes more uniform through the wall thickness. For very thick wall, or in other words for high values of ratio δ, there are high gradients of the ratio σ_t/p_i, and thus of stress σ_t, in the area adjacent to the inner radius, while the same stress drops significantly in moving away from the inner radius and towards the outer radius.

Consequently, if the circular cylinder in question consists of shallow-hardening steel, it is possible that the effect of heat treatment will not penetrate to a sufficient depth into the wall thickness. As it is thus very likely that yield stress will be exceeded locally, a plastic flow may be generated, with an elastic–plastic boundary that advances from the inner radius towards the outer radius, even though the structure is still stressed predominantly in the elastic field.

It also follows that, as the highest stresses are located around the hole, it is not necessary to have a monolithic structure consisting of a high yield strength material. It is more rational to employ a composite two-component structure consisting of a fairly thin sleeve (liner) made of a high strength material, and a thicker outer tube consisting of a low strength and then less costly material.

Another design consideration can be drawn from relations (4.59), which give the displacements at the inner and outer radius of the cylinder in question or, equivalently, the associated tangential strains. The second relation (4.59) yields the following expression for tangential strain at the outer radius:

$$\varepsilon_{t,r=r_e} = \frac{p_i}{E}\left(\frac{2-\nu}{\delta^2-1}\right). \tag{4.69}$$

Both relations (4.59) lead to the following relation that correlates tangential strain at the inner and outer radii in the elastic field:

$$\varepsilon_{t,r=r_i} = \varepsilon_{t,r=r_e}\frac{\left[(1-2\nu)+\delta^2(1+\nu)\right]}{2-\nu}. \tag{4.70}$$

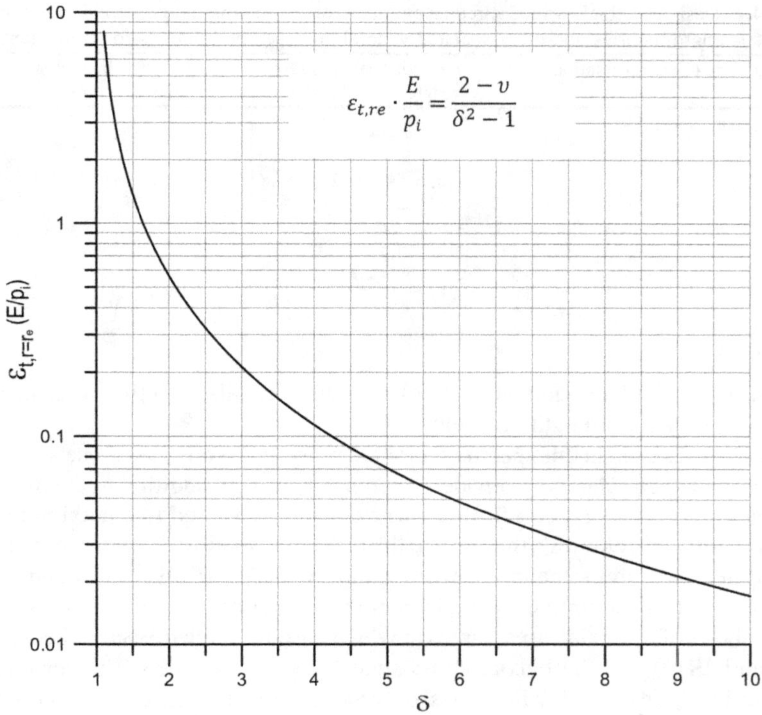

Fig. 4.8 Distribution of $E\varepsilon_{t,r=r_e}/p_i$ versus δ, for $\nu = 0.3$

Figure 4.8 shows tangential strain $\varepsilon_{t,r=r_e}$ at the outer radius given by relation (4.69), but expressed in form $E\varepsilon_{t,r=r_e}/p_i$, as a function of δ, for $\nu = 0.3$, in a thick-walled circular cylinder under internal pressure only. As can be seen from the curve, $\varepsilon_{t,r=r_e}$ decreases rapidly as δ increases, for any given value of ratio p_i/E. As can also be seen, there is no point in having ratios δ in excess of (5–6) in order to minimize this strain, which can be accomplished by using a composite two-component structure consisting of a fairly thin liner made of a high elastic modulus material, and an outer portion consisting of a less costly material with a lower elastic modulus.

4.7 Design of Thick-Walled Circular Cylinders

As indicated in the foregoing sections, the stress state in thick-walled circular cylinders under internal and external pressure is triaxial (σ_r, σ_t and σ_z all other than zero) or biaxial (σ_r and σ_t other than zero and $\sigma_z = 0$), depending on whether they have closed or open ends. In the design analysis of these structures, in any case, it is

Table 4.1 Design of thick-walled circular cylinders under internal and external pressure: summary of strength theories and the associated general and specific relations for equivalent stress

Strength theory	σ_e (General relation)	σ_e for triaxial stress state, with $\sigma_z = A$	σ_e for biaxial stress state, with $\sigma_z = 0$
σ_{max}	$\sigma_e = \sigma_{max} = \sigma_t$	$\sigma_e = A + \frac{B}{r^2}$	$\sigma_e = A + \frac{B}{r^2}$
ε_{max}	$\sigma_e = E\varepsilon_{max} = \sigma_{max}$ $= \sigma_t - \nu(\sigma_r + \sigma_z)$	$\sigma_e = A(1 - 2\nu) + \frac{B}{r^2}(1 + \nu)$	$\sigma_e = A(1 - \nu) + \frac{B}{r^2}(1 + \nu)$
τ_{max}	$\sigma_e = \sigma_{max} - \sigma_{min} = \sigma_t - \sigma_r$	$\sigma_e = \frac{2B}{r^2}$	$\sigma_e = \frac{2B}{r^2}$
ED_{max}	$\sigma_e^2 = \sigma_t^2 + \sigma_z^2 + \sigma_r^2$ $- (\sigma_t\sigma_z + \sigma_t\sigma_r + \sigma_z\sigma_r)$	$\sigma_e = \sqrt{3}\frac{B}{r^2}$	$\sigma_e = \sqrt{A^2 + 3\frac{B^2}{r^4}}$

necessary to calculate equivalent stress σ_e, which entails selecting an appropriate strength theory (or strength criterion).

The theories used in this specific sector are more numerous than those adopted for the thin-walled circular cylinders described in Chap. 1. In addition to the maximum distortion energy theory, two further energy theories are used (the maximum strain energy theory, or Beltrami's theory, and the maximum modified strain energy theory, or modified Beltrami's theory, see Giovannozzi 1965, Iurzolla 1981) and, for the maximum normal strain theory, a distinction is made between the two cases mentioned above of a triaxial stress state (in which case it is called Bach's theory, Bach 1909 and 1910) and the biaxial stress state (where it is called Winkler's theory, Winkler 1867). In the following pages, only the strength theories described in Chap. 1 for the thin-walled circular cylinder will be used. For the use of these additional strength theories or other theories, the reader is thus referred to specialized texts on the subject.

For the four strength theories considered here for use both in the design analysis and in the response (or verification) analysis of thick-walled circular cylinders, the summary given in Table 4.1—like its equivalent in Table 1.1 for thin-walled cylinders—shows the general relations that express the equivalent stress for a triaxial stress state for each of these theories in the second column, and the corresponding specific relations for the equivalent stresses obtained by substituting relations (4.21) and (4.30) or (4.31) or (4.33) in the general relations, again for a triaxial stress state, in the third column. The fourth column shows the same relations as in the third column, but specialized for the biaxial stress state, i.e., with σ_r and σ_t other than zero and $\sigma_z = 0$.

The explicit general relations which express equivalent stress σ_e according to the four strength theories considered for the two cases with $\sigma_z \neq 0$ or $\sigma_z = 0$ will not be detailed here. These relations can be readily obtained from those given in Table 4.1 by substituting the expressions for integration constants A and B given

Table 4.2 Numerical factors a and b appearing in relation (4.71)

Strength theory	σ_{max}	ε_{max}	τ_{max}	ED_{max}
a	1	$1 - 2\nu$	0	0
b	1	$1 + \nu$	2	$\sqrt{3}$

by relations (4.28) for the case of a circular cylinder under internal or external pressure only. Assuming that stresses are in the linear elastic field, the general case in which we have $p_i \neq 0$ and $p_e \neq 0$ can obviously be addressed using the method of superposition.

However, it should be emphasized that, for the four strength theories indicated above and a triaxial stress state with the longitudinal stress given by the arithmetic average of the radial and hoop stresses, for which relation (4.32) applies, the equivalent stress can be expressed by a single relation, viz.:

$$\sigma_e = \frac{a+b}{2}\sigma_t + \frac{a-b}{2}\sigma_r, \tag{4.71}$$

where a and b are numerical factors that depend on the selected strength theories. They are shown in Table 4.2.

Introducing relations (4.21) in relation (4.71) yields:

$$\sigma_e = aA + b\frac{B}{r^2}. \tag{4.72}$$

Substituting relations (4.28) in the latter then yields:

Table 4.3 Design of thick-walled circular cylinders under internal pressure only: summary of expressions for equivalent stress and the response and design relations for the four strength theories

Strength theory		Stresses $\sigma_{e,i}$ and response analysis relations	Design analysis relations
σ_{max}		$\sigma_{e,i} = p_i\frac{\delta^2+1}{\delta^2-1} \leq \sigma_a$	$s = r_i\left(\sqrt{\frac{\sigma_a+p_i}{\sigma_a-p_i}} - 1\right)$
ε_{max}	$\sigma_z \neq 0$	$\sigma_{e,i} = \frac{p_i[(1-2\nu)+\delta^2(1+\nu)]}{\delta^2-1} \leq \sigma_a$	$s = r_i\left(\sqrt{\frac{\sigma_a+p_i(1-2\nu)}{\sigma_a-p_i(1+\nu)}} - 1\right)$
	$\sigma_z = 0$	$\sigma_{e,i} = \frac{p_i[(1-\nu)+\delta^2(1+\nu)]}{\delta^2-1} \leq \sigma_a$	$s = r_i\left(\sqrt{\frac{\sigma_a+p_i(1-\nu)}{\sigma_a-p_i(1+\nu)}} - 1\right)$
τ_{max}		$\sigma_{e,i} = p_i\frac{2\delta^2}{\delta^2-1} \leq \sigma_a$	$s = r_i\left(\sqrt{\frac{\sigma_a}{\sigma_a-2p_i}} - 1\right)$
ED_{max}	$\sigma_z \neq 0$	$\sigma_{e,i} = p_i\frac{\sqrt{3}\delta^2}{\delta^2-1} \leq \sigma_a$	$s = r_i\left(\sqrt{\frac{\sigma_a}{\sigma_a-\sqrt{3}p_i}} - 1\right)$
	$\sigma_z = 0$	$\sigma_{e,i} = p_i\frac{\sqrt{1+3\delta^4}}{\delta^2-1} \leq \sigma_a$	$s = r_i\left(\sqrt{\frac{\sigma_a^2+p_i\sqrt{4\sigma_a^2-3p_i^2}}{\sigma_a^2-3p_i^2}} - 1\right)$

$$\begin{aligned}\sigma_e &= \frac{p_i r_i^2}{r_e^2 - r_i^2}\left(a + b\frac{r_e^2}{r^2}\right) - \frac{p_e r_e^2}{r_e^2 - r_i^2}\left(a + b\frac{r_i^2}{r^2}\right) \\ &= \frac{p_i}{\delta^2 - 1}\left(a + b\frac{r_e^2}{r^2}\right) - \frac{\delta^2 p_e}{\delta^2 - 1}\left(a + b\frac{r_i^2}{r^2}\right).\end{aligned} \tag{4.73}$$

This relation can be readily specialized for the cases of circular cylinders with closed ends under internal pressure only or external pressure only.

4.7.1 Thick-Walled Circular Cylinders Under Internal Pressure

For thick-walled circular cylinders under internal pressure, the terms in p_e disappear from the general relations (4.28) expressing the integration constants. As indicated in Section 4.4.1, we will thus have $A = B/r_e^2 = K'$, with K' given by the third relation (4.39). As the material is at greatest risk at inner radius ($r = r_i$), where σ_t and σ_r assume their absolute maximum and minimum values respectively, while axial stress σ_z is intermediate (and is other than zero for a cylinder with closed ends, and zero for a cylinder with open ends), the equivalent stress to be used for both design analysis and response analysis is that calculated at this radius, viz., $\sigma_e = \sigma_{e,i}$.

Table 4.3 summarizes the relations obtained from those in Table 4.1 using this procedure. The second column shows the expressions for equivalent stress $\sigma_{e,i}$ at the inner radius and the response analysis expressions given by the condition $\sigma_{e,i} \leq \sigma_a$ (this is the material's allowable stress or working stress), while the third column shows the corresponding design analysis relations. Obviously, the σ_{max} and τ_{max} strength theories lead to the same response and design relations, independently of whether we have $\sigma_z \neq 0$ or $\sigma_z = 0$. The other two strength theories, for which the equivalent stress depends on all three principal stresses, lead to response and design relations that differ according to whether we consider the case in which we have $\sigma_z \neq 0$ or that in which we have $\sigma_z = 0$.

It should be noted that, for $\sigma_z = (\sigma_r + \sigma_t)/2 = A$, the response and design relations given in Table 4.3 can also be obtained from relation (4.73). The latter, for $r = r_i$, yields:

$$\sigma_{e,i} = \frac{p_i}{\delta^2 - 1}\left(a + b\delta^2\right) \tag{4.74}$$

from which the following expression of ratio δ can be obtained:

Table 4.4 Parameters for comparing strength theories for thick-walled circular cylinders under internal pressure only

	Strength theory		$p_i/\sigma_{e,i}$	p_{cr}	$\sigma_{e,e}$	$\Delta\sigma = \sigma_{e,i} - \sigma_{e,e}$	$R = \sigma_{e,i}/\sigma_{e,e}$
1	σ_{max}		$\frac{p_i}{\sigma_{e,i}} = \frac{\delta^2 - 1}{\delta^2 + 1}$	$p_{cr} = \sigma_a$	$\sigma_{e,e} = \frac{2p_i}{\delta^2 - 1}$	$\Delta\sigma = p_i$	$R = \frac{\delta^2+1}{2}$
2	ε_{max}	$\sigma_z \neq 0$	$\frac{p_i}{\sigma_{e,i}} = \frac{\delta^2 - 1}{[(1-2v) + \delta^2(1+v)]}$	$p_{cr} = \frac{\sigma_a}{(1+v)}$	$\sigma_{e,e} = \frac{(2-v)p_i}{\delta^2 - 1}$	$\Delta\sigma = p_i(1+v)$	$R = \frac{[(1-2v)+\delta^2(1+v)]}{(2-v)}$
2′		$\sigma_z = 0$	$\frac{p_i}{\sigma_{e,i}} = \frac{\delta^2 - 1}{[(1-v) + \delta^2(1+v)]}.$	$p_{cr} = \frac{\sigma_a}{(1+v)}$	$\sigma_{e,e} = \frac{2p_i}{\delta^2 - 1}$	$\Delta\sigma = p_i(1+v)$	$R = \frac{[(1-v)+\delta^2(1+v)]}{(2-v)}$
3	τ_{max}		$\frac{p_i}{\sigma_{e,i}} = \frac{\delta^2 - 1}{2\delta^2}$	$p_{cr} = \frac{\sigma_a}{2}$	$\sigma_{e,e} = \frac{2p_i}{\delta^2 - 1}$	$\Delta\sigma = 2p_i$	$R = \delta^2$
4	ED_{max}	$\sigma_z \neq 0$	$\frac{p_i}{\sigma_{e,i}} = \frac{\delta^2 - 1}{\sqrt{3}\delta^2}$	$p_{cr} = \frac{\sigma_a}{\sqrt{3}}$	$\sigma_{e,e} = \frac{\sqrt{3}p_i}{\delta^2 - 1}$	$\Delta\sigma = \sqrt{3}p_i$	$R = \delta^2$
4′		$\sigma_z = 0$	$\frac{p_i}{\sigma_{e,i}} = \frac{\delta^2 - 1}{\sqrt{1 + 3\delta^4}}$	$p_{cr} = \frac{\sigma_a}{\sqrt{3}}$	$\sigma_{e,e} = \frac{2p_i}{\delta^2 - 1}$	$\Delta\sigma = \frac{p_i}{\delta^2-1}(\sqrt{1+3\delta^4} - 2)$	$R = \frac{\sqrt{1+3\delta^4}}{2}$
5	σ_{max} thin-walled cylinder (Hamburg formula)		$\frac{p_i}{\sigma_{e,i}} = \delta - 1$	–	$\sigma_{e,e} = \frac{p_i}{\delta - 1}$	$\Delta\sigma = 0$	$R = 1$
6	τ_{max} thin-walled cylinder (ISO formula)		$\frac{p_i}{\sigma_{e,i}} = 2\frac{\delta - 1}{\delta + 1}$	$p_{cr} = 2\sigma_a$	$\sigma_{e,e} = \frac{p_i}{\delta - 1}$	$\Delta\sigma = \frac{p_i}{2}$	$R = \frac{\delta+1}{2}$

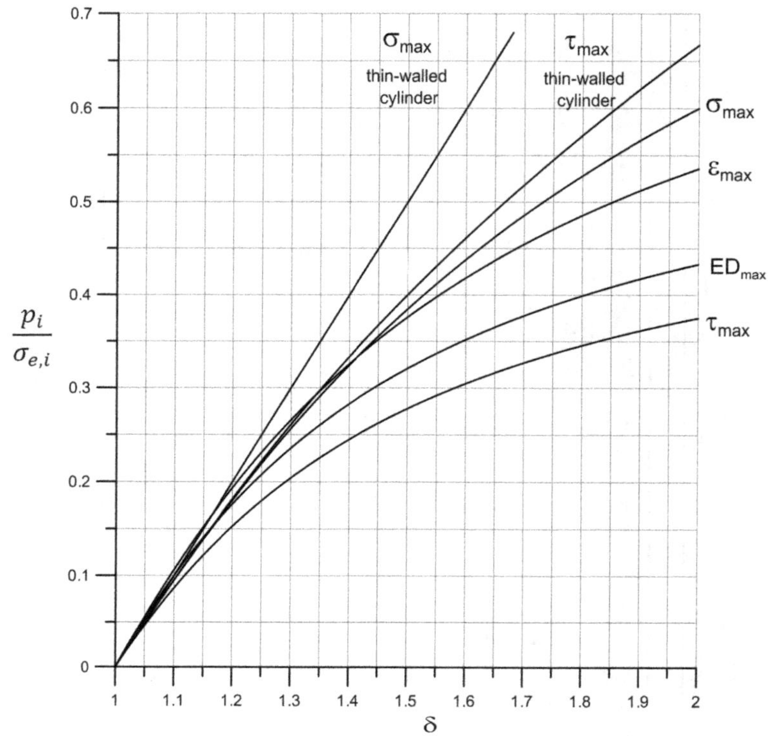

Fig. 4.9 Ratios $p_i/\sigma_{e,i} = f(\delta)$ for the strength theories considered herein

$$\delta = \sqrt{\frac{\sigma_{e,i} + p_i a}{\sigma_{e,i} - p_i b}}. \tag{4.75}$$

Noting that thickness is expressed by $s = r_i(\delta - 1)$, the relations given in Table 4.3 can be obtained from relations (4.74) and (4.75) by substituting allowable stress σ_a for $\sigma_{e,i}$, as well as the values of factors a and b appearing in Table 4.2.

As for thin-walled circular cylinders, to compare the results of using the four strength theories considered here for thick-walled circular cylinders under internal pressure only, it is advisable to introduce the dimensionless ratio $p_i/\sigma_{e,i}$ obtained from the equality relations given in the second column of Table 4.3 (these equality relations are those preceding the inequality sign). For the strength theories considered here, the expressions of the ratios $p_i/\sigma_{e,i}$determined in this way are shown in the second column of Table 4.4.

In Fig. 4.9, the same ratios are plotted versus δ, with δ varying in the range $1 \leq \delta \leq 2$. For the ε_{max} and maximum distortion energy theories, only the plots for $p_i/\sigma_{e,i}$ relating to a Poisson's ratio of 0.3 and condition $\sigma_z \neq 0$ are shown. For purposes of comparison, the figure also shows the $p_i/\sigma_{e,i}$ diagrams for the σ_{max} and τ_{max} theories applying to thin-walled circular cylinders. For $\delta \to \infty$, the four curves in Fig. 4.9 for the thick-walled circular cylinder exhibit a horizontal asymptote of ordinate $p_i/\sigma_{e,i} = 1/b$. Moreover, it can be concluded from relation (4.75) that the derivative $d\delta/d(p_i/\sigma_{e,i})$, if calculated for $p_i/\sigma_{e,i} = 0$, is in all cases equal to the arithmetic average of $(a + b)$. Consequently, not only do some of these four curves intersect each other, but they also intersect the straight line representing the σ_{max} theory for the thin-walled circular cylinder.

As is immediately apparent from Fig. 4.9, as ratio δ decreases (remember that by convention, thin-walled circular cylinders fall in the range $1.00 < \delta \leq 1.10$), the relations for the thick-walled cylinders as well as the two for the thin-walled cylinders result in $p_i/\sigma_{e,i}$ curves that are superimposed and thus lead to results that practically coincide. By contrast, the differences in the values for δ, and thus of thickness s for any given ratio $p_i/\sigma_{e,i}$, that depend on which strength theory is used, tend to increase along with δ. Figure 4.9 highlights the fact that the most conservative strength theory, or in other words that which for any given ratio $p_i/\sigma_{e,i}$ results in maximum thickness, is the τ_{max} theory. The figure also shows what happens up to the value $\delta = 2$ considered here (in this connection, it should be noted that $\delta = 2$ is a diametral ratio that is unlikely to be reached in actual practical applications, even though there are actual cases in which it has been exceeded). The differences that depend on which strength theory is used are even greater for $\delta > 2$.

On the basis of the relations given in the last column of Table 4.3, it can be readily demonstrated that we have $\delta_{\tau_{max}} > \delta_{ED_{max}} > \delta_{\varepsilon_{max}}$ in all cases. The location of the points where the curve for the σ_{max} strength theory intersect the remaining three curves can also be readily identified. These points are of undoubted interest, as they delimit the areas where prudence would dictate the use of one theory rather than another.

An enlargement of Fig. 4.9 adjacent to abscissa $\delta = 1.10$ shows that the non-linear $p_i/\sigma_{e,i} = f(\delta)$ curve for the maximum distortion energy strength theory intersects the line $p_i/\sigma_{e,i} = (\delta - 1)$ representing the so-called Hamburg formula for calculating the thickness of cylinders under internal pressure (this formula derives from using the σ_{max} strength theory for thin-walled circular cylinders), at the point whose coordinates are (1.101;0.101). For any given value of ratio $p_i/\sigma_{e,i}$ and for $\delta > 1.101$, the Hamburg formula results in wall thicknesses that are smaller than those calculated with the maximum distortion energy theory, while the contrary is true for $\delta < 1.101$. This is the justification behind the international convention that sets the thickness limit for a thin-walled circular cylinder, after

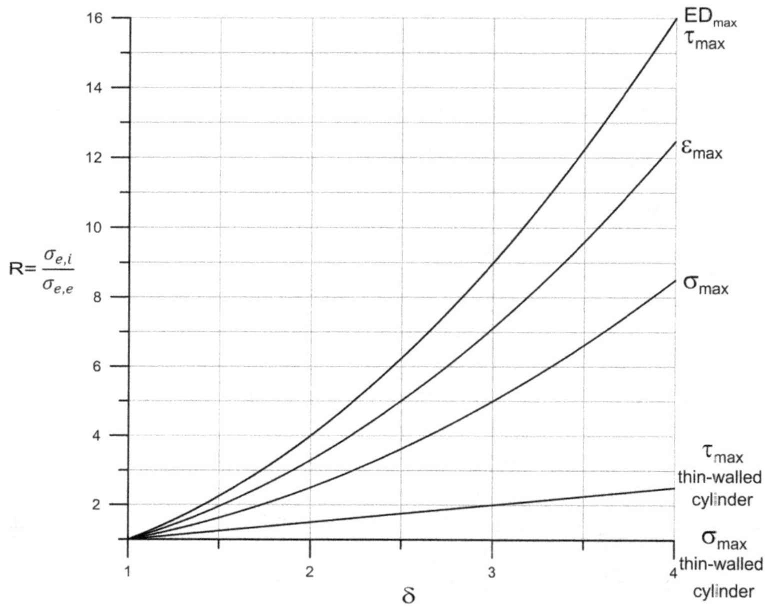

Fig. 4.10 Ratios $R = f(\delta)$ for the strength theories considered herein

which it is considered a thick-walled circular cylinder, at $\delta = 1.10$. As was noted in Chap. 1, Sect. 1.1, moreover, the curve expressing the Hamburg formula and that for the ϵ_{max} strength theory intersect at point $\delta = 1.16$ on the abscissa, which defines the demarcation between thick-walled circular cylinder and thin-walled circular cylinder on the basis of this theory.

Many other values have been introduced for the ratio δ establishing the dividing line between the two families of thin-walled and thick-walled circular cylinders as a function of the points of intersection between the curves shown in Fig. 4.9 and other curves based on further strength theories (maximum strain energy, modified maximum strain energy, etc.) which will not be given here for reasons of space. In this connection, the reader is directed to texts covering a broader scope.

The results obtained using the strength theories considered above can be compared on the basis of other parameters in addition to dimensionless ratio $p_i/\sigma_{e,i}$. Among these additional parameters, specific mention should be made of the critical pressure p_{cr}, defined as the pressure for which the wall thickness of the circular cylinder becomes infinite (this pressure is obtained by equaling the denominator of the fraction under the square root to zero in the relations given in the third column of Table 4.3), the difference $\Delta\sigma = \sigma_{e,i} - \sigma_{e,e}$, where $\sigma_{e,e}$ is the equivalent stress calculated at the outer radius, and the ratio $R = \sigma_{e,i}/\sigma_{e,e}$. These

Table 4.5 Design of thick-walled circular cylinders under external pressure only: summary of expressions for equivalent stress and the response and design relations for the four strength theories

Strength theory		Stresses $\sigma_{e,i}$ and response analysis relations	Design analysis relations
σ_{max}		$\sigma_{e,i} = -p_e \frac{2\delta^2}{\delta^2 - 1} \leq \sigma_a$	$s = r_i\left(\sqrt{\frac{\sigma_a}{\sigma_a - 2p_e}} - 1\right)$
ε_{max}	$\sigma_z \neq 0$	$\sigma_{e,i} = -p_e \frac{[\delta^2(2 - \nu)]}{\delta^2 - 1} \leq \sigma_a$	$s = r_i\left(\sqrt{\frac{\sigma_a}{[\sigma_a - p_e(2 - \nu)]}} - 1\right)$
	$\sigma_z = 0$	$\sigma_{e,i} = -p_e \frac{2\delta^2}{\delta^2 - 1} \leq \sigma_a$	$s = r_i\left(\sqrt{\frac{\sigma_a}{\sigma_a - 2p_e}} - 1\right)$
τ_{max}	$\sigma_z \neq 0$	$\sigma_{e,i} = -p_e \frac{2\delta^2}{\delta^2 - 1} \leq \sigma_a$	$s = r_i\left(\sqrt{\frac{\sigma_a}{\sigma_a - 2p_e}} - 1\right)$
	$\sigma_z = 0$	$\sigma_{e,i} = -p_e \frac{2\delta^2}{\delta^2 - 1} \leq \sigma_a$	$s = r_i\left(\sqrt{\frac{\sigma_a}{\sigma_a - 2p_e}} - 1\right)$
ED_{max}	$\sigma_z \neq 0$	$\sigma_{e,i} = -p_e \frac{\sqrt{3}\delta^2}{\delta^2 - 1} \leq \sigma_a$	$s = r_i\left(\sqrt{\frac{\sigma_a}{\sigma_a - \sqrt{3}p_e}} - 1\right)$
	$\sigma_z = 0$	$\sigma_{e,i} = -p_e \frac{2\delta^2}{\delta^2 - 1} \leq \sigma_a$	$s = r_i\left(\sqrt{\frac{\sigma_a}{\sigma_a - 2p_e}} - 1\right)$

parameters are shown in Table 4.4, together with the equivalent stress $\sigma_{e,e}$ which characterizes parameters $\Delta\sigma$ and R. By way of comparison, the two lines at the bottom of the table show the corresponding parameters resulting from the two σ_{max} and τ_{max} strength theories applied to the thin-walled circular cylinder under internal pressure only.

As the relations given in the two last columns in Table 4.1 indicate, the equivalent stress varies along the radius, or in other words through the circular cylinder's wall thickness. Whatever strength theory is used, this stress is at its maximum at the inner radius, and minimum at the outer radius, while the maximum and minimum values differ according to the theory employed. Consequently, the material is much more highly stressed at the inner radius than at the outer radius, and its tensile strength properties are not put to good use. Quantitatively, this is borne out by the two parameters $\Delta\sigma$ and R. The latter is shown versus δ in Fig. 4.10, from which it can be seen that, for any given wall thickness, or in other words for any given δ, the theories that results in the maximum value of ratio R (and thus the maximum difference $\Delta\sigma$) is the τ_{max} theory (as well as the maximum distortion energy theory which is superposed on it). For any given R, on the other hand, the theory resulting in the maximum thickness is the σ_{max} theory. It goes without saying that the axis of the abscissa, characterized by $R = 1$, represents the Hamburg formula.

Table 4.6 Parameters for comparing strength theories for thick-walled circular cylinders under external pressure only

	Strength theory		$p_e/\sigma_{e,i}$	p_{cr}	$\sigma_{e,e}$	$\Delta\sigma = \sigma_{e,i} - \sigma_{e,e}$	$R = \sigma_{e,i}/\sigma_{e,e}$
1	σ_{max}		$\frac{p_e}{\sigma_{e,i}} = -\frac{\delta^2-1}{2\delta^2}$	$p_{cr} = \frac{\sigma_a}{2}$	$\sigma_{e,e} = -p_e\frac{\delta^2-1}{\delta^2-1}$	$\Delta\sigma = -p_e$	$R = \frac{2\delta^2}{\delta^2+1}$
2	ε_{max}	$\sigma_z \neq 0$	$\frac{p_e}{\sigma_{e,i}} = -\frac{\delta^2-1}{\delta^2(2-v)}$	$p_{cr} = \frac{\sigma_a}{2-v}$	$\sigma_{e,e} = -\frac{p_e[\delta^2(1-2v)+(1+v)]}{\delta^2-1}$	$\Delta\sigma = -p_e(1+v)$	$R = \frac{\delta^2(2-v)}{\delta^2(1-2v)+(1+v)}$
2′		$\sigma_z = 0$	$\frac{p_e}{\sigma_{e,i}} = -\frac{\delta^2-1}{2\delta^2}$	$p_{cr} = \frac{\sigma_a}{2}$	$\sigma_{e,e} = -\frac{p_e[\delta^2(1-v)+(1+v)]}{\delta^2-1}$	$\Delta\sigma = -p_e(1+v)$	$R = \frac{2\delta^2}{\delta^2(1-v)+(1+v)}$
3	τ_{max}	$\sigma_z \neq 0$	$\frac{p_e}{\sigma_{e,i}} = -\frac{\delta^2-1}{2\delta^2}$	$p_{cr} = \frac{\sigma_a}{2}$	$\sigma_{e,e} = -p_e\frac{2}{\delta^2-1}$	$\Delta\sigma = -2p_e$	$R = \delta^2$
3′		$\sigma_z = 0$	$\frac{p_e}{\sigma_{e,i}} = -\frac{\delta^2-1}{2\delta^2}$	$p_{cr} = \frac{\sigma_a}{2}$	$\sigma_{e,e} = -p_e\frac{\delta^2+1}{\delta^2-1}$	$\Delta\sigma = -p_e$	$R = \frac{2\delta^2}{\delta^2+1}$
4	ED_{max}	$\sigma_z \neq 0$	$\frac{p_e}{\sigma_{e,i}} = -\frac{\delta^2-1}{\sqrt{3}\delta^2}$	$p_{cr} = \frac{\sigma_a}{\sqrt{3}}$	$\sigma_{e,e} = -p_e\frac{\sqrt{3}}{\delta^2-1}$	$\Delta\sigma = -\sqrt{3}p_e$	$R = \delta^2$
4′		$\sigma_z = 0$	$\frac{p_e}{\sigma_{e,i}} = -\frac{\delta^2-1}{2\delta^2}$	$p_{cr} = \frac{\sigma_a}{2}$	$\sigma_{e,e} = -p_e\frac{\sqrt{3+\delta^4}}{\delta^2-1}$	$\Delta\sigma = -\frac{p_e}{\delta^2-1}(2\delta^2 - \sqrt{3+\delta^4})$	$R = \frac{2\delta^2}{2\delta^2 - \sqrt{3+\delta^4}}$
5	σ_{max} thin-walled cylinders for $p_e \neq 0$ and $p_i = 0$		$\frac{p_e}{\sigma_{e,i}} = -\frac{\delta-1}{\delta}$	$p_{cr} = \sigma_a$	$\sigma_{e,e} = -p_e\frac{\delta}{\delta-1}$	$\Delta\sigma = 0$	$R = 1$

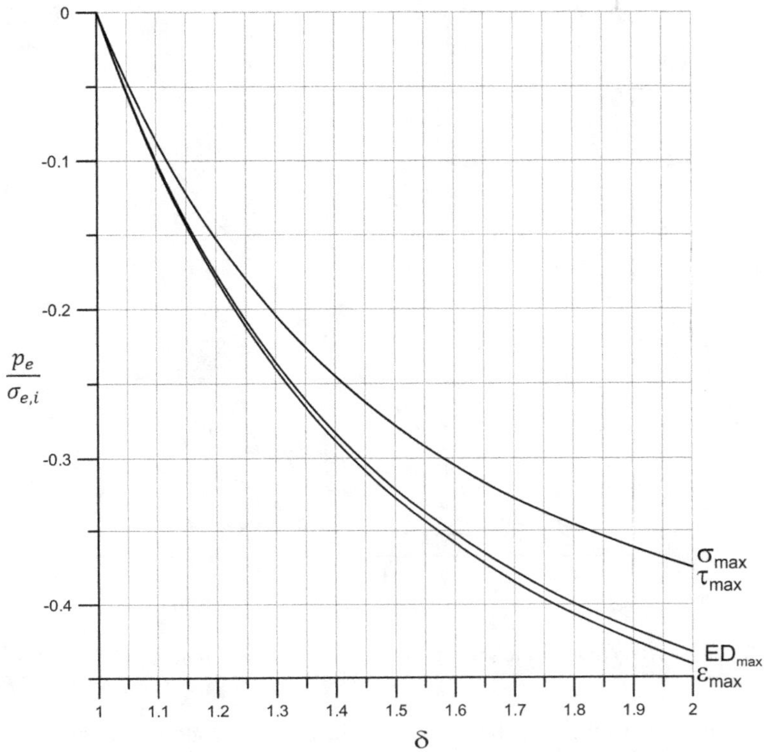

Fig. 4.11 Ratios $p_e/\sigma_{e,i} = f(\delta)$ for the strength theories considered herein

4.7.2 Thick-Walled Circular Cylinders Under External Pressure

For thick-walled circular cylinders under external pressure only, the terms in p_i disappear from the general relations (4.28) expressing the integration constants. As indicated in Sect. 4.4.2, we will thus have $A = B/r_i^2 = K''$, with K'' given by the third relation (4.46). Here again, the material is at greatest risk at the inner radius, where hoop stress σ_t, which in absolute value represents the maximum stress, assumes the highest value of the stress field.

By contrast with the situation in the thick-walled circular cylinder under internal pressure only, however, the minimum stress here differs in the two cases with $\sigma_z \neq 0$ and $\sigma_z = 0$: in the first case, the minimum stress, again in absolute value, is given by σ_r, whereas in the second case it is represented by stress σ_z, which is zero. In relation to the strength theory considered, this obviously has repercussions on the equivalent stress calculation. This affects the τ_{max} theory, for which the equivalent stress depends on the maximum and minimum of the three

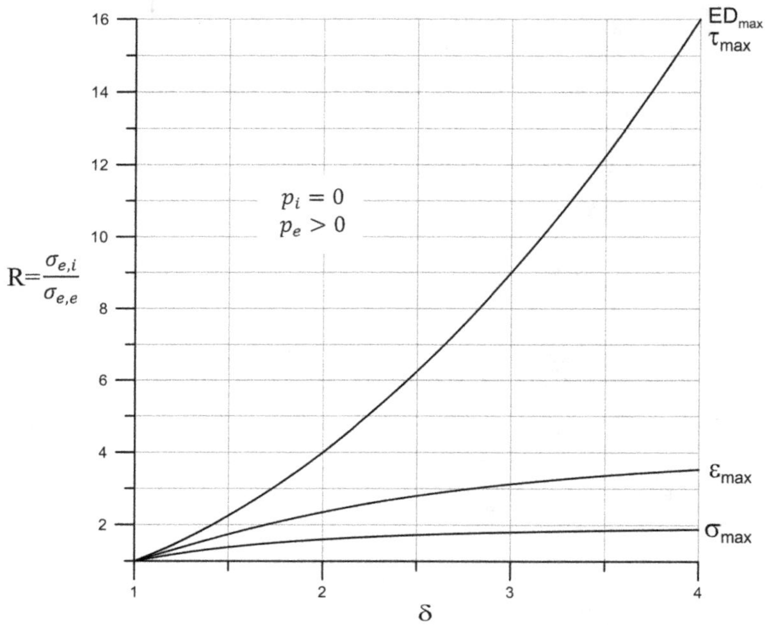

Fig. 4.12 Ratios $R = f(\delta)$. for the strength theories considered herein

principal stresses. In view of the curves for these stresses along the wall thickness (see Fig. 4.5), in fact, it can be concluded that the equivalent stress $\sigma_{e,i}$ calculated at the inner radius has the same expression in both cases, while the equivalent stress $\sigma_{e,e}$ calculated at the outer radius is given by two different relations, depending on whether we have $\sigma_z \neq 0$ or $\sigma_z = 0$ (see Table 4.6).

The equivalent stress to be used for both response analysis and design analysis is that corresponding to the inner radius, i.e., $\sigma_e = \sigma_{e,i}$. Table 4.5 summarizes the relations obtained from the general relations given in Table 4.1 using this procedure. The second column shows the expressions for equivalent stress $\sigma_{e,i}$ and the response analysis expressions given by condition $\sigma_{e,i} = \sigma_a$, while the third column shows the corresponding design analysis relations. Here again, it should be noted that, for $\sigma_z = (\sigma_r + \sigma_t)/2 = A$, the response and design relations given in Table 4.5 can also be obtained from relation (4.73), using the same procedure described in the preceding section.

Table 4.6 shows the parameters that are commonly used to compare the results of using the four strength theories considered herein in designing thick-walled circular cylinders under external pressure only. These are the companion parameters to those given in Table 4.4 for the thick-walled cylinder under internal pressure only. As for the latter, the first comparison parameter is ratio $p_e/\sigma_{e,i}$, which is obtained from the equality relations given in the second column of Table 4.5. In Fig. 4.11, the $p_e/\sigma_{e,i} = f(\delta)$ ratios for each strength theory are

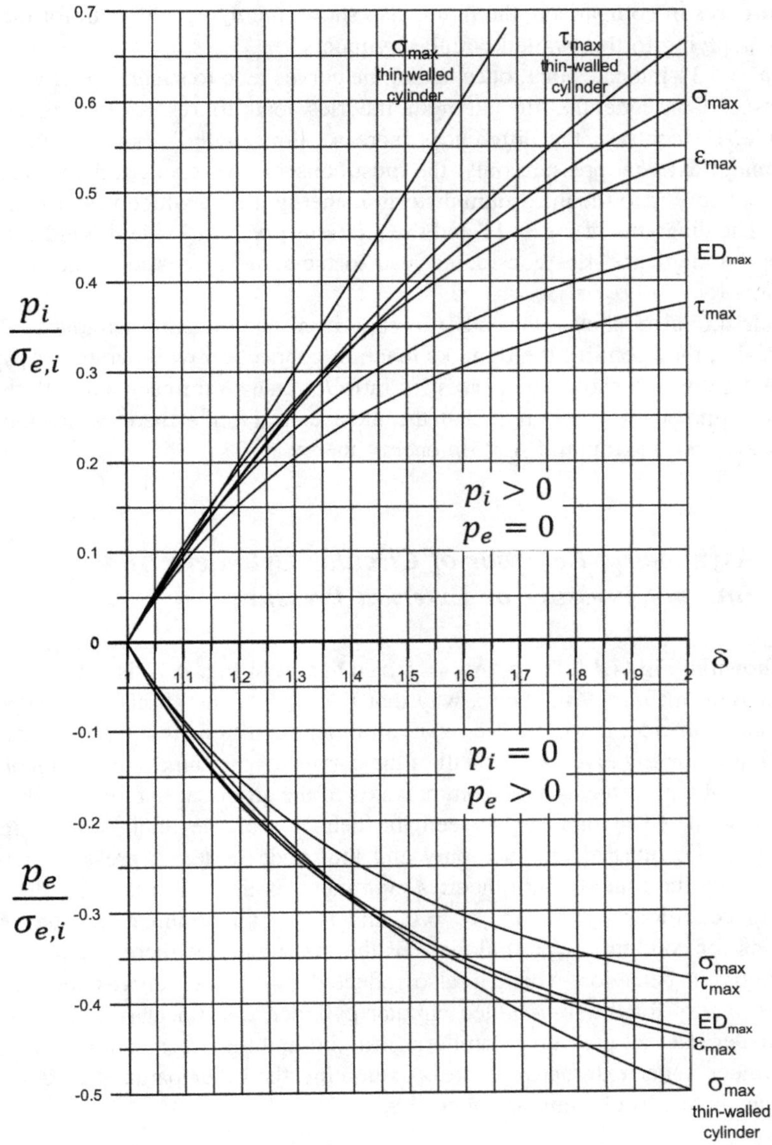

Fig. 4.13 Diagrams for $p_i/\sigma_{e,i} = f(\delta)$ and $p_e/\sigma_{e,i} = f(\delta)$, demonstrating the asymmetric behavior shown by cylinders under internal and external pressure

plotted versus δ, which here again varies in the range $1.0 \leq \delta \leq 2.0$. Likewise, for the only theory for which this ratio depends on Poisson's ratio ν (the ε_{max} theory), the plot shown is that for $\nu = 0.3$. All diagrams shown refer to condition $\sigma_z \neq 0$.

For purposes of comparison, the figure also shows the $p_e/\sigma_{e,i}$ diagram for the σ_{max} theory applying to thin-walled circular cylinders.

Figure 4.11 indicates that, once again, the curves tend to superimpose as ratio δ decreases. Consequently, the strength theories lead to results that practically coincide. By contrast, the differences increase along with δ. For the cylindrical body under external pressure only, the most conservative strength theory is again the τ_{max} theory (and the maximum distortion energy theory which is superimposed on it). The diagrams in Fig. 4.11 and the relations given in Table 4.6 indicate that the four strength theories considered lead to the same expressions and the same diagrams for ratio $p_e/\sigma_{e,i}$ if $\sigma_z = 0$.

Table 4.6 also shows the other comparison parameters mentioned above. Essentially, they confirm the remarks made in connection with parameter $p_e/\sigma_{e,i}$. Figure 4.12, which shows the curves for ratio R versus δ for each strength theory, makes it immediately apparent that the most conservative theories are the τ_{max} theory and the maximum distortion energy theory.

4.7.3 Asymmetric Behavior of Circular Cylinders Under Internal Pressure or External Pressure

If as shown in Fig. 4.13, the $p_e/\sigma_{e,i} = f(\delta)$ diagram in Fig. 4.11 is placed beneath the diagram in Fig. 4.9 in such a way that the axes of the abscissa are superimposed and coincide, it will be seen that, all other conditions remaining equal (and in particular for any given σ_z), not all of the curves for the same strength theory are symmetrical with respect to the common axis of the abscissa. In other words, there is no generalized symmetry between the behavior of the thick-walled circular cylinder under internal pressure only and that under external pressure only, as described by the four strength theories. Symmetry is limited to the τ_{max} diagrams for both conditions $\sigma_z \neq 0$ and $\sigma_z = 0$, and to the maximum distortion energy diagrams for condition $\sigma_z \neq 0$ alone. For the other theories there is a substantial asymmetry in behavior, which is also reflected by the two curves for the σ_{max} strength theory for the thin-walled circular cylinder, one linear (that for the cylinder under internal pressure, Hamburg formula) and the other nonlinear (that for the cylinder under external pressure). Examining the other parameters described above leads to entirely similar conclusions.

Chapter 5
Concentric Circular Cylinders with Radial Interference Fit and Multilayer Circular Cylindrical Structures

5.1 Introduction

The increasing need for structural assemblies and components with a high strength-to-weight ratio has led to the development of composite materials and composite structures. Today, there is no technological sector where composite materials, and even advanced composite materials, are not used. This is because the directionality of their mechanical properties can be put to effective use in designing high-performance structures that would not otherwise be feasible. Here, however, we will not deal with structures consisting of composite materials, calculation methods for which are covered in a number of specialized texts.

In the following pages, our focus will be on the equally numerous cases of composite structures (in detail, composite concentric cylinders), in which different components consisting of homogeneous and isotropic materials are assembled in order to maximize the advantages that each can provide in reaching design objectives. Specifically, we will analyze composite structures consisting of concentric thick-walled cylindrical bodies assembled with a radial interference fit (Prosciutto 1952, Faupel 1964, Giovannozzi 1965, Juvinall 1967, Belluzzi 1971, Caboni 1971, Feodosyev 1973, Ford and Alexander 1977, Iurzolla 1981, Burr 1982, Manna 1983, Dubbel 1983).

Components can be assembled with an interference fit either by forcing them together with a press, in which case the terms *force fit*, or *press fit* are used, or by means of a *shrink fit* (*shrinking*), where the outer member is preheated so that it expands to fit over the inner member, or the latter is precooled so that it contracts. Mixed technologies are also used in which two or all three of these processes are combined.

For shrink fits employing heat, the inner cylinder is maintained at the initial temperature T_i and the outer cylinder must be brought to a final temperature $T_{f,e} > T_i$ which is sufficient to expand its diameter so that the inner cylinder can be slid into it. The magnitudes involved in this process are linked by the following relation:

$$T_{f,e} - T_i = (d_{c,f} - d_{c,i})/\alpha d_{c,i} \tag{5.1}$$

V. Vullo, *Circular Cylinders and Pressure Vessels*, Springer Series in Solid and Structural Mechanics 3, DOI: 10.1007/978-3-319-00690-1_5,

where $d_{c,f}$ and $d_{c,i}$ are the final and initial shrink-fit diameters of the outer cylinder and α is the coefficient of linear thermal expansion of the cylinder material. Note that the two shrink-fit diameters (the inner diameter of the outer cylinder and the outer diameter of the inner cylinder) do not differ, as they are equal to the nominal dimensions of the two cylindrical surfaces assembled in an interference fit. In general, the initial temperature coincides with the ambient temperature T_a, as the shrink-fit process is carried out starting from the latter.

To increase the total difference in temperature or to lower the temperature of the outer member, which could reach excessive levels, the assembly process can also employ simultaneous cooling. This brings the inner member to a temperature $T_{f,i}$ below the initial temperature T_i by an amount sufficient to ensure that its diameter contracts. In this case, the relation linking the various magnitudes is as follows:

$$-(T_{f,i} - T_i) = (d_{c,f} - d_{c,i})/\alpha d_{c,i}. \tag{5.2}$$

5.2 Two Concentric Circular Cylinders Assembled with an Interference Fit

We will first examine the simplest case of two concentric circular cylinders of annular cross-section consisting of homogeneous and isotropic material and assembled with a radial interference fit (Fig. 5.1).

To analyze the stress state generated in these two circular cylinders as a result of the shrink-fit pressure (or contact pressure) at the interface between the

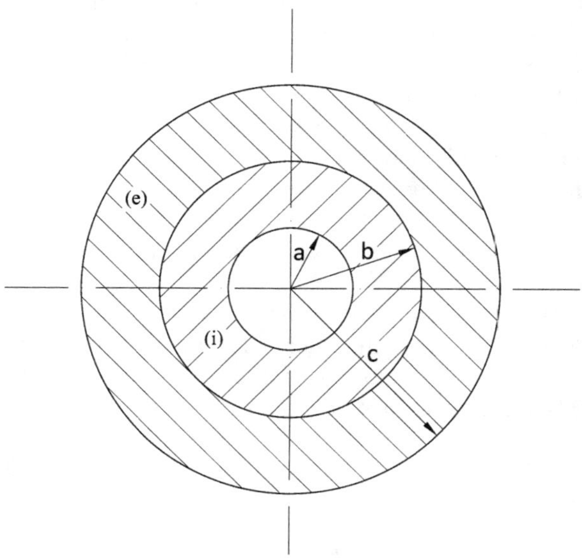

Fig. 5.1 Coaxial circular cylinders of equal length assembled with a radial interference fit

members, we will first assume that this pressure is uniformly distributed and such as to stress the two circular cylinders in the elastic field. We will also assume that both cylinders to be assembled are of the same length (if one of the two circular cylinders is shorter than the other, as in the case of shrink-fit shaft/hub assemblies, there will be a higher stress state at the end of the shorter circular cylinder that is generally accounted for by introducing a stress concentration factor, whose value may be 2 or higher).

Let (a, b) and (b, c) be the inner and outer radii of the inner cylinder (often referred to as the liner) and the outer cylinder, whose magnitudes are here designated with the subscripts i and e, respectively. The common radius b of the interface between the two circular cylinders is that of the assembled structure. Before the shrink fit, the outer radius of the inner member and the inner radius of the outer member will obviously be greater than and lesser than b respectively, but the difference between them will be eliminated once the members are shrunk together. As the cylinders are made of different materials, let (E_i, ν_i) and (E_e, ν_e) be the Young's modulus and Poisson's ratios for the inner and outer cylinder materials respectively.

At the shrink-fit interface of radius b (Fig. 5.1), a contact or shrink-fit pressure p_f, is generated between the two mating members assembled with a radial interference. Consequently, the inner cylinder will be under external pressure, while the outer cylinder will be under internal pressure. To calculate the stress and strain states and radial displacements in the two cylinders, we must first determine the hyperstatic unknown, i.e., the shrink-fit pressure p_f deriving from an assigned diametral interference $2i$ (i is the radial interference). Naturally, determining the function $p_f = p_f(i)$ that links the shrink-fit pressure p_f to radial interference i entails taking into consideration the stresses and strains, and thus the displacements, at radius b of both cylinders. Though the subscript b should be used to designate these magnitudes, it will be omitted below in order not to overload the symbols with subscripts (as indicated earlier, the subscripts i and e designate the magnitudes referring to the inner and outer cylinders).

Figure 5.2a, b shows the curves for radial and hoop stresses correlated with the hyperstatic unknown p_f for the inner and outer cylinders respectively. The figures also show the characteristic parameters of interest as well as the local values of the stresses at each radius, including those for the shrink-fit interface. These local values are also shown with reference to the unknown p_f.

Given that $u = r\varepsilon_t$, and assuming that $\sigma_z = 0$, the following expressions for radial displacements $u_{b,i}$ and $u_{b,e}$ at the shrink-fit interface subjected to shrink-fit pressure p_f and regarded as part of the inner and outer cylinder respectively are readily obtained:

$$u_{b,i} = b(\varepsilon_t)_{r=b} = \frac{b}{E_i}(\sigma_t - \nu\sigma_r)_{r=b} = -\frac{p_f b}{E_i}\left(\frac{\delta_i^2 + 1}{\delta_i^2 - 1} - \nu_i\right) = -\frac{p_f b}{E_i}\left(\frac{b^2 + a^2}{b^2 - a^2} - \nu_i\right) \tag{5.3}$$

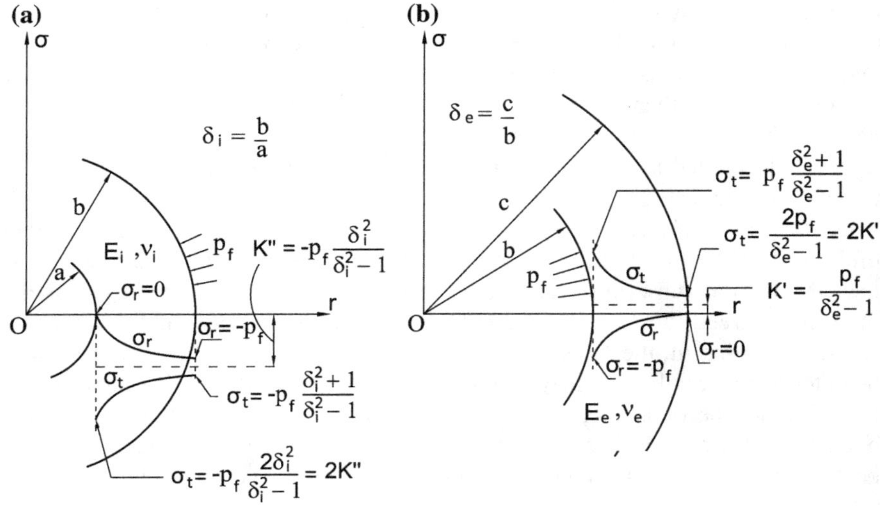

Fig. 5.2 Distribution curves of stresses σ_r and σ_t, magnitudes and parameters of interest in two circular cylinders assembled with an interference fit: **a** inner cylinder; **b** outer cylinder

and

$$u_{b,e} = b(\varepsilon_t)_{r=b} = \frac{b}{E_e}(\sigma_t - \nu\sigma_r)_{r=b} = \frac{p_f b}{E_e}\left(\frac{\delta_e^2 + 1}{\delta_e^2 - 1} + \nu_e\right) = \frac{p_f b}{E_e}\left(\frac{c^2 + b^2}{c^2 - b^2} + \nu_e\right). \tag{5.4}$$

These relations, where the minus sign indicates a radial displacement in the centripetal direction (i.e., in contrast with the positive direction of r), are used to calculate total radial interference i, which is obviously equal to the algebraic difference $\left(u_{b,e} - u_{b,i}\right)$. In the general case, we thus have:

$$i = u_{b,e} - u_{b,i} = \frac{p_f b}{E_e}\left(\frac{c^2 + b^2}{c^2 - b^2} + \nu_e\right) + \frac{p_f b}{E_i}\left(\frac{b^2 + a^2}{b^2 - a^2} - \nu_i\right) = p_f b k, \tag{5.5}$$

where the parameter k is given by the following relation:

$$k = \frac{1}{E_e}\left(\frac{c^2 + b^2}{c^2 - b^2} + \nu_e\right) + \frac{1}{E_i}\left(\frac{b^2 + a^2}{b^2 - a^2} - \nu_i\right). \tag{5.6}$$

This parameter summarizes the geometric magnitudes of the shrink fit and the elastic magnitudes of the materials used for the two cylinders in the composite structure considered here.

In the particular case in which the two circular cylinders consist of the same material ($E_i = E_e = E$; $\nu_e = \nu_i = \nu$), relation (5.5) becomes:

$$i = \frac{2p_f b^3 (c^2 - a^2)}{E(c^2 - b^2)(b^2 - a^2)}. \tag{5.7}$$

It follows that the shrink-fit pressure caused by an assigned diametral interference $2i$ is given by the relation:

$$p_f = \frac{Ei(c^2 - b^2)(b^2 - a^2)}{2b^3 (c^2 - a^2)}. \tag{5.8}$$

Obviously, in the general case of cylinders made of different materials, it will be necessary to use relation (5.5) and solve for p_f. It is equally obvious that if a hollow outer cylinder is assembled together with a solid inner cylinder, as is the case of a shrink-fit shaft/hub assembly, we must put $a = 0$ in all of the foregoing relations.

Once the hyperstatic unknown p_f has been determined, relations (4.46) can be used to arrive at the following relations defining the stress state in the inner cylinder, which is subjected to external pressure (remember that we have assumed $\sigma_z = 0$):

$$\begin{aligned} \sigma_{r,i} &= -\frac{p_f b^2}{b^2 - a^2}\left(1 - \frac{a^2}{r^2}\right) \\ \sigma_{t,i} &= -\frac{p_f b^2}{b^2 - a^2}\left(1 + \frac{a^2}{r^2}\right). \end{aligned} \tag{5.9}$$

If the inner cylinder is solid, as in a shrink-fit shaft/hub assembly, we will have $a = 0$. Consequently:

$$\sigma_{r,i} = \sigma_{t,i} = -p_f = \text{const.} \tag{5.10}$$

In other words, the radial and hoop stresses, both of which are compressive, are constant along the entire radius of the solid inner cylinder and are equal in value to the shrink-fit pressure.

The stress state in the outer cylinder, which is under internal pressure, is obtained from relations (4.18), which enable us to write:

$$\begin{aligned} \sigma_{r,e} &= \frac{p_f b^2}{c^2 - b^2}\left(1 - \frac{c^2}{r^2}\right) \\ \sigma_{t,e} &= \frac{p_f b^2}{c^2 - b^2}\left(1 + \frac{c^2}{r^2}\right). \end{aligned} \tag{5.11}$$

Figure 5.3 shows two typical stress distributions in two shrink-fit assemblies: the first involves two hollow cylinders, while in the second, a hollow cylinder and a solid cylinder are assembled together, as in the case of a shrink-fit shaft/hub assembly.

The distributions of stresses σ_r and σ_t produced by shrink-fit pressure p_f can be obtained very simply by representing the two stresses as a function of variable $-(1/r^2)$. For a double-walled structure consisting of two hollow cylinders,

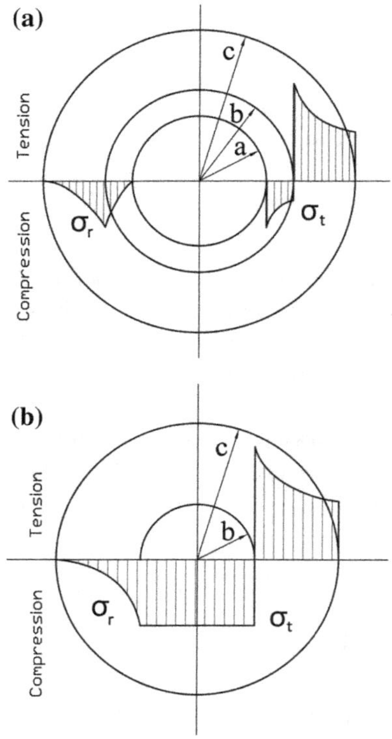

Fig. 5.3 Distribution curves of stresses σ_r and σ_t in two shrink-fit assemblies between: **a** two hollow cylinders; **b** one hollow cylinder and one solid cylinder

for example, Fig. 5.4a shows how to plot the two diagrams for stresses σ_r and σ_t when shrink-fit pressure p_f is known. In fact, once points 1, 2 and 3 whose coordinates are $-(1/a^2)$, $-(1/b^2)$ and $-(1/c^2)$ respectively on the negative abscissa semi-axis have been identified, it is sufficient to trace a line joining 2 on the abscissa with value $-p_f$ on the ordinate and join point 4 thus found with points 1 and 3 (both having zero radial stress) with two straight lines: we will thus obtain the diagrams for stresses σ_r in the inner and outer cylinder.

We then extend these two straight lines to intersect the axis of the ordinate, i.e., the σ axis, at points A and A', and trace another two straight lines passing through these points and symmetrical to the previous two straight lines with respect to the two horizontals for A and A' (note that the abscissa of point O and thus those of points A and A' correspond to $r = \infty$). The line segments intercepted on these two straight lines by the verticals through points 1, 2 and 3 make up the diagrams for stresses σ_t in the inner and outer cylinder. Figure 5.4b, on the other hand, shows the distributions of stresses σ_r and σ_t in the two cylinders versus radius r. We can proceed in the same way to represent stresses σ_r and σ_t in a shrink fit between a hollow outer cylinder and a solid inner cylinder.

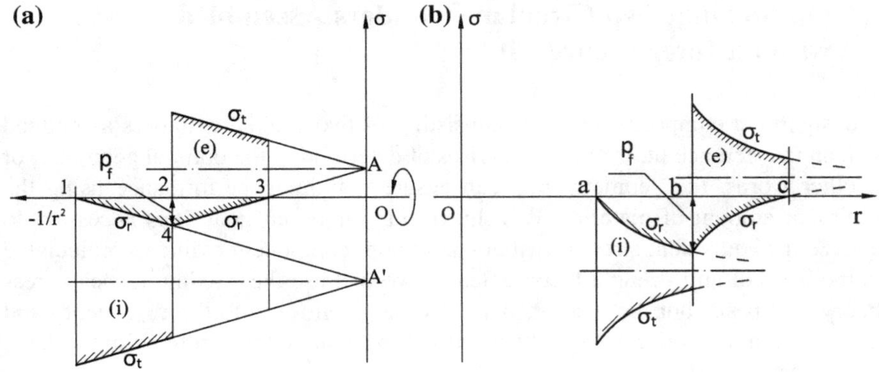

Fig. 5.4 Distribution curves of stresses σ_r and σ_t in two hollow cylinders assembled with an interference fit: **a** linear diagrams; **b** nonlinear diagrams (cubic hyperbolas)

It is essential that several fundamental concepts and considerations be borne in mind when designing shrink-fit assemblies of the kind described above.

First, it is necessary to know the effective initial state of the circular cylinders involved. The circular cylinders, in fact, may be affected by appreciable residual stresses resulting from heat treatment and/or machining processes, which are inevitably added to those deriving from service loads and the shrink-fit process. It must also be noted that, for shrink fits employing heat, temperatures above 260 °C can cause annealing effects in certain steels, especially if they have been heat-treated to a high hardness level.

In addition, high thermal gradients resulting from the contact thermal resistances arising at the interface when the two circular cylinders are shrunk together can cause high thermal stresses and even plastic flow if the material's yield stress is exceeded. Nor should we forget the brittleness problems associated with shrink fits that employ cooling, or the fact that extreme care must be taken in calculating dimensional tolerances and in carrying out the processes controls needed to achieve them, as the shrink-fit assembly performance in service will depend on them.

A wide range of practical problems can be solved by means of relation (5.8) or, in the more general case of dissimilar materials, by means of relation (5.5) and the relations deriving from it. Among the latter, special mention should be made of those used in calculations for multilayer vessels used in the chemical and nuclear industries, and for shaft/hub assemblies. All of the foregoing relations apply to simple shrink-fit assembly, where only the contact pressure associated with interference is at work. Very often, however, the composite structure thus obtained is affected by other loads, such as internal and/or external pressures, body forces when the structure is subject to angular acceleration or centrifugal load, or thermal stresses resulting from heat flow. Obviously, all of the stresses acting on the structure must be considered in defining its overall stress and strain state.

5.3 Optimizing Two Circular Cylinders Assembled with an Interference Fit

In designing a composite structure consisting of two coaxial cylinders assembled with an interference fit, the problem arises of determining the optimal geometry, or in other words, the geometry that can ensure maximum performance using the minimum amount of material. To achieve this important goal, it is necessary to analyze the equivalent stress distributions in both component cylinders, calculated on the basis of any strength theory. Below, we will use the maximum shear stress theory as a basis, but the relations will also be given for both the τ_{max} theory and the maximum distortion energy theory. It will be assumed that both cylinders have open ends ($\sigma_z = 0$).

As can be seen from Fig. 5.2a, the maximum hoop stress in the inner cylinder occurs at the inner radius ($r = a$), while the maximum radial stress is at the outer radius ($r = b$). Both these stresses are negative, i.e., compressive. In addition, the inequality $|\sigma_t| > |\sigma_r|$ occurs at any value of the radius. The maximum shear stress $\tau_{max} = \sigma_t/2$ (remember that at radius $r = a$, we have $\sigma_r = \sigma_z = 0$) will thus be at the inner radius of the inner cylinder and, recalling relation (5.5), will be:

$$\tau_{i,max} = \pm p_f \frac{\delta_i^2}{\delta_i^2 - 1} = \pm \frac{i\delta_i^2}{bk\left(\delta_i^2 - 1\right)} = \pm \frac{ib}{k(b^2 - a^2)}. \tag{5.12}$$

For the outer cylinder, on the other hand, it can be seen from Fig. 5.2b that both the maximum hoop stress and the maximum radial stress occur at the contact interface ($r = b$), where the former is a tensile stress and the latter is compressive. Maximum shear stress $\tau_{max} = (\sigma_t - \sigma_r)/2$ is thus at the same interface and, once again recalling relation (5.5), is:

$$\tau_{e,max} = \pm p_f \frac{\delta_e^2}{\delta_e^2 - 1} = \pm \frac{i\delta_e^2}{bk\left(\delta_e^2 - 1\right)} = \pm \frac{ic^2}{bk(c^2 - b^2)}. \tag{5.13}$$

For an optimized design, the inner and the outer cylinder must simultaneously have the same maximum equivalent stresses and, consequently, the same maximum shear stresses. It is thus not necessary to plot the complete diagrams representing the variability along the wall thickness of the two circular cylinders. To make best use of the material and its strength properties, this maximum shear stress value common to the two assembled components must be equal to the materials' elastic limit, i.e., $\sigma_s/2$ (here again, the proportional limit, the elastic limit and the yield point are assumed to coincide). With the optimization criterion thus defined, the equality of relations (5.12) and (5.13) yields:

$$b = \sqrt{ac}. \tag{5.14}$$

In other words, the optimal value for the radius at interface ($r = b$) is the geometric mean of the inner and outer radii of the inner and outer cylinder respectively. Substituting relation (5.14) in relations (5.12) and (5.13) and noting that the maximum shear stress in the uniaxial tensile test is one half of the yield stress, we obtain:

$$\tau_{i,max} = \frac{\sigma_{s,i}}{2} = \frac{i\sqrt{ac}}{ka(c-a)} \tag{5.15}$$

and

$$\tau_{e,max} = \frac{\sigma_{s,e}}{2} = \frac{ic}{k\sqrt{ac}(c-a)}. \tag{5.16}$$

The values of radial interference i required for the shrink-fit assembly can be determined by means of relation (5.15) or relation (5.16). If the materials used for the two inner and outer cylinders have the same yield strength, we will have $\sigma_{s,i} = \sigma_{s,e} = \sigma_s$. All the other magnitudes (interface pressure, etc.) are determined using the foregoing relations which, as will be recalled, are based on the τ_{max} strength theory.

If we wish to use the maximum distortion energy strength theory, which, as will be discussed in Chap. 7, best interprets the experimental results, at least for ductile materials, care must be taken to substitute $\sigma_s/\sqrt{3}$ for $\sigma_s/2$ in the foregoing relations. With this theory, relation (5.16) gives the following value for the optimal radial interference i_{opt}:

$$i_{opt} = \sigma_s\sqrt{\frac{ac}{3}}\left(\frac{c-a}{c}\right)\left[\frac{1}{E_e}\left(\frac{c+a}{c-a}+\nu_e\right)+\frac{1}{E_i}\left(\frac{c+a}{c-a}-\nu_i\right)\right]. \tag{5.17}$$

Recalling relation (5.5), relation (5.17) gives the following expression for the optimal interface pressure:

$$p_{f,opt} = \frac{\sigma_s}{\sqrt{3}}\left(\frac{c-a}{c}\right). \tag{5.18}$$

In the case where $E_i = E_e = E$ and $\nu_i = \nu_e = \nu$, or in other words when the two cylindrical bodies of annular cross-section consist of the same material, relation (5.17) is specialized as:

$$i_{opt} = \frac{2\sigma_s}{E\sqrt{3}}\sqrt{ac}\left(\frac{c+a}{c}\right). \tag{5.19}$$

In these conditions, as long as the two components have the same yield stress, the optimal geometry is defined by relation (5.14), which means that both cylinders must have the same ratio of outer to inner radius, i.e., $\delta_i = \delta_e$. This result is completely general and extends to any multilayer structure consisting of any number of circular cylinders.

As can be concluded from relations (5.12) and (5.13), if the inner cylinder is solid and both members of the shrink-fit assembly consist of the same material treated so as to have the same yield stress, the equivalent stress in the solid cylinder is always less than that in the member heated during the shrink fit process, viz., the hollow circular cylinder. In these conditions, it is not possible to identify a specific optimal geometric configuration. As will be demonstrated below, an optimal design configuration can be identified only if the strength of the material used for the solid inner cylinder is less than that of the hub material.

If the two members of the shrink-fit assembly are both hollow and are made of different materials with different yield stresses σ_s, it will be necessary to use relations (5.12) and (5.13), with k given by relation (5.6). For an optimal design, both members must have the same maximum shear stress, i.e., the tautological relation $\tau_{i,max}(\tau_{e,max}/\tau_{i,max}) = \tau_{e,max}$ must be satisfied. Introducing the ratio $\phi = \tau_{e,max}/\tau_{i,max}$ which appears in this relation and expressing it in view of relations (5.12) and (5.13), we arrive at the following expression whereby the optimum value of the interface radius b can be calculated:

$$b^2 = \frac{c^2(\phi - 1) \pm c\sqrt{c^2(\phi - 1)^2 + 4\phi a^2}}{2\phi}. \tag{5.20}$$

Relation (5.12), setting $\tau_{i,max} = \tau_{e,max}/\phi = \sigma_{s,e}/\phi\sqrt{3}$, yields the following optimum value for radial interference:

$$i_{opt} = \frac{k\sigma_{s,e}}{\phi b\sqrt{3}}\left(b^2 - a^2\right). \tag{5.21}$$

The latter, recalling relation (5.5), then gives:

$$p_{f,opt} = \frac{\sigma_{s,e}}{\phi b^2\sqrt{3}}\left(b^2 - a^2\right). \tag{5.22}$$

Obviously, the value of b given by relation (5.20) must be introduced in relations (5.21) and (5.22). Here again, the foregoing relations can be specialized for the various particular cases. If the inner member is hollow, and the two members are made of the same material, but the yield stresses differ because different heat treatments have been used, we will have the following relation instead of (5.21):

$$i_{opt} = \frac{2\sigma_{s,e} b(c^2 - a^2)}{\sqrt{3}\phi E(c^2 - b^2)}. \tag{5.23}$$

Equaling this latter relation with (5.5) and noting that $E_e = E_i$ and $\nu_e = \nu_i$, we can see that $p_{f,opt}$ is once again given by relation (5.22).

By contrast, if the inner cylinder is solid (a solid shaft), the materials of the two members are different and the yield stresses also differ, relation (5.20) with $a = 0$ will give:

$$b = c\sqrt{(\phi - 1)/\phi}. \tag{5.24}$$

This is valid only for $\phi > 1$. This means that if the solid shaft consists of a material whose yield strength is less than that of the shrunk-on hub, an optimal design can be achieved. In this case, the optimal shrink-fit pressure is obtained from relation (5.22), where $a = 0$ [here, as $p_{f,opt}$ is independent of b, it is not necessary to substitute the expression of b given by (5.24) in relation (5.22)]; we thus obtain the following relation:

$$p_{f,opt} = \frac{\sigma_{s,e}}{\sqrt{3}\phi}. \tag{5.25}$$

For a solid shaft, and if the two members consist of the same material but have different yield strengths, the optimum configuration is obtained with an interference given by relation (5.23) where $a = 0$, substituting relation (5.24) in it. We thus have:

$$i_{opt} = \frac{2\sigma_{s,e}c}{\sqrt{3}E}\sqrt{\frac{\phi - 1}{\phi}}. \tag{5.26}$$

5.4 Optimizing Two Circular Cylinders Assembled with an Interference Fit and Subjected to Internal Pressure

We will now consider a double-walled structure consisting of two coaxial circular cylinders, one of which is shrunk onto the other, with p_f as the initial shrink-fit pressure. In actual service conditions, the structure is under internal pressure p_i. Assuming that stresses are in the linear elastic field, the overall stress state will be given by the superposition of the stresses correlated with pressures p_f and p_i. Figure 5.5 shows the distribution of stresses σ_r and σ_t in the two component cylinders due respectively to the working pressure p_i (these are the same in the monolithic structure and the double-walled structure of identical total thickness, provided that

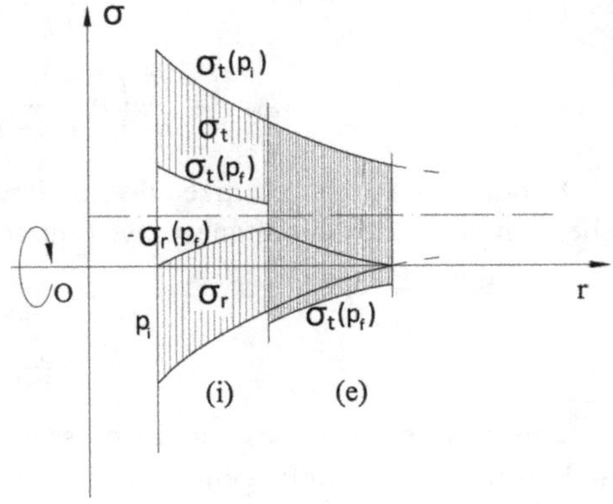

Fig. 5.5 Distribution curves of total stresses σ_r and σ_t resulting from the simultaneous action of shrink-fit pressure p_f and working pressure p_i in a double-walled composite structure

the absolute value of p_f is greater than zero) and to the shrink-fit pressure p_f, where the latter change sign, i.e., are rotated by 180° relative to the axis of the abscissa (these are a mirror image relative to the r axis of those shown in Fig. 5.4b).

In this way, the total stresses given by the superposition of the two effects, i.e., $\sigma_r = \sigma_r(p_i) + \sigma_r(p_f)$ and $\sigma_t = \sigma_t(p_i) + \sigma_t(p_f)$, are represented radius by radius by the vertical segments intercepted between the homologous curves for the stresses due to p_i and p_f. The vertically shaded areas in Fig. 5.5 thus provide a graphical representation of the stress state.

Given these premises, we will now consider the two cylinders assembled with an interference fit examined in the preceding section, assuming that the inner cylinder is under internal pressure p_i. The cases involving a solid inner cylinder presented above thus do not apply here. As regards the maximum shear stress strength theory (and the maximum distortion energy strength theory, if we assume that $\sigma_z = A$), four cases will be examined below.

5.4.1 Case 1 (Same Material and Same Yield Strength)

In the first case, both assembled components consist of the same material (and their Young's moduli and Poisson's ratios are thus identical) and their yield strength is the same. Using the method of superposition (as usual, it is assumed that the material exhibits linear elastic behavior up to yield, i.e., that the proportional limit, the elastic limit and the yield point coincide), and recalling relations (4.39) and (4.46), we arrive at the following relations which express the maximum shear stress in the inner component and outer component respectively (it should be noted that the double-walled composite structure's behavior as regards internal pressure p_i is that of a single-walled structure, while as regards contact pressure p_f, its behavior reflects its actual two-component construction):

$$\tau_{i,max} = p_i \frac{c^2}{c^2 - a^2} - p_f \frac{b^2}{b^2 - a^2} \tag{5.27}$$

and

$$\tau_{e,max} = \frac{c^2}{b^2} \left(p_i \frac{a^2}{c^2 - a^2} + p_f \frac{b^2}{c^2 - b^2} \right). \tag{5.28}$$

Here again, for an optimized design, both components must reach the maximum shear stress values simultaneously. Equaling relations (5.27) and (5.28) yields:

$$p_f = p_i \frac{\frac{c^2 (b^2 - a^2)}{b^2 (c^2 - a^2)}}{\left(\frac{c^2}{c^2 - b^2} + \frac{b^2}{b^2 - a^2} \right)}. \tag{5.29}$$

This relation expresses contact pressure p_f as a function of working pressure p_i and shrink-fit assembly geometry. Substituting relation (5.29) in either relation

(5.27) or (5.28) and noting that with the maximum distortion energy strength theory, we have $\tau_{max} = \sigma_s/\sqrt{3}$, the following relation is obtained:

$$p_{i,max} = \frac{2\sigma_s}{\sqrt{3}}\left[1 - \frac{1}{2}\left(\frac{b^2}{c^2} + \frac{a^2}{b^2}\right)\right]. \tag{5.30}$$

This relation gives the maximum allowable internal pressure that the composite structure in question can withstand in the elastic range. Equaling the derivative of relation (5.30) with respect to b to zero, we find that the optimal value of b is once again given by relation (5.14). Substituting this optimal value of b in relation (5.30) then gives:

$$p_{i,opt} = \frac{2\sigma_s}{\sqrt{3}}\left(1 - \frac{a}{c}\right). \tag{5.31}$$

Knowing the optimal value for the radius at the shrink-fit interface, however, is not sufficient to ensure that the structure is correctly designed. It is also necessary to calculate the optimal interference in order to select the most appropriate machining processes for assembling the two cylinders. By introducing relation (5.14) in relation (5.8), we obtain:

$$p_{f,opt} = \frac{Ei(c-a)}{2\sqrt{ac}(c+a)}. \tag{5.32}$$

Relations (5.29), (5.31) and (5.32) then yield:

$$p_{f,opt} = \frac{\sigma_s(c-a)^2}{c\sqrt{3}(c+a)} = \frac{\sigma_s(\delta-1)^2}{\delta\sqrt{3}(\delta+1)}. \tag{5.33}$$

and

$$i_{opt} = \frac{2\sigma_s}{E\sqrt{3}}\sqrt{ac}\left(\frac{c-a}{c}\right) = \frac{2\sigma_s a}{\sqrt{\delta}}\left(\frac{\delta-1}{E\sqrt{3}}\right), \tag{5.34}$$

where $\delta = c/a$.

Figure 5.6 shows the elastic stress distribution in a composite structure consisting of two coaxial circular cylinders (the inner cylinder is designated with i, and the outer cylinder with e) assembled with an interference fit and under internal pressure p_i, in the general case where the structure is not optimized. Detail (a) shows the stresses resulting from the shrink fit alone, i.e., from shrink-fit pressure p_f, with $p_i = 0$ (these stresses are indicated with one superscript). Detail (b) shows stresses due to internal pressure p_i alone (indicated with two superscripts; it should be noted that the structure behaves as if it consisted of a single component, providing that interference i is greater than or, at minimum, equal to zero). Detail (c) shows the total stresses (indicated without superscripts), or in other words those obtained from the superposition of the effects of the shrink-fit

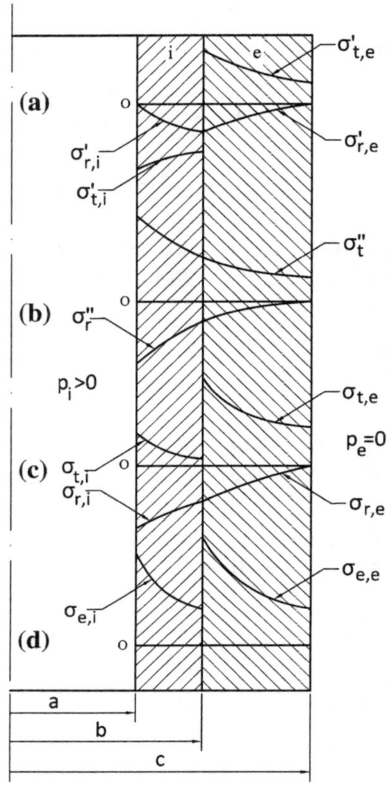

Fig. 5.6 Stress distributions in composite structure consisting of two circular cylinders assembled with an interference fit: **a** stresses resulting from shrink fit; **b** stresses resulting from internal pressure; **c** stresses resulting from the superposition of the two effects; **d** equivalent stresses in the two cylinders

pressure and the internal pressure. Detail (d) shows the equivalent stresses σ_e calculated with any of the strength theories.

It can be concluded from this figure that, for a specified geometry (i.e., for a given ratio $\delta = c/a$), and for an assigned external load given by the internal pressure p_i, the initial shrink-fit interference resulting from shrink-fit pressure p_f reduces the hoop stress in the inner cylinder even in a composite structure which is not optimized, but increases this stress in the outer cylinder. As the hoop stress is that which makes the greatest contribution to the equivalent stress calculated on the basis of any strength theory, the favorable influence of the initial shrink-fit interference is as clear as it is important.

Obviously, if the composite structure had been optimized using the procedure described above, the two maximum values for equivalent stress would have been the same. The relations expressing the optimized magnitudes in this case are named Gadolin's conditions after the Russian scientist who first derived them (see Feodosyev 1973).

5.4.2 Case 2 (Same Material and Different Yield Strength)

In the second case, both components again consist of the same material, but their yield strengths differ. Here, instead of relation (5.29), the following optimization relation is obtained from relations (5.27) and (5.28), recalling that $\tau_{e,max}/\tau_{i,max} = \phi$ and introducing the ratio $\delta = c/a$:

$$\left(p_i\frac{\delta^2}{\delta^2-1} - p_f\frac{\delta_i^2}{\delta_i^2-1}\right)\phi = \frac{\delta^2}{\delta_i^2}\left(p_i\frac{1}{\delta^2-1} + p_f\frac{\delta_i^2}{\delta^2-\delta_i^2}\right). \tag{5.35}$$

Solving this relation for shrink-fit pressure p_f (the latter will be a function of p_i, of geometry and of ϕ) and substituting the expression thus obtained in either relation (5.27) or (5.28) and once again noting that with the maximum distortion energy theory, we have $\tau_{max} = \sigma_s/\sqrt{3}$, we obtain the following expression for the maximum internal pressure that the structure can withstand in the elastic range:

$$p_{i,max} = \frac{\sigma_{s,i}}{\sqrt{3}}\left(1+\phi-\frac{1}{\delta_i^2}-\frac{\delta_i^2}{\delta^2}\phi\right). \tag{5.36}$$

Equaling the derivative of relation (5.36) with respect to δ_i to zero, we find that the optimal value is defined by one of the two equivalent relations given below:

$$\delta_i^2 = \frac{\delta}{\sqrt{\phi}} \quad \text{or} \quad b^2 = \frac{ac}{\sqrt{\phi}}. \tag{5.37}$$

Substituting the first relation (5.37) in relation (5.36) gives the following expression for the optimal internal pressure at the yield limit:

$$p_{i,opt} = \frac{2\sigma_{s,i}}{\sqrt{3}}\left(\frac{1+\phi}{2}-\frac{\sqrt{\phi}}{\delta}\right). \tag{5.38}$$

Relations (5.38), (5.35) and (5.5) then yield the following relations which express the optimal shrink-fit pressure and the corresponding radial interference respectively:

$$p_{f,opt} = \frac{\sigma_{s,i}\phi}{\delta\sqrt{3}(\delta^2-1)}\left(\delta_i^2\phi-1\right)^2\left(\delta-\sqrt{\phi}\right) \tag{5.39}$$

and

$$i_{opt} = \frac{2\sigma_{s,i}b\phi}{E\delta^2\sqrt{3}}\left[\frac{1+\delta(\delta\phi-2\sqrt{\phi})}{\sqrt{\phi}-1}\right]. \tag{5.40}$$

5.4.3 Case 3 (Different Material and Same Yield Strength)

In the third case, where the components consist of different materials but have the same yield strength, the optimal design is once again defined by relation (5.14), while we have the following relations in place of (5.36), (5.39) and (5.40):

$$p_{i,max} = \frac{2\sigma_s}{\sqrt{3}}\left(1 - \frac{a}{c}\right) = \frac{2\sigma_s}{\sqrt{3}}\left(\frac{\delta - 1}{\delta}\right) \tag{5.41}$$

$$p_{f,opt} = \frac{\sigma_s(c-a)^2}{c\sqrt{3}(c+a)} = \frac{\sigma_s(\delta-1)^2}{\delta\sqrt{3}(\delta+1)} \tag{5.42}$$

$$i_{opt} = \frac{\sigma_s}{c\sqrt{3}}\sqrt{ac}\left[\frac{(c-a)^2}{c+a}\right]\left[\frac{1}{E_e}\left(\frac{c+a}{c-a} + \nu_e\right) + \frac{1}{E_i}\left(\frac{c+a}{c-a} - \nu_i\right)\right]. \tag{5.43}$$

5.4.4 Case 4 (Different Material and Different Yield Strength)

In the fourth case, with components made of different materials with different yield strengths, the optimal design is defined by relations (5.37), while we have the following relations in place of (5.36), (5.39) and (5.40):

$$p_{i,opt} = \frac{2\sigma_{s,i}}{\sqrt{3}}\left(\frac{1+\phi}{2} - \frac{\sqrt{\phi}}{\delta}\right). \tag{5.44}$$

$$p_{f,opt} = \frac{\sigma_{s,i}}{\delta\sqrt{3}(\delta^2-1)}\left(\delta_i^2\phi - 1\right)^2\left(\delta - \sqrt{\phi}\right) \tag{5.45}$$

$$i_{opt} = \frac{\sigma_{s,i}bk}{\delta\sqrt{3}(\delta^2-1)}\left(\delta_i^2\phi - 1\right)^2\left(\delta - \sqrt{\phi}\right). \tag{5.46}$$

5.4.5 Concluding Remarks

In conclusion, it should first be noted that all of the foregoing relations have been extensively confirmed experimentally, as demonstrated by the results given in the literature for systematic test campaigns on two types of structure: the single-walled structure (monobloc-type construction) and the double-walled structure consisting of two coaxial cylindrical bodies, both of annular cross-section, assembled with a shrink fit. Both types of structure were subjected to an internal pressure sufficient to bring them to the onset of the plastic state at the inner radius. The differences between the numerical values calculated theoretically and determined experimentally for each magnitude, moreover, can be regarded as insignificant from the standpoint of practical applications, as they are within measurements errors.

From the designer's perspective, it is interesting to compare the elastic responses up to the yield point (once again, it should be recalled that the proportional limit, the elastic limit and the yield point are here assumed to coincide) of these two types of structure, with δ, and thus geometry, material and the latter's mechanical strength properties remaining the same. As a comparison parameter, we will take the ratio $p_{i,i}/\sigma_s$ of the internal pressure that, on the basis of the maximum distortion energy theories, leads to the onset of the plastic state at the structure's inner radius [calculated with relation (7.18) given in Chap. 7 for the single-walled structure, and relation (5.31) for the composite structure consisting of two cylinders] to the yield stress σ_s of the material used for both structures, expressed as a function of δ (Fig. 5.7). The curve for the monobloc-type structure shows that, for $\delta > (3.5–4.0)$, there is a relatively modest increase in the pressure at which the fibers at the inner radius begin to yield. By contrast, the curve for the composite shrink-fit structure shows that there are significant advantages over the single-walled structure for any given δ. For this composite structure, in fact, ratio $p_{i,i}/\sigma_s$ is significantly higher for an assigned value of δ and tends to be twice as high for $\delta \rightarrow \infty$.

Another useful design comparison between the two structures is shown in Fig. 5.8, where ratio δ for the monobloc-type structure is plotted versus that of the two-walled composite structure, with the same internal pressure bringing the inner radius to the onset of the plastic state. The relation that links the two values of δ is obtained by equaling relation (7.18) as mentioned earlier to relation (5.38) written for $\phi = 1$ (as is immediately apparent, relation (5.38) is the generalization of relation (5.31) to the case where the two cylinders of the composite structure are made of materials with different mechanical strength properties).

By way of example, Fig. 5.8 shows that for any given level of performance in terms of internal pressure $p_{i,i}$ capable of bringing the inner radius to the onset of

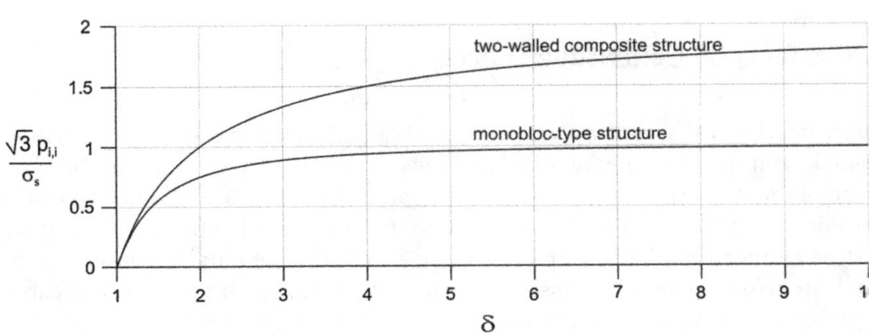

Fig. 5.7 Ratio $p_{i,i}/\sigma_s$ versus δ for the monobloc-type structure and for the composite structure consisting of two cylinders with optimized geometry assembled with a shrink fit

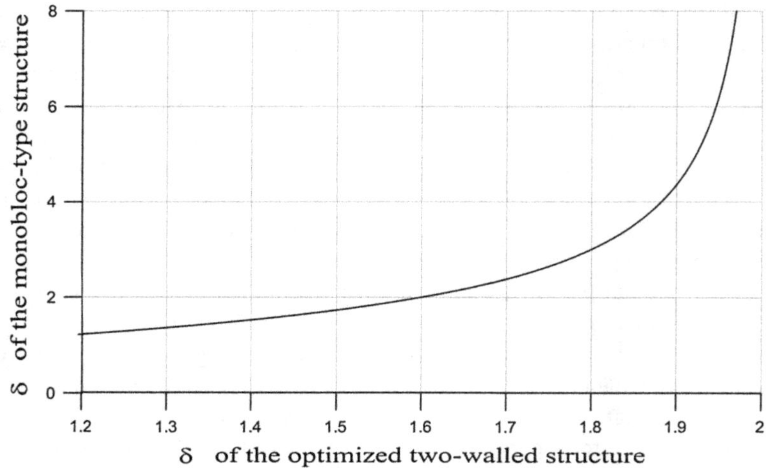

Fig. 5.8 Equivalence of ratios δ for the monobloc-type structure and the composite structure consisting of two cylinders assembled with a shrink fit

the plastic state, a monobloc-type structure such as a vessel consisting of a single cylinder with $\delta = 6$ can be replaced with a composite structure consisting of two cylinders assembled with a shrink fit, with $\delta = 1.95$.

With the same procedure, curves similar to those shown in Fig. 5.7 can be constructed that apply to the case where the two cylinders of the composite structure consist of different materials or of the same material but with different tensile properties, i.e., because different heat treatments have been used. In this way, the internal pressure that leads to the onset of yield at the inner radius can be increased by one-third, by raising the outer cylinder's yield strength $\sigma_{s,e}$ from $\sigma_{s,i}/2$ to $1.25\sigma_{s,i}$. Once again, these curves are obtained by equaling relation (7.18) with relation (5.38), where, however, $\phi \neq 1$.

5.5 Effects of Centrifugal Force

Many interference shrink fits are often used at high rotational speeds. Centrifugal loading can produce radial displacements that tend to loosen the shrink-fit assembly and, if loading is very intense, cause it to loosen. This effect must be considered when calculating the interference to be achieved in the shrink fit. It will first be assumed that both members of the shrink-fit assembly shown in Fig. 5.1 can be regarded as disks of constant thickness, for which a plane stress state can be hypothesized. The radial displacement $u = u(r)$ at the generic radius of a disk of inner radius r_i and outer radius r_e, subjected only to centrifugal load, is given by the relation:

$$u(r) = \frac{(3+\nu)\gamma\omega^2 r_e^2 r}{8E}\left[\left(1+\frac{r_i^2}{r_e^2}\right)(1-\nu)+(1+\nu)\frac{r_i^2}{r^2}-\frac{r^2(1-\nu^2)}{r_e^2(3+\nu)}\right] \quad (5.47)$$

where ω and γ are the angular velocity and the material's density respectively. Where both disk-shaped members of the shrink-fit assembly are of annular cross-section (Fig. 5.1) and consist of the same material, the radial displacements at the interface of radius b considered as part of the inner cylinder and of the outer cylinder are given by the following relations:

$$u_{b,i} = \frac{b\gamma\omega^2}{4E}\left[b^2(1-\nu)+a^2(3+\nu)\right] \quad (5.48)$$

and

$$u_{b,e} = \frac{b\gamma\omega^2}{4E}\left[b^2(1-\nu)+c^2(3+\nu)\right]. \quad (5.49)$$

The reduction in radial interference $\Delta i = (u_{b,e} - u_{b,i})$ is thus:

$$\Delta i = \frac{b\gamma\omega^2 c^2}{4E}(3+\nu)\left(1-\frac{a^2}{c^2}\right) \quad (5.50)$$

here $c\omega$ is the linear velocity (or peripheral velocity) at the outer radius of the outer member.

If the geometry is still that of the shrink-fit assembly shown in Fig. 5.1, but the materials of the two disk-shaped members have different properties ($E_i \neq E_e$; $\nu_i \neq \nu_e; \gamma_i \neq \gamma_e$), instead of relation (5.50) we will have:

$$\Delta i = \frac{\omega^2 c^2 b}{4}\left\{\frac{\gamma_e}{E_e}\left[(3+\nu_e)+(1-\nu_e)\frac{b^2}{c^2}\right]-\frac{\gamma_i}{E_i}\left[(3+\nu_i)\frac{a^2}{c^2}+(1-\nu_i)\frac{b^2}{c^2}\right]\right\}. \quad (5.51)$$

In the particular case where the inner member is solid (the classic shrink-fit shaft/hub assembly), we will have $a = 0$. Consequently, relations (5.50) and (5.51) respectively become:

$$\Delta i = \frac{b\gamma\omega^2 c^2}{4E}(3+\nu) \quad (5.52)$$

and

$$\Delta i = \frac{\omega^2 c^2 b}{4}\left\{\frac{\gamma_e}{E_e}\left[(3+\nu_e)+(1-\nu_e)\frac{b^2}{c^2}\right]-\frac{\gamma_i}{E_i}(1-\nu_i)\frac{b^2}{c^2}\right\}. \quad (5.53)$$

From the standpoint of design optimization, it is obvious that the centrifugal force makes its influence felt. Consequently, the purely static calculation must be revised to check that optimal conditions can be also maintained at the operating speed. In other words, it is necessary to check that the variation in shrink-fit pressure, and then in contact forces resulting from centrifugal load is compatible with the requirements for an optimal shrink fit discussed above.

If the inner member consists of a material with a lower elastic modulus than that of the outer member, the shrink-fit pressure will increase as a result of centrifugal load. It is thus possible to assemble the members initially using a shrink fit with less interference, so that optimal conditions are reached at the rotational speed used in actual service. Conversely, if the material of the outer member has a lower elastic modulus, the two members will tend to separate as angular velocity increases, and the shrink-fit pressure will drop. Optimizing the design for the operating speed will thus be impossible, as the increase in the initial shrink-fit interference required to compensate for loosening at operating speed could be high enough to lead to plastic deformation in the two members, resulting in redistributions of the stresses associated with significant departures from the optimal conditions.

Obviously, when the total displacement of the interface regarded as part of the inner member (this displacement is the sum of the displacements defined by relations (5.3) and (5.48), due to shrink-fit pressure p_f and angular velocity ω respectively), equals the total displacement of the same interface regarded as part of the outer member (the sum of the displacements defined by relations (5.4) and (5.49), correlated with shrink-fit pressure p_f and angular velocity ω respectively), the shrink-fit connection will loosen.

Conversely, where the axial length of the two members of the shrink-fit assembly shown in Fig. 5.1 is such that they must be regarded as circular cylinders of finite length with free ends, for which the strain state is known to be plane, the relation given below must be used instead of the basic relation (5.47), as it takes the triaxial stress state into account (in this connection, see: Vullo and Vivio 2013). In this case, we will have the following relations instead of relations (5.47)–(5.53) respectively::

$$u(r) = \frac{(3-2\nu)\gamma\omega^2 r_e^2 r}{8E(1-\nu)}\left[\left(1+\frac{r_i^2}{r_e^2}\right)\frac{(3+5\nu)}{(3-2\nu)} + (1+\nu)\frac{r_i^2}{r^2} - \frac{r^2(1+\nu)(1-2\nu)}{r_e^2(3-2\nu)}\right] \tag{5.54}$$

$$u_{b,i} = \frac{b\gamma\omega^2}{4E(1-\nu)}\left[b^2\left(1-2\nu+\nu^2\right) + a^2\left(3-2\nu-\nu^2\right)\right] \tag{5.55}$$

$$u_{b,e} = \frac{b\gamma\omega^2}{4E(1-\nu)}\left[b^2\left(1-2\nu+\nu^2\right) + c^2\left(3-2\nu-\nu^2\right)\right] \tag{5.56}$$

$$\Delta i = \frac{b\gamma\omega^2 c^2}{4E(1-\nu)}\left(3-2\nu-\nu^2\right)\left(1-\frac{a^2}{c^2}\right) \tag{5.57}$$

$$\Delta i = \frac{\omega^2 c^2 b}{4(1-v)} \left\{ \frac{\gamma_e}{E_e} \left[\left(3 - 2v_e - v_e^2\right) + \left(1 - 2v_e + v_e^2\right) \frac{b^2}{c^2} \right] - \frac{\gamma_i}{E_i} \left[\left(3 - 2v_i - v_i^2\right) \frac{a^2}{c^2} + \left(1 - 2v_i + v_i^2\right) \frac{b^2}{c^2} \right] \right\} \tag{5.58}$$

$$\Delta i = \frac{b\gamma\omega^2 c^2}{4E(1-v)} \left(3 - 2v - v^2\right) \tag{5.59}$$

$$\Delta i = \frac{\omega^2 c^2 b}{4(1-v)} \left\{ \frac{\gamma_e}{E_e} \left[\left(3 - 2v_e - v_e^2\right) + \left(1 - 2v_e + v_e^2\right) \frac{b^2}{c^2} \right] - \frac{\gamma_i}{E_i} \left(1 - 2v_i + v_i^2\right) \frac{b^2}{c^2} \right\}. \tag{5.60}$$

5.6 Shrink-Fit Shaft/Hub Assemblies

The general relation (5.5) and the more specific relation (5.7) correlate radial interference i to shrink-fit pressure p_f. For the shrink-fit shaft/hub assemblies of interest to us here, it is clear that this shrink-fit pressure must be commensurate with the torque and/or the axial load to be transmitted. We will consider a typical shrink-fit shaft/hub assembly, assuming that the materials used for both components have identical elastic properties. As we are dealing with a solid shaft, we have $a = 0$, so that relations (5.7) and (5.8) become, in order:

$$i = \frac{2p_f bc^2}{E(c^2 - b^2)} \tag{5.61}$$

and

$$p_f = \frac{Ei(c^2 - b^2)}{2bc^2}. \tag{5.62}$$

If l is the length of the hub and f_t is the friction coefficient in torsion at shaft/hub contact interface, the transmittable torque M_t at the slip limit is the product of the active surface $2\pi bl$ by the tangential force per unit of surface area $f_t p_f$ and by radius b. Consequently:

$$M_t = (2\pi bl)\left(f_t p_f\right) b = 2\pi f_t p_f b^2 l. \tag{5.63}$$

For the typical design problem in which the torque and shaft dimensions are assigned, relation (5.63) can be used to determine shrink-fit pressure p_f after assuming a low friction coefficient in torsion by introducing an appropriate safety factor. Subsequently, the required radial interference i is calculated using relation (5.61) or (5.7), depending on whether the shaft is solid or hollow. The diametral interference $2i$ to be indicated on the manufacturing drawings of the two members

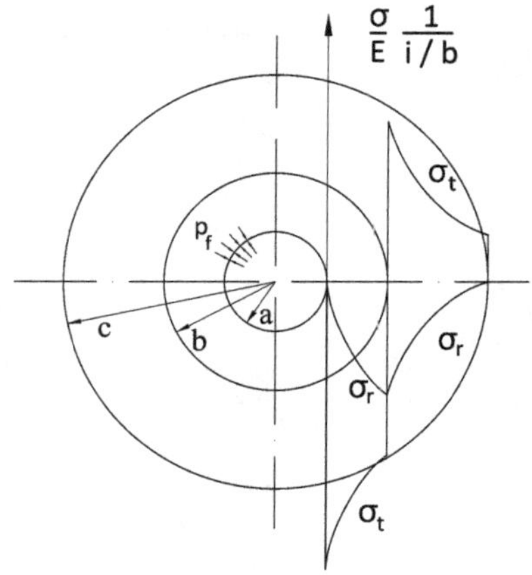

Fig. 5.9 Distribution curves of hoop stress σ_t and radial stress σ_r in dimensionless form in a shrink-fit shaft/hub assembly

of the shrink-fit assembly is then calculated, given that, in reality, only the diameters can be measured and thus controlled. As a further step once the shrink-fit pressure is known, the stress and strain states of the two components of the shaft/hub assembly are calculated. A typical distribution of the hoop and radial stresses in these two components is shown in Fig. 5.9, where the stresses are represented in dimensionless form by means of parameter $(\sigma/E)/(i/b)$.

The axial load that the shrink-fit assembly defined above can withstand before loosening is given by the following relation:

$$F_a = 2\pi f p_f l b, \tag{5.64}$$

where f is the coefficient of the friction.

5.7 Multilayer Circular Cylindrical Structures

Applications involving rather demanding technological requirements employ multilayer structures consisting of three or more concentric circular cylinders shrunk together with a specified radial interference.

In the most general case as illustrated in Fig. 5.10, the multilayer structure consists of n concentric circular cylinders made of different materials. The figure shows the geometric magnitudes, the elastic properties of each material, and the pressures acting at the radii that delimit the component cylinders, i.e., at the shrink-fit interfaces between them. For the generic mth layer, the diametral ratio is $\delta_m = r_m/r_{m-1}$.

In current practice, the outer layers are assembled first, and the resulting subsystem is then shrunk onto the innermost cylinder, or liner. To analyze this type of structure, the data and knowledge discussed in the preceding sections are sufficient. Thus, the stress state in the volume element in the mth layer between inner

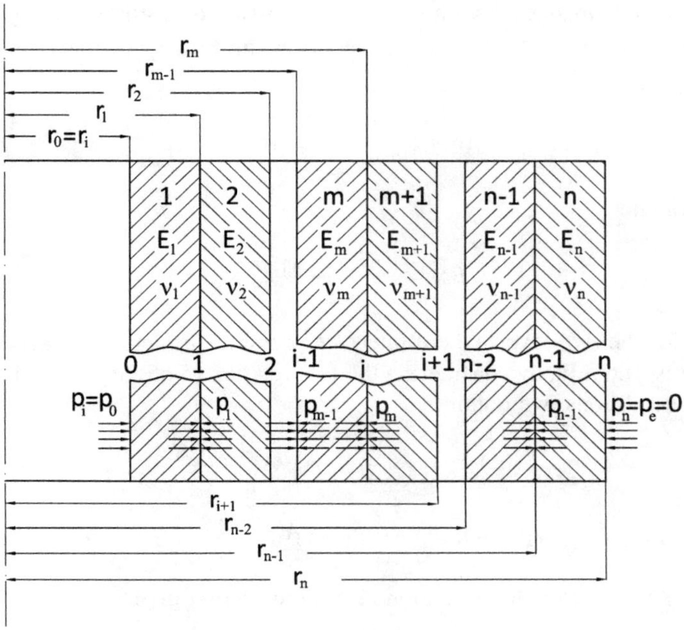

Fig. 5.10 Multilayer structure: geometry, elastic properties of materials and pressures at the various radii

radius $r_{m\text{-}1}$ and outer radius r_m, is, if triaxial with $\sigma_z = A$, defined by the following set of three relations:

$$\begin{aligned}
\left|\sigma_{r,m}\right| &= A_m - \frac{B_m}{r^2} \\
\left|\sigma_{t,m}\right| &= A_m + \frac{B_m}{r^2} \\
\left|\sigma_{z,m}\right| &= A_m
\end{aligned} \tag{5.65}$$

where

$$A_m = \frac{p_{m-1} r_{m-1}^2 - p_m r_m^2}{r_m^2 - r_{m-1}^2} \quad \text{and} \quad B = \frac{(p_{m-i} - p_m) r_{m-i}^2 r_m^2}{r_m^2 - r_{m-1}^2}. \tag{5.66}$$

Local values of radial and hoop stresses at the inner radius r_{m-1} of the mth layer are given by:

$$\begin{aligned}
\left|(\sigma_r)_{r=r_{m-1}}\right| &= A_m - \frac{B_m}{r_{m-1}^2} \\
\left|(\sigma_t)_{r=r_{m-1}}\right| &= A_m + \frac{B_m}{r_{m-1}^2}.
\end{aligned} \tag{5.67}$$

For a triaxial stress state with $\sigma_z = A$, the maximum distortion energy strength theory gives $(\sigma_t - \sigma_r) = (2/\sqrt{3})\sigma_e$. Applying this theory to the two foregoing relations gives:

$$\left|(\sigma_t)_{r=r_{m-1}}\right| - \left|(\sigma_r)_{r=r_{m-1}}\right| = \frac{2}{\sqrt{3}}\left|(\sigma_e)_{r=r_{m-1}}\right| = 2\frac{B_m}{r_{m-1}^2}. \tag{5.68}$$

We thus obtain

$$B_m = \frac{\left|(\sigma_e)_{r=r_{m-1}}\right|}{\sqrt{3}} r_{m-1}^2. \tag{5.69}$$

Again for the mth layer, where interface pressures p_m and p_{m-1} act at radii r_m and r_{m-1} respectively, by considering the first relation (5.65) and imposing the two boundary conditions, we obtain:

$$\begin{aligned}(\sigma_r)_{r=r_{m-1}} &= -p_{m-1} = A_m - \frac{B_m}{r_{m-1}^2}\\ (\sigma_r)_{r=r_m} &= -p_m = A_m - \frac{B_m}{r_m^2}.\end{aligned} \tag{5.70}$$

Subtracting the first of these relations from the second gives:

$$p_{m-1} - p_m = B_m\left(\frac{1}{r_{m-1}^2} - \frac{1}{r_m^2}\right). \tag{5.71}$$

Substituting relation (5.69) in the latter and introducing the diametral ratio δ_m defined above yields the relation

$$p_{m-1} - p_m = \frac{\left|(\sigma_e)_{r=r_{m-1}}\right|}{\sqrt{3}} \frac{\delta_m^2 - 1}{\delta_m^2} \tag{5.72}$$

which, with the notation

$$\frac{\delta_m^2 - 1}{\sqrt{3}\delta_m^2} = k_m, \tag{5.73}$$

then becomes

$$p_{m-1} - p_m = k_m\left|(\sigma_e)_{r=r_{m-1}}\right|. \tag{5.74}$$

To optimize the multilayer structure examined here, the same equivalent stress $\sigma_{e,i}$ at the inner surface of the component cylinders must be the same for all layers. The treatment presented below applies only to the case in which all component layers consist of the same material with the same tensile properties, i.e., the case where $E_1 = E_2 = \cdots = E_n$, $\nu_1 = \nu_2 = \cdots = \nu_n$ and $\sigma_{s,1} = \sigma_{s,2} = \cdots = \sigma_{s,n}$. Here, the optimization condition is as follows:

$$\left|(\sigma_{e,i})_m\right| = \left|(\sigma_{e,i})_{r=r_m}\right| = \text{const} = \sigma_{e,i}. \tag{5.75}$$

Applying this optimization relation to (5.74) and noting that $m = 0, 1, 2, \ldots, n$, yields the following system of equations

$$\begin{aligned} & p_i - p_1 = p_0 - p_1 = k_1 \sigma_{e,i} \\ & p_1 - p_2 = k_2 \sigma_{e,i} \\ & \ldots\ldots\ldots\ldots\ldots\ldots \\ & p_{m-1} - p_m = k_m \sigma_{e,i} \\ & \ldots\ldots\ldots\ldots\ldots\ldots \\ & p_{n-1} - p_n = p_{n-1} - p_e = k_n \sigma_{e,i}, \end{aligned} \tag{5.76}$$

from which, adding member by member, we obtain the relation

$$p_i - p_e = \sigma_{e,i}(k_1 + k_2 + \cdots + k_n) = \sigma_{e,i} \sum_{m=1}^{n} k_m \tag{5.77}$$

and thus

$$\sigma_{e,i} = \frac{p_i - p_e}{\sum_{m=1}^{n} k_m}. \tag{5.78}$$

For optimum structure design on the basis of relation (5.75), the equivalent stress $\sigma_{e,i}$ at the inner radius of any component layer must be minimum for any given pressure difference $(p_i - p_e)$. Consequently, the summation appearing in the denominator in relation (5.78) must be maximum. This maximum can be found using the Lagrange multipliers method, which makes it possible to identify the constrained maxima and minima of a function.

To apply this method to the case of interest to us here, we designate the summation function to be maximized as $f(x)$, identify a constraint condition $\phi(x)$ and introduce the new function $F(x)$ defined by the relation:

$$F(x) = f(x) + \lambda \phi(x) \tag{5.79}$$

where λ is a parameter other than zero and independent of variable x. To determine the maximum of $f(x)$, we must have:

$$\frac{\partial F(x)}{\partial x} = \frac{\partial f(x)}{\partial x} + \lambda \frac{\partial \phi(x)}{\partial x} = 0. \tag{5.80}$$

In the case examined here, recalling relation (5.73), we have

$$f(x) = \sum_{m=1}^{n} k_m = \sum_{m=1}^{n} \frac{x_m - 1}{\sqrt{3} x_m} \tag{5.81}$$

where, for convenience, the notation $x_m = \delta_m^2$ was introduced. The constraint equation $\phi(x)$ can be inferred from simple geometric considerations regarding the ratios of the radii delimiting the structure's individual layers. In this connection (see Fig. 5.10), we must have:

$$\prod_{m=1}^{n} \left(\frac{r_m}{r_{m-1}} \right)^2 = \prod_{m=1}^{n} \delta_m^2 = \prod_{m=1}^{n} x_m = \left(\frac{r_e}{r_i} \right)^2 = \delta^2 = K = \text{const.} \tag{5.82}$$

Consequently, the constraint equation can be written in the following form:

$$\phi(x) = \prod_{m=1}^{n} x_m - K = 0. \tag{5.83}$$

Deriving relations (5.81) and (5.83) with respect to x_m, we obtain:

$$\begin{aligned} \frac{\partial f(x)}{\partial x_m} &= \frac{1}{\sqrt{3}x_m^2} \\ \frac{\partial \phi(x)}{\partial x_m} &= \frac{\prod_{m=1}^{n} x_m}{x_m} = \frac{\phi(x) + K}{x_m}. \end{aligned} \tag{5.84}$$

By substituting the latter in relation (5.80) and considering constraint equation (5.83), we obtain the following system:

$$\begin{aligned} 1 + \lambda x_m \sqrt{3}[\phi(x) + K] &= 0 \\ x_1 x_2 x_3 \ldots x_n - K &= 0, \end{aligned} \tag{5.85}$$

which is satisfied for:

$$\begin{aligned} x_1 &= x_2 = x_3 = \cdots = x_n = x = \text{const} \\ &= \delta_1^2 = \delta_2^2 = \delta_3^2 = \cdots = \delta_n^2 = \delta^2 = \text{const} \\ &= k_1 = k_2 = k_3 = \cdots = k_n = k = \text{const}. \end{aligned} \tag{5.86}$$

We can thus conclude that the optimal configuration for a multilayer structure, with the restrictive assumptions described above, is that in which all component layers have the same diametral ratio δ. It is thus possible to generalize the result obtained with a double-walled structure whose components consist of the same material having the same tensile properties (case 1 discussed in Sect. 5.4.1), represented by relation (5.14).

Recalling relations (5.86) together with notation $x_m = \delta_m^2$, relation (5.78) becomes:

$$\sigma_{e,i} = \frac{p_i - p_e}{nk} = \frac{\delta^2\sqrt{3}(p_i - p_e)}{n(\delta^2 - 1)}. \tag{5.87}$$

Establishing that $\sigma_{e,i} \leq \sigma_a$, this is to be considered a response analysis relation. If we set $\sigma_{e,i} = \sigma_a$ in relation (5.87) and solving the equality thus obtained for δ, we obtain the following design analysis relation:

$$\delta = \sqrt{\frac{\sigma_a}{\sigma_a - \frac{\sqrt{3}(p_i - p_e)}{n}}}. \tag{5.88}$$

We can conclude from this relation that there is also a critical value of the internal pressure in a multilayer structure given by the relation:

$$p_{cr} = \frac{n\sigma_a}{\sqrt{3}} + p_e. \tag{5.89}$$

At this critical pressure, the thickness becomes infinite.

If the maximum shear stress strength theory is used, we obtain the following relations in place of (5.87)–(5.89):

$$\sigma_{e,i} = \frac{2\delta^2(p_i - p_e)}{n(\delta^2 - 1)} \tag{5.90}$$

$$\delta = \sqrt{\frac{\sigma_a}{\sigma_a - \frac{2(p_i - p_e)}{n}}} \tag{5.91}$$

$$p_{cr} = \frac{n\sigma_a}{2} + p_e. \tag{5.92}$$

For a multilayer structure under internal pressure only, Fig. 5.11 shows the ratios $p_i/\sigma_{e,i}$ obtained with the maximum distortion energy and τ_{max} strength theories versus δ (results obtained with n varying from 1 to 8 are shown). The curves indicate that, all other conditions remaining equal, the structure's load bearing capacity increases significantly along with the number of layers, as well as the greater severity associated with the τ_{max} strength theory as compared to the maximum distortion energy strength theory.

For a further comparison of the results obtained using these two strength theories, it is advisable to introduce parameters $\Delta\sigma = \sigma_{e,i} - \sigma_{e,e}$ and $R = \sigma_{e,i}/\sigma_{e,e}$ as was done for the single-walled cylinder. The local values of radial and hoop stresses at outer radius r_m of the mth layer are given by the relations:

$$\begin{aligned} \left|(\sigma_r)_{r=r_m}\right| &= A_m - \frac{B_m}{r_m^2} \\ \left|(\sigma_t)_{r=r_m}\right| &= A_m + \frac{B_m}{r_m^2}. \end{aligned} \tag{5.93}$$

As for the inner radius of the same layer, the maximum distortion energy strength theory gives the following for a triaxial stress state ($\sigma_z = A$):

$$\left|(\sigma_t)_{r=r_m}\right| - \left|(\sigma_r)_{r=r_m}\right| = \frac{2}{\sqrt{3}}\left|(\sigma_e)_{r=r_m}\right| = 2\frac{B_m}{r_m^2}. \tag{5.94}$$

Consequently, dividing relation (5.69) by the latter member by member gives:

$$\frac{\left|(\sigma_t)_{r=r_{m-1}}\right|}{\left|(\sigma_t)_{r=r_m}\right|} = \frac{(\sigma_{e,i})_m}{(\sigma_{e,e})_m} = \frac{r_m^2}{r_{m-1}^2} = \delta_m^2. \tag{5.95}$$

However, relations (5.75) and (5.86) show that both $\sigma_{e,i}$ and δ are constant. As $\sigma_{e,e}$ must thus also be constant, we have:

$$R = \frac{\sigma_{e,i}}{\sigma_{e,e}} = \delta^2. \tag{5.96}$$

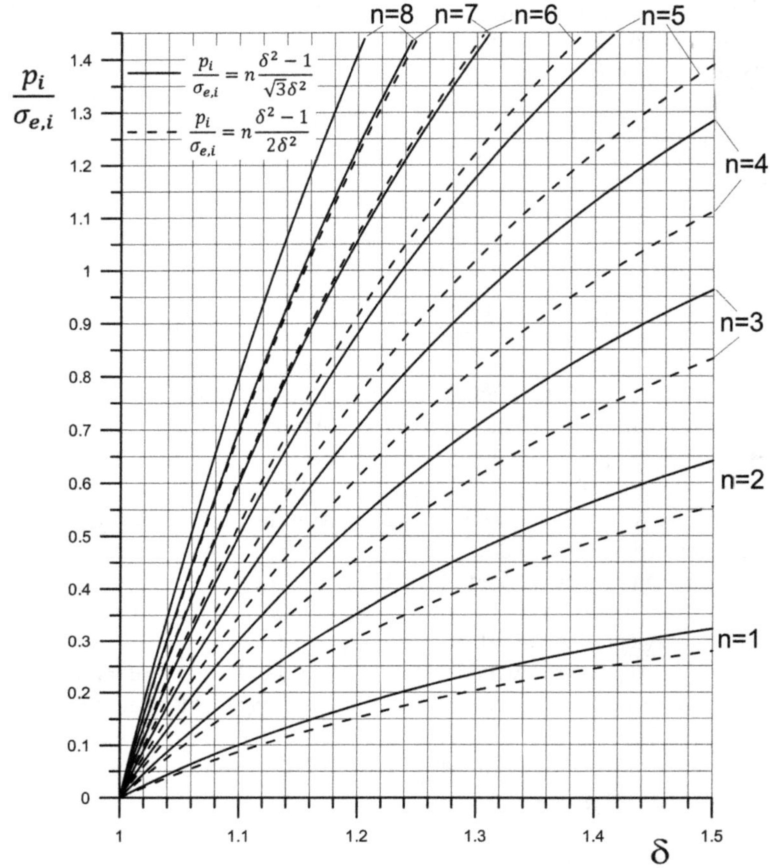

Fig. 5.11 Ratios $p_i/\sigma_{e,i} = f(\delta)$ for a multilayer structure with n varying from 1 to 8, calculated with ED_{max} and τ_{max} strength theories

From the latter relation, and considering relation (5.87), we obtain:

$$\sigma_{e,e} = \frac{\sqrt{3}(p_i - p_e)}{n(\delta^2 - 1)}. \tag{5.97}$$

Accordingly,

$$\Delta\sigma = \sigma_{e,i} - \sigma_{e,e} = \frac{\sqrt{3}(p_i - p_e)}{n}. \tag{5.98}$$

Table 5.1 Main characteristic parameters of the multilayer structure consisting of n layers of the same material treated so as to have the same tensile properties and under internal pressure only, on the basis of two different strength theories for a triaxial stress state ($\sigma_z = A$)

Strength theory	$p_i/\sigma_{e,i}$	δ	p_{cr}	$\sigma_{e,e}$	$\Delta\sigma = \sigma_{e,i} - \sigma_{e,e}$	$R = \sigma_{e,i}/\sigma_{e,e}$
ED_{max}	$\frac{p_e}{\sigma_{e,i}} = \frac{n(\delta^2-1)}{\sqrt{3}\delta^2}$	$\delta = \sqrt{\frac{\sigma_a}{\sigma_a - \frac{\sqrt{3}p_i}{n}}}$	$p_{cr} = \frac{n\sigma_a}{\sqrt{3}}$	$\sigma_{e,e} = \frac{\sqrt{3}p_i}{n(\delta^2-1)}$	$\Delta\sigma = \frac{\sqrt{3}p_i}{n}$	$R = \delta^2$
τ_{max}	$\frac{p_e}{\sigma_{e,i}} = \frac{n(\delta^2-1)}{2\delta^2}$	$\delta = \sqrt{\frac{\sigma_a}{\sigma_a - \frac{2p_i}{n}}}$	$p_{cr} = \frac{n\sigma_a}{2}$	$\sigma_{e,e} = \frac{2p_i}{n(\delta^2-1)}$	$\Delta\sigma = \frac{2p_i}{n}$	$R = \delta^2$

If the τ_{max} strength theory is used, we obtain the following relations in place of (5.97) and (5.98):

$$\sigma_{e,e} = \frac{2(p_i - p_e)}{n(\delta^2 - 1)} \tag{5.99}$$

and

$$\Delta\sigma = \sigma_{e,i} - \sigma_{e,e} = \frac{2(p_i - p_e)}{n}. \tag{5.100}$$

Table 5.1 summarizes the main characteristic parameters for the multilayer structure under internal pressure only (this, in fact, is the condition encountered in most industrial applications) correlated with the maximum distortion energy and τ_{max} strength theories.

For a multilayer structure under internal pressure only, Fig. 5.12 shows the ratios $p_i/\sigma_{e,e}$ obtained with the maximum distortion energy and τ_{max} strength theories versus δ. As in Fig. 5.11, the results obtained with n varying from 1 to 8 are shown. The function $R = R(\delta)$ given by relation (5.96) is represented at the upper left hand side of the figure.

The pressure p_{m-1} arising under working conditions between two adjoining layers of the multilayer structure [the (m-1)th and mth layers] is determined from relation (5.74), substituting the expression for $\sigma_{e,i}$ given by relation (5.87) in it and noting that, in the optimized condition, relation (5.86) indicates that we must have $k_m = k$ and, consequently $\delta_m^2 = \delta^2$. We will thus have:

$$p_{m-1} - p_m = \frac{p_i - p_e}{nk} k_m = \frac{p_i - p_e}{n} = \Delta p = \text{const.} \tag{5.101}$$

For the entire multilayer structure, we will have $(p_i - p_e) = n\Delta p$ (while for the mth interface we have $p_i - p_m = m\Delta p$). It can thus be concluded that:

$$p_m = p_i - m\Delta p = \frac{p_i(n - m) + mp_e}{n}. \tag{5.102}$$

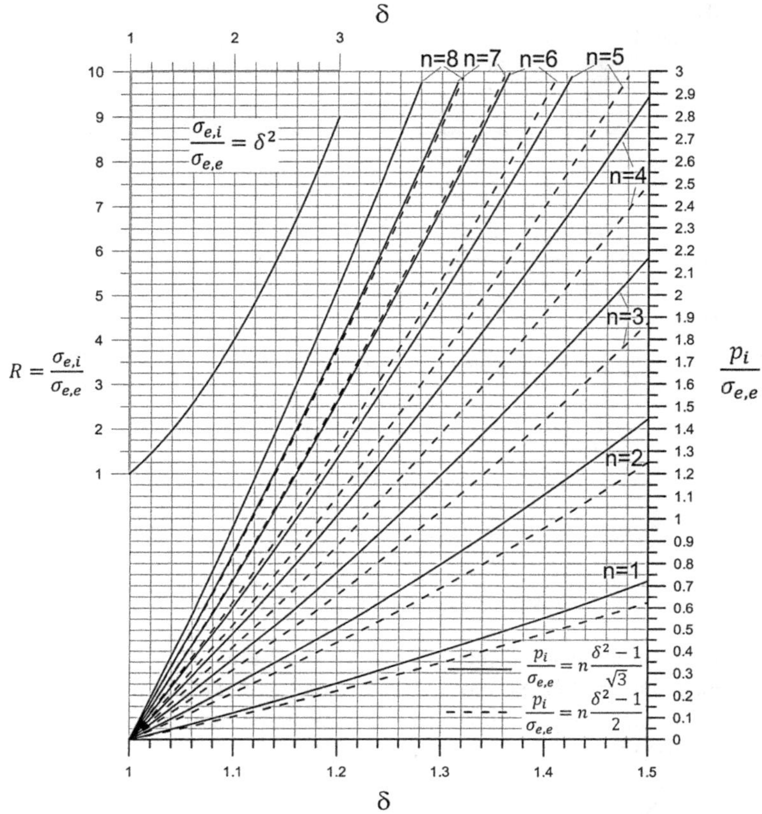

Fig. 5.12 Ratio $R = \delta^2$ and ratios $p_i/\sigma_{e,e} = f(\delta)$ for a multilayer structure with n varying from 1 to 8, calculated with ED_{max} and τ_{max} strength theories

In the particular case with zero external pressure, relation (5.102) becomes:

$$p_m = \frac{p_i(n-m)}{n}. \tag{5.103}$$

From the designer's standpoint, it is necessary to calculate the contact pressure in the generic interface between two adjoining layers. For the outer cylindrical surface of the mth layer, the method of superposition can be used since it is assumed that the material exhibits linear elastic behavior. Accordingly, the total radial stress acting on this surface can be expressed as the sum of the radial stress resulting from internal pressure p_i and that deriving from contact pressure $p_{f,m}$. We thus have:

$$\left|(\sigma_t)_{r=r_m}\right|_{tot} = \left|(\sigma_r)_{r=r_m}\right|_{p_i} + \left|(\sigma_r)_{r=r_m}\right|_{p_{f,m}}; \tag{5.104}$$

but

$$\begin{aligned} \left|(\sigma_t)_{r=r_m}\right|_{tot} &= -p_m \\ \left|(\sigma_r)_{r=r_m}\right|_{p_{f,m}} &= -\left|p_{f,m}\right| \\ \left|(\sigma_r)_{r=r_m}\right|_{p_i} &= -\left|(p_m)_{p_i}\right| = A - \frac{B}{r_m^2}, \end{aligned} \tag{5.105}$$

where A and B are given by relations (4.28). Considering relations (5.105), (5.102) and (4.28), relation (5.104) yields:

$$\left|p_{f,m}\right| = \frac{p_i(n-m) + mp_e}{n} + \frac{p_i - p_e\left(\frac{r_e}{r_i}\right)^2}{\left(\frac{r_e}{r_i}\right)^2 - 1} - \frac{p_i - p_e}{\left(\frac{r_e}{r_i}\right)^2 - 1}\left(\frac{r_e^2}{r_m^2}\right). \tag{5.106}$$

However, given that $(r_e/r_i) = \delta^n$ and $(r_e/r_m) = \delta^{n-m}$, relation (5.106) becomes:

$$\left|p_{f,m}\right| = \frac{p_i(n-m) + mp_e}{n} - \frac{p_i\left[\delta^{2(n-m)} - 1\right] + p_e\delta^{2n}\left[1 - \delta^{-2m}\right]}{\delta^{2n} - 1}. \tag{5.107}$$

This expresses the contact pressure at radius r_m. For $p_e = 0$, relation (5.107) becomes:

$$\left|p_{f,m}\right| = p_i\left[\frac{n-m}{n} - \frac{\delta^{2(n-m)} - 1}{\delta^{2n} - 1}\right]. \tag{5.108}$$

Chapter 6
Circular Cylinders Subjected to a Radial Temperature Gradient and Stressed in the Elastic Range

6.1 General Considerations

Circular cylinders, whether they feature very small axial dimensions and can thus be regarded as rings, or have axial dimensions that are of the same order of magnitude as their cross-sectional dimension or are much larger, are frequently subjected to complex thermal loading of varying intensity, and often with a high gradient, as a result of heat flow (Giovannozzi 1965, Zudans, Yen and Steigelmann 1965, Baker et al. 1968, Sim 1973, Tabakman and Lin 1978, Iurzolla 1981, Burr 1982, Boley and Weiner 1997, Cook and Young 1999, Vullo and Vivio 2013). In designing these structures, the problem thus arises of determining the stress and strain state once the temperature distributions occurring in them have been evaluated using considerations based on the laws of heat transfer (Carslaw and Jaeger 1959).

To address this intriguing problem, it is necessary to introduce several limiting and simplifying assumptions, in addition to those that have already been made, regarding the stress and strain state resulting from surface forces. The general equations for heat transfer contain a term for the heat generated within the structure (as occurs, for instance, in the fuel bars used in nuclear reactors). Phenomena of this kind will not be considered here, as they would complicate the approach to the problem, given that the equations for thermoelasticity and those for heat conduction would be mutually dependent and would influence each other. By decoupling the two types of equation, the simplifying assumption introduced here clearly leads to a more straightforward approach.

Under transient thermal load conditions, moreover, the displacements and the resulting strains become time-dependent magnitudes (Zudans, Yen and Steigelmann 1965, Belloni and Lo Conte 2002). In this case, the inertia forces appear in the equilibrium equations in addition to the internal forces. Here, we will assume steady state conditions and thus not consider the contribution of inertia forces, which would not in any case be very high. It will also be assumed that material behaves elastically (linear elastic range), and that the Young's modulus E and the Poisson's ratio ν as well as the material's coefficient of thermal expansion α are independent of temperature. Should it be necessary to regard this latter thermophysical constant as

V. Vullo, *Circular Cylinders and Pressure Vessels*, Springer Series in Solid and Structural Mechanics 3, DOI: 10.1007/978-3-319-00690-1_6,

varying with temperature, as it is always multiplied by the temperature, the problem can be addressed by considering the product αT as a variable. In addition, it will be assumed that displacements and deformations are small (and we are thus dealing with Lagrangian or engineering strains), that the material's behavior is homogeneous and isotropic in response to both mechanical and thermal loads, and that the temperature can be determined independently of the deformations of the structure. Finally, generalized axisymmetry will be assumed.

To provide a clearer understanding of the question, we will deal in sequence with cylinders of annular cross-section, with small axial dimension (rings) for which a plane stress state ($\sigma_z = 0$) can be assumed, with long circular cylinders with clamped ends or indefinitely extended in the direction of their axis, for which a plane strain state ($\varepsilon_z = 0$) can be assumed, and with circular cylinders of finite length with free ends, for which the strain state is characterized by a translation of the generic cross-section of abscissa z ($\varepsilon_z = const$) parallel to itself along the axis.

6.2 Cylinders of Annular Cross-Section with Small Axial Dimension (Rings)

We will consider a cylindrical structure of annular cross-section and any radial thickness (see Fig. 2.1, where however the wall is assumed thick), having a geometry of revolution and an axial dimension h which is small compared to the outer radius (r_i and r_e are the inner and outer radii respectively), consisting of homogeneous and isotropic material stressed in the linear elastic field by an axisymmetric thermal load, where the temperature variation with the radius is expressed by the entirely general function $T = T(r)$, and which is invariant in the direction of its z axis. This cylindrical body, which can thus be regarded as a thick-walled ring, is thus subject to pure radial heat flow.

In this case, the displacement v in the direction of the tangent to the generic circumference of radius r is zero, the stress state is plane, i.e., $\sigma_z = 0$, σ_r and σ_t as well as the radial displacement u are functions only of coordinate r and $\tau_{rt} = 0$. The equilibrium equation is given by (4.7), which is shown below for convenience together with the geometric relations (4.8) and the compatibility Eq. (4.9), all of which still apply here:

$$\frac{d}{dr}(\sigma_r \cdot r) - \sigma_t = 0 \tag{6.1}$$

$$\varepsilon_r = \frac{du}{dr}; \qquad \varepsilon_t = \frac{u}{r} \tag{6.2}$$

$$\varepsilon_r = \frac{d}{dr}(r \cdot \varepsilon_t) \tag{6.3}$$

The constitutive law which links the stress and strain components under thermal load and expresses the thermoelasticity relations for the plane strain state is as follows:

$$\begin{Bmatrix} \varepsilon_r \\ \varepsilon_t \\ \gamma_{rt} \end{Bmatrix} = \begin{bmatrix} 1/E & -\nu/E & 0 \\ -\nu/E & 1/E & 0 \\ 0 & 0 & 1/G \end{bmatrix} \begin{Bmatrix} \sigma_r \\ \sigma_t \\ \tau_{rt} \end{Bmatrix} + \alpha T \begin{Bmatrix} 1 \\ 1 \\ 0 \end{Bmatrix}. \tag{6.4}$$

This relation is the extension of Eq. (4.10) to thermal load. Developing Eq. (6.4) yields the thermoelasticity relations:

$$\begin{aligned} \varepsilon_r &= \frac{1}{E}(\sigma_r - \nu\sigma_t) + \alpha T \\ \varepsilon_t &= \frac{1}{E}(\sigma_t - \nu\sigma_r) + \alpha T \\ \gamma_{rt} &= \frac{1}{G}\tau_{rt}. \end{aligned} \tag{6.5}$$

Given the assumption of axisymmetry, it is necessary that $\gamma_{rt} = 0$; consequently, only the first two of these relations apply here. Solving the system of two equations consisting of the first two relations (6.5) for σ_r and σ_t gives:

$$\begin{aligned} \sigma_r &= \frac{E}{1-\nu^2}[(\varepsilon_r - \alpha T) + \nu(\varepsilon_t - \alpha T)] \\ \sigma_t &= \frac{E}{1-\nu^2}[(\varepsilon_t - \alpha T) + \nu(\varepsilon_r - \alpha T)]. \end{aligned} \tag{6.6}$$

Expanding the equilibrium Eq. (6.1), substituting relations (6.6) in it together with the derivative $d\sigma_r/dr$ obtained from the first relation (6.6) and considering geometric relations (6.2) yields the following non-homogeneous second order differential equation in terms of u:

$$\frac{d^2u}{dr^2} + \frac{1}{r}\frac{du}{dr} - \frac{u^2}{r^2} = \frac{d}{dr}\left[\frac{1}{r}\frac{d}{dr}(ur)\right] = \alpha(1+\nu)\frac{dT}{dr}. \tag{6.7}$$

This equation governs the radial displacement field of a ring subjected to steady state radial heat flow.

Integrating Eq. (6.7) twice in succession and introducing a constant C_1 which represents the constant $C_1'/2$ obtained by integrating the associated homogeneous equation yields:

$$u = \frac{\alpha(1+\nu)}{r}\int_{r_i}^{r} Tr dr + C_1 r + \frac{C_2}{r} \tag{6.8}$$

where C_1 and C_2 are integration constants to be determined by setting boundary conditions. By deriving Eq. (6.8) for r, we then obtain:

$$\frac{du}{dr} = \alpha(1+\nu)T - \frac{\alpha(1+\nu)}{r^2}\int_{r_i}^{r} Tr dr + C_1 - \frac{C_2}{r^2}. \tag{6.9}$$

Recalling Eq. (6.2) and substituting relations (6.8) and (6.9) in Eq. (6.7) yields:

$$\begin{cases} \sigma_r = -\dfrac{\alpha E}{r^2}\displaystyle\int_{r_i}^{r} Trdr + \dfrac{EC_1}{(1-\nu)} - \dfrac{EC_2}{(1+\nu)r^2} \\ \sigma_t = \alpha E\left(\dfrac{1}{r^2}\displaystyle\int_{r_i}^{r} Trdr - T\right) + \dfrac{EC_1}{(1-\nu)} + \dfrac{EC_2}{(1+\nu)r^2}\,. \end{cases} \tag{6.10}$$

Assuming that the inner and outer radii are not affected by surface forces (which will be accounted for separately through the method of superposition), the two boundary conditions to be set are:

$$\begin{aligned} \sigma_r &= 0 \text{ for } r = r_i \\ \sigma_r &= 0 \text{ for } r = r_e. \end{aligned} \tag{6.11}$$

With these boundary conditions, the following relations are obtained from the system of two equations in two unknowns given by the first Eq. (6.10):

$$\begin{aligned} C_1 &= \frac{\alpha(1-\nu)}{r_e^2 - r_i^2}\int_{r_i}^{re} Trdr \\ C_2 &= \frac{\alpha(1-\nu)r_i^2}{r_e^2 - r_i^2}\int_{r_i}^{re} Trdr. \end{aligned} \tag{6.12}$$

Substituting the latter relations in Eq. (6.10) and in Eq. (6.8) then yields the following set of three solving equations:

$$\begin{aligned} \sigma_r &= \frac{\alpha E}{r_e^2 - r_i^2}\left(1 - \frac{r_i^2}{r^2}\right)\int_{r_i}^{re} Trdr - \frac{\alpha E}{r^2}\int_{r_i}^{r} Trdr \\ \sigma_t &= \frac{\alpha E}{r_e^2 - r_i^2}\left(1 + \frac{r_i^2}{r^2}\right)\int_{r_i}^{re} Trdr + \alpha E\left(\frac{1}{r^2}\int_{r_i}^{r} Trdr - T\right) \\ u &= \frac{\left[r^2(1-\nu) + r_i^2(1+\nu)\right]}{r(r_e^2 - r_i^2)}\alpha\int_{r_i}^{re} Trdr + \frac{\alpha(1+\nu)}{r}\int_{r_i}^{r} Trdr. \end{aligned} \tag{6.13}$$

The first two of these relations define the stress state, while the third defines the radial displacement from which the strain state is determined using Eq. (6.2). From these relations, it can be concluded that in the particular case in which temperature is constant, the two stresses σ_r and σ_t are zero, while the radial displacement will be $u = \alpha Tr$.

The corresponding relations for solid circular cylinders featuring small axial thickness, and which are thus disk-shaped, can be deduced from those presented above for cylindrical bodies of annular cross-section by setting $r_i = 0$ in Eq. (6.13). This yields the following three relations:

$$\begin{aligned}
\sigma_r &= \alpha E\left(\frac{1}{r_e^2}\int_0^{re} Trdr - \frac{1}{r^2}\int_0^{r} Trdr\right)\\
\sigma_t &= \alpha E\left(\frac{1}{r_e^2}\int_0^{re} Trdr + \frac{1}{r^2}\int_0^{r} Trdr - T\right)\\
u &= \frac{\alpha(1+\nu)}{r}\int_0^{r} Trdr - \frac{\alpha(1-\nu)}{r_e^2}r\int_0^{re} Trdr.
\end{aligned} \tag{6.14}$$

Relations (6.14) are indeterminate for $r = 0$ and thus give rise to a singularity at the axis. However, if the temperature here has a well-defined value (in other words, there must be no concentrated heat sources at the axis), these relations can be used to calculate stresses and radial displacement at the center of these disk-shaped cylindrical bodies, i.e., at their axes. In fact, with a Taylor series expansion of function $T = T(r)$, such that

$$T(r) = T(0) + T'(0)r + \frac{1}{2}T''(0)r^2 + \ldots \tag{6.15}$$

and noting that

$$\begin{aligned}
&\lim_{r\to 0}\frac{1}{r^2}\int_0^{r} Trdr = \frac{1}{2}T(0)\\
&\lim_{r\to 0}\frac{1}{r}\int_0^{r} Trdr = 0,
\end{aligned} \tag{6.16}$$

we arrive at the conclusion, which is in any case fairly intuitive and can be demonstrated using other methods, that at the center of this cylindrical body we have:

$$(\sigma_r)_{r=0} = (\sigma_t)_{r=0} = \alpha E\left[\frac{1}{r_e^2}\int_0^{re} Trdr - \frac{1}{2}T(0)\right];\ (u)_{r=0} = 0. \tag{6.17}$$

6.3 Long Cylinders of Annular Cross-Section

We will now consider a circular cylinder of any radial thickness (Fig. 6.1a) having theoretically infinite axial length and geometry of revolution, consisting of homogeneous and isotropic material stressed in the linear elastic field by surface

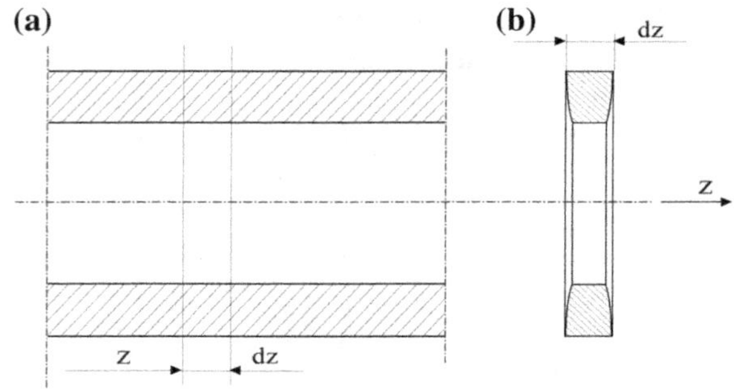

Fig. 6.1 **a** Thick-walled circular cylinder extended indefinitely in the direction of the z axis; **b** isolated small element of axial dimension dz and its axial strain

forces and thermal loads which are all axisymmetric and invariant in the direction of axis z. In these conditions, taking for example a circular cylinder having large radial thickness, each elementary annular portion of this cylinder located between two cross-sections having coordinates z and $z + dz$ (Fig. 6.1a) would, if considered in isolation, behave as a ring, with a resulting strain state that would show an axial strain ε_z which is a function of the radius (Fig. 6.1b).

However, in long (or infinitely long) circular cylinders subjected to surface forces at the inner and outer radii and to thermal load, it is not conceivable that longitudinal unit elongations (axial strains) and axial displacements can vary with the radius, except at the ends, which are assumed to be free. The fact that it is impossible to have a longitudinal unit elongation ε_z that differs with radius r (expansion cannot proceed freely in a so loaded continuous structure) gives rise to a state of longitudinal coaction characterized by a longitudinal normal stress σ_z which is a function of the radius. By comparison with a ring subjected to the same load conditions, where a plane stress state is found, the problem examined here entails either a plane strain state ($\varepsilon_z = 0$) if the cylindrical body is constrained at its ends or indefinitely extended in the direction of its axis, or a strain state characterized by a uniform axial translation of a generic cross-section ($\varepsilon_z = const$), if the circular cylinder has finite axial length and is free at the ends. Both cases of strain characterized by $\varepsilon_z = 0$ and by $\varepsilon_z = const$ will be examined here in turn.

The general treatment of this case assumes that the end sections of the circular cylinder are not loaded. The stress state resulting from any longitudinal external loads acting on the ends can be evaluated separately and then added to those analyzed below, as the principle of superposition applies in the linear elastic field assumed here. As for the ring, the radial, tangential and axial directions are the principal directions of both stresses and strains, as the cylindrical body is loaded axisymmetrically in the radial direction and uniformly in the axial direction.

As a general premise, it should first be borne in mind that the equilibrium equation in the radial direction (6.1), the geometric relations (6.2) and the compatibility Eq. (6.3) continue to apply. To solve the problem, two further relations must be added to those indicated above, viz.:

$$\varepsilon_z = \frac{dw}{dz} = \varepsilon_0 = const \tag{6.18}$$

and

$$F_z = \int_A \sigma_z dA = 0. \tag{6.19}$$

Relation (6.18) expresses the plane strain state condition, in cases where the constant is zero, or the strain state characterized by a uniform axial translation of a generic cross-section of the circular cylinder in cases where the constant is not zero; w denotes the displacement of the generic cross-section of abscissa z, in the direction of the z axis. Relation (6.19), which applies to the cylinder with free unloaded ends, expresses the global equilibrium condition of a generic cross-section A in the direction of the z axis; dA is the area element of this cross-section, while F_z is the axial load.

The thermoelasticity relations which link the strain components (normal strain components and shear strain components) to the stress components (normal stress components and shear stress components) and express Hooke's constitutive laws for the triaxial stress state in the linear elastic range with thermal loading are as follows:

$$\begin{Bmatrix} \varepsilon_r \\ \varepsilon_t \\ \varepsilon_z \\ \gamma_{rt} \\ \gamma_{tz} \\ \gamma_{zr} \end{Bmatrix} = \begin{bmatrix} 1/E & -\nu/E & -\nu/E & 0 & 0 & 0 \\ -\nu/E & 1/E & -\nu/E & 0 & 0 & 0 \\ -\nu/E & -\nu/E & 1/E & 0 & 0 & 0 \\ 0 & 0 & 0 & 1/G & 0 & 0 \\ 0 & 0 & 0 & 0 & 1/G & 0 \\ 0 & 0 & 0 & 0 & 0 & 1/G \end{bmatrix} \begin{Bmatrix} \sigma_r \\ \sigma_t \\ \sigma_z \\ \tau_{rt} \\ \tau_{tz} \\ \tau_{zr} \end{Bmatrix} + \alpha T \begin{Bmatrix} 1 \\ 1 \\ 1 \\ 0 \\ 0 \\ 0 \end{Bmatrix}. \tag{6.20}$$

These relations can be written in the following form:

$$\begin{aligned} \varepsilon_r &= \frac{1}{E}[\sigma_r - \nu(\sigma_t + \sigma_z)] + \alpha T \\ \varepsilon_t &= \frac{1}{E}[\sigma_t - \nu(\sigma_z + \sigma_r)] + \alpha T \\ \varepsilon_z &= \frac{1}{E}[\sigma_z - \nu(\sigma_r + \sigma_t)] + \alpha T \\ \gamma_{rt} &= \frac{1}{G}\tau_{rt} \\ \gamma_{tz} &= \frac{1}{G}\tau_{tz} \\ \gamma_{zr} &= \frac{1}{G}\tau_{zr}. \end{aligned} \tag{6.21}$$

The last three of these relations, i.e., those linking shear strain components to shear stress components, are clearly not influenced by temperature, as unconstrained thermal expansion does not cause angular distortions in a homogeneous and isotropic material.

From the first three thermoelasticity relations (6.20) or (6.21), introducing the first invariant of the strain tensor, i.e., volume expansion

$$e = \varepsilon_r + \varepsilon_t + \varepsilon_z = \frac{(1-2\nu)}{E} J_1 + 3\alpha T, \tag{6.22}$$

with $J_1 = (\sigma_r + \sigma_t + \sigma_z)$ being the first invariant of the stress tensor, we obtain the following general relations, which are also called thermoelasticity relations and express the three principal radial, hoop and axial stresses as a function of the principal strains:

$$\begin{cases} \sigma_r = \dfrac{E}{1+\nu}\left(\dfrac{\nu E}{1-2\nu} + \varepsilon_r\right) - \dfrac{E}{1-2\nu}\alpha T \\ \sigma_t = \dfrac{E}{1+\nu}\left(\dfrac{\nu E}{1-2\nu} + \varepsilon_t\right) - \dfrac{E}{1-2\nu}\alpha T \\ \sigma_z = \dfrac{E}{1+\nu}\left(\dfrac{\nu E}{1-2\nu} + \varepsilon_z\right) - \dfrac{E}{1-2\nu}\alpha T. \end{cases} \tag{6.23}$$

These relations can also be written in the following more compact form:

$$\begin{cases} \sigma_r = \lambda e + 2G\varepsilon_r - \dfrac{E}{1-2\nu}\alpha T \\ \sigma_t = \lambda e + 2G\varepsilon_t - \dfrac{E}{1-2\nu}\alpha T \\ \sigma_z = \lambda e + 2G\varepsilon_z - \dfrac{E}{1-2\nu}\alpha T \end{cases} \tag{6.24}$$

where $\lambda = \nu E/[(1+\nu)(1-2\nu)]$ is Lamé's constant and $G = E/[2(1+\nu)]$ is the modulus of elasticity in shear or modulus of rigidity.

Relation (6.18) must be respected for any constraint condition at the ends of the circular cylinder considered in the direction of the z axis, along which the temperature gradient is assumed to be zero, and axisymmetry is also assumed. Consequently, deriving the first of relations (6.23) or of relations (6.24) by respect to the radius gives:

$$\frac{d\sigma_r}{dr} = \frac{E}{1+\nu}\left[\frac{\nu}{1-2\nu}\left(\frac{d^2u}{dr^2} + \frac{1}{r}\frac{du}{dr} - \frac{u^2}{r^2}\right) + \frac{d^2u}{dr^2}\right] - \frac{E}{1-2\nu}\alpha\frac{dT}{dr}. \tag{6.25}$$

Expanding the equilibrium Eq. (6.1) which holds for axisymmetric circular cylinders and introducing the first and second of relations (6.23) or of relations (6.24) as well as relation (6.25) yields the following second order differential equation in terms of displacement $u = u\ (r)$:

$$\frac{d^2u}{dr^2} + \frac{1}{r}\frac{du}{dr} - \frac{u^2}{r^2} - \alpha\left(\frac{1+\nu}{1-\nu}\right)\frac{dT}{dr} = 0. \tag{6.26}$$

This can also be obtained from Eq. (6.7) after substituting (as will be demonstrated in the following section) α with $\alpha(1+\nu)$ and ν with $\nu/(1-\nu)$ in the latter.

Without assuming any law for temperature distribution along the radius in order to ensure the most general possible treatment for the problem, two successive integrations of relation (6.26) lead to the relation:

$$u = \alpha\left(\frac{1+\nu}{1-\nu}\right)\frac{1}{r}\int_{r_i}^{r} Trdr + C_1 r + \frac{C_2}{r} \tag{6.27}$$

and then, through derivation, to:

$$\frac{du}{dr} = \alpha\frac{1+\nu}{1-\nu}T - \alpha\left(\frac{1+\nu}{1-\nu}\right)\frac{1}{r^2}\int_{r_i}^{r} Trdr + C_1 - \frac{C_2}{r^2}\,. \tag{6.28}$$

Clearly, relations (6.27) and (6.28) can also be obtained from relations (6.8) and (6.9) respectively, substituting α and ν with $\alpha(1 + \nu)$ and $\nu/(1-\nu)$ in them.

6.4 Long Circular Cylinder with Clamped Ends or Indefinitely Extended in the Direction of Its Axis and Subjected to Thermal Load Defined by a Generic Function $T = T(r)$

For a circular cylinder of finite length clamped at the ends and for a circular cylinder indefinitely extended in the direction of its axis, we necessarily have $\varepsilon_z = 0$. Accordingly, the third relation (6.21) gives:

$$\sigma_z = \nu(\sigma_r + \sigma_t) - E\alpha T. \tag{6.29}$$

Substituting the σ_z thus determined in the first two thermoelasticity relations given by Eq. (6.21) yields the following expressions for unit normal elongations or normal strain components ε_r and ε_t:

$$\begin{cases} \varepsilon_r = \dfrac{1-\nu^2}{E}\left(\sigma_r - \dfrac{\nu}{1-\nu}\sigma_t\right) + (1+\nu)\alpha T \\ \varepsilon_t = \dfrac{1-\nu^2}{E}\left(\sigma_t - \dfrac{\nu}{1-\nu}\sigma_r\right) + (1+\nu)\alpha T\,. \end{cases} \tag{6.30}$$

If we compare these latter relations with the equivalents applying to the ring (where the stress state is plane, i.e., $\sigma_z = 0$) given by Eq. (6.5), we can see that passing from the ring to the long circular cylinder subjected to the same stress state (surface forces and thermal load) involves the following variations, which concern only the material's thermomechanical magnitudes:

$$\begin{array}{ccc} \sigma_z = 0 & \rightarrow & \varepsilon_z = 0 \\ \alpha & \rightarrow & \alpha\cdot(1+\nu) \\ 1/E & \rightarrow & (1-\nu^2)/E \\ \nu & \rightarrow & \nu/(1-\nu)\,. \end{array} \tag{6.31}$$

To take advantage of the foregoing relations, we will now return to Eq. (6.10) which, omitting integration constants C_1 and C_2, express the stress state in a ring subjected to surface forces at the inner and outer radii and to a temperature gradient along the radius with a plane stress state ($\sigma_z = 0$). Introducing variations (6.31) in Eq. (6.10) yields the following relations:

$$\begin{cases} \sigma_r = -\dfrac{\alpha E}{(1-\nu)r^2}\displaystyle\int_{r_i}^{r} Trdr + \dfrac{EC_1}{(1+\nu)(1-2\nu)} - \dfrac{EC_2}{(1+\nu)r^2} \\ \sigma_t = \dfrac{\alpha E}{(1-\nu)r^2}\displaystyle\int_{r_i}^{r} Trdr - \dfrac{\alpha ET}{(1-\nu)} + \dfrac{EC_1}{(1+\nu)(1-2\nu)} + \dfrac{EC_2}{(1+\nu)r^2}. \end{cases} \tag{6.32}$$

Obviously, relations (6.32) can be obtained directly from the first two thermoelasticity relations (6.23) or (6.24) by substituting relations (6.27) and (6.28) in them and merging the constant $\varepsilon_z = \varepsilon_0 = const$ in constant C_1.

By introducing the new constants

$$\begin{aligned} A &= \frac{EC_1}{(1+\nu)(1-2\nu)} \\ B &= \frac{EC_2}{(1+\nu)}, \end{aligned} \tag{6.33}$$

relations (6.32) take the following more compact form:

$$\begin{cases} \sigma_r = A - \dfrac{B}{r^2} - \dfrac{\alpha E}{(1-\nu)r^2}\displaystyle\int_{r_i}^{r} Trdr \\ \sigma_t = A + \dfrac{B}{r^2} + \dfrac{\alpha E}{(1-\nu)}\left(\dfrac{1}{r^2}\displaystyle\int_{r_i}^{r} Trdr - T\right). \end{cases} \tag{6.34}$$

For a hollow circular cylinder subjected only to a temperature gradient along the radius (the surface forces at the inner and outer radii are thus assumed to be zero; the effects of any such forces will be evaluated separately and are added to those resulting from the temperature, as the method of superposition is assumed to hold true), the boundary conditions to be set are once again those given by relations (6.11). Accordingly, the system of two equations in two unknowns obtained from the first relation (6.34) yields:

$$\begin{aligned} A &= \frac{\alpha E}{(1-\nu)\left(r_e^2 - r_i^2\right)}\int_{r_i}^{re} Trdr \\ B &= \frac{\alpha E r_i^2}{(1-\nu)\left(r_e^2 - r_i^2\right)}\int_{r_i}^{re} Trdr. \end{aligned} \tag{6.35}$$

Substituting the constants thus determined in relations (6.34) yields the following expressions for stresses σ_r and σ_t :

$$\sigma_r = \frac{\alpha E}{1-\nu}\left[\frac{1}{r_e^2 - r_i^2}\left(1-\frac{r_i^2}{r^2}\right)\int_{r_i}^{re} Trdr - \frac{1}{r^2}\int_{r_i}^{r} Trdr\right]$$
$$\sigma_t = \frac{\alpha E}{1-\nu}\left[\frac{1}{r_e^2 - r_i^2}\left(1+\frac{r_i^2}{r^2}\right)\int_{r_i}^{re} Trdr + \frac{1}{r^2}\int_{r_i}^{r} Trdr - T\right]. \quad (6.36)$$

Subsequently, substituting relations (6.36) in relation (6.29) gives the following general expression of axial stress σ_z:

$$\sigma_z = \frac{\alpha E}{1-\nu}\left(\frac{2\nu}{r_e^2 - r_i^2}\int_{r_i}^{re} Trdr - T\right). \quad (6.37)$$

On the other hand, the radial displacement $u = u(r)$ is linked to tangential strain ε_t by relation $u = r\varepsilon_t$; given the second relation (6.21), we thus have:

$$u = r\varepsilon_t = \frac{r}{E}[\sigma_t - \nu(\sigma_r + \sigma_z) + \alpha ET]. \quad (6.38)$$

Substituting (6.36) and (6.37) in this relation then yields:

$$u = \frac{\alpha(1+\nu)}{r(1-\nu)}\left[\frac{(1-2\nu)r^2 + r_i^2}{r_e^2 - r_i^2}\int_{r_i}^{re} Trdr + \int_{r_i}^{r} Trdr\right]. \quad (6.39)$$

In conclusion, the three principal stresses that define the stress state and radial displacement in a circular cylinder with clamped ends or indefinitely extended in the direction of its axis, with no load at either the inner or outer radius and subjected to a temperature gradient along the radius according to any function $T = T(r)$ are as follows:

$$\sigma_r = \frac{\alpha E}{1-\nu}\left[\frac{1}{r_e^2 - r_i^2}\left(1-\frac{r_i^2}{r^2}\right)\int_{r_i}^{re} Trdr - \frac{1}{r^2}\int_{r_i}^{r} Trdr\right]$$
$$\sigma_t = \frac{\alpha E}{1-\nu}\left[\frac{1}{r_e^2 - r_i^2}\left(1+\frac{r_i^2}{r^2}\right)\int_{r_i}^{re} Trdr + \frac{1}{r^2}\int_{r_i}^{r} Trdr - T\right]$$
$$\sigma_z = \frac{\alpha E}{1-\nu}\left(\frac{2\nu}{r_e^2 - r_i^2}\int_{r_i}^{re} Trdr - T\right) \quad (6.40)$$
$$u = \frac{\alpha(1+\nu)}{r(1-\nu)}\left[\frac{(1-2\nu)r^2 + r_i^2}{r_e^2 - r_i^2}\int_{r_i}^{re} Trdr + \int_{r_i}^{r} Trdr\right].$$

Obviously, in the case considered here of a circular cylinder constrained at the ends or indefinitely extended in the direction of its axis, the axial displacement w of any cross-section of abscissa z will be zero. It is also obvious that the first two of the relations given above can be deduced from the first two relations (6.13) after substituting α with $\alpha(1 + \nu)$, ν with $\nu/(1-\nu)$ and E with $E/(1 - \nu^2)$ in them.

If the same circular cylinder is also subjected to surface force distributions acting at the inner and outer radii, the stress state and the resulting radial displacement are given by the corresponding relations given in Chap. 4. It goes without saying that the overall stress state and the total radial displacement can be determined by means of the method of superposition, that is as the sum of the contributions made by the various loads, separately acting on the structure, as it has been assumed that Hooke's law holds (linear elastic field). For the hollow cylindrical body, for $r_i \to 0$, the triaxial problem also gives rise to the singularity encountered in the biaxial problem; here again, in calculating the values at the inner radius for radial and hoop stresses and radial displacement, the singularity requires us to consider the ratio r_i^2/r^2 as unity.

As above, the corresponding relations for the solid circular cylinder constrained at the ends or indefinitely extended in the direction of its axis, with no load at the outer radius and subjected to a temperature gradient along the radius according to any law function $T = T(r)$ can be derived from those given earlier for the cylinder of annular cross-section, setting $r_i = 0$ in relations (6.40). We thus obtain the following four relations:

$$\begin{aligned}
\sigma_r &= \frac{\alpha E}{1-\nu}\left[\frac{1}{r_e^2}\int_0^{re} Trdr - \frac{1}{r^2}\int_0^{r} Trdr\right]\\
\sigma_t &= \frac{\alpha E}{1-\nu}\left[\frac{1}{r_e^2}\int_0^{re} Trdr + \frac{1}{r^2}\int_0^{r} Trdr - T\right]\\
\sigma_z &= \frac{\alpha E}{1-\nu}\left(\frac{2\nu}{r_e^2}\int_0^{re} Trdr - T\right)\\
u &= \frac{\alpha(1+\nu)}{r(1-\nu)}\left[\frac{(1-2\nu)r^2}{r_e^2}\int_0^{re} Trdr + \int_0^{r} Trdr\right].
\end{aligned} \tag{6.41}$$

These relations can also be deduced from the general relations given earlier, noting that $r_i = 0$ and that radial displacement u must be zero at the axis. It can be concluded from Eq. (6.27) that, as the integral appearing in it converges on zero for $r \to 0$, we must necessarily have $C_2 = B = 0$. If the cylindrical bar (a solid cylindrical body is nothing more than a rod) is not subjected to surface forces at the outer radius, the second constant A is determined from the first relation (6.34) by establishing that radial stress σ_r is zero for $r = r_e$. We thus obtain:

$$A = \frac{\alpha E}{(1 - \nu) r_e^2} \int_0^{re} Trdr, \tag{6.42}$$

which, substituted in relations (6.34), yields the first two relations (6.41). The other two relations (6.41) are found by following the same procedure used for the cylinder of annular cross-section.

6.4.1 Long Circular Cylinder with Clamped Ends or Indefinitely Extended in the Direction of Its Axis and Subjected to Thermal Load Defined by Fourier's Law

We will now assume that the function $T = T(r)$ of temperature distribution with the radius is Fourier's law (Fourier 1822) of heat-flow by conduction in cylindrical bodies of circular or annular cross-section having inside radius r_i and outside radius r_e. For long circular cylinders such as heat exchanger tubes, the tubing and piping in chemical, petrochemical and nuclear plants that carry fluids at high temperatures to the plant's subsystems, the circular cylindrical liners used in reciprocating machines (engines and compressors), smokestacks and so forth, heat flow is essentially radial, except at the ends. In the steady-state conditions considered here, the heat flow passes through a generic cylindrical small element having unitary axial length, inner radius r and outer radius $r + dr$, and is expressed by Fourier's equation:

$$Q = -2\pi kr \frac{dT}{dr}, \tag{6.43}$$

where Q is the heat flow, that is instantaneous rate at which heat is transferred in direction normal to flow surface (Q is considered positive when in the centrifugal direction), k is the material's coefficient of thermal conductivity, $2\pi r$ is the flow surface having unitary axial length, and dT/dr is the instantaneous temperature gradient in the direction normal to flow surface (the minus sign indicates that heat flow is in the direction of decreasing temperatures, from the inner to the outer radius).

Thermal conductivity is a thermo-physical magnitude of the material and in general depends on temperature. For anisotropic materials, k also depends on the direction of heat flow. In the case of isotropic materials, the coefficient of thermal conductivity k can be expressed by means of a power series in temperature T and, in many cases of design interest, can be approximated with a constant or a linear function of T.

For one-dimensional heat flow under steady-state conditions in a homogeneous material where there are no energy sources and no energy sinks and for which k is a known function of temperature and direction, Q is constant and Eq. (6.43) can be integrated provided that wall temperatures T_i and T_e at the cylinder's inner and outer radii are known. Barring exceptional cases, which will be briefly discussed in Sect. 6.7, steady-state heat flow conditions are almost always respected in design practice. Indeed, in the problems of greatest interest to the structural designer, variations of temperature over time can almost always be ignored, and the temperature profile can be considered as a succession of states, each under steady-state conditions, for which the relations given below apply (Zudans, Yen and Steigelmann 1965).

By separating the variables and integrating Eq. (6.43) between the inner and the outer radii, with the general assumption that k is a function of temperature, we obtain:

$$\frac{Q}{2\pi}\ln\frac{r_e}{r_i} = -\int_{T_i}^{T_e} k dT. \tag{6.44}$$

Multiplying and dividing the right-hand side member of relation (6.44) by the difference between wall temperatures $(T_i - T_e)$, this relation becomes:

$$\frac{Q}{2\pi}\ln\frac{r_e}{r_i} = -\frac{\int_{T_i}^{T_e} kdT}{(T_i - T_e)}(T_i - T_e). \tag{6.45}$$

The fractional term, indicated with

$$\bar{k} = \frac{\int_{T_i}^{T_e} kdT}{(T_i - T_e)}, \tag{6.46}$$

represents the average value of k in the temperature range $(T_i - T_e)$. When this range is relatively small, a constant coefficient of thermal conductivity k, equal to $\bar{k}$ and calculated as the arithmetic mean of the values of k at temperatures T_i and T_e is usually introduced in Fourier's Eq. (6.43). This is the assumption that will be made henceforth.

With this assumption, integrating Fourier's equation between the generic radius r and the outer radius (the same result can be reached by integrating Eq. (6.43) and calculating the integration constant with the condition that $T = T_e$ at the outer radius, i.e., at $r = r_e$) gives the following function $T = T(r)$:

$$T = T_e + \frac{Q}{2\pi k}\ln\frac{r_e}{r} \tag{6.47}$$

where T_e is the temperature at the outer radius.

For $r = r_i$, i.e., for $T = T_i$, where T_i is the temperature at the inner radius, relation (6.47) gives:

$$\frac{Q}{2\pi k} = \frac{T_i - T_e}{\ln r_e/r_i}. \tag{6.48}$$

Consequently, relation (6.47) can be rewritten in the following form:

$$T = T_e + \frac{T_i - T_e}{\ln r_e/r_i}\ln\frac{r_e}{r}. \tag{6.49}$$

It is now necessary to introduce relation (6.49), which expresses the temperature distribution with the radius according to Fourier's law as specialized for the case in question, in relations (6.40) and proceed with integrations. As regards the radial and hoop stresses given by the first two relations (6.40), it should be noted that the latter derive from integrating the non-homogeneous differential Eq. (6.26), where the first derivative of temperature *dT*/*dr* appears in the temperature term. It follows that a constant temperature resulting from a slow, uniform heating or cooling of the cylindrical structure does not produce stresses, but only expansion. This observation is valid in general, regardless of the type of cylindrical structure concerned, provided that the material is isotropic and its elastic and thermo-physical properties are independent of the radius. For this reason, the term T_e appearing in relation (6.49) can be neglected when performing integrations if the distance from the clamped ends is large enough that the perturbations at the edges contemplated by Saint–Venant's principle will not be encountered. At a sufficient distance from the clamped ends, in fact, radial and tangential expansions are not constrained, and term T_e thus does not contribute to stresses σ_r and σ_t in any way.

By introducing relation (6.49) in relations (6.40) and performing integrations, we then obtain the following relations:

$$\begin{aligned}
\sigma_r &= \frac{\alpha E(T_i - T_e)}{2(1-\nu)(\ln r_e/r_i)}\left[-\ln\frac{r_e}{r} + \frac{r_i^2}{r_e^2 - r_i^2}\left(\frac{r_e^2}{r^2} - 1\right)\ln\frac{r_e}{r_i}\right]\\
\sigma_t &= \frac{\alpha E(T_i - T_e)}{2(1-\nu)(\ln r_e/r_i)}\left[1 - \ln\frac{r_e}{r} - \frac{r_i^2}{r_e^2 - r_i^2}\left(\frac{r_e^2}{r^2} + 1\right)\ln\frac{r_e}{r_i}\right].\\
\sigma_z &= \frac{\alpha E}{2(1-\nu)}\left\{\frac{(T_i - T_e)}{(\ln r_e/r_i)}\left[\nu - \frac{2\nu r_i^2}{r_e^2 - r_i^2}\ln\frac{r_e}{r_i} - 2\ln\frac{r_e}{r}\right] - 2T_e(1-\nu)\right\}.\\
u &= \alpha(1+\nu)r\left\{\frac{(T_i - T_e)}{2(1-\nu)(\ln r_e/r_i)}\left[(1-\nu) + \ln\frac{r_e}{r} - \frac{r_i^2}{r_e^2 - r_i^2}\ln\frac{r_e}{r_i}\left(\frac{r_e^2}{r^2} + 1 - 2\nu\right)\right] + T_e\right\}.
\end{aligned} \tag{6.50}$$

These relations describe the stress state and radial displacement in the long hollow cylindrical body with clamped ends, subject to a thermal load whose temperature distribution with the radius follows Fourier's law.

The term T_e does not appear in the parentheses in the first two relations (6.50): as indicated above, this is because radial and tangential expansions are unconstrained except at the clamped ends, and the term T_e appearing in relation (6.49) makes no contribution to stresses σ_r and σ_t. In the third relation (6.50), on the other hand, a term in T_e appears between the braces, given that the axial expansion, unlike the other expansions, is constrained. A constant temperature in a circular cylinder with clamped ends will thus give rise to a stress state with $\sigma_r = \sigma_t = 0$ and with $\sigma_z = \alpha E T_e \neq 0$ if the temperature $T_e = const$ is not the same as the assembly temperature taken as a reference. A term in T_e also appears between the braces in the last relation (6.50), because it is assumed that the temperature at the outer radius differs from reference temperature T_0, which is in general assumed to be the assembly temperature. If T_e were to be assumed as the reference temperature, the associated term between braces would disappear. The value of T_e to be introduced in calculations is the difference between the temperature measured on the outer surface and reference temperature T_0.

Figure 6.1 shows distributions of temperature and stresses in a steel circular cylinder ($E = 210$ GPa; $\nu = 0.3$; $\alpha = 12 \times 10^{-6}$ °C^{-1}), with $\delta = 2$ ($r_e = 2r_i$) in dimensionless form and as a function of radius r, which is likewise made dimensionless relative to the outside radius r_e. Stresses are made dimensionless by dividing them by the factor $\bar{\sigma} = aE(T_i - T_e)/2(1 - \nu)$ which, for steels, assumes a value of around 1.80 MPa for each °C of difference between the temperatures at the inner and outer radii. Temperature is also shown in dimensionless form, as the ratio $\bar{T} = (T - T_e)/(T_i - T_e)$, and thus assumes values of 1 and 0 at the inner radius and outer radius respectively. Its behavior between these two extremes is obviously logarithmic.

As is immediately clear from the figure, the most dangerous stress state as regards strength capacity is that at the inner radius, where stresses σ_z and σ_t, both compressive, reach their highest values, viz., 3.68 and 2.20 MPa respectively for each °C of difference between the temperatures at the inner and outer radii. For a temperature differential ($T_i - T_e$) of 100 °C, these two stresses would be 368 and 220 MPa, or in other words sizeable fractions of the material's yield stress σ_s. For steady-state heat flow, i.e., for static thermal load, and a structure consisting of AISI 1060 steel, whose yield stress is 480 MPa, this stress state would correspond to safety factors of 1.30, 1.30 and 1.50 respectively, determined using the σ_{max}, τ_{max} and maximum distortion energy strength theories.

The end constraints thus result in a stress state that can be doubly dangerous, as it not only influences strength conditions, but can also trigger global and local buckling if the circular cylinder's wall is thin. It goes without saying that in designing long circular cylinders (tubing and piping used in the chemical, petrochemical and nuclear industries, for example), it is advisable to avoid axial constraints at the ends whenever possible.

Passing from the inner radius to the outer, both stresses σ_z and σ_t drop in absolute value. However, while longitudinal stress remains negative, reaching its minimum absolute value at the outer radius, the hoop stress changes sign for a value of r which can be readily determined from the second relation (6.50), becoming a tensile stress and, at the outer radius, assuming a value of 1.40 MPa for each °C of difference between the temperatures at the inner and outer radii. The radial stress, zero at inner and outer radii, is always negative and reaches its maximum value, which is in any case not very high and can be readily calculated, for a value of r that can once again be determined by establishing that the derivative $d\sigma_r/dr$ obtained from the first relation (6.50) is zero.

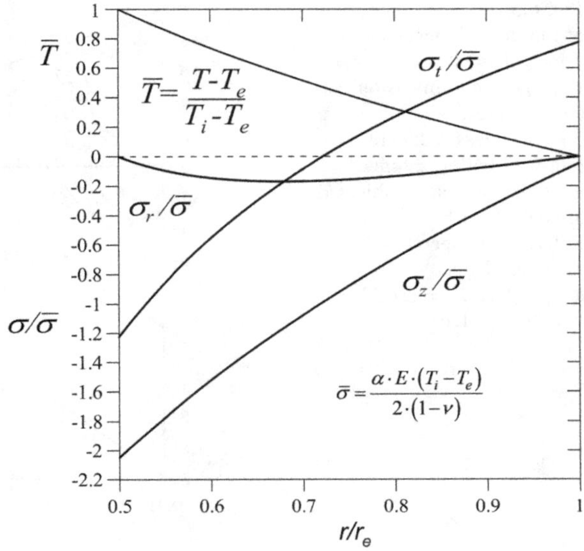

Fig. 6.2 Distribution curves of the three dimensionless principal stresses $\sigma_r/\bar{\sigma}$, $\sigma_t/\bar{\sigma}$ and $\sigma_z/\bar{\sigma}$ and dimensionless temperature $\bar{T}$ along the dimensionless radius of a hollow cylinder with clamped ends subjected to centrifugal heat flow. Reference temperature $T_0 = 20$ °C; outer temperature $T_e = 20$ °C; $\delta = 2$

Figure 6.2 applies for centrifugal heat flow ($T_i > T_e$). For centripetal heat flow ($T_i < T_e$), the distribution of the three principal stresses through the wall thickness is obtained from that indicated in Fig. 6.2 as the mirror image of the curves shown therein, i.e., by rotating the plots for $\sigma_r/\bar{\sigma}$, $\sigma_t/\bar{\sigma}$ and $\sigma_z/\bar{\sigma}$ by 180° relative to the axis of the abscissa. The temperature distribution is obtained by rotating the $\bar{T} = [(T - Te)/(Ti - Te)]$ curve shown in Fig. 6.2, again by 180°, but around a vertical axis, so that the dimensionless temperature assumes values of 0 and 1 at the inner radius and outer radius respectively.

When the circular cylinder has thin walls, as in the case of a cylinder liner (e.g., cast-in liners, as a cast iron liner in a aluminum engine block, and dry or wet pressed-in liners, i.e., replaceable tubular sleeves inside cylinders of an internal combustion engine or a reciprocating compressor in which the pistons travel, see Gibbs 1971, Garro 1992, Giacosa 2000), the temperature distribution through the wall thickness may be assumed to be linear (Fig. 6.3), as will be demonstrated below, and corresponds to an equally linear distribution of hoop and longitudinal stresses. Radial stress is constantly zero. The maximum values of stress σ_t are found at the inner radius and outer radius, where they equal the factor $\bar{\sigma} = aE(T_i - T_e)/2(1 - \nu)$, as they are respectively compressive and tensile stresses for centrifugal heat flow. Stress σ_z is zero at the outer radius and reaches a maximum value of $2\bar{\sigma}$ at the inner radius. Figure 6.3 shows the curves for the three principal stresses in a circular cylinder with a ratio $\delta = r_e/r_i = 1.10$ subjected to centrifugal heat flow; once again, stresses are made dimensionless by dividing them by the factor $\bar{\sigma}$. This value of δ corresponds to a thickness/diameter ratio of 1/20, which as mentioned in Chap. 1 is by convention the upper limit for a thin wall.

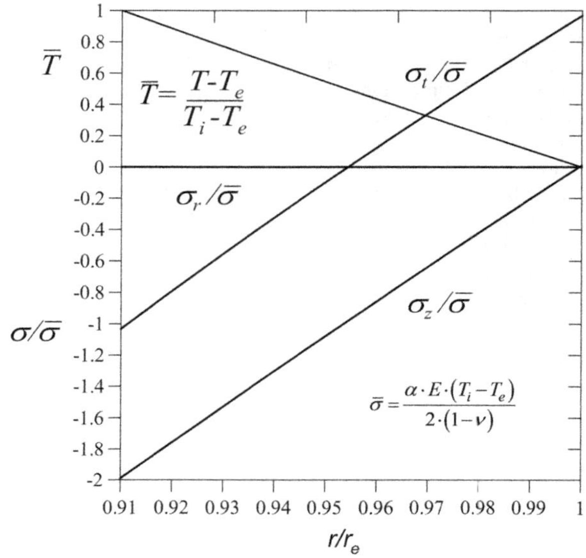

Fig. 6.3 Distribution curves of the three dimensionless principal stresses $\sigma_r/\bar{\sigma}$, $\sigma_t/\bar{\sigma}$ and $\sigma_z/\bar{\sigma}$ and dimensionless temperature $\bar{T}$ along the dimensionless radius of a thin-wall hollow cylinder with clamped ends subjected to centrifugal heat flow. Reference temperature $T_0 = 20$ °C; outer temperature $T_e = 20$ °C; $\delta = r_e/r_i = 1.10$

Stresses σ_t and σ_z are compressive at the inner radius and assume their maximum negative values of 1.80 MPa and 3.60 MPa respectively for each degree of temperature difference $(T_i - T_e)$. Passing from the inner radius to the outer, hoop stress drops linearly in absolute value, changes sign to become a tensile stress, and at the outer radius assumes a value equal and opposite to that at the inner radius. Axial stress also drops in absolute value, again following a linear distribution, remaining negative at all times until reaching zero at the outer radius. Figure 6.3 shows temperature in dimensionless form as the ratio $\bar{T} = (T - T_e)/(T_i - T_e)$; as can be seen, the temperature assumes values of 1 and 0 at the inner radius and outer radius respectively.

Figure 6.3 applies for centrifugal heat flow $(T_i > T_e)$. For centripetal heat flow $(T_i < T_e)$, the distribution of the two non-zero principal stresses through the wall thickness is obtained from that indicated in Fig. 6.3 as the mirror image of the curves shown therein, i.e., by rotating the plots for $\sigma_t/\bar{\sigma}$ and $\sigma_z/\bar{\sigma}$ by 180° relative to the axis of the abscissa. The temperature distribution is obtained by rotating the $\bar{T} = (T - T_e)/(T_i - T_e)$ curve shown in Fig. 6.3, again by 180°, but around a vertical axis, so that the dimensionless temperature assumes values of 0 and 1 at the inner radius and outer radius respectively.

Given axisymmetry and compatibility, the radius of curvature in the circumferential direction remains constant, so bending stresses are zero and the only stresses are membrane stresses. Near the ends of the cylinder, however, the structure's interactions with the constraint come into play. In particular, if there are clamped edges, localized bending effects arise which must be taken into account. The increases in the stress state can be quantified on the basis of Coates' theory (1930). Specifically, if the structure's material is steel, this increase is approximately 25 %.

The first two relations (6.50) can also be written in the following more compact form:

$$\begin{aligned}\sigma_r &= \frac{\alpha E}{2(1-\nu)}(T_i - T_e)H = \bar{\sigma}H \\ \sigma_t &= \frac{\alpha E}{2(1-\nu)}(T_i - T_e)K = \bar{\sigma}K,\end{aligned} \tag{6.51}$$

where $H = H(r_e/r_i,\ r_e/r)$ and $K = K(r_e/r_i,\ r_e/r)$ are functions of r_e/r_i and of r_e/r given respectively by:

$$\begin{aligned}H &= \frac{1}{(\ln r_e/r_i)}\left[-\ln\frac{r_e}{r} + \frac{r_i^2}{r_e^2 - r_i^2}\left(\frac{r_e^2}{r^2} - 1\right)\ln\frac{r_e}{r_i}\right] \\ K &= \frac{1}{(\ln r_e/r_i)}\left[1 - \ln\frac{r_e}{r} - \frac{r_i^2}{r_e^2 - r_i^2}\left(\frac{r_e^2}{r^2} + 1\right)\ln\frac{r_e}{r_i}\right].\end{aligned} \tag{6.52}$$

The behavior of these functions, and thus the distribution of the correlated radial and hoop stresses, is shown versus the dimensionless radial coordinate $x = (r-r_i)/(r_e-r_i)$ in Figs. 6.4 and 6.5.

As can be seen from these figures, the maximum values of H and K and, consequently, the maximum values of radial stress σ_r and hoop stress σ_t are heavily influenced by ratio δ. For $\delta \leq 1.10$ (thin-walled circular cylinder), the maximum values assumed by function K are an order of magnitude higher than those assumed by function H: when values of K are higher in absolute value than 1, H assumes a maximum value, again in absolute value, of around 0.023, which for steels corresponds to values of radial stress $\sigma_r = \bar{\sigma}H$ and hoop stress $\sigma_t = \bar{\sigma}K$ which are higher and lower by 0.04 MPa and 1 MPa respectively for a temperature difference of 1 °C. Accordingly, it can be maintained, with an error that is entirely negligible from the design standpoint, that radial stress is zero at any radius whatsoever. By way of example, we can compare the maximum values assumed by radial and hoop stresses in a steel circular cylinder, with $\delta = 1.10$, i.e., at the upper limit of thickness for a thin wall, for a temperature difference $(T_i - T_e) = 100$ °C. We will have $\sigma_{r,max} = \bar{\sigma}H_{max} = 180 \cdot 0.023 = 4.14$ MPa and $\sigma_{t,max} = \bar{\sigma}K_{max} = 180 \cdot 1.45 = 261$ MPa; there is thus a difference of almost two orders of magnitude between the hoop stress and the radial stress.

Both Figs. 6.4 and 6.5 apply for centrifugal heat flow ($T_i > T_e$). For centripetal heat flow ($T_i < T_e$), the distributions of the functions H and K versus the dimensionless radial coordinate x are obtained from those indicated in Figs. 6.4 and 6.5 as the mirror image of the family of curves shown therein, i.e., by rotating the plots for H and K by 180° relative to the axis of the abscissa.

The relations describing the stress state and radial displacement in a long solid circular cylinder (i.e., in a circular bar) with clamped ends subject to a thermal load whose temperature distribution with the radius follows Fourier's law are obtained from relations (6.50), with $r_i = 0$. We thus have:

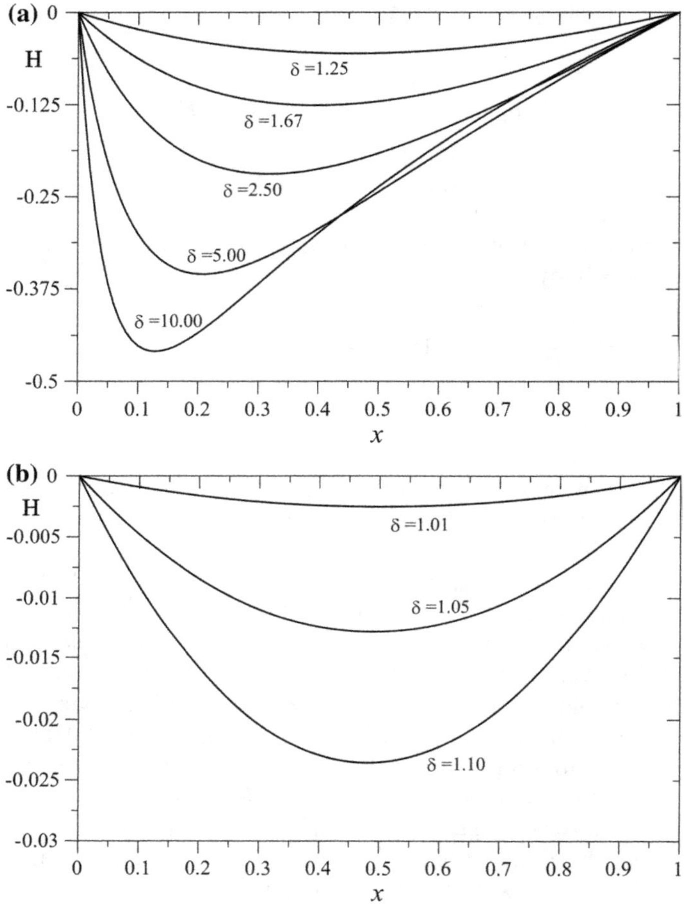

Fig. 6.4 Function H versus dimensionless radial coordinate $x = (r-r_i)/(r_e-r_i)$, for different values of parameter δ: **a** $1.25 < \delta < 10.00$; **b** $1.01 < \delta < 1.10$

$$
\begin{aligned}
\sigma_r &= -\frac{\alpha E(T_i - T_e)}{2(1-\nu)(\ln r_e/r_i)} \ln\frac{r_e}{r} \\
\sigma_t &= \frac{\alpha E(T_i - T_e)}{2(1-\nu)(\ln r_e/r_i)} \left(1 - \ln\frac{r_e}{r}\right) \\
\sigma_z &= \frac{\alpha E}{2(1-\nu)} \left[\frac{(T_i - T_e)}{(\ln r_e/r_i)} \left(\nu - 2\ln\frac{r_e}{r}\right) - 2T_e(1-\nu)\right] \\
u &= \alpha(1+\nu) r \left[\frac{(T_i - T_e)}{2(1-\nu)(\ln r_e/r_i)} \left(1 - \nu + \ln\frac{r_e}{r}\right) + T_e\right].
\end{aligned}
\tag{6.53}
$$

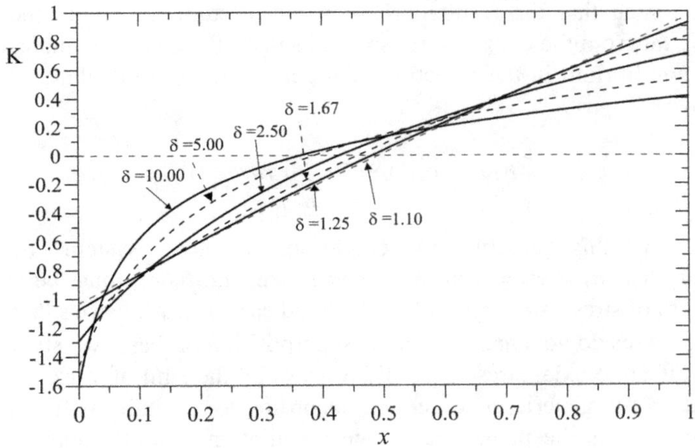

Fig. 6.5 Function K versus dimensionless radial coordinate $x = (r - r_i)/(r_e - r_i)$ for different values of parameter δ, with $1.10 < \delta < 10.00$

6.5 Circular Cylinder of Finite Length with Free Ends and Subjected to Thermal Load Defined by a Generic Function $T = T(r)$

We will now consider the somewhat more complex case of a hollow circular cylinder with free ends whose length is finite but large compared to its cross-section diameter, and subjected to thermal load. As there is no external axial force to balance the resultant of axial stress σ_z given by the third relation (6.40) and having the distribution shown in Fig. 6.6b, the latter cannot represent the actual axial stress state. The effective axial stress, which we will call σ'_z, should have a

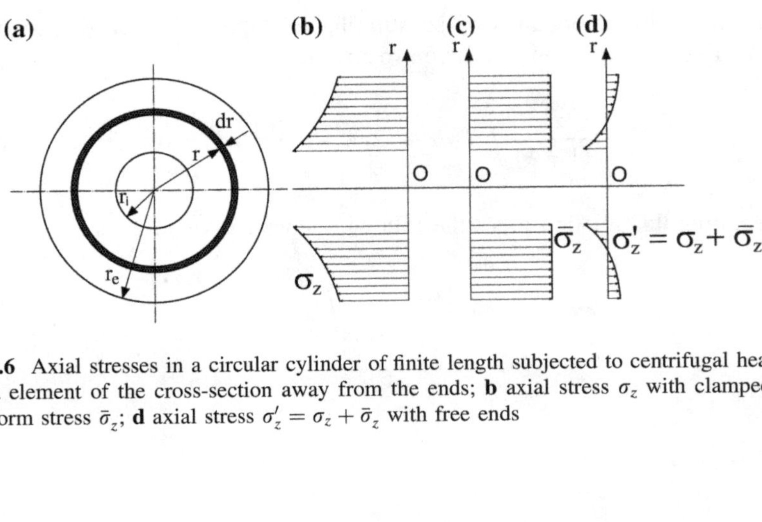

Fig. 6.6 Axial stresses in a circular cylinder of finite length subjected to centrifugal heat flow: **a** area element of the cross-section away from the ends; **b** axial stress σ_z with clamped ends; **c** uniform stress $\bar{\sigma}_z$; **d** axial stress $\sigma'_z = \sigma_z + \bar{\sigma}_z$ with free ends

distribution such that the resultant F'_z, on any section normal to the axis and sufficiently far from the ends is zero, so as to satisfy the condition indicated above of global equilibrium in the direction of the axis; because of relation (6.19), we should have:

$$F'_z = \int_A \sigma'_z dA = 2\pi \int_{r_i}^{r_e} \sigma'_z r dr = 0. \tag{6.54}$$

To satisfy this equilibrium condition, a new system of stresses $\bar{\sigma}_r$, $\bar{\sigma}_t$ and $\bar{\sigma}_z$, with $\bar{\sigma}_r = \bar{\sigma}_t = 0$ and $\bar{\sigma}_z = const$ (see Fig. 6.6c), must be superposed to the system of stresses σ_r, σ_t and σ_z calculated earlier. It is obvious that the radial and hoop stresses do not change with the superposition of these two stress systems, while the effective axial stress σ'_z will be given by the sum of stress σ_z, which is distributed on the generic cross-section according to a nonlinear function of the radius and satisfies the plane strain state assumption ($\varepsilon_z = 0$), and of stress $\bar{\sigma}_z$, which is distributed uniformly on the same section, so that the latter can remain plane and perpendicular to the axis in compliance with the assumption (6.18) of a strain state characterized by a uniform axial translation of a generic cross-section. In this case, this assumption enables us to write $\varepsilon'_z = \varepsilon_z + \bar{\varepsilon}_z = \bar{\varepsilon}_z = const$.

To determine $\bar{\sigma}_z$, we can use condition (6.54), setting $\sigma'_z = \sigma_z + \bar{\sigma}_z$. We thus have:

$$\int_{r_i}^{r_e} (\sigma_z + \bar{\sigma}_z) r dr = 0. \tag{6.55}$$

To facilitate calculating stress $\bar{\sigma}_z$ distributed uniformly on the cross-section, we prefer here to use the following expression of axial stress σ_z, which is also entirely general, instead of the third relation (6.40):

$$\sigma_z = -\frac{\alpha E T}{(1-\nu)} + 2\nu A. \tag{6.56}$$

This expression is obtained by substituting relations (6.34) in relation (6.29). Introducing this expression in relation (6.55) yields:

$$-\frac{\alpha E}{(1-\nu)} \int_{r_i}^{r_e} T r dr + (2\nu A + \bar{\sigma}_z) \int_{r_i}^{r_e} r dr = 0. \tag{6.57}$$

Integrating this relation gives the following constant value of $\bar{\sigma}_z$:

$$\bar{\sigma}_z = \frac{2\alpha E}{(1-\nu)(r_e^2 - r_i^2)} \int_{r_i}^{r_e} T r dr - 2\nu A. \tag{6.58}$$

Adding relation (6.58) and relation (6.56) gives the following expression of effective axial stress σ'_z:

$$\sigma'_z = \sigma_z + \bar{\sigma}_z = \frac{\alpha E}{(1-\nu)}\left(\frac{2}{r_e^2 - r_i^2}\int_{r_i}^{r_e} Trdr - T\right). \tag{6.59}$$

Figure 6.6d shows the distribution of effective axial stress σ'_z in a circular cylinder of finite length with free ends subjected to thermal load, as the sum of the distributions of σ_z and $\bar{\sigma}_z$. Naturally, the effective axial stress σ'_z is correlated with a uniform axial strain ε'_z as well as with an axial displacement w which is independent of radius r in the generic cross-section of abscissa z. It is equally obvious that this expression of σ'_z does not apply at the outer edges, as there can be no reaction on free ends and σ'_z is thus zero at all points (at any radius). It goes in any case without saying that, in accordance with Saint–Venant's principle, relation (6.59) represents the actual axial stress state of a circular cylinder with free ends starting from a distance from the edges which is of the same order of magnitude as the outer radius.

In conclusion, the three principal stresses which define the stress state in a circular cylinder of finite length with free ends, with no surface force distributions at either the inner or outer radii and subjected to a temperature gradient along the radius according to any function $T = T(r)$ are as follows:

$$\begin{aligned}
\sigma_r &= \frac{\alpha E}{1-\nu}\left[\frac{1}{r_e^2 - r_i^2}\left(1 - \frac{r_i^2}{r^2}\right)\int_{r_i}^{re} Trdr - \frac{1}{r^2}\int_{r_i}^{r} Trdr\right] \\
\sigma_t &= \frac{\alpha E}{1-\nu}\left[\frac{1}{r_e^2 - r_i^2}\left(1 + \frac{r_i^2}{r^2}\right)\int_{r_i}^{re} Trdr + \frac{1}{r^2}\int_{r_i}^{r} Trdr - T\right] \\
\sigma_z &= \frac{\alpha E}{1-\nu}\left(\frac{2}{r_e^2 - r_i^2}\int_{r_i}^{re} Trdr - T\right).
\end{aligned} \tag{6.60}$$

Introducing relations (6.60) in relation (6.38), yields the following relation expressing radial displacement $u = u(r)$:

$$u = \frac{\alpha}{r(1-\nu)}\left\{\left[\frac{(1-3\nu)r^2 + r_i^2(1+\nu)}{r_e^2 - r_i^2}\right]\int_{r_i}^{re} Trdr + (1+\nu)\int_{r_i}^{r} Trdr\right\}. \tag{6.61}$$

As relation (6.18) gives $\varepsilon_z = dw/dz = const$ and, consequently, $w = \varepsilon_z \cdot z$, given that $\varepsilon_z = const$ and the integration constant, with $w = 0$ for $z = 0$, is zero, substituting relations (6.60) in the third Eq. (6.21) and multiplying by z gives the following relation expressing axial displacement $w = w(z)$ which, assuming a strain state characterized by a uniform axial translation of a generic cross-section of the circular cylinder, is independent of variable r:

$$w = z\varepsilon_z = \frac{z}{E}[\sigma_z - \nu(\sigma_r + \sigma_t) + E\alpha T] = \frac{2\alpha z}{r_e^2 - r_i^2}\int_{r_i}^{re} Trdr. \quad (6.62)$$

Relations (6.60)–(6.62) fully describe the stress state and displacements of the hollow circular cylinder with free ends considered here. If the same circular cylinder is also subjected to surface force distributions acting at the inner and outer radii as well as at the free ends, the overall stress state and radial and axial displacements can once again be determined as the sum of the contributions made by the various loads acting on the structure. Relations (6.60)–(6.62) also apply if the circular cylinder is not hollow; in this case, it is sufficient to set $r_i = 0$. As has been repeatedly specified, the problem regarding singularity remains unchanged.

To sum up, the general relations expressing the three principal stresses σ_r, σ_t and σ_z, radial displacement u and axial displacement w in a circular cylinder of finite length with free ends subjected to a thermal load defined by a generic radial temperature distribution $T = T(r)$ are as follows:

$$\sigma_r = \frac{\alpha E}{1-\nu}\left[\frac{1}{r_e^2 - r_i^2}\left(1 - \frac{r_i^2}{r^2}\right)\int_{r_i}^{re} Trdr - \frac{1}{r^2}\int_{r_i}^{r} Trdr\right]$$

$$\sigma_t = \frac{\alpha E}{1-\nu}\left[\frac{1}{r_e^2 - r_i^2}\left(1 + \frac{r_i^2}{r^2}\right)\int_{r_i}^{re} Trdr + \frac{1}{r^2}\int_{r_i}^{r} Trdr - T\right]$$

$$\sigma_z = \frac{\alpha E}{1-\nu}\left(\frac{2}{r_e^2 - r_i^2}\int_{r_i}^{re} Trdr - T\right) \quad (6.63)$$

$$u = \frac{\alpha}{r(1-\nu)}\left\{\left[\frac{(1-3\nu)r^2 + r_i^2(1+\nu)}{r_e^2 - r_i^2}\right]\int_{r_i}^{re} Trdr + (1+\nu)\int_{r_i}^{r} Trdr\right\}$$

$$w = z\varepsilon_z = \frac{z}{E}[\sigma_z - \nu(\sigma_r + \sigma_t) + E\alpha T] = \frac{2\alpha z}{r_e^2 - r_i^2}\int_{r_i}^{re} Trdr.$$

As was specified earlier, the first two relations (6.63) and the first two relations (6.40) coincide. From the first three relations (6.63), moreover, it can also be concluded that $\sigma_z = (\sigma_r + \sigma_t)$ in a circular cylinder with free ends subjected to a radial temperature gradient, i.e., the longitudinal stress is the sum of the radial and hoop stresses. Here again, the corresponding relations for a solid circular cylinder with free ends, with no surface force distribution at the outer radius and subjected to a temperature gradient along the radius according to any function $T = T(r)$, can be derived from those given above for a circular cylinder of annular cross-section by setting $r_i = 0$ in relations (6.63). We thus obtain the following five relations:

$$\sigma_r = \frac{\alpha E}{1-v}\left(\frac{1}{r_e^2}\int_0^{re} Trdr - \frac{1}{r^2}\int_0^{r} Trdr\right)$$

$$\sigma_t = \frac{\alpha E}{1-v}\left(\frac{1}{r_e^2}\int_0^{re} Trdr + \frac{1}{r^2}\int_0^{r} Trdr - T\right)$$

$$\sigma_z = \frac{\alpha E}{1-v}\left(\frac{2}{r_e^2}\int_0^{re} Trdr - T\right) \qquad (6.64)$$

$$u = \frac{\alpha}{r(1-v)}\left[\left(\frac{(1-3v)r^2}{r_e^2}\right)\int_0^{re} Trdr + (1+v)\int_0^{r} Trdr\right]$$

$$w = z\varepsilon_z = \frac{z}{E}[\sigma_z - v(\sigma_r+\sigma_t) + E\alpha T] = \frac{2\alpha z}{r_e^2}\int_0^{re} Trdr.$$

6.5.1 Circular Cylinder of Finite Length with Free Ends and Subjected to Thermal Load Defined by Fourier's Law

Here again, introducing relation (6.49) in relations (6.63) and performing integrations gives the following:

$$\sigma_r = \frac{\alpha E(T_i - T_e)}{2(1-v)(\ln r_e/r_i)}\left[-\ln\frac{r_e}{r} + \frac{r_i^2}{r_e^2 - r_i^2}\left(\frac{r_e^2}{r^2} - 1\right)\ln\frac{r_e}{r_i}\right]$$

$$\sigma_t = \frac{\alpha E(T_i - T_e)}{2(1-v)(\ln r_e/r_i)}\left[1 - \ln\frac{r_e}{r} - \frac{r_i^2}{r_e^2 - r_i^2}\left(\frac{r_e^2}{r^2} + 1\right)\ln\frac{r_e}{r_i}\right]$$

$$\sigma_z = \frac{\alpha E(T_i - T_e)}{2(1-v)(\ln r_e/r_i)}\left[1 - 2\ln\frac{r_e}{r} - \frac{2r_i^2}{r_e^2 - r_i^2}\ln\frac{r_e}{r_i}\right]$$

$$u = \frac{\alpha(T_i - T_e)}{2r(1-v)(\ln r_e/r_i)}\left\{r^2\left[(1-v) + (1+v)\ln\frac{r_e}{r}\right] - \frac{\left[(1-3v)r^2 + (1+v)r_e^2\right]}{r_e^2 - r_i^2} r_i^2 \ln\frac{r_e}{r_i}\right\}$$

$$w = \frac{\alpha(T_i - T_e)}{(\ln r_e/r_i)} z\left(\frac{1}{2} - \frac{r_i^2}{r_e^2 - r_i^2}\ln\frac{r_e}{r_i}\right).$$

(6.65)

These relations describe the stress state and the radial and axial displacements in hollow circular cylinders of finite length with free ends, subject to a thermal load whose temperature distribution along the radius follows Fourier's law.

The first two relations (6.65) coincide with the corresponding relations (6.50); from the standpoint of stresses σ_r and σ_t there is thus no difference between the circular cylinder with free ends and the circular cylinder clamped at its ends or indefinitely extended in the direction of its axis. From the first three relations (6.65), we find that the longitudinal stress σ_z in the same circular cylinder is equal to the sum $(\sigma_r + \sigma_t)$. These relations also show that a constant temperature in this structure with free ends results in a stress state characterized by $\sigma_r = \sigma_t = \sigma_z = 0$; unlike the corresponding structure constrained at its ends and subjected to the same temperature distribution $T = const$, the longitudinal stress will be zero, as the structure is free to expand in the axial direction.

It should also be noted that here, in contrast to the situation for the circular cylinder constrained at its ends or indefinitely extended in the direction of its axis, the radial and longitudinal expansions are both unconstrained, and terms in T_e consequently do not appear in the last two relations (6.65). This conclusion is reached using relations (6.38) and (6.62), which express radial and longitudinal displacements respectively, and assuming that displacements are evaluated starting from the configuration for temperature T_e appearing in relation (6.49) and taken as the reference (or assembly) temperature. If the latter differs from T_e, the additional terms $r\alpha T_e$ and $z\alpha T_e$ would appear in the expressions $u = u(r)$ and $w = w\,(z)$ respectively, which are given by the last two relations (6.65), while the value of T_e to be introduced in calculations is the difference between the temperature measured on the outer surface and reference temperature T_0.

It can be concluded from the first relation (6.65) that the radial stress σ_r is zero at the inner radius ($r = r_i$) and at the outer radius ($r = r_e$), while it retains the same sign for every other value of r, as it is negative for centrifugal heat flow ($T_i > T_e$) and positive for centripetal heat flow ($T_i < T_e$). The second and third relations (6.65) show that stresses σ_t and σ_z assume the same values at the inner and outer radii, where they are respectively:

$$
\begin{aligned}
(\sigma_t)_{r=r_i} = (\sigma_z)_{r=r_i} &= \frac{\alpha E}{2(1-\nu)} \frac{T_i - T_e}{\ln(r_e/r_i)} \left(1 - \frac{2r_e^2}{r_e^2 - r_i^2} \ln\frac{r_e}{r_i}\right) \\
(\sigma_t)_{r=r_e} = (\sigma_z)_{r=r_e} &= \frac{\alpha E}{2(1-\nu)} \frac{T_i - T_e}{\ln(r_e/r_i)} \left(1 - \frac{2r_i^2}{r_e^2 - r_i^2} \ln\frac{r_e}{r_i}\right).
\end{aligned}
\tag{6.66}
$$

These are compression stresses towards the inner radius and tensile stresses towards the outer radius for centrifugal heat flow, and the converse for centripetal heat flow. For all other values of r differing from $r = r_i$ and from $r = r_e$, we have $\sigma_t \neq \sigma_z$.

Figure 6.7 shows distribution curves of temperature and principal stresses in dimensionless form in a steel circular cylinder ($E = 210$ GPa; $\nu = 0.3$; $\alpha = 12 \times 10^{-6}$ °C^{-1}), with $\delta = 2$ ($r_e = 2r_i$), free at the ends and subjected to centrifugal heat flow. Stresses and temperature are made dimensionless in the same way illustrated for the circular cylinder clamped at the ends.

As can be seen from the figure, the absolute maximum values of the stresses are those for longitudinal and hoop stresses at the inner radius, where the two values are equal, negative (i.e., are compressive), and equal to 2.20 MPa for each degree of difference between the temperatures at the inner and outer radii. For a temperature difference $(T_i - T_e)$ of 100 °C, both of these stresses would be 220 MPa. In this case, for steady-state heat flow, i.e., for static stress, for the same AISI 1060 steel considered for the circular cylinder clamped at the ends, and again using the three σ_{max}, τ_{max} and maximum distortion energy strength theories, this stress state would correspond to a safety factor of 2.18 for all three theories. All other conditions remaining equal, the circular cylinder with free ends is much more resistant to thermal load than the circular cylinder clamped at the ends.

Passing from the inner radius to the outer, these stresses drop in absolute value, change sign to become a tensile stress, and at the outer radius reach the same value of 1.34 MPa for each °C of difference between the temperatures at the inner and outer radii. For steel or metals in general, the designer's attention must be focused on the inner radius, where the material is at greatest risk. For reinforced concrete or masonry construction as used in certain smokestacks, this tensile stress state at the outer radius often causes more or less generalized fractures, as these materials have low tensile strength.

In the entire range of variability for r, the distribution curve of axial stress remains below that for hoop stress except at the ends, where the two values coincide.

Figure 6.7 applies for centrifugal heat flow $(T_i > T_e)$. For centripetal heat flow $(T_i < T_e)$, here as for the cylindrical structure clamped at the ends, the distribution

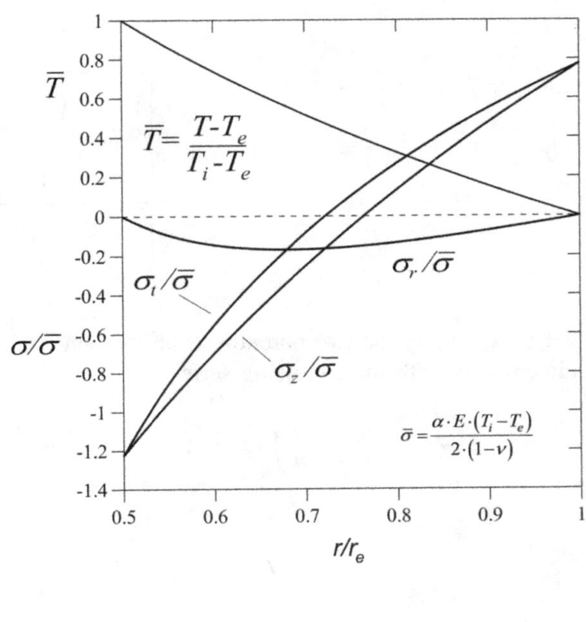

Fig. 6.7 Distribution curves of the three dimensionless principal stresses $\sigma_r/\bar{\sigma}$, $\sigma_t/\bar{\sigma}$ and $\sigma_z/\bar{\sigma}$ and dimensionless temperature $\bar{T}$ along the dimensionless radius of a hollow circular cylinder with free ends subjected to centrifugal heat flow: reference temperature $T_0 = 20$ °C; outer temperature $T_e = 20$ °C; $\delta = 2$

of the three principal stresses and the temperature distribution along the dimensionless radius are obtained from those indicated in Fig. 6.7 by rotating the curves for $\sigma_r/\bar{\sigma}$, $\sigma_t/\bar{\sigma}$ and $\sigma_z/\bar{\sigma}$ and the $\bar{T} = (T - T_e)/(T_i - T_e)$ curve by 180° relative to the axis of the abscissa and to a vertical axis respectively.

When the circular cylinder has thin wall, as in the case of a cylinder liner for an internal combustion engine or a reciprocating compressor, the temperature distribution along the dimensionless radius may be assumed to be linear, and corresponds to an equally linear distribution of hoop and longitudinal stresses whose curves overlap for the entire range of variability for r. Radial stress is constantly zero.

To see that this is in fact true, it is sufficient to consider relations (6.66), which express hoop and axial stresses at the inner and outer radii as a function of ratio r_e/r_i. These relations can be rewritten to introduce the circular cylinder's wall thickness s, as $r_e/r_i = (1 + s/r_i)$. If s is small compared to r_i, a power series expansion of $\ln(r_e/r_i)$ yields:

$$\ln\frac{r_e}{r_i} = \frac{s}{r_i}\left[1 - \frac{1}{2}\left(\frac{s}{r_i}\right) + \frac{1}{3}\left(\frac{s}{r_i}\right)^2 - \frac{1}{4}\left(\frac{s}{r_i}\right)^3 + \cdots\right]. \tag{6.67}$$

Thus, by stopping the expansion at the third term in the square brackets, we find that the expression

$$\frac{1}{\ln(r_e/r_i)} \cdot \left(1 - \frac{2 \cdot r_e^2}{r_e^2 - r_i^2} \cdot \ln\frac{r_e}{r_i}\right), \tag{6.68}$$

which appears in the second member of the first relation (6.66) can be reduced to the following

$$\begin{aligned}\frac{1}{\ln(r_e/r_i)} \cdot \left(1 - \frac{2 \cdot r_e^2}{r_e^2 - r_i^2} \cdot \ln\frac{r_e}{r_i}\right) &= \frac{1}{\frac{s}{r_i} \cdot \left[1 - \frac{1}{2} \cdot \left(\frac{s}{r_i}\right) + \frac{1}{3} \cdot \left(\frac{s}{r_i}\right)^2\right]} - \frac{2 \cdot \left(1 + \frac{s}{r_i}\right)^2}{\left(1 + \frac{s}{r_i}\right)^2 - 1} \\ &= \frac{1}{\frac{s}{r_i} \cdot \left[1 - \frac{1}{2} \cdot \left(\frac{s}{r_i}\right) + \frac{1}{3} \cdot \left(\frac{s}{r_i}\right)^2\right]} - \frac{\left(1 + \frac{s}{r_i}\right)^2}{\frac{s}{r_i} \cdot \left(1 + \frac{s}{2 \cdot r_i}\right)}.\end{aligned} \tag{6.69}$$

By expanding the two parentheses in the denominator of the last expression of this equality with the following series

$$(1 + x)^a = \sum_{n=0}^{\infty} \binom{a}{n} \cdot x^n = 1 + a \cdot x + \frac{a \cdot (a - 1)}{2!} x^2 + \ldots, \tag{6.70}$$

with $a = -1$ (the two parentheses are in the denominator), stopping the expansion at the second degree terms, whereby:

$$(1+x)^{-1} = 1 - x + x^2 + \ldots \tag{6.71}$$

and setting x equal to $-(s/2r_i - s^2/3r_i^2)$ in the square brackets of relation (6.69) and equal to $s/2r_i$ in the parentheses, we can then obtain the following from relation (6.69):

$$\begin{aligned}\frac{1}{\ln(r_e/r_i)} \cdot \left(1 - \frac{2 \cdot r_e^2}{r_e^2 - r_i^2} \cdot \ln\frac{r_e}{r_i}\right) &= \frac{r_i}{s} \cdot \left[\left(1 + \frac{s}{2 \cdot r_i} - \frac{s^2}{3 \cdot r_i^2}\right) - \left(1 + \frac{s}{r_i}\right)^2 \cdot \left(1 - \frac{s}{2 \cdot r_i} - \frac{s^2}{4 \cdot r_i^2}\right)\right] \\ &= -\left(1 + \frac{s}{3 \cdot r_i}\right).\end{aligned} \tag{6.72}$$

In the series expansion of the square brackets appearing in relation (6.69), as in the elaboration of the square brackets appearing in relation (6.72), it should be noted that the powers $(s/r_i)^3$ and $(s/r_i)^4$ by respect to the linear and quadratic terms of this ratio were ignored.

Proceeding in the same way with the expression

$$\frac{1}{\ln r_e/r_i} \cdot \left(1 - \frac{2 \cdot r_i^2}{r_e^2 - r_i^2} \cdot \ln\frac{r_e}{r_i}\right), \tag{6.73}$$

which appears in the second relation (6.66), we obtain:

$$\frac{1}{\ln(r_e/r_i)} \cdot \left(1 - \frac{2 \cdot r_i^2}{r_e^2 - r_i^2} \cdot \ln\frac{r_e}{r_i}\right) = 1 - \frac{s}{3 \cdot r_i}. \tag{6.74}$$

Substituting expressions (6.72) and (6.74) in relations (6.66) gives:

$$\begin{aligned}(\sigma_t)_{r=r_i} &= (\sigma_z)_{r=r_i} = \frac{\alpha E(T_i - T_e)}{2(1-\nu)}\left(1 + \frac{s}{3r_i}\right) \\ (\sigma_t)_{r=r_e} &= (\sigma_z)_{r=r_e} = \frac{\alpha E(T_i - T_e)}{2(1-\nu)}\left(1 - \frac{s}{3r_i}\right).\end{aligned} \tag{6.75}$$

If $s \ll r_i$, i.e., in the case of a very thin wall, relations (6.75) can be further simplified to become:

$$\begin{aligned}(\sigma_t)_{r=r_i} &= (\sigma_z)_{r=r_i} = \frac{\alpha E(T_i - T_e)}{2(1-\nu)} \\ (\sigma_t)_{r=r_e} &= (\sigma_z)_{r=r_e} = \frac{\alpha E(T_i - T_e)}{2(1-\nu)}.\end{aligned} \tag{6.76}$$

Consequently, the values assumed by the hoop and axial stresses at the inner radius, which equal the factor $\bar{\sigma}$, are equal and opposite to those assumed by the same stresses at the outer radius. With the same degree of approximation, we can say that temperature variation with the radius is linear, and thus

$$T = T_i - (T_i - T_e)\frac{r - r_i}{r_e - r_i} = T_e + (T_i - T_e)\frac{r_e - r}{r_e - r_i}. \tag{6.77}$$

Introducing this relation in the first three relations (6.63) yields:

$$\begin{aligned} \sigma_r &= 0 \\ \sigma_t = \sigma_z &= \frac{\alpha E(T_i - T_e)}{2(1 - \nu)}\left(1 - 2\frac{r_e - r}{r_e - r_i}\right). \end{aligned} \tag{6.78}$$

For $r = r_i$ and for $r = r_e$, the second relation (6.78) yields relations (6.76).

Figure 6.8 shows the distribution curves of the three principal stresses, which here again are made dimensionless by dividing them by the factor $\bar{\sigma}$, in a circular cylinder with a ratio $\delta = r_e/r_i = 1.10$ subjected to centrifugal heat flow. This value of δ corresponds to a thickness/diameter ratio of 1/20, which, as has been mentioned several times, is by convention the upper limit for a thin wall. In this condition (where the thickness is small, but not very small), the maximum values for stresses σ_t and σ_z at the inner and outer radii are only approximately equal to the factor $\bar{\sigma} = a \cdot E \cdot (T_i - T_e) / 2 \cdot (1 - \nu)$, but the differences are not significant for design purposes.

Stresses σ_t and σ_z are compressive at the inner radius and assume their maximum negative value of around 1.80 MPa for each degree of temperature difference $(T_i - T_e)$. Passing from the inner radius to the outer, these stresses drop linearly in absolute value, change sign to become tensile stresses, and at the outer radius

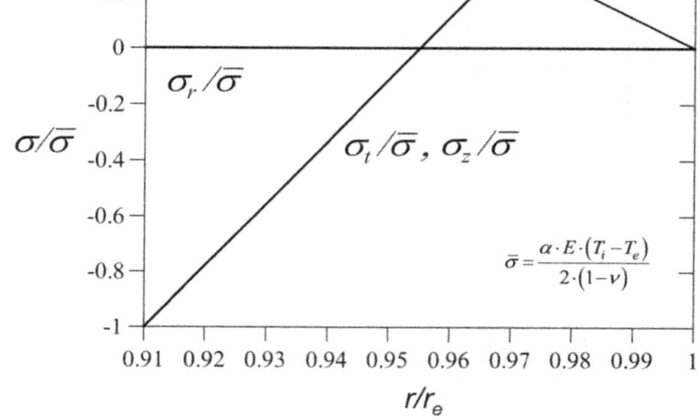

Fig. 6.8 Distribution curves of the three dimensionless principal stresses $\sigma_r/\bar{\sigma}$, $\sigma_t/\bar{\sigma}$ and $\sigma_z/\bar{\sigma}$ and dimensionless temperature $\bar{T}$ along the dimensionless radius of a thin-walled hollow circular cylinder with free ends subjected to centrifugal heat flow: reference temperature $T_0 = 20$ °C; outer temperature $T_e = 20$ °C; $\delta = 1.10$

assume a value approximately equal and opposite to that at the inner radius. Figure 6.8 shows temperature in dimensionless form as the ratio $\bar{T} = (T - T_e)/(T_i - T_e)$.

These stress values correspond to the radial and hoop bending stresses found at the upper face and lower face of a circular plate clamped at the outer radius and subject to a temperature gradient through the thickness, with upper face temperature T_i and lower face temperature T_e. It is thus clear that the stresses in the thin-walled circular cylinder and those in the plate are comparable only as regards the numerical values and their behavior in the direction of the temperature gradient, while the stress field is different, as it is characterized by σ_t and σ_z in the circular cylinder and by σ_r and σ_t in the circular plate.

Figure 6.8 applies for centrifugal heat flow ($T_i > T_e$). For centripetal heat flow ($T_i < T_e$), the distribution of the two non-zero principal stresses along the dimensionless radius is obtained from that indicated in Fig. 6.8 as the mirror image of the curves shown therein, i.e., by rotating the plots for $\sigma_t/\bar{\sigma}$ and $\sigma_z/\bar{\sigma}$ by 180° relative to the axis of the abscissa. The temperature distribution is obtained by rotating the $\bar{T} = (T - T_e)/(T_i - T_e)$ curve shown in Fig 6.8, again by 180°, but around a vertical axis, so that the dimensionless temperature assumes values of 0 and 1 at the inner radius and outer radius respectively.

6.6 Edge Effects on the Free Ends of a Thin-Walled Circular Cylinder Subjected to Thermal Load

What we have said so far for the circular cylinder of finite length with free ends and subjected to thermal load holds as long as it can be assumed that there is a plane strain state characterized by a uniform axial translation of a generic cross-section ($\varepsilon_z = \varepsilon_0 = const$). For a cylinder of finite length, this assumption is respected at a sufficient distance from the ends, but as this distance is reduced, the stress state should vary given that there should be a plane stress state at the ends ($\sigma_z = 0$).

The stress and strain state at the free ends can be rigorously analyzed by using the relations governing heat flow and those for thermoelasticity simultaneously. With the rather rough assumption that the finite length does not have an appreciable influence on temperature distribution through the wall thickness and that the temperature varies according to a linear function along the radius, an approximate solution for the thin-walled circular cylinder can be reached which consists of calculating the local stress state as the superposition of that calculated above assuming a plane strain state characterized by a uniform axial translation of a generic cross-section and that determined by a localized bending moment which generates stresses σ_z equal and opposite to these stresses at the free ends. In this way, we will have $\sigma_z = 0$ at the free ends. This approach, though it respects the equilibrium conditions, leads to results that are highly approximate (Belloni and Lo Conte 2002).

Bearing relations (6.76) and thin cylindrical shell theory in mind, it is obvious that the longitudinal bending moment M_{z0} per unit length of the edge to be applied to the free ends of the circular cylinder (for example, the end having coordinate $z = 0$) must satisfy the relation $\sigma_z = 6 \cdot M_{z0}/s^2$, so that

$$M_{z0} = \frac{\alpha E(T_i - T_e)s^2}{12(1-\nu)}. \tag{6.79}$$

For centrifugal heat flow, it is also obvious that M_z must have a direction such that the fibers at the inner radius will be in tension and those at the outer radius will be compressed (this is the direction assumed as positive), with the converse being true for centripetal heat flow.

From the general bending theory of thin cylindrical shells (Timoshenko and Woinowsky-Krieger 1959) we have:

$$\begin{aligned} M_z &= D\frac{d^2u}{dz^2} = M_{z0}e^{-\beta z}(\cos\beta z + \sin\beta z) \\ M_t &= \nu M_z = \nu M_{z0}e^{-\beta z}(\cos\beta z + \sin\beta z) \\ u &= \frac{M_{z0}}{2D\beta^2}e^{-\beta z}(\cos\beta z - \sin\beta z), \end{aligned} \tag{6.80}$$

where, with the meaning of the other symbols remaining the same, M_t is the anticlastic tangential bending moment, $D = Es^3/12(1-\nu^2)$ is the flexural rigidity of the cylindrical shell, $\beta = \sqrt[4]{3(1-\nu^2)/R^2s^2}$ is a characteristic parameter of the cylindrical shell having the dimension $[length]^{-1}$, and $R = (r_i + r_e)/2$ is the average radius. It should be noted that β here does not have the same meaning as the $(\beta = 1/\delta = r_i/r_e)$ used so far. In the edge effects of thin cylindrical shell bending theory, β denotes a magnitude linked to the thickness, the flexural rigidity and the elastic properties of the shell material.

Figure 6.9a shows the distribution curves of the function $T = T(r)$ through the wall thickness for centrifugal heat flow and of the corresponding axial stress σ_z on the end surface with axial coordinate $z = 0$. Figure 6.9b, on the other hand, shows the distribution curve of the radial displacement $u = u(z)$ of the mean cylindrical surface's generating line (i.e., the line bisecting the thickness of the cylindrical shell) and thus of the deflection resulting from applying the edge bending moment M_{z0}, uniformely distributed along the edge $z = 0$.

At the inner edge, the axial and hoop stresses due to the localized bending moment will be:

$$\begin{aligned} \sigma_{z,M} &= \frac{6M_z}{s^2} = \frac{6M_{z0}}{s^2}e^{-\beta z}(\cos\beta z + \sin\beta z) \\ \sigma_{t,M} &= \frac{E}{R}u + \frac{6M_t}{s^2} = M_{z0}e^{-\beta z}\left[\frac{E}{2RD\beta^2}(\cos\beta z - \sin\beta z) + \frac{6\nu}{s^2}(\cos\beta z + \sin\beta z)\right]. \end{aligned} \tag{6.81}$$

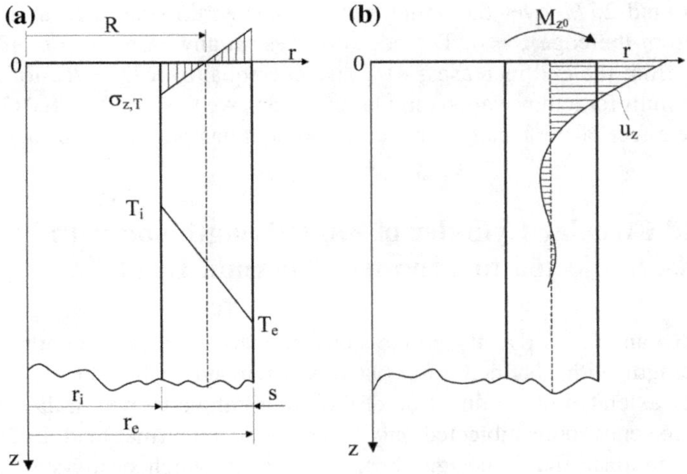

Fig. 6.9 **a** Distribution curves of $T = T(r)$ and the correlated axial stress σ_z; **b** deflection $u = u(z)$ resulting from localized bending moment M_{z0}

At the outer edge, these stresses will be:

$$\sigma_{z,M} = -\frac{6M_z}{s^2} = -\frac{6M_{z0}}{s^2}e^{-\beta z}(\cos\beta z + \sin\beta z)$$
$$\sigma_{t,M} = \frac{E}{R}u - \frac{6M_t}{s^2} = M_{z0}e^{-\beta z}\left[\frac{E}{2RD\beta^2}(\cos\beta z - \sin\beta z) - \frac{6\nu}{s^2}(\cos\beta z + \sin\beta z)\right]. \tag{6.82}$$

At the end of the circular cylinder ($z = 0$), the overall axial and hoop stresses, which are the sum of those due to thermal load and those caused by the localized bending moment, will thus be as follows at the inner edge:

$$\sigma_z = 0$$
$$\sigma_t = \sigma_{t,T} + \sigma_{t,M} = -\frac{\alpha E(T_i - T_e)}{2(1-\nu)}\left(1 - \sqrt{\frac{1-\nu^2}{3}} - \nu\right) = \sigma_{t,T}\left(1 - \sqrt{\frac{1-\nu^2}{3}} - \nu\right). \tag{6.83}$$

At the outer edge, they will be:

$$\sigma_z = 0$$
$$\sigma_t = \sigma_{t,T}\left(1 - \sqrt{\frac{1-\nu^2}{3}} - \nu\right). \tag{6.84}$$

For centrifugal heat flow, it can thus be concluded that hoop stress decreases at the inner radius and increases at the outer radius. For $\nu = 0.3$ (steels), the increase in σ_t is around 25 % over the value with a plane strain state, i.e., at a sufficient distance from the edge $z = 0$. The edge stresses, in any case, drop rapidly as the distance z from the end increases. At a distance equal to R ($z = R$), for $R/s = 10$ (the upper limit for a thin-walled circular cylinder), we have $e^{-\beta z} = 0.017$; in other words, the effect of the localized bending moment has practically disappeared.

6.7 Solid Circular Cylinder of Finite Length and with Free Ends, Subjected to Transient Thermal Load

In Sects. 6.4 and 6.5, we analyzed the stress and strain states in circular cylinders of finite length with clamped ends (as was discussed earlier, circular cylinders indefinitely extended in the direction of their axis can be treated in the same way) and with free ends, both subjected only to steady-state thermal load. In Sect. 6.4 it was also mentioned that in design, there are cases in which steady-state heat flow conditions are not respected. One of these cases is that of the solid cylindrical rod of finite length with free ends. In this case, the thermal transient must always be considered, as the entire mass will be at the same temperature after the transient has ended and there will be no stresses due to thermal loading.

We will consider a solid circular cylinder (solid cylindrical rod) with free ends, initially at reference temperature T_0, and we will suppose that, starting from instant $t = 0$, the temperature at the outer radius ($r = r_e$) is brought to and maintained at $T_e \neq T_0$. After a theoretically infinite period of time, the temperature at all points of the cylindrical rod's mass should be $T = T_e$.

At the generic instant of time t starting from the initial time $t = 0$, temperature $T = T(r, t)$ is given by solving the following partial differential equation of first order:

$$2\pi rk\frac{\partial T}{\partial r}dt - 2\pi(r+dr)k\frac{\partial}{\partial r}\left(T + \frac{\partial T}{\partial r}dr\right)dt - 2\pi\gamma c\frac{\partial T}{\partial t}rdrdt = 0. \tag{6.85}$$

For the annular element of the cylindrical rod having unitary axial length with inner radius r and outer radius $r + dr$, and bearing in mind that in accordance with Fourier's law, the heat flow moving through the generic cylindrical surface of radius r is proportional to the temperature gradient, this equation expresses the thermal balance between the heat entering the cylinder of radius r, the heat exiting from the cylinder of radius $r + dr$ and the heat accumulated in the element, expressed respectively by the three monomials appearing in Eq. (6.85). In this equation, with the meaning of the other symbols remaining the same, c denotes the specific heat of the rod material.

Expanding Eq. (6.85) and ignoring the higher-order infinitesimals in the first member yields the following partial differential equation of second order:

$$kr\frac{\partial^2 T}{\partial r^2} + k\frac{\partial T}{\partial r} + c\gamma r\frac{\partial T}{\partial t} = 0. \tag{6.86}$$

The solution of this latter equation, which satisfies the boundary conditions and the assigned initial conditions, is given by the series expansion:

$$T = (T_e - T_0)\sum_{n=1}^{\infty} A_n J_0\left(B_n \frac{r}{r_e}\right) e^{-\frac{k\beta_n^2 t}{\gamma c r^2}}, \tag{6.87}$$

where J_0 and J_1 are the Bessel functions of the first kind, respectively of order zero and of order one, β_n are the zeros of the equation $J_0(x) = 0$ and $A_n = 2/[\beta_n \cdot J_1(\beta_n)]$. In this connection, it should be borne in mind that the Bessel functions of the first kind $J_{\pm n}(x)$, with n being a positive or negative integer and, in particular, zero, are the solutions of the second order differential equation:

$$x^2 \frac{d^2 w}{dx^2} + x\frac{dw}{dx} + (x^2 - n^2)w = 0, \tag{6.88}$$

which is the Bessel differential equation. It should also be borne in mind that, for every $n = p$, Eq. (6.88) admits other solutions: these are the Bessel functions of the second kind $Y_n(x)$, also called Weber functions, and of the third kind $H_n^{(1)}(x)$ and $H_n^{(2)}(x)$, also called Hankel functions(Abramovitz and Stegun 1972; Smirnov 1970).

From Fig. 6.10, which shows the curves for the Bessel functions $J_0(x)$ and $J_1(x)$ as well as for the Weber functions $Y_0(x)$ and $Y_1(x)$, we can see that the first three zeros of equation $J_0(x)$ are $\beta_1 = 2.4048$, $\beta_2 = 5.5201$ and $\beta_3 = 8.6537$, while the first three zeros of equation $J_1(x)$ are $\beta_1' = 3.8317$, $\beta_2' = 7.0156$ and $\beta_3' = 10.1735$.

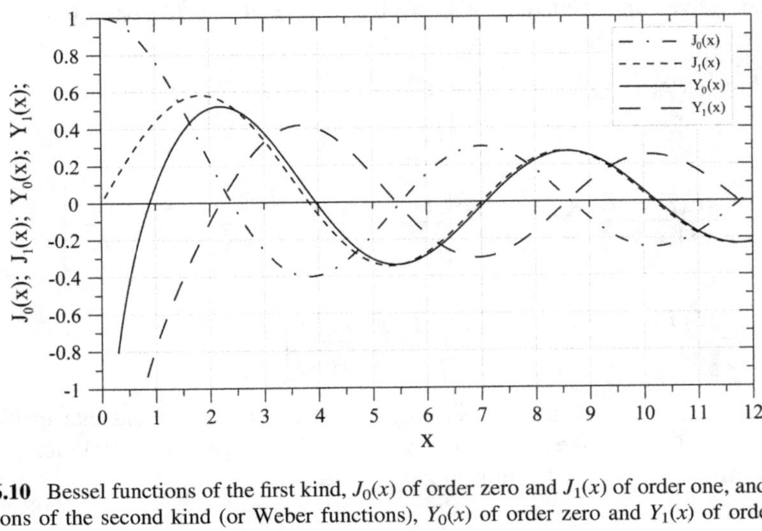

Fig. 6.10 Bessel functions of the first kind, $J_0(x)$ of order zero and $J_1(x)$ of order one, and Bessel functions of the second kind (or Weber functions), $Y_0(x)$ of order zero and $Y_1(x)$ of order one

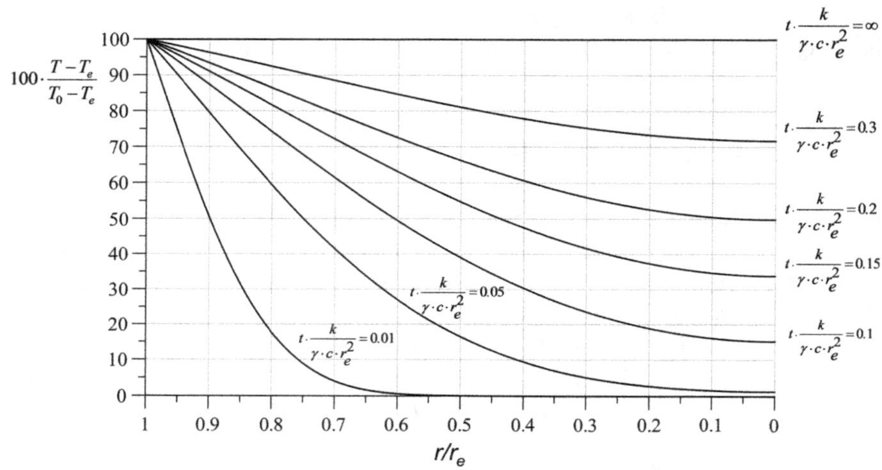

Fig. 6.11 Temperature distributions with time and along the radius in a steel cylindrical rod

Figure 6.11 shows temperature distribution along the radius in a steel solid cylindrical rod at different instants of time $tk/\gamma c r_e^2$ starting from the initial instant $t = 0$; temperature is given in dimensionless form, as the percentage ratio $[(T–T_e)/(T_0–T_e)] \cdot 100$.

Once function $T = T\ (r,\ t)$ at a given instant of time t is known, familiar numerical analysis methods can be used to calculate the integrals appearing in relations (6.63) written for $r_i = 0$ and thus determine the rod's stress and strain state at that instant. It is also possible to employ an analytical approach based on series expansions together with some of the relations linking the Bessel functions, their derivatives and their integrals. We thus arrive at the following relations:

$$\sigma_r = \frac{2\alpha E(T_0 - T_e)}{(1-\nu)} \sum_{n=1}^{\infty} e^{-\frac{k\beta_n^2 t}{\gamma c r_e^2}} \left[\frac{1}{\beta_n^2} - \left(\frac{r_e}{\beta_n^2 r} \right) \frac{J_1(\beta_n \frac{r}{r_e})}{J_1(\beta_n)} \right]$$

$$\sigma_t = \frac{2\alpha E(T_0 - T_e)}{(1-\nu)} \sum_{n=1}^{\infty} e^{-\frac{k\beta_n^2 t}{\gamma c r_e^2}} \left[\frac{1}{\beta_n^2} + \left(\frac{r_e}{\beta_n^2 r} \right) \frac{J_1\left(\beta_n \frac{r}{r_e}\right)}{J_1(\beta_n)} - \frac{J_0\left(\beta_n \frac{r}{r_e}\right)}{\beta_n J_1(\beta_n)} \right] \tag{6.89}$$

$$\sigma_z = \frac{2\alpha E(T_0 - T_e)}{(1-\nu)} \sum_{n=1}^{\infty} e^{-\frac{k\beta_n^2 t}{\gamma c r_e^2}} \left[\frac{2}{\beta_n^2} - \frac{J_0(\beta_n \frac{r}{r_e})}{\beta_n J_1(\beta_n)} \right].$$

When t is very short, the temperature on the outer surface and that inside the cylindrical rod can still be considered to be T_e and T_0 respectively. Relations (6.63) written for $r_i = 0$, with t tending to zero, yield:

$$\sigma_r = 0$$
$$\sigma_t = \sigma_z = \frac{\alpha E(T_0 - T_e)}{(1 - \nu)}. \tag{6.90}$$

We arrive at the same results using the series expansions (6.89), given that

$$\sum_{n=1}^{\infty} \frac{1}{\beta_n^2} = \frac{1}{4}. \tag{6.91}$$

The second relation (6.90) gives the maximum values of the stresses characterizing the stress state in a cylindrical rod subjected to heat flow along the radius. This stress is compressive during heating and tensile during cooling, and equals that needed to annul thermal expansion. To reduce this stress state in shafts and rods, it is necessary to heat or cool gradually, or to increase the heating time in proportion to the square of the diameter.

Chapter 7
Thick-Walled Circular Cylinders in the Linear Elastic-Perfectly Plastic State After Loading Beyond the Elastic Range

7.1 General and Scope

Tests carried out on thick-walled circular cylinders under internal pressure demonstrate unequivocally that failure occurs at pressures corresponding to maximum nominal elastic stresses that are not only much above the material's yield strength, but also exceed its failure strength. This is due to the fact that if the equivalent stress calculated with a yield theory at the inner radius, where the material is at greatest risk, locally exceeds the material's elastic limit, the stress field will be profoundly changed. In particular, there will be a more uniform redistribution of stresses, together with a reduction in their concentrations near geometric discontinuities such as notches, holes, flanged connections, etc. In addition, once the pressure has been released, a circular cylinder stressed beyond yielding will have residual internal stresses that in the subsequent pressure cycles will be superposed on the stresses resulting from the pressure loads. As these residual stresses are generally compressive, the designer can use them to advantage in reducing the maximum tensile stresses due to the working pressure (Faupel and Furbeck, 1953, Faupel 1955).

Analyzing the stress state in a circular cylinder loaded beyond yielding until the material reaches the fully plastic state is of considerable importance for highly stressed tubular structures, for proof testing to failure and interpreting the results, and for reducing stress intensification effects in machine parts and structures by creating an appropriate residual state of coaction in the material. This analysis, conducted with the usual assumptions of axisymmetry and a generalized plane strain state, so that the problem is mono-dimensional, leads to closed form solutions even without making rough simplifications concerning the material's stress–strain curve (Crossland and Bones 1955 and 1958, Faupel 1956, Gormann 1967, Capurso 1971, Franciosi 1979, Liguori 1983).

However, to gain a better grasp of the strain mechanics occurring in a circular cylinder under a pressure that stresses it beyond the elastic range, from the onset of yielding up to ultimate stress, it is best to proceed step by step. Accordingly, it will first be assumed that the stress–strain curve is that of a linear elastic-perfectly plastic material (material with behavior perfectly elastic up to yield point and

V. Vullo, *Circular Cylinders and Pressure Vessels*, Springer Series in Solid and Structural Mechanics 3, DOI: 10.1007/978-3-319-00690-1_7,

perfectly plastic up to failure). Subsequently, we will renounce this restrictive assumption and take this curve to be the far more general one for an elasto-plastic material with nonlinear strain hardening (Batdorf and Budiansky 1949, Westergard 1952, Hoffmann and Sachs 1953, Goodier and Hodge 1958, Smith and Sidebottom 1965, Colladine 1985, Alexandrov and Potatov 1990, François, Pineau and Zaoui 1998, Dowling 1999, Haddad 2000, Lubliner 2006).

In line with this stepwise approach, we will assume in this chapter that the stress–strain curve $\sigma = \sigma(\varepsilon)$ obtained from tensile tests on a standard specimen is as shown in Fig. 7.1, which corresponds to linear elastic-perfectly plastic behavior, that is with no strain hardening, as occurs in ductile metals and alloys. The material shows perfectly elastic behavior up to the yield-point stress σ_s (in this case, the proportional limit, the elastic limit and the yield-point stress coincide), followed by perfectly plastic behavior up to failure.

It will also be assumed that, regardless of the material's condition (i.e., virgin or otherwise), the $\sigma = \sigma(\varepsilon)$ curve for a specimen subjected to a compression test is the mirror image of the origin of the tensile test curve. Here and in the rest of the text, in other words, we will ignore the Bauschinger effect (Bauschinger 1886), whereby there will be a reduction in the compressive (or tensile) yield strength of a specimen whose material has undergone prior non-zero plastic tensile (or compressive) strain. The lowering of the yield-point stress when deformation in one direction is followed by deformation in the opposite direction is called the Bauschinger effect.

It is then assumed that, starting from a generic point *A* (Fig. 7.1) located beyond *P* on the horizontal straight-line of the stress–strain curve $\sigma = \sigma(\varepsilon)$ and reducing the load, the representative point will move along path *AB* exactly parallel to the linear elastic line segment *OP*. Consequently, when the load reaches zero, there will be a permanent plastic strain given by line segment *OB* on the abscissa. Finally, it will be assumed that, starting from point *B* and increasing load again, the representative point will move along path *BA* in the opposite direction to unloading path *AB*. In other words, it will be assumed that the hysteresis cycle associated with the energy losses per unit volume occurring in the unloading and loading cycle will have zero area. A further increase in load beyond point *A* will result in a movement towards the right along the horizontal straight-line of the

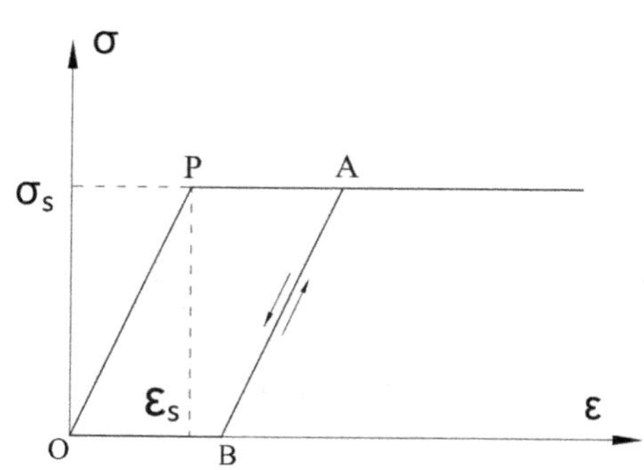

Fig. 7.1 Stress–strain curve $\sigma = \sigma(\varepsilon)$ for a linear elastic-perfectly plastic material

stress–strain curve $\sigma = \sigma(\varepsilon)$. Obviously, this model of the material's behavior is highly approximate (for a treatment of the problem using models that are closer to reality, the reader is directed to Chap. 9 as well as to specialized texts).

As was indicated in Sect. Purpose and Basic Assumptions of the Introduction, in a circular cylinder stressed symmetrically around its axis and loaded uniformly along the latter (a condition satisfied by all three of the cases considered in Sect. 4.3 and characterized by $\sigma_z = 0$, $\sigma_z = A$ and $\sigma_z = 2\nu A$ respectively), the radial, tangential (or circumferential) and axial directions will be the principal directions of both stresses and strains. It was also mentioned that, from the practical standpoint, most interest focuses on cases where strains are infinitesimal (small strains or small deformations assumption) and those where strains have a non-negligible finite value (large strains or large deformations assumption).

In the following treatment, it will be assumed that the strains are small, and that the associated strain state is thus described by conventional strains (nominal or Lagrangian or engineering strains) ε_r, ε_t and ε_z. In cases involving large strains (see, e.g., Sect. 3.6), it will clearly be necessary to introduce the true strains (logarithmic or natural strains) $\overline{\varepsilon_r} = \ln(1 + \varepsilon_r)$, $\overline{\varepsilon_t} = \ln(1 + \varepsilon_t)$ and $\overline{\varepsilon_z} = \ln(1 + \varepsilon_z)$. Here, however, problems involving large strains will not be addressed.

Here, the subject discussed involves small deformations. For this reason, the strains of interest to us are only the conventional strains ε_r, ε_t and ε_z. As is known, the assumption that strains are small is fully verified for materials that show well defined yielding in the effective $\sigma = \sigma(\varepsilon)$ curve. In such cases, there is a gradual passage from an elastic strain state to the plastic strain state, with a plastic-elastic interface (plastic boundary) that, from the moment permanent deformation first appears, shifts to mark the boundary between two regions, one plastic and one elastic. Because of the two regions' continuity and contiguity, strains in the plastic region must be of comparable magnitude with those in the elastic region, or in other words must be small because those of the elastic region are small.

7.2 Yield Theories

In dealing with the problem of a circular cylinder stressed beyond the elastic range, it is necessary to consider a yield theory (or yield criterion). In this area, the maximum shear stress yield theory (Tresca yield theory or τ_{max} yield theory) and maximum distortion energy yield theory (von Mises yield theory or ED_{max} yield theory) are normally used, but we should not forget that the maximum normal stress yield theory (σ_{max} yield theory) and maximum shear strain yield theory (ε_{max} yield theory) have also been used, and continue to be employed today. Because of their importance, reference will also be made to the Mohr's yield theory or internal friction yield theory.

For thick-walled circular cylinders under internal and external pressure which, in the most general case, are characterized by a triaxial stress state described by the three principal stresses σ_t, σ_r and σ_z, the specific formulation of the first four yield theories (τ_{max}, ED_{max}, σ_{max} and ε_{max} yield theories) is as follows (the relations given below are those for the most frequent case, viz., that in which the maximum principal stress is the hoop stress):

Maximum shear stress yield theory, τ_{max} (in order of time, Coulomb-Tresca-Guest yield theory):

$$\sigma_t - \sigma_r = \pm\,\sigma_s \tag{7.1}$$

or

$$\sigma_t - \sigma_z = \pm\,\sigma_s \tag{7.2}$$

depending on whether $\sigma_t > \sigma_z > \sigma_r$ or $\sigma_t > \sigma_r > \sigma_z$ (σ_s is obviously the yield-point stress).

Maximum distortion energy yield theory, ED_{max} (in order of time, Maxwell-Huber-von Mises-Hencky yield theory):

$$\sigma_t^2 + \sigma_r^2 + \sigma_z^2 - \sigma_t\sigma_r - \sigma_r\sigma_z - \sigma_z\sigma_t = \sigma_s^2. \tag{7.3}$$

Maximum normal stress yield theory, σ_{max} (in order of time, Lamé-Rankine yield theory):

$$\sigma_t = \pm\sigma_s. \tag{7.4}$$

Maximum normal strain yield theory, ε_{max} (Saint–Venant's yield theory):

$$\sigma_t - \nu(\sigma_r + \sigma_z) = \pm\sigma_s. \tag{7.5}$$

To understand which yield theory best interprets the experimental results, a few words are in order concerning the latter. Except for very limited data from tests on thick-walled cylinders with closed ends under internal pressure, most available experimental results involve biaxial stress states. Given these circumstances, it is necessary to consider the plane with coordinate axes (σ_r, σ_t) when attempting to compare data calculated with theoretical models and experimental data. For the sake of convenience, it is also advisable to make stresses dimensionless by relating them to the yield stress σ_s, so that plane (σ_r, σ_t) becomes plane $(\sigma_r/\sigma_s, \sigma_t/\sigma_s)$.

On the basis of relation (7.3) written for $\sigma_z = 0$, the formulation of the maximum distortion energy yield theory, ED_{max}, becomes:

$$\sigma_t^2 + \sigma_r^2 - \sigma_t\sigma_r = \sigma_s^2. \tag{7.6}$$

In plane (σ_r, σ_t), this is the equation of an ellipse (the ellipse of plasticity), with major and minor semi-axes equal to $\sigma_s\sqrt{2}$ and $\sigma_s\sqrt{2/3}$, positioned so that they bisect dihedra σ_r, σ_t and $(-\sigma_r, -\sigma_t)$, and $(-\sigma_r, \sigma_t)$ and $(\sigma_r, -\sigma_t)$ respectively. The plot of relation (7.6) is called *yield locus*; its correlated ellipse is represented in Fig. 7.2 in plane $(\sigma_r/\sigma_s, \sigma_t/\sigma_s)$.

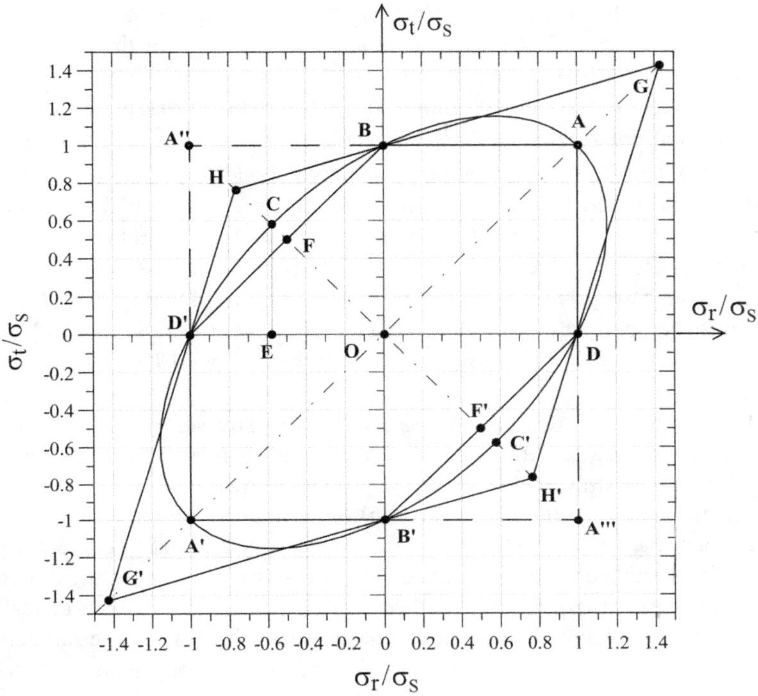

Fig. 7.2 *Ellipse*, *hexagon*, *rhombus* and *square* of plasticity in plane $(\sigma_r/\sigma_s, \sigma_t/\sigma_s)$; $\overline{OB} = \overline{OD} = 1; \overline{OA} = \sqrt{2}; \overline{OE} = \overline{CE} = 1/\sqrt{3}; \overline{OC} = \sqrt{2/3}$; for the ε_{max} yield theory, $\nu = 0.3$

As can be seen from the dimensionless stresses diagram shown in Fig. 4.4, stress σ_t is always tensile in a thick-walled circular cylinder under internal pressure alone, while stress σ_r is always compressive except at the outer radius, where it is zero. At any radius, the values assumed by hoop stress are always greater than the absolute value of radial stress, i.e., the ratio $\sigma_t/|\sigma_r|$ is above unity at all times. Consequently, the arc of ellipse of interest to us here is that between the line bisecting dihedron $(-\sigma_r/\sigma_s, \sigma_t/\sigma_s)$ and the axis of stresses σ_t/σ_s, i.e., the arc $\widehat{BC}$ whose correspondent in the fourth quadrant is arc $\widehat{B'C'}$.

By contrast, as the dimensionless stresses diagram in Fig. 4.5 shows, both stresses σ_r and σ_t in a thick-walled circular cylinder under external pressure alone are compressive (an exception is radial stress at the inner radius, which is zero). Here again, at any radius, the absolute value of hoop stress is always greater than the absolute value of radial stress, i.e., the ratio $|\sigma_t|/|\sigma_r|$ is above unity at all times. In this case, the arc of ellipse of interest to us here is that between the line bisecting dihedron $(-\sigma_r/\sigma_s, -\sigma_t/\sigma_s)$ and the negative part of the σ_t/σ_s axis, i.e., the arc $\widehat{A'B'}$, whose correspondent in the first quadrant is arc $\widehat{AB}$.

If the yield theory used is that of maximum shear stress (the Tresca yield criterion), instead of having the yield ellipse in plane (σ_r, σ_t), we will have a different yield locus, i.e., the hexagon of plasticity (yield hexagon) as shown in plane $(\sigma_r/\sigma_s, \sigma_t/\sigma_s)$ in Fig. 7.2. With the same approach as used for the maximum distortion energy yield theory, we can conclude that the parts of this yield hexagon of interest in the cases of a circular cylinder subject only to internal pressure or subject only to external pressure are respectively the line segment $\overline{BF}$, which corresponds to line segment $\overline{B'F'}$, and line segment $\overline{A'B'}$ corresponding to line segment $\overline{AB}$.

It should be noted that the maximum distortion energy yield theory can be regarded as the generalization of the maximum shear stress yield theory, in that it also considers the influence exerted by the third principal stress on yielding of the material. To be handled as a generalization of the τ_{max} yield theory, the ED_{max} yield theory must be seen as a maximum octahedral shear stress yield theory $\tau_{ott,max}$, independently due to Ros and Eichinger 1949, and Nadai 1950 (Roš–Eichinger–Nadai yield theory).

As we know from continuum mechanics, the stresses acting on the eight octahedral planes (each of the octahedral planes cuts across one of the corners of a principal element, stressed by the three principal stresses $\sigma_1, \sigma_2, \sigma_3$, so that the eight planes together form an octahedron) have interesting and significant properties. In fact, the same normal stress acts on each of the eight octahedral planes delimiting the octahedron thus obtained. This stress state, a hydrostatic stress state, tends to compress or expand the octahedron homothetically, without distorting it, with a consequent change in volume but not in shape. The shear stresses acting on each of the eight octahedral planes are also equal, but they tend to distort the octahedron without changing its volume. Although the octahedral shear stress is smaller than the maximum principal shear stress (in this connection, it should be borne in mind that in Mohr's plane $\tau = \tau(\sigma)$, the octahedral stresses are represented by the points outside the two smaller Mohr's circles and inside the principal Mohr's circle, i.e., lie in the area between these circles), it is a single value which depends on all three principal stresses $\sigma_1, \sigma_2, \sigma_3$.

The following relations express the octahedral normal stress and the octahedral shear stress as a function both of these three principal stresses, and of the first and second tensor invariants for principal axes, given by $I_1 = \sigma_1 + \sigma_2 + \sigma_3$ and $I_2 = -(\sigma_1\sigma_2 + \sigma_1\sigma_3 + \sigma_2\sigma_3)$ respectively. The octahedral shear stress is also expressed as a function of the second deviator invariant for principal axes J_2, linked to I_1 and I_2 by the relation $J_2 = I_2 + (1/3)I_1^2$.

$$\sigma_{ott} = \frac{1}{3}(\sigma_1 + \sigma_2 + \sigma_3) = \frac{1}{3}I_1 \tag{7.7}$$

$$\tau_{ott} = \frac{1}{3}\left[(\sigma_1 - \sigma_2)^2 + (\sigma_2 - \sigma_3)^2 + (\sigma_3 - \sigma_1)^2\right]^{1/2} = \frac{\sqrt{2}}{3}\sqrt{I_1^2 + 3I_2} = \sqrt{\frac{2}{3}}J_2. \tag{7.8}$$

The octahedral shear stress can also be expressed as a function of the three principal shear stresses $\tau_{1,2}, \tau_{1,3}, \tau_{2,3}$ [i.e., the maximum values of the shear stresses correlated with the three Mohr's circles in Mohr's plane (σ, τ)] as follows:

$$\tau_{ott} = \frac{2}{3}(\tau_{1,2}^2 + \tau_{1,3}^2 + \tau_{2,3}^2)^{1/2}. \tag{7.9}$$

According to the maximum octahedral shear stress yield theory, the yield point will be reached when

$$(\sigma_1 - \sigma_2)^2 + (\sigma_2 - \sigma_3)^2 + (\sigma_3 - \sigma_1)^2 = 2\sigma_s^2. \tag{7.10}$$

This relation, where $\sigma_1 = \sigma_t$, $\sigma_2 = \sigma_z$ and $\sigma_3 = \sigma_r$, is linked to Eq. (7.3), corresponding to the maximum distortion energy yield theory.

Both the maximum distortion energy yield theory and the maximum octahedral shear stress yield theory take the influence of all three principal stresses into account. In addition, these yield theories assume that the material's behavior is symmetrical in tension and compression, and put no limitations on hydrostatic stress states which tend to compress or enlarge the principal element but not distort it. This latter assumption, though providing a good interpretation of the material's mechanical behavior in compression, does not give a realistic picture of failure mechanics under triaxial tension, when failure is known to occur through *brittle fracture* or *cleavage* (separation or fragmentation of a structural member into two or more parts when the strained cohesive bonds between grain boundaries break down, permitting sudden cleavage, with the broken halves revealing little if any distortion) rather than *distortion* (or *ductile fracture* with plastic deformations correlated to shear stresses, and dislocations moving along slip planes and resulting in the formation of parallel slip-bands in the microstructure and Lüders' bands, also called Piobert's bands or Hartmann's lines or stretcher strains, in the macrostructure, because such bands were first observed by Piobert (1842), Lüders (1860) and Hartmann (1896); this type of deformation is also referred to as the Piobert effect, see Dieter 1988).

Though these two theories lead to the same expression for equivalent stress (also called ideal stress or comparison stress), and to the same yield locus in plane (σ_r, σ_t) or in plane $(\sigma_r/\sigma_s, \sigma_t/\sigma_s)$, they are conceptually very different. The maximum octahedral shear stress theory, which is based on the stress state, can like the less general τ_{max} theory be used beyond the proportional range to analyze both yielding and strain hardening in a multiaxial stress state. By contrast, the maximum distortion energy theory, based as it is on the strain energy associated with changes in shape without changes in volume calculated using linear elastic relations, establishes that yielding begins when this energy reaches a critical level, which from the standpoint of the stress state occurs when the second deviator invariant for principal axes J_2 reaches a critical limit value. This conceptual clarification was made by Nadai, who showed that the material's plasticization is governed by the stress state that depends on the equilibrium conditions and not on the distortion energy, the latter being calculated by means of a linear elastic relation. Despite the fact that this concept was made clear long ago, however, there is still a tendency to speak of the Von Mises criterion in the plastic range as well.

Figure 7.2 also shows the yield loci which represent the other two yield theories considered, viz., the σ_{max} and ε_{max} yield theories, in plane $(\sigma_r/\sigma_s; \sigma_t/\sigma_s)$. For the material we have assumed (which exhibits the same behavior in tensile and in compression tests), the maximum normal stress yield theory translates in this plane into the square of plasticity (yield square) obtained by extending, pairwise, sides $\overline{AB}$ and $\overline{A'B'}$ and sides $\overline{AD}$ and $\overline{A'D'}$ until they meet. Lastly, the maximum normal strain yield theory translates into the rhombus of plasticity (yield rhombus) whose major and minor diagonals are superimposed on the major and minor semi-axes of the ellipse of plasticity, and whose four sides pass through points B, D, B', and D' and converge pairwise on the points lying on the major diagonal having coordinates $[1/(1-\nu), 1/(1-\nu)]$ and $[-1/(1-\nu), -1/(1-\nu)]$.

Mohr's yield theory assumes that yielding begins when the stress state is represented by the points in Mohr's plane $\tau = \tau(\sigma)$ that lie on the envelope of the infinite Mohr's circles for stress states at the limit of yield onset. In Mohr's plane, this envelope is represented by two nonlinear curves that are symmetrical with respect to the axis of the abscissas (σ axis) and converge on a point of the latter with a positive abscissa, where the material breaks down as a result of lack of cohesion (Fig. 7.3), in which the tangent is parallel to the axis of the ordinates (τ axis). These two curves form the so-called *limiting curve*. The osculating circumference to this curve at the point where the latter intersects the σ axis represents the locus of the limiting curve points for stress states in which failure occurs through cleavage, whereas the limiting curve points outside the osculating circumference represent stress states in which failure occurs through distortion.

These two curves become two straight lines (simplified Mohr's yield criterion), again symmetrical with respect to the σ axis and converging on a point of the latter having coordinates ($\sigma > 0, \tau = 0$), in the case where the experimental data used are provided by two uniaxial tests, the first in pure tension and the second in pure compression (Fig. 7.4). From this standpoint, Mohr's yield criterion can be considered as a modification of the τ_{max} yield theory. For this latter theory, the envelopes of all principal Mohr's circles at the yielding onset (usually uniaxial tension and compression data are available and there are thus only two principal Mohr's circles) on Mohr's coordinates are two horizontal lines having ordinate $\tau = \pm\sigma_s/2$.

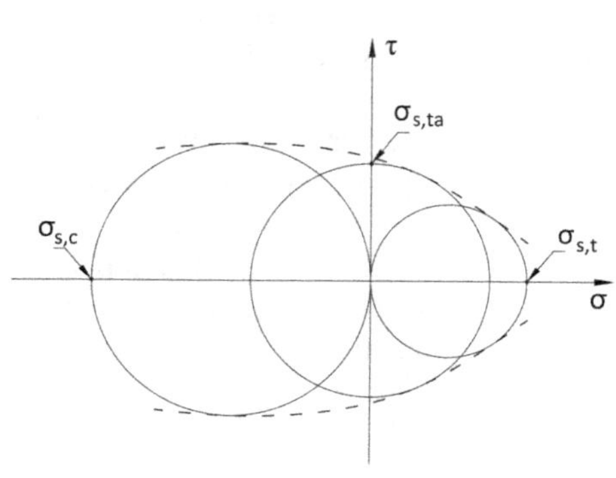

Fig. 7.3 *Plasticity curves* for Mohr's yield theory, in plane $\tau = \tau(\sigma)$

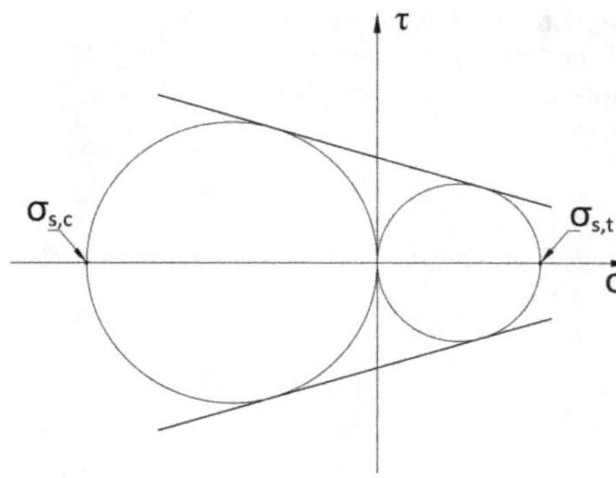

Fig. 7.4 *Plasticity lines* for the simplified Mohr's yield theory, in plane $\tau = \tau(\sigma)$

In plane (σ_t, σ_r), Mohr's yield criterion for the biaxial stress state results in two nonlinear curves that are symmetrical with respect to the diagonal bisecting the dihedra of the first and third quadrants. The points where these two curves (Fig. 7.5) intersect the coordinate axes and the shear diagonal (the straight line bisecting the dihedra of the second and fourth quadrants) can be readily identified by means of uniaxial tension, uniaxial compression and pure shear tests. The upper curve will thus intersect the axis of the abscissa at the point whose coordinates are $\left(-\sigma_{s,c}, 0\right)$, the axis of the ordinate at the point whose coordinates are $\left(0, \sigma_{s,t}\right)$ and the shear diagonal at the point whose coordinates are $\left(-\sigma_{s,ta}, \sigma_{s,ta}\right)$, where $\sigma_{s,t}$, $\sigma_{s,c}$ and $\sigma_{s,ta}$ are the yield stresses in uniaxial tension test, uniaxial compression test and pure shear test respectively. For the lower curve, the points of intersection with the three straight lines can be identified through the symmetry indicated above.

Just as the two straight lines in Fig. 7.4, which are symmetrical with respect to the axis of the abscissa, are an approximate representation of the two curves in Fig. 7.3 for Mohr's yield criterion in Mohr's plane, in plane (σ_t, σ_r), the two curves in Fig. 7.5 can be represented by a piecewise-linear closed curve having the shape of an irregular hexagon, whose sides differ in the first and third quadrants

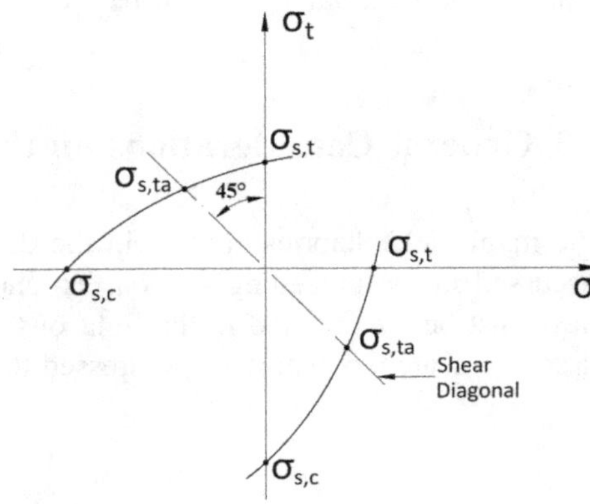

Fig. 7.5 *Plasticity curves* for Mohr's yield theory, in plane (σ_t, σ_r)

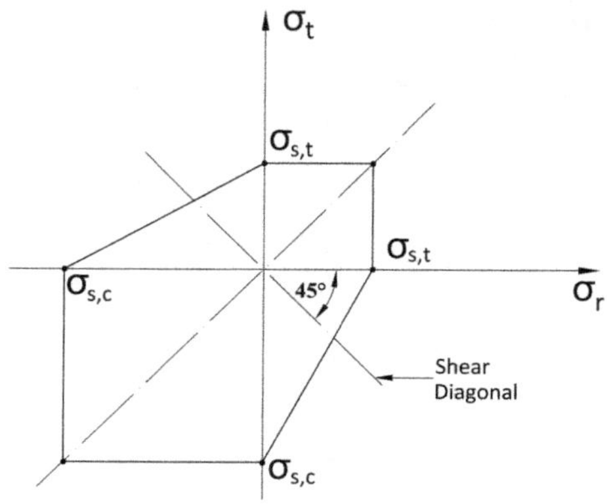

Fig. 7.6 *Irregular hexagon* of plasticity for the simplified Mohr's yield theory, in plane (σ_t, σ_r)

(Fig. 7.6). This is the hexagon of plasticity (yield hexagon) for Mohr's simplified yield criterion. From this standpoint as well, Mohr's yield criterion can be considered as a modification of the τ_{max} yield theory.

In other words, the $\tau = \tau(\sigma)$ and $\sigma_t = \sigma_t(\sigma_r)$ curves for Mohr's yield criterion can be obtained starting from the corresponding curves for the τ_{max} theory, considering and including the effect of internal friction. For this reason, Mohr's yield criterion is also called the internal friction yield theory. Proposed in 1897 by Reyto (see Juvinall 1967) and subsequently developed by Marin (1962), this yield criterion assumes that sliding along the slip planes is inhibited by the friction forces associated with the compressive forces acting orthogonally to said slip planes.

In plane (σ_t, σ_r), Mohr's yield criterion in its generalized formulation, i.e., that using comprehensive experimental data, leads to a yield locus having shape of a closed and oval-like curve similar to the ellipse of plasticity, but asymmetrical with respect to the shear diagonal. It is thus an oval (oval of plasticity), larger in the third quadrant and smaller in the first quadrant, but still symmetrical with respect to the straight-line bisecting the first and third quadrants.

Mohr's yield criterion is thus more general than the maximum octahedral shear stress and maximum distortion energy yield theories, in that it can account for and simulate the material's different behavior in compression and tension.

7.3 General Considerations on the Use of Yield Theories

The main yield theories applied in the design sector of interest to us here were discussed in the preceding section. To choose the best theory for use, however, they must be reexamined in the light of the experimental results obtained by the various researchers who have addressed this topic. To achieve the desired biaxial stress state, many of these researchers have used thin-walled steel, copper and aluminum tubes subjected simultaneously to internal pressure and to axial and torsional load, or tubes under combined tensile and torsional load. Most of the results obtained, all from tests at ambient temperature, consist of points representing stress states in the first and fourth quadrants, and which are denser around the yield ellipse or, at most, fall in the zones between the straight lines of the yield hexagon and the corresponding curve segments of the yield ellipse. The location of these experimental points, which are not shown in Fig. 7.2 to avoid over-complication, demonstrates unequivocally that the yield theory that best interprets experimental evidence is that of maximum octahedral shear stress and the formally equivalent (formally, but not conceptually) maximum distortion energy theory.

For certain load combinations, several of the four yield theories (σ_{max}, τ_{max}, ε_{max} and ED_{max}) provide the same results, and are equally advantageous. In other cases, the results of the theoretical models differ, and the problem thus arises of selecting the criterion that bests interprets the actual risk of failure in the material. Generally speaking, it must be recognized that there is no single theory: we can only identify the theory that is best for a particular purpose. The same theory applied in different situations may result in errors. For example, repeated experimental tests have demonstrated that yield strength in torsion is about 12–15 % higher than the equivalent shear strength determined through tensile testing. This discrepancy, though it can be explained theoretically, would not arise if the τ_{max} yield theory were correct. The maximum distortion energy yield theory (the term will be employed in order to be consistent with the usage that is normally encountered in textbooks on plasticity, though beyond the elastic limit it would in fact be more correct to speak only of the maximum octahedral shear stress yield theory. As this conceptual inaccuracy is now firmly enshrined in the technical literature, however, we will continue henceforth to refer to this criterion only as the maximum distortion energy theory) does not suffer from this limitation, and its accuracy is borne out both by theoretical considerations and experimental evidence, making it the optimal choice among the different yield criteria.

A number of tests carried out on specimens consisting of cast iron or other brittle materials have led to the conclusion that the σ_{max} strength theory (and hence yield theory) is better for brittle materials than the other criteria. However, data from more accurate tests conducted in recent years have clearly demonstrated that the behavior of brittle materials can be effectively interpreted by the maximum distortion energy theory, providing that account is made for the effects of stress concentrations resulting from the inclusions that often cause a structural component to lose ductility.

Lastly, the ε_{max} yield theory, widely used in the nineteenth century and the first half of the twentieth, particularly in the design of artillery and of porcelain and reinforced concrete components, has been largely abandoned today. Here, it is used only for purposes of comparison in determining the pressure that leads to the onset of the plastic state at the inner radius of a thick-walled circular cylinder under internal pressure.

From the general design standpoint, in any case, the σ_{max} yield theory can also be used with success, providing that the stress state is represented by points falling in the first and third quadrants of Fig. 7.2. In the first quadrant, both stresses are positive and the σ_{max} and τ_{max} yield theories lead to the same results. When the principal stresses are all equal, the three σ_{max}, τ_{max} and maximum distortion energy yield theories lead to the same results. If the stress state is represented by points falling in the second and third quadrants, it is obvious that only the τ_{max} and maximum distortion energy yield theories can be used.

In mechanical design, the material's yield criterion is selected and used on the basis of more general considerations. In these pages, we are dealing with operating conditions that do not involve excessive elastic deflections, buckling, creep, impact and fatigue, all of which call for further and more specific considerations. Notch effects and stress concentrations are considered here only as regards the triaxialty of the stress state they produce; for the fatigue aspects they involve, the reader is directed to specialized textbooks. Attention focuses here on static cases, for which the following three situations should be borne firmly in mind:

1. Stress concentrations are of minimum practical significance when the material's yield theory is that of initial ductile yielding. At the notch, a certain amount of initial yield takes place which gives rise to a favorable redistribution of stresses. However, it must be recognized that stress concentrations, together with other synergistic factors such as impact loads, low temperatures, marginally ductile behavior of the material, etc., can play a significant role as regards to the possibility of a brittle failure.
2. The contribution that stress concentrations make to the brittle failure of homogeneous materials is extremely important, and the designer must make every effort to minimize the stress concentration factor K_t (think, for example, of cutting a pane of glass, when a small load is sufficient to break the glass along the line scored by the cutting tool).
3. For brittle failure of non-homogeneous materials such as gray cast iron, stress concentrations may be of little or no importance. In gray cast iron, the graphite flakes dispersed in the metal matrix cause notch effects that can be more significant than those associated with geometrical notches. This is why a cast iron bar with a small transverse hole subjected to tensile stress will often break at a location other than that of the hole.

Where a preliminary review of actual design conditions indicates that component failure may take place during service with modes that differ from those found for standard specimens, it is obvious that the yield theories discussed above cannot be applied. To give an actual example, if it is suspected that brittle fracture could occur in a component consisting of normally ductile steel, none of these yield theories is capable of linking this condition to the data obtained from standard tensile tests. In such a situation, it will be necessary to use more advanced theoretical considerations and experimental data from specific tests (impact tests, for instance) carried out under conditions that approximate those encountered during actual service to the greatest possible extent.

We will now focus attention on cases where the material yield theories can be correctly applied, starting from their use in predicting ductile yielding. Here, the yield theory that interprets experimental evidence most accurately is that of the maximum distortion energy. It is thus recommended that it be used, though with some reservations. For the biaxial stress state, when good experimental data are available that go beyond those from simple tensile tests, Mohr's yield criterion is to be preferred. In this connection, however, a possible weakness of Mohr's yield criterion should be noted: as it is difficult to obtain highly accurate compression and torsion data (i.e., one of the more general stress states conceivable in experimental conditions), the errors may be greater than those associated with the maximum distortion energy yield theory.

For triaxial stress states, the amount of experimental data available is relatively modest. The maximum distortion energy yield theory can nevertheless be applied with good results, but there is also no lack of grounds for recommending that the more conservative internal friction yield theory be used (see Marin 1962). For both biaxial and triaxial stress states, it is still fairly common to use the simple and slightly conservative τ_{max} theory together with data from standard tensile tests. Many design codes and calculation standards are based on this criterion.

The predictive theories for ductile fracture coincide in general with the predictive theories for ductile yielding. However, the experimental data available in this connection are not only more limited numerically, but also show greater scatter. For these reasons, it is sometimes recommended that the internal friction yield theory be used instead of the maximum distortion energy yield theory.

To predict the fracture strength of materials that can be affected by brittle fractures in tension and compression, it is once again necessary to have reliable experimental data. For biaxial stress states, Mohr's criterion can be used. The internal friction theory can be employed if uniaxial tension and compression data are available, and is reasonably accurate. It is often recommended for this reason, but the results obtained are in general conservative. The σ_{max} theory is occasionally used, but it often leads to results that, for biaxial tension–compression stress states, can be affected by significant errors, as they are heavily skewed towards the non-conservative side. Experimental results obtained by a number of researchers on the basis of tests conducted on gray cast iron are clearly included between the theoretical curves for the σ_{max} theory and the internal friction criterion, which are thus in agreement when a biaxial stress state is considered.

Selecting the most appropriate yield theory calls for considerable caution, as it is heavily influenced by the type of material and its mechanical behavior up to failure, as well as by the stress state of the structure in question. In general, the following conclusions can be drawn.

- If the structure fails as a result of yielding of the material or excessive permanent set, if the material has good plastic properties, with behavior that is roughly symmetrical in uniaxial compression and tension, and if the stress state is triaxial or biaxial, the criterion providing the best fit with experimental results is the maximum octahedral shear stress yield theory or the equivalent maximum distortion energy yield theory. In the same conditions, all other considerations being equal, the τ_{max} yield theory results in a higher safety factor, while Mohr's yield criterion leads to the minimum possible value for this safety factor.

- Again where the structure fails as a result of yielding or excessive permanent set, but the material has poor plastic properties, with asymmetric behavior in uniaxial compression and tension, and for triaxial or biaxial stress states, Mohr's yield criterion best approximates the experimental results, with the limiting curve for true stresses. In the same conditions and again with all other considerations being equal, Mohr's simplified yield theory provides a higher safety factor.
- If the structure fails through fracture with large deformations (which occurs for metals with good plastic properties in both uniaxial tension and uniaxial compression) and triaxial or biaxial stress states, two situations may arise, depending on whether failure entails relative slip of the two faces of the fracture surface, or separation of the faces. According to many researchers, the τ_{max} yield theory is to be preferred in the first case, and the σ_{max} yield theory in the second. However, there is no unanimous consensus regarding this conclusion, especially when it is based on stresses calculated for the initial cross-sectional area of the undeformed structure. From the design standpoint, the problem of which yield theory should be selected is of limited practical importance: with very few exceptions, the need to prevent excessive strains is much more restrictive than the fracture considerations.
- Lastly, where the structure fails through fracture but with limited deformations (as occurs for metals or nonmetals exhibiting brittle behavior under certain service conditions or partially plastic behavior under other service conditions) and for triaxial or biaxial stress states, the theory that provides the best fit with experimental data is once again Mohr's criterion, especially if the stresses are true stresses.

7.4 Elastic-Breakdown Pressure

When a circular cylinder that is unaffected by residual stresses due to heat treatment, manufacturing processes, prior loading histories beyond yielding or other causes is subjected to sufficiently high internal or external pressures at the inner radius where the material is at greatest risk, plastic flow begins. The material stops behaving elastically and, as indicated in Fig. 7.1, the innermost fibers begin to yield when the equivalent stress calculated with the selected yield theory reaches σ_s at the inner radius.

To determine the elastic-breakdown pressures (the pressures that trigger the onset of plastic flow at the inner radius of the circular cylinder), we will here use all four of the yield theories (ED_{max}, τ_{max}, ε_{max}, σ_{max}) described above. Mohr's yield criterion will not be employed, given that it should, for the reasons outlined in Sect. 7.3, be based on regression curves from experimental data, which are not available. For each of the four yield theories considered, we will examine the three cases where $\sigma_z = 0$, $\sigma_z = A = (1/2)(\sigma_t + \sigma_r)$ and $\sigma_z = 2\nu A = \nu(\sigma_t + \sigma_r)$ or $\varepsilon_z = 0$, corresponding to the cases of circular cylinders with free open ends, with free closed ends and with clamped ends respectively, designated by the letters (a), (b) and (c) in front of the associated expressions.

In these three cases, the maximum distortion energy yield theory given by Eq. (7.3) can be specialized in the following three relations:

$$\text{(a)}\quad \sqrt{\sigma_t^2 + \sigma_r^2 - \sigma_t\sigma_r} = \sigma_s \tag{7.11}$$

$$\text{(b)}\quad \frac{\sqrt{3}}{2}(\sigma_t - \sigma_r) = \sigma_s \tag{7.12}$$

$$\text{(c)}\quad \sqrt{\sigma_t^2 + \sigma_r^2 - \sigma_t\sigma_r + \sigma_z[\sigma_z - (\sigma_r + \sigma_t)]} = \sigma_s. \tag{7.13}$$

Clearly, relation (7.13) is simply relation (7.3) rewritten in order to facilitate the calculation procedure. Though it is obvious that we should set $\sigma_z = \nu(\sigma_t + \sigma_r)$ in this relation, doing so would lengthen it somewhat, and we have thus opted to retain the compact form here. Substituting the general relations for σ_r and σ_t given by equations (4.21) in all three relations and $\sigma_z = 2\nu A$ in the third, recalling the general expressions for the known terms A and B given by equations (4.28), introducing the dimensionless diameter ratio δ and considering inner radius $r = r_i$ will give the following relations which apply to the circular cylinder under simultaneous internal and external pressure:

$$\text{(a)}\quad \frac{1}{\delta^2 - 1}\sqrt{\left(p_i - p_e\delta^2\right)^2 + 3\delta^4(p_i - p_e)^2} = \sigma_s \tag{7.14}$$

$$\text{(b)}\quad \frac{\sqrt{3}\delta^2}{\delta^2 - 1}(p_i - p_e) = \sigma_s \tag{7.15}$$

$$\text{(c)}\quad \frac{1}{\delta^2 - 1}\sqrt{\left(p_i - p_e\delta^2\right)^2[1 - 4\nu(1 - \nu)] + 3\delta^4(p_i - p_e)^2} = \sigma_s. \tag{7.16}$$

For a circular cylinder under internal pressure alone, these general relations give the following expressions which in the three cases considered define the internal pressure $p_{i,i}$ at which plastic flow begins at the bore and the innermost fibers (the second subscript i thus denotes the inner radius) enter the plastic state:

$$\text{(a)}\quad p_{i,i} = \sigma_s \frac{\delta^2 - 1}{\sqrt{3\delta^4 + 1}} \tag{7.17}$$

$$\text{(b)}\quad p_{i,i} = \frac{\sigma_s}{\sqrt{3}}\frac{\delta^2 - 1}{\delta^2} \tag{7.18}$$

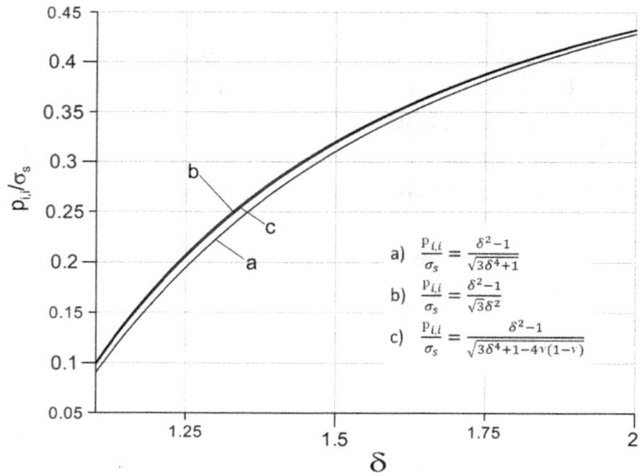

Fig. 7.7 Ratios $p_{i,i}/\sigma_s$ versus δ, for the three cases: **a** ($\sigma_z = 0$), **b** ($\sigma_z = A$), **c** ($\sigma_z = 2\nu A$), with $\nu = 0.3$

$$\text{(c)} \quad p_{i,i} = \sigma_s \frac{\delta^2 - 1}{\sqrt{3\delta^4 + 1 - 4\nu(1 - \nu)}}. \tag{7.19}$$

The results obtained from these three relations in terms of elastic-breakdown pressure are substantially equivalent, as there is very little difference in their numerical values. The differences decrease as diameter ratio δ increases, and even at only $\delta = 2$ are less than 2 %. The circular cylinder with free open ends ($\sigma_z = 0$) has the lowest elastic-breakdown pressure. In Fig. 7.7, the ratios $p_{i,i}/\sigma_s$ obtained from the three relations (7.17), (7.18) and (7.19) are plotted versus δ; curve c applies for $\nu = 0.3$. As can be inferred from extending these curves to very high values of δ, in the elastic range there is no point in going beyond values of (4–5) for this dimensionless ratio. For lower ratios δ, it is preferable to use materials with better strength properties. As can also be seen, curves b and c (where the latter is calculated for $\nu = 0.3$) are practically superimposed.

For a circular cylinder under external pressure alone, the general relations (7.14), (7.15) and (7.16) give the following expressions which in the three cases considered define the external pressure $p_{e,i}$ at which plastic flow begins at the inner radius (here as above, the second subscript i denotes the inner radius):

$$\text{(a)} \quad p_{e,i} = \sigma_s \frac{\delta^2 - 1}{2\delta^2} \tag{7.20}$$

$$\text{(b)} \quad p_{e,i} = \sigma_s \frac{\delta^2 - 1}{\sqrt{3}\delta^2} \tag{7.21}$$

$$\text{(c)} \quad p_{e,i} = \sigma_s \frac{\delta^2 - 1}{2\delta^2\sqrt{1 - \nu + \nu^2}}. \tag{7.22}$$

Obviously, if there were residual stresses of any kind in the circular cylinder before the working pressure was applied, the elastic-breakdown pressure will be higher or lower, depending on whether these stresses are compressive or tensile. The residual stresses must in any case be borne in mind, as they are added to those deriving from service loads.

We will now apply the τ_{max} yield theory to the three cases indicated above, with $\sigma_z = 0$, $\sigma_z = A$ and $\sigma_z = 2\nu A$ respectively. In the case where σ_t and σ_r are the maximum and minimum principal stresses respectively and σ_z is the intermediate stress, we will have:

$$\tau_{max} = \frac{\sigma_t - \sigma_r}{2} = \frac{\sigma_s}{2}. \tag{7.23}$$

This occurs in all three of the cases considered ($\sigma_z = 0$; $\sigma_z = A$; $\sigma_z = 2\nu A$) for the circular cylinder under internal pressure alone and, for the circular cylinder under external pressure alone, in the two cases of structures with free closed ends ($\sigma_z = A$) and clamped ends ($\varepsilon_z = 0$, i.e., $\sigma_z = 2\nu A$). In all these cases, the τ_{max} yield theory given by Eq. (7.1) with the plus sign in the second member, together with relations (4.21) expressing principal stresses σ_r and σ_t, leads to the following general relation for the onset of plastic flow at the inner radius:

$$\frac{2\delta^2}{\delta^2 - 1}(p_i - p_e) = \sigma_s. \tag{7.24}$$

By contrast, where σ_t and $\sigma_z = 0$ are maximum and minimum principal stresses respectively and σ_r is the intermediate principal stress, the τ_{max} yield theory leads to the following relation:

$$\tau_{max} = \frac{\sigma_t - \sigma_z}{2} = \frac{\sigma_t}{2} = \frac{\sigma_s}{2}. \tag{7.25}$$

This occurs only for the circular cylinder under external pressure with free open ends ($\sigma_z = 0$). In this case, noting that the yield theory is expressed by Eq. (7.2) with $\sigma_z = 0$ and considering the plus sign in the second member, repeating the procedure indicated above yields the following general relation for the onset of plastic flow at the innermost fibers:

$$\frac{\delta^2 + 1}{\delta^2 - 1}(p_i - p_e) = \sigma_s. \tag{7.26}$$

For the circular cylinder under internal pressure only, relation (7.24) must be used in all three cases considered here. It can be inferred from this relation that the expression for the elastic-breakdown pressure is as follows:

$$p_{i,i} = \sigma_s \frac{\delta^2 - 1}{2\delta^2}. \tag{7.27}$$

For the circular cylinder under external pressure only, relation (7.24) must still be used in the two cases where the structure has free closed ends ($\sigma_z = A$) or clamped ends ($\varepsilon_z = 0$ and, consequently, $\sigma_z = 2\nu A$). However, relation (7.26) must be used if the structure has free open ends, i.e., $\sigma_z = 0$. In the first two cases and in the third, the external pressure which leads to the onset of plastic flow at the inner radius will respectively be:

$$\text{(b) and (c)} \quad p_{e,i} = \sigma_s \frac{\delta^2 - 1}{2\delta^2} \tag{7.28}$$

$$\text{(a)} \quad p_{e,i} = \sigma_s \frac{\delta^2 - 1}{\delta^2 + 1}. \tag{7.29}$$

If the σ_{max} yield theory expressed by relation (7.4) is used with the plus sign in the second member, we will obtain relation (7.26) for the inner radius in all three cases. Thus, for both the circular cylinder under internal pressure only and the circular cylinder under external pressure only, the internal or external pressures for the onset of plastic flow at the inner radius are expressed by the same relation, viz.:

$$p_{i,i} = p_{e,i} = \sigma_s \frac{\delta^2 - 1}{\delta^2 + 1}. \tag{7.30}$$

Lastly, we will apply the ε_{max} yield theory given by relation (7.5), as usual with the plus sign in the second member. If we consider the inner radius and relations (4.21) and (4.28), we will obtain the following general relations in the three cases defined above (case *a*, with $\sigma_z = 0$; case *b*, with $\sigma_z = A$; case *c*, with $\sigma_z = 2\nu A$):

$$\text{(a)} \quad \frac{1}{\delta^2 - 1}\left\{p_i\left[(1-\nu) + \delta^2(1+\nu)\right] - 2p_e\delta^2\right\} = \sigma_s \tag{7.31}$$

$$\text{(b)} \quad \frac{1}{\delta^2 - 1}\left\{p_i\left[(1-2\nu) + \delta^2(1+\nu)\right] - p_e\delta^2(2-\nu)\right\} = \sigma_s \tag{7.32}$$

$$\text{(c)} \quad \frac{1-\nu}{\delta^2 - 1}\left\{p_i\left[(1-2\nu) + \delta^2\right] - 2p_e\delta^2(1-\nu)\right\} = \sigma_s. \tag{7.33}$$

For the circular cylinder under internal pressure only, relations (7.31), (7.32) and (7.33) can be specialized for the three cases to give the following expressions for the internal pressure that triggers the onset of plastic flow at the inner radius:

$$\text{(a)} \quad p_{i,i} = \sigma_s \frac{\delta^2 - 1}{\left[(1-\nu) + \delta^2(1+\nu)\right]} \tag{7.34}$$

$$\text{(b)} \quad p_{i,i} = \sigma_s \frac{\delta^2 - 1}{\left[(1-2\nu) + \delta^2(1+\nu)\right]} \tag{7.35}$$

$$\text{(c)} \quad p_{i,i} = \sigma_s \frac{\delta^2 - 1}{(1+\nu)\left[(1-2\nu) + \delta^2\right]}. \tag{7.36}$$

For the circular cylinder under external pressure only, relations (7.31), (7.32) and (7.33) provide the following expressions for the external pressure that triggers the onset of plastic flow, again at the inner radius:

$$\text{(a)} \quad p_{e,i} = \sigma_s \frac{\delta^2 - 1}{2\delta^2} \tag{7.37}$$

$$\text{(b)} \quad p_{e,i} = \sigma_s \frac{\delta^2 - 1}{\delta^2(2 - \nu)} \tag{7.38}$$

$$\text{(c)} \quad p_{e,i} = \sigma_s \frac{\delta^2 - 1}{2\delta^2(1 - \nu^2)}. \tag{7.39}$$

To compare the results obtained with the four yield theories, we can once again use the dimensionless ratio $p_{i,i}/\sigma_s$ or $p_{e,i}/\sigma_s$, depending on whether we are dealing with a circular cylinder subject only to internal pressure or only to external pressure. By way of example, we will illustrate the results obtained for a circular cylinder under internal pressure only and subject to a stress state characterized by $\sigma_t \neq 0$, $\sigma_r \neq 0$. and $\sigma_z = 0$. Figure 7.8 shows the $p_{i,i}/\sigma_s = f(\delta)$ curves for the onset of plastic flow at the inner radius associated with the four yield theories considered. The asymptotic values for each of the four curves can be readily

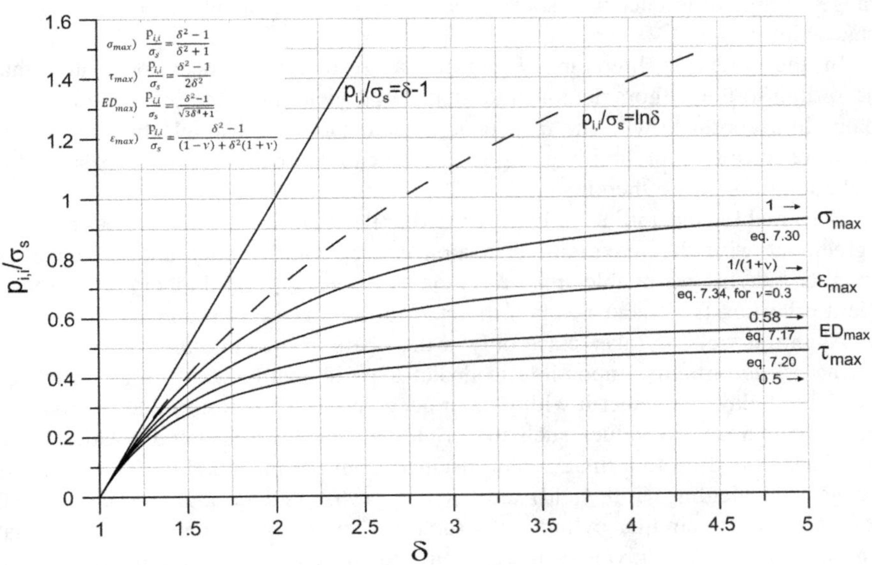

Fig. 7.8 Ratios $(p_{i,i}/\sigma_s) = f(\delta)$ at the onset of plastic flow for the four yield theories considered, with $\sigma_z = 0$; ratio $(p_{i,i}/\sigma_s) = \delta - 1$ according to the Hamburg formula in fully plastic state of the wall thickness; ratio $(p_{i,i}/\sigma_s) = f(\delta)$ in fully plastic state for the maximum shear stress yield theory (*dashed line*)

determined on the basis of the simple consideration that where δ tends to very high values, and at the limit to infinity, maximum hoop stres $\sigma_{t,max}$ will tend to p_i, while maximum radial stress $\sigma_{r,max}$ remains equal to $-p_i$. Thus, as $\sigma_z = 0.$, we will have a pure shear plane stress state (pure shear) for which, in plane $\tau = \tau(\sigma)$, the principal Mohr's circle will be centered at the origin of the coordinate axes.

Where the dimensionless diameter ratio δ tends to unity, i.e., at the extreme left of its theoretical range of variability $(1,0 \leq \delta < \infty)$, the stress state is close to that for uniaxial tension and the principal Mohr's circle representing in plane $\tau = \tau(\sigma)$ is almost tangent at the origin of the axis of the ordinate. This is the stress state we have in a thin-walled circular cylinder. It is represented by the fifth solid-line curve shown in Fig. 7.8, i.e., the uppermost curve corresponding to relation $(p_{i,i}/\sigma_s) = \delta - 1$, or in other words the Hamburg formula.

It should also be noted that all four of the yield theories considered here lead to the same result for a uniaxial stress state, i.e., with $\sigma_t \neq 0$ and $\sigma_r = \sigma_z = 0$. It can also be deduced that, as a consequence of the assumption that stresses are uniformly distributed through the wall thickness, once yielding has arisen at the inner radius, it will propagate instantaneously through the wall. In other words, the structure's entire cross-section will yield.

Although Fig. 7.8 applies to the case where $\sigma_z = 0$, the concepts it illustrates can be generalized to the case with non-zero axial stress. The considerations to be made in this case are left to the reader. In this connection, it should be borne in mind that certain yield theories permit principal stress σ_z to vary within a specified range with no difference whatsoever compared to the results obtained with a stress state with $\sigma_z = 0$.

In the scientific literature, experimental results have been presented that according to the authors demonstrate that initial yielding at the inner radius takes place in accordance with the σ_{max} or ε_{max} yield theories. Though this contradicts what we have seen in the foregoing section, it can be explained on the basis of the considerations made therein.

Presumably, the initial yielding of a ductile material occurs in approximate agreement with the maximum distortion energy yield theory and, as an even rougher but still reasonable approximation, with the τ_{max} yield theory (for Mohr's yield criterion, the reader is again directed to the preceding section).

As initial yielding takes place only at the fibers of the circular cylinder's inner surface, it is virtually impossible to detect experimentally with the test methods available today. To detect it with greater accuracy, it is necessary that the internal pressure increase to values such that yielding and the associated elastic/plastic interface extend up to a finite radial distance from the inner radius. There will be no general yielding (full yield) until it has propagated through the entire wall thickness of the circular cylindrical structure. This involves applying an internal pressure such as that given by the dashed-line curve in Fig. 7.8, which applies to the pressure bringing the entire wall thickness to the fully plastic state according to the τ_{max} yield theory.

It can thus be expected that initial yielding will in fact occur at pressures midway between those on the curves for the onset of plastic flow and the fully plastic state with a given yield theory, depending on the sensitivity of the measurement instruments used. In Fig. 7.8, these two limiting curves for the τ_{max} yield theory are the lowermost curve and the dashed-line curve. These considerations, moreover, demonstrate another important point that the designer should keep in mind: a structural member can continue to function normally under service loads that are significantly higher than those causing initial yielding.

For a circular cylinder made of a brittle material, the σ_{max} yield theory can be used, but taking care to refer to the ultimate stress rather than the yield stress. For such materials, it is probably more conservative to use the internal friction yield theory. A circular cylinder made of a brittle material must obviously be designed with a higher safety factor than that introduced for a ductile material, as brittle structures fail through complete fracture of the entire cross section, whereas with ductile materials it is possible to rely on a plastic reserve between initial yielding (or small yielding) and fracture.

7.5 Circular Cylinder Loaded Beyond Initial Yielding

The stress state of a thick-walled circular cylinder under an internal pressure sufficient to load it beyond initial yielding will here be analyzed in the most general condition, viz., that of a triaxial stress state. This, in any case, is the condition that normally obtains in the equipment used for the *autofrettage* or *self-hooping* process.

7.5.1 Autofrettage Procedures

The autofrettage process is carried out with procedures that either apply pressure to the entire length of the structure at once, or to successive portions of limited length until the entire structure is self-hooped and so left with a compressive residual stress. In the former case (Fig. 7.9), two seals are used to close both ends of the circular cylinder and enable the entire length to be pressurized by hydraulic fluid delivered through an appropriate supply line. In the latter case, a pair of mobile seals are used in what could be termed a moving mandrel configuration, and hydraulic fluid is introduced between the seals and pressurized by means of an external circuit or the mechanical action of a ram (Fig. 7.10).

To enable the autofrettage process to be carried out more quickly, the internal pressure acting on the entire length of the structure is sometimes accompanied by the external pressure generated by a so-called container consisting of side-by-side rings of sufficient thickness, and hence sufficient stiffness, to prevent the radial expansion of the structure's outer radius and generate a pressure distribution p_e on

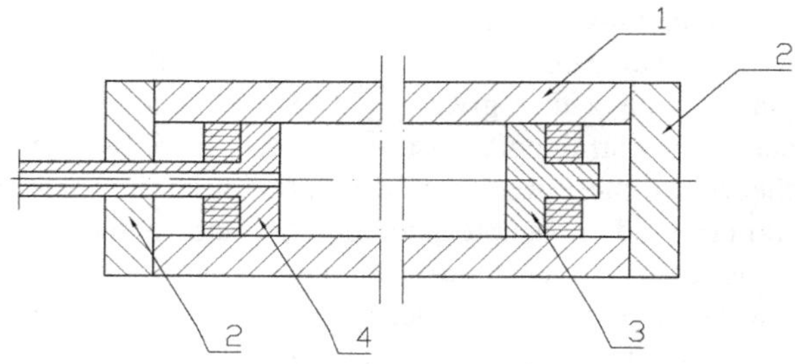

Fig. 7.9 Functional pattern of hydraulic autofrettage in closed-end conditions: *1* Circular cylinder to be autofrettaged, *2* End closures, *3* Plug with fixed seal, *4* Plug with pressurized fluid supply port and fixed seal

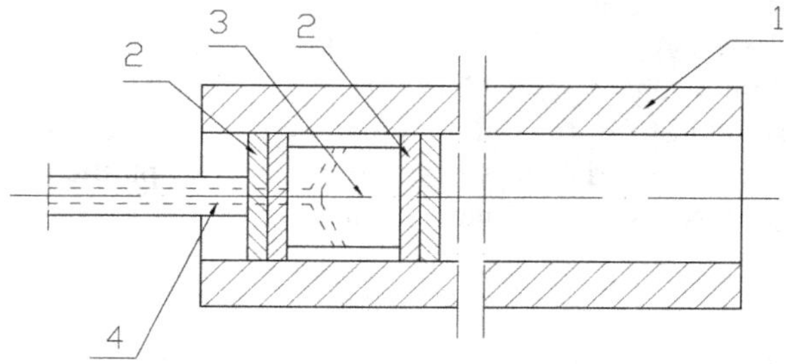

Fig. 7.10 Autofrettage with mobile seal equipment: *1* Circular cylinder to be autofrettaged, *2* Seals, *3* Moving mandrel, *4* Ram

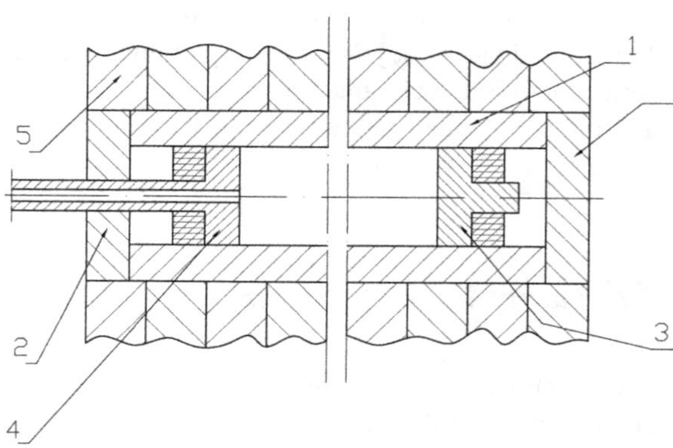

Fig. 7.11 Autofrettage in a container enclosing the entire length of the circular cylinder; arrangement as shown in Fig. 7.9, with the addition of container *5* consisting of rigid side-by-side rings

structure's outer surface which adds to the action of the internal pressure produced by the hydraulic fluid (Fig. 7.11).

Another type of autofrettage called swage autofrettage is based on the mechanical action of an oversized mandrel whose outer radius is larger than the bore of the circular cylinder to be autofrettaged, and which is forced axially through the structure by hydraulic fluid pressurized and supplied by an external pump (Fig. 7.12).

In all of these autofrettage processes, the stress state is triaxial, with three non-zero principal stresses σ_t, σ_r, and σ_z. In the zones contacted by the fixed or moving seals, stress distribution σ_z is not constant through the cross-section. It becomes

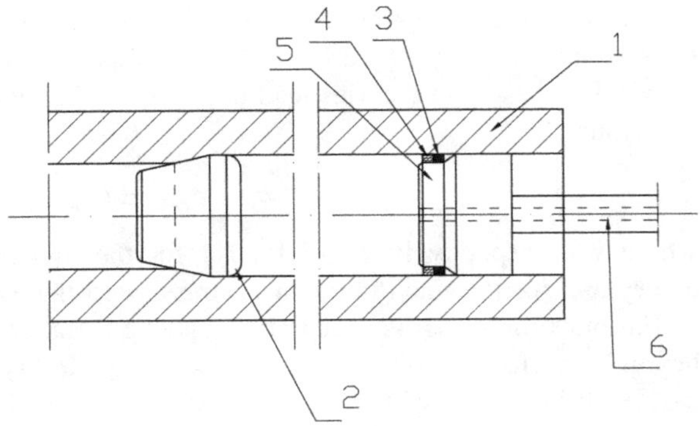

Fig. 7.12 Swage autofrettage: *1* Circular cylinder to be autofrettaged, *2* Mandrel, *3* Mild steel seals, *4* Rubber seals, *5* Plunger, *6* Pressurized fluid supply port

constant at a distance from the perturbed zones which is of the same order of magnitude as the cross-section, in accordance with Saint–Venant's principle. These are the conditions which apply for the considerations discussed below.

7.5.2 *Analysis of Circular Cylinders Loaded Beyond Initial Yielding*

In the context described in the preceding section, we will consider a thick-walled circular cylinder under internal pressure producing a triaxial stress state and consisting of an ideal material whose stress–strain curve $\sigma = \sigma(\varepsilon)$ is linear elastic-perfectly plastic (Fig. 7.1). We will gradually increase internal pressure p_i until the material begins to enter the plastic state at inner radius r_i, where on the basis of relations (4.39) the stress state is characterized by the following maximum radial, hoop and axial stresses:

$$\sigma_{r,r_i} = -p_i; \quad \sigma_{t,r_i} = p_i \frac{\delta^2 + 1}{\delta^2 - 1}; \quad \sigma_z = \frac{p_i}{\delta^2 - 1} = const. \tag{7.40}$$

From the stress distribution curves shown in Fig. 4.4, in fact, it is reasonable to expect the material's failure to begin at the inner radius, where stresses reach the maximum value. In this connection, and as regards the pressure leading to the onset of the plastic state, all four of the yield theories presented in Sect. 7.4 agree in stating that yielding will start at the innermost fibers.

Here, however only the maximum distortion energy ED_{max} and maximum shear stress τ_{max} yield theories will be used in dealing with circular cylinders loaded beyond initial yielding and in analyzing the stress and strain states generated in them. This choice is motivated not only by the considerations regarding the yield theories that can potentially be used given in Sect. 7.3, but also by the fact that the two selected yield theories lead to similar and very simple relations. In the triaxial stress state considered here ($\sigma_z = A$), a comparison of relations (7.10) and (7.23) shows that these two theories lead to the following equation representing the yield criterion:

$$\sigma_t - \sigma_r = \pm k\sigma_s, \tag{7.41}$$

where k is respectively $2/\sqrt{3}$ and 1 for the maximum distortion energy yield theory and for the maximum shear stress yield theory.

For our problem, it is necessary to use both Eq. (7.41), which, for stress values beyond the elastic limit, substitutes the compatibility equation (obviously, in the case examined here, which is that of the first loading beyond yielding as a result of internal pressure in a circular cylinder consisting of virgin material, Eq. (7.41) must be considered with the plus sign in the second member) and the equilibrium Eq. (4.7). Expanding the latter and substituting Eq. (7.41) in the expanded relation gives the following first order differential equation with separable variables:

$$\frac{\mathrm{d}\sigma_r}{\mathrm{d}r} r = k\sigma_s. \tag{7.42}$$

Integrating this equation enables us to write

$$\sigma_r = k\sigma_s \ln r + C, \tag{7.43}$$

where C is an integration constant to be determined by imposing boundary conditions at the inner radius or the outer radius. Here, we prefer to use the boundary condition at the inner radius $r = r_i$, where it is necessary that $\sigma_{r,r_i} = -p_i$, so that Eq. (7.43) gives:

$$C = -(p_i + k\sigma_s \ln r_i). \tag{7.44}$$

Substituting the latter in Eq. (7.43) yields:

$$\sigma_r = -p_i + k\sigma_s \ln\frac{r}{r_i}. \tag{7.45}$$

By introducing this last relation in Eq. (7.41), we then obtain:

$$\sigma_t = -p_i + k\sigma_s\left(1 + \ln\frac{r}{r_i}\right). \tag{7.46}$$

Lastly, we assume that the plastic region has a stress state in which the third principal stress σ_z is the arithmetic average of the other two principal stresses (radial stress and hoop stress). On the basis of this assumption, which will be demonstrated in more general terms in Chap. 9, we will have:

$$\sigma_z = -p_i + k\sigma_s\left(\frac{1}{2} + \ln\frac{r}{r_i}\right). \tag{7.47}$$

The stress state in the entire plastic region, wherever the plastic-elastic interface (or plastic boundary) defined by radius r_p, in the domain $r_i \leq r_p \leq r_e$, is located, will thus be that described by the following set of three equations:

$$\begin{aligned}\sigma_r &= -p_i + k\sigma_s \ln\frac{r}{r_i}\\ \sigma_t &= -p_i + k\sigma_s\left(1 + \ln\frac{r}{r_i}\right)\\ \sigma_z &= -p_i + k\sigma_s\left(\frac{1}{2} + \ln\frac{r}{r_i}\right)\end{aligned} \tag{7.48}$$

The plastic-elastic interface, which consists of the cylindrical surface of radius r_p, thus divides the structure's effective cross-sectional area into two distinct adjoining regions: the inner region, where the material works in the plastic range, and the outer region, where the material still works in the elastic range (Fig. 7.13). It should be noted that equations (7.48) are entirely general, as the relations (7.41) and (4.7) from which they derive are equally general.

As was mentioned earlier, the constant C appearing in Eq. (7.43) can also be determined by imposing the boundary condition at the outer radius r_e, where $\sigma_{r,r_e} = p_e$ or $\sigma_{r,r_e} = 0$, depending on whether autofrettage is performed with or without a container; thus, the constant will be, respectively:

$$C = -(p_e + k\sigma_s \ln r_e); \tag{7.49}$$

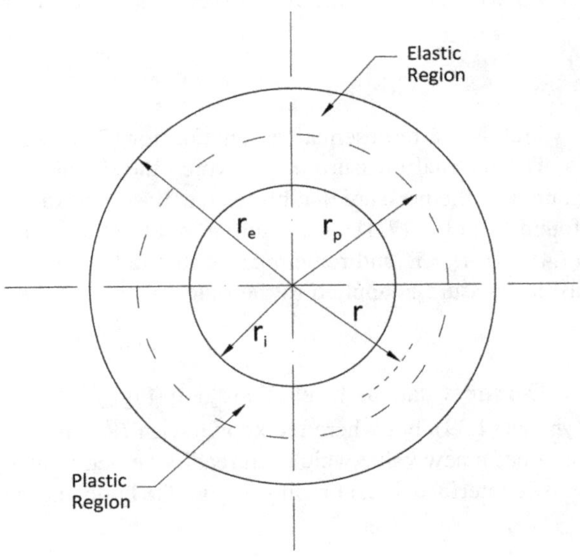

Fig. 7.13 Plastic-elastic interface (*plastic boundary*), with plastic region (*inner*) and elastic region (*outer*)

and

$$C = -k\sigma_s \ln r_e. \tag{7.50}$$

In place of the set of three equations (7.48), we will have the following sets of three equations for the two respective cases:

$$\begin{aligned} \sigma_r &= -p_e + k\sigma_s \ln\frac{r}{r_e} \\ \sigma_t &= -p_e + k\sigma_s\left(1 + \ln\frac{r}{r_e}\right) \\ \sigma_z &= -p_e + k\sigma_s\left(\frac{1}{2} + \ln\frac{r}{r_e}\right) \end{aligned} \tag{7.51}$$

and

$$\begin{aligned} \sigma_r &= k\sigma_s \ln\frac{r}{r_e} \\ \sigma_t &= k\sigma_s\left(1 + \ln\frac{r}{r_e}\right) \\ \sigma_z &= k\sigma_s\left(\frac{1}{2} + \ln\frac{r}{r_e}\right). \end{aligned} \tag{7.52}$$

The following considerations are based on the set of three equations (7.48). As only the case of autofrettage performed without a container will be examined, we will have $p_e = 0$.

The internal pressure $p_{i,i}$ which leads to the onset of plastic flow at the bore of our circular cylinder is found by substituting the first two relations (7.40) in Eq. (7.41) with the plus sign in the second member, thereby obtaining the equation

$$p_{i,i} = k\sigma_s \frac{\delta^2 - 1}{2\delta^2} \tag{7.53}$$

which is the representative equation for (7.18) and (7.27).

The internal pressure $p_{i,e}$ at which the circular cylinder is in the fully plastic state, i.e., the pressure that brings the plastic-elastic interface to the outer radius, is found from Eq. (7.41), substituting in it the hoop stress σ_t given by the second Eq. (7.48) for $r = r_e$ and remembering that radial stress σ_r at the outer radius is zero, as no pressure is applied to the outer surface of the structure. We thus obtain:

$$p_{i,e} = k\sigma_s \ln\delta. \tag{7.54}$$

The stress state in the elastic region (Fig. 7.13) is given by relations of the same type as (4.39), but where the known term K' is no longer equal to $p_i/(\delta^2 - 1)$, but assumes a new value which can readily be determined by noting that at the plastic-elastic interface, i.e., at radius $r = r_p$, the hoop and radial stresses given by the first two equations (4.39) and specialized for this radius must satisfy Eq. (7.41) representing the yield criterion. We thus obtain:

$$K' = k\sigma_s \frac{r_p^2}{2r_e^2}. \tag{7.55}$$

Accordingly, the stress state in the elastic region is given by the following relations:

$$\begin{aligned}\sigma_r &= k\sigma_s \frac{r_p^2}{2r_e^2}\left(1 - \frac{r_e^2}{r^2}\right)\\ \sigma_t &= k\sigma_s \frac{r_p^2}{2r_e^2}\left(1 + \frac{r_e^2}{r^2}\right)\\ \sigma_z &= k\sigma_s \frac{r_p^2}{2r_e^2}.\end{aligned} \tag{7.56}$$

The internal pressure $p_{i,p}$ which brings the plastic-elastic interface to radius $r = r_p$ is determined by establishing that because of continuity, the hoop stress at this radius calculated with the second relation (7.56) is equal to that calculated with the second relation (7.48). We thus obtain:

$$p_{i,p} = k\sigma_s \left[\frac{1}{2}\left(1 - \frac{r_p^2}{r_e^2}\right) + \ln \frac{r_p}{r_i}\right]. \tag{7.57}$$

It should be pointed out that we arrive at the same relation (7.57) if we impose continuity of σ_r rather than σ_t at the plastic-elastic interface. For $r_p = r_i$ and $r_p = r_e$, relation (7.57) gives the values of internal pressure at the onset of plastic flow and at the fully plastic state, given by relations (7.53) and (7.54) respectively. The radial stress $\sigma_{r,p}$ at the plastic-elastic interface is determined from the first relation (7.56) for $r = r_p$; it is given by the relation:

$$\sigma_{r,p} = -k\sigma_s \frac{r_e^2 - r_p^2}{2r_e^2}. \tag{7.58}$$

The strain state in the elastic region of our circular cylinder is found from the general equations for thermoelasticity given by relations (1.17) after omitting the temperature terms which are of no interest to us here. Substituting relations (7.56) in them gives the following relations for the three principal unit elongations (principal strains), which as will be recalled apply to the case of the circular cylinder with closed ends:

$$\varepsilon_r = \frac{k\sigma_s}{E}\frac{r_p^2}{2r_e^2}\left[(1 - 2\nu) - \frac{r_e^2}{r^2}(1 + \nu)\right]$$

$$\varepsilon_t = \frac{k\sigma_s}{E}\frac{r_p^2}{2r_e^2}\left[(1-2\nu) + \frac{r_e^2}{r^2}(1+\nu)\right] \tag{7.59}$$
$$\varepsilon_z = \frac{k\sigma_s}{E}\frac{r_p^2}{2r_e^2}(1-2\nu).$$

Calculating the radial displacement in this region is extremely straightforward: it is sufficient to remember that $u = r\varepsilon_t$ and substitute in the latter the second relation (7.59). We thus obtain:

$$u = \frac{k\sigma_s r}{E}\frac{r_p^2}{2r_e^2}\left[(1-2\nu) + \frac{r_e^2}{r^2}(1+\nu)\right]. \tag{7.60}$$

The stress state in the plastic region of our circular cylinder is described by the following set of three relations:

$$\sigma_r = k\sigma_s\left[-\frac{1}{2}\left(1-\frac{r_p^2}{r_e^2}\right) + \ln\frac{r}{r_p}\right]$$
$$\sigma_t = k\sigma_s\left[\frac{1}{2}\left(1+\frac{r_p^2}{r_e^2}\right) + \ln\frac{r}{r_p}\right] \tag{7.61}$$
$$\sigma_z = k\sigma_s\left[\frac{r_p^2}{2r_e^2} + \ln\frac{r}{r_p}\right].$$

These relations are obtained from equations (7.48) after substituting p_i in the latter with the pressure $p_{i,p}$ given by relation (7.57).

Figure 7.14 shows the distribution of the three principal stresses, made dimensionless relative to yield stress σ_s, in the plastic and elastic regions of a mild steel circular cylinder ($\sigma_s = 240\,\text{MPa}$) with closed ends, featuring a dimensionless diameter ratio $\delta = 3$ and autofrettaged by applying an internal pressure which brings the plastic-elastic interface to a radius $r_p = 1,75r_i$. The stress distribution curves shown in the figure were calculated with the three relations (7.56) for the elastic region, and with the three relations (7.61) for the plastic region, applying the maximum distortion energy strength theory or in other words setting $k = 2/\sqrt{3}$ in all relations.

Correctly calculating the strains in the plastic region is quite complex. What is of interest in analyzing the autofrettage process, however, is the unit elongation at the inner radius r_i, which can be calculated as a rough approximation under the assumption that the volume remains constant during deformation (material incompressible and volume expansion will thus be zero). From this perspective, once radial displacement u_{r_p} at the plastic-elastic interface has been calculated by means of relation (7.60) with $r = r_p$, the radial displacement at the inner radius can be assumed to be equal to:

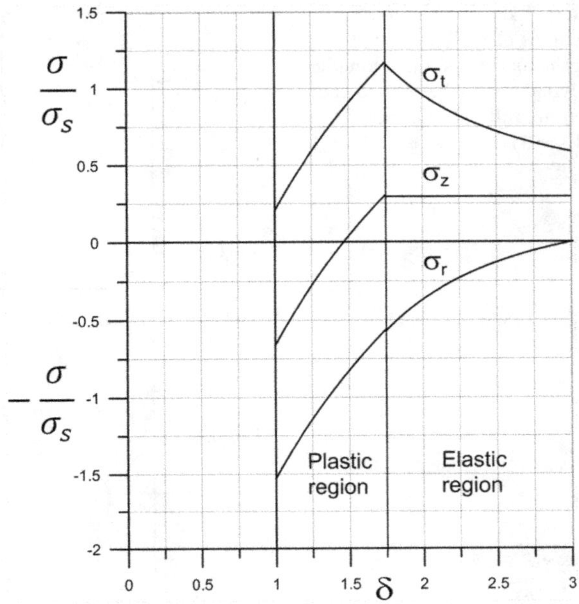

Fig. 7.14 Distribution curves of the dimensionless principal stresses in a circular cylinder with closed ends, under an internal pressure just sufficient to cause reverse yielding ($\delta = 3$; $\sigma_s \cong 240$ MPa)

$$u_{r_i} = u_{r_p}\left(r_p/r_i\right). \tag{7.62}$$

Correspondingly, in terms of circumferential principal strain, we will have

$$\varepsilon_{t,r_i} = \varepsilon_{t,r_p}\left(r_p/r_i\right)^2. \tag{7.63}$$

When the circular cylinder's effective cross-sectional area is in the fully plastic state, the plastic-elastic interface moves to the outer radius, i.e., $r_p = r_e$, as a result of an internal pressure given by relations (7.54). The stress state in the fully plastic structure is determined from relations (7.61), setting $r_p = r_e$. This gives the following relations:

$$\begin{aligned}\sigma_r &= k\sigma_s \ln\frac{r}{r_e}\\ \sigma_t &= k\sigma_s\left(1+\ln\frac{r}{r_e}\right)\\ \sigma_z &= k\sigma_s\left(\frac{1}{2}+\ln\frac{r}{r_e}\right).\end{aligned} \tag{7.64}$$

For the sake of completeness, the equations whereby the unit elongations at the inner and outer radii of the circular cylinder in the fully plastic state are given below:

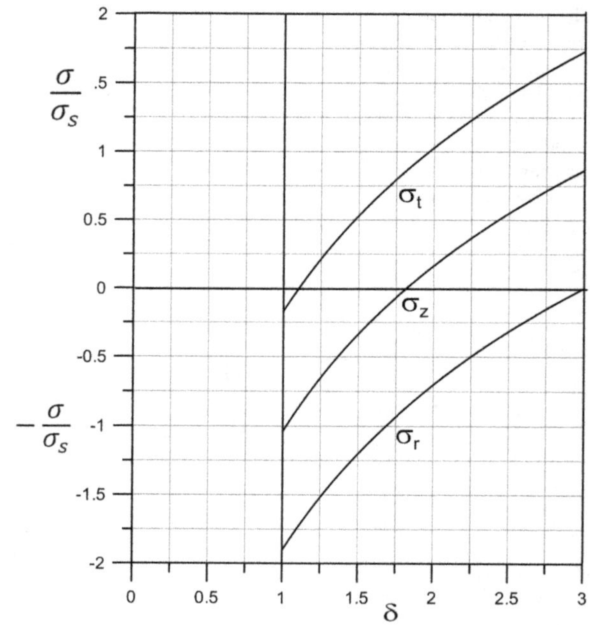

Fig. 7.15 Distribution curves of the dimensionless principal stresses in a circular cylinder with free closed ends in the fully plastic state ($\delta = 3$)

$$
\begin{aligned}
\varepsilon_{t,re} &= \frac{k\sigma_s}{2E}(2-\nu)\\
\varepsilon_{t,r_i} &= \frac{k\sigma_s}{2E}(2-\nu)\delta^2\\
\varepsilon_{z,r_i,r_e} &= \frac{k\sigma_s}{2E}(1-2\nu)\\
\varepsilon_r &= -(\varepsilon_t+\varepsilon_z).
\end{aligned} \tag{7.65}
$$

Figure 7.15 shows the distribution curves of the three principal stresses, made dimensionless relative to the yield stress, through the wall thickness of a circular cylinder in the fully plastic state with a diameter ratio $\delta = 3$. These stresses were calculated using the maximum distortion energy yield theory, or in other words with relations (7.64) in which $k = 2/\sqrt{3}$.

As can be deduced from the stress distribution curves in Figs. 7.14 and 7.15, the stress distribution in both the elasto-plastic range and the plastic range is substantially different from that in the linear elastic range. The differences are even more noticeable if we consider the equivalent stress σ_e calculated for the sake of consistency with the maximum distortion energy yield theory. In the linear elastic range, i.e., for $p_i \le p_{i,i}$, this stress is highest at the bore and decreases nonlinearly as we move from the inner radius towards the outer radius. In the elasto-plastic range, by contrast, the equivalent stress remains constantly equal to σ_s in the entire plastic region, i.e., for $r_i \le r \le r_p$, and then decreases nonlinearly in the outer elastic region, i.e., for $r_p \le r \le r_e$. Lastly, for the structure in the fully plastic state, the equivalent stress is constantly equal to σ_s at any radius, i.e., for $r_i \le r \le r_p$. The dimensionless ratio p_i/σ_s also increases nonlinearly, from the value given by relation (7.53) to that defined by relation (7.54), as r_p increases, i.e., as the plastic-elastic interface moves from the inner to the outer radius.

7.5.3 Plastic Reserve

Figure 7.16 shows the distribution curves of the dimensionless ratio $p_{i,i}/\sigma_s = f(\delta)$ curve for the onset of plastic flow at the bore of a circular cylinder with free closed ends ($\sigma_z = A$) and the ratio $p_{i,e}/\sigma_s = f(\delta)$ for the structure's effective cross-sectional area in the fully plastic state, calculated in pairs with the maximum shear stress and maximum distortion energy yield theories. An examination of these distribution curves and those given for the circular cylinder with free open ends shown in Fig. 7.8 leads to several interesting design considerations regarding the material's plastic reserve capacity, which can be quantified through the *limit-design factor*

$$L = \frac{\Lambda_{cp}}{\Lambda_{ip}} \tag{7.66}$$

defined, in general, as the ratio of the load Λ_{cp} needed to bring the material to the fully plastic state and the load Λ_{ip} at the onset of plastic flow (the elastic-breakdown in the case for which proportional limit, elastic limit and yield-point stress

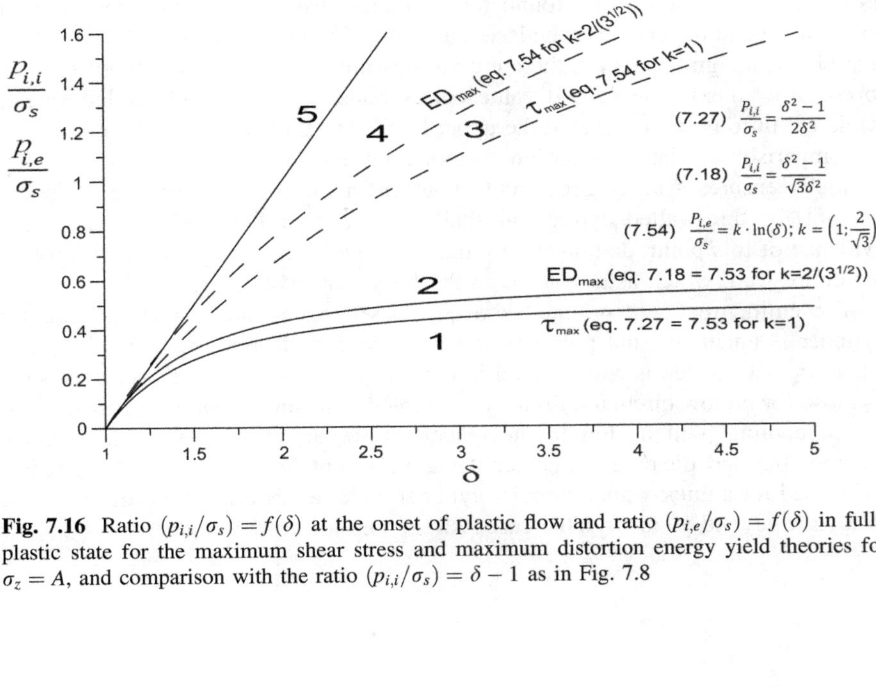

Fig. 7.16 Ratio $(p_{i,i}/\sigma_s) = f(\delta)$ at the onset of plastic flow and ratio $(p_{i,e}/\sigma_s) = f(\delta)$ in fully plastic state for the maximum shear stress and maximum distortion energy yield theories for $\sigma_z = A$, and comparison with the ratio $(p_{i,i}/\sigma_s) = \delta - 1$ as in Fig. 7.8

coincide). It should be pointed out that the term "load" is used here to denote whatever magnitude is characteristic of the plasticization phenomena being investigated. In analyzing the formation of a plastic hinge in a deflected beam, for instance, the load consists of the bending moment applied to the beam, whereas if we are examining the plasticization of a rotor resulting from overspeeding, the load will be the rotor's angular velocity.

In the case of interest to us here, the load is clearly the internal pressure. Accordingly, recalling relations (7.54) and (7.53) which respectively define the pressures leading to the fully plastic state and the onset of plastic flow in a circular cylinder with free closed ends, relation (7.66) becomes:

$$L = \frac{p_{i,e}}{p_{i,i}} = \frac{2\delta^2}{\delta^2 - 1} \ln\delta. \tag{7.67}$$

With reference to Fig. 7.16, the limit-design factor is simply the ratio, for a given value of the dimensionless diameter ratio δ, of the ordinates of the curves for fully plastic flow and the onset of the plastic state plotted for a given yield theory, or in other words, the ratio of the ordinates of curves 4 and 2 correlated to the maximum distortion energy yield theory or of curves 3 and 1 correlated to the maximum shear stress yield theory for an assigned value of δ. Both relation (7.67) and the curves given in Fig. 7.16 clearly show that the limit-design factor, and thus the plastic reserve capacity of the material of an autofrettaged circular cylinder, increase with δ according to a logarithmic function. From this standpoint, moreover, there is no difference between the two yield theories considered.

This factor assumes values of 3.69, 1.85 and 1.10 for $\delta = 6$, $\delta = 2$ and $\delta = 1.1$ respectively. The first of these values is given here by way of comparison with the corresponding value of 1.44 found for a rotating disk of the same cross section stressed beyond the elastic range [see Vullo and Vivio (2013)]. However, it has no real practical significance, as there are no pressure vessels with $\delta = 6$ nor reasons to construct them. The second value corresponds to a very thick-walled vessel, while the third is for a vessel at the upper limit for a thin-walled circular cylinder. A comparison of the limit-design factors for these latter two cases shows that plasticity ensures a fairly good safety margin for thick-walled circular cylinders but not for thin-walled cylindrical shells. Curve 5 in Fig. 7.16 provides clear evidence of this point, demonstrating that once yield starts, it propagates through the entire wall, which as a result is in the fully plastic state.

The entire treatment outlined above is based on assuming that the circular cylinder is under internal pressure and has free closed ends ($\sigma_z = A$). The case where $\sigma_z = 0$, which is encountered in other types of autofrettage processes such as those for hollow circular cylinders assembled with shrink fits involving heating and/or cooling, will be left to the reader. However, the fundamental relations whereby this specific case can be handled are given in Sect. 7.4, as are the relations to be used for a thick-walled circular cylinder under an external pressure sufficient to load it beyond yielding. This case will also be left to the reader.

Chapter 8
Autofrettage of Thick-Walled Circular Cylinders

8.1 General Considerations on Autofrettage

A circular cylinder with free closed ends working in the elastic range has limited resistance to loads arising from internal pressure (Faupel 1964, Juvinall 1967, Brownell and Young 1968, Iurzolla 1981). As was demonstrated in Sect. 7.5.2, the internal pressure at which yielding begins at this structure's bore is given by relation (7.53). It can be deduced from this relation that, for very large diameter ratios δ (and, at the limit, for $\delta \rightarrow \infty$), the maximum pressure in the elastic range that the circular cylinder can withstand is $\sigma_s/\sqrt{3}$ and $\sigma_s/2$, depending on whether the maximum distortion energy or the maximum shear stress yield theory is used.

One of the methods of increasing the elastic range of operation of a circular cylinder under internal pressure is *autofrettage*, also known as *self-hooping* or *overstressing*. This process, which was briefly described in Sect. 7.5.1, is as old as it is interesting from the structural standpoint. Though first developed to increase the strength of artillery pieces (cannons, guns, missile launchers, rocket launchers, torpedo launchers and so forth), autofrettage is now used for many highly stressed thick-walled cylinders employed in a wide variety of applications in the chemical, petrochemical, petroleum and nuclear industries (Jorgensen 1958, Kalnins and Updike 2001).

As was mentioned earlier, the autofrettage process consists of inducing plastic flow in the circular cylinder by subjecting it at a particular point of the production cycle to an internal pressure above that given by relation (7.53): this is the autofrettage pressure, also known as the self-hooping or overstressing pressure. In the structure's material, plastic flow begins at the bore and progresses through the wall thickness as the autofrettage pressure increases. As a result, the plastic-elastic interface, or plastic boundary, between the inner region showing plastic behavior and the outer region whose behavior is still elastic, moves from the inner to the outer radius.

This process gives rise to a state of coaction in the wall thickness of the circular cylinder deriving from the fact that the permanent set produced in the inner plastic region by the pressure acting at the inner radius prevents the material in the outer elastic region from contracting once pressure is released. The elastic outer layer thus has the effect of hooping the plastic internal layer; with this overstressing

V. Vullo, *Circular Cylinders and Pressure Vessels*, Springer Series in Solid and Structural Mechanics 3, DOI: 10.1007/978-3-319-00690-1_8,

procedure, we thus obtain an effect similar but stronger than that achieved with the interference fit procedure. This plastic flow, which for a linear elastic-perfectly plastic material proceeds in accordance with relation (7.57), is such that a residual stress state characterized around the bore by a compressive hoop stress distribution remains in the wall after the pressure is released. The circular cylinder is then said to be autofrettaged. As the material's behavior is assumed to be linear elastic-perfectly plastic, subsequently applying an internal pressure which does not exceed the autofrettage pressure will cause the circular cylinder to react elastically.

Over time, our understanding of the autofrettage process has grown, and now encompasses reverse yielding (and, more in general, successive yield cycles, first in tension, then in compression, then again in tension, and so on) of circular cylinders with a high diameter ratio δ, subjected to autofrettage until the entire wall thickness is in the plastic state in order to produce strong, high performance structures such as those used in hypervelocity launchers for naval ordnance. While further details of the autofrettage process will be provided in Sect. 8.4, the preliminary considerations given below should be mastered first (Dawson 1960, Dawson and Seigel 1963, Moradi et al. 2001, Zhao and Wang 2010).

8.2 Residual and Working Stresses

Attention will focus here on calculating both the residual stresses that will remain in the autofrettaged circular cylinder after the autofrettage pressure is released, and the overall stress state in the autofrettaged structure subjected to the working pressure, and thus in actual service conditions. For this purpose, the relations presented in the preceding chapter must be kept firmly in mind, as they are fundamental.

Figure 8.1 illustrates how the dimensionless hoop and radial stress distributions vary through the circular cylinder's wall thickness as internal pressure p_i increases from the level $p_{i,i}$ leading to the onset of plastic flow at the bore and given by relation (7.53) to the pressure $p_{i,e}$ given by relation (7.54) at which the entire wall thickness of the structure is in the plastic state and the plastic-elastic boundary is thus located at the outer radius. The circular cylinder considered here has a diameter ratio $\delta = 2$. Curves 1 and 2 for the onset of plastic flow are calculated with elastic-field relations (4.39), setting $p_i = p_{i,i}$. The dashed-line curves for the partially plastic state, i.e., for the plastic-elastic range, are calculated with relations (7.56) for the elastic region and relations (7.61) for the plastic region, with $k = 2/\sqrt{3}$ (the maximum distortion energy yield theory is thus applied), for two plastic-elastic interfaces defined by $r_p = 1.4\, r_i$ and $r_p = 1.8\, r_i$ respectively. Curves 3 and 4 for the fully plastic state are calculated with relations (7.64), again with $k = 2/\sqrt{3}$.

Curves 1 and 2, which define the structure's stress state when initial yielding occurs at the bore radius, are calculated with the elastic-field equations in accordance with the stress–strain curve $\sigma = \sigma(\varepsilon)$ for a linear elastic-perfectly plastic material such as that shown in Fig. 7.1. As internal pressure p_i increases from the value $p_{i,i}$ at the onset of plastic flow, radial stress σ_r will vary, though it will still

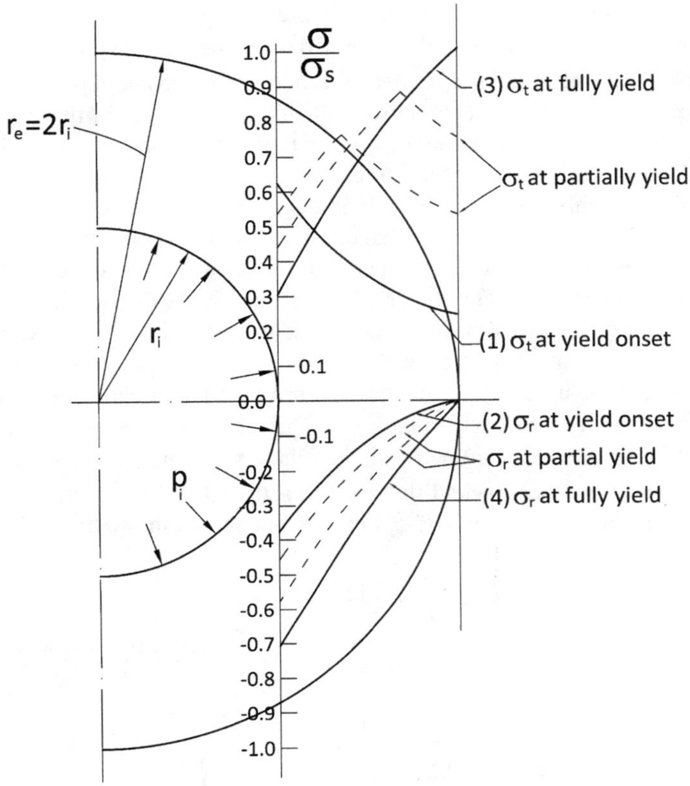

Fig. 8.1 Distribution *curves* of the dimensionless hoop and radial stresses calculated with the maximum distortion energy yield theory, for: the onset of plastic flow (*curves 1* and *2*); two partially plastic states (*dashed line curves*); the fully plastic state (*curves 3* and *4*)

have to satisfy the boundary conditions, i.e., must be $\sigma_r = -p_i$ and $\sigma_r = 0$ at the inner and outer radii respectively. As p_i increases, hoop stress σ_t must drop at the inner radius in accordance with the yield theory given by Eq. (7.41); in addition, it must satisfy Eq. (7.41) regardless of the location of the plastic-elastic boundary as it moves from the inner towards the outer radius. In other words, the algebraic difference $(\sigma_t - \sigma_r)$ in the plastic region, i.e., the distance between the two curves representing the hoop stress and the radial stress measured along a straight-line parallel to the ordinate axis, must be equal to $\left(2/\sqrt{3}\right)\sigma_s$ at all times.

To calculate the residual stress state that will remain in the circular cylinder when the autofrettage pressure is released, it will first be assumed that the equivalent residual stress σ_{e,res,r_i} at inner radius r_i is less than the compressive yield-point stress of the material and that the latter is equal to the tensile yield-point stress, so that there is no Bauschinger effect. The residual stress state in the material is thus such that it does not cause further compressive yielding at the inner radius, i.e., where tensile yielding took place at the time the autofrettage pressure was applied.

Figure 8.2 shows the distribution curves representing the residual hoop and radial stresses ($\sigma^p_{t,\,res}$ and $\sigma^p_{r,\,res}$ respectively) obtained when the internal pressure $p_{i,\,e}$ which fully overstressed the circular cylinder by bringing the plastic-elastic boundary to the outer radius is released. The entire wall thickness is thus in the plastic state, hence the superscript p designating the plastic region (the superscript e will be used below to denote the elastic region). Curves 1 and 2 show the hoop and radial stresses corresponding to this pressure, and obviously coincide with curves 3 and 4 in Fig. 8.1. Curves 3 and 4, on the other hand, represent the stress state, whose sign has changed, that pressure $p_{i,\,e}$ would produce if the material were to continue to exhibit perfectly elastic behavior. The material's behavior while the internal pressure is released is in line with the assumptions advanced in Sect. 7.1 (see Fig. 7.1, unloading path AB). To give an idea of how this works, it is as if a pressure—$p_{i,\,e}$ were superimposed at the unloading stage on pressure $p_{i,\,e}$. These curves are thus calculated using relations (4.39), where $p_i = p_{i,\,e}$. Curves 5 and 6 represent the residual hoop and radial stresses determined with the relations given below and applying the maximum distortion energy yield theory, i.e., setting $k = 2/\sqrt{3}$ in these relations (to avoid over-complication, the $\sigma^p_{z,\,res}$ curve is not shown in the figure).

$$\begin{aligned}
\sigma^p_{r,\,res} &= k\sigma_s \ln\frac{r}{r_e} - \frac{p_{i,e}}{\delta^2 - 1}\left(1 - \frac{r_e^2}{r^2}\right)\\
\sigma^p_{t,\,res} &= k\sigma_s\left(1 + \ln\frac{r}{r_e}\right) - \frac{p_{i,e}}{\delta^2 - 1}\left(1 + \frac{r_e^2}{r^2}\right)\\
\sigma^p_{z,\,res} &= k\sigma_s\left(\frac{1}{2} + \ln\frac{r}{r_e}\right) - \frac{p_{i,e}}{\delta^2 - 1}.
\end{aligned} \tag{8.1}$$

Relations (8.1) are obtained by consistently subtracting (i.e., radial stress from radial stress, etc.) relations (4.39) written for $p_i = p_{i,\,e}$ from relations (7.64). In other words, curves 5 and 6 are the algebraic sum of curves 1 and 3 and of curves 2 and 4 respectively. Here, the superscript p indicates that the plastic region extends throughout the whole wall thickness.

It is clear that, in accordance with the phenomena illustrated in Fig. 7.1, there will be no yield of the material either during unloading from $p_{i,\,e}$ to zero pressure or during reloading from zero pressure to $p_{i,\,e}$ providing that the diameter ratio δ is below the threshold at which *reverse yielding* as defined below takes place. It is equally clear that curves 3 and 4 in Fig. 8.2 can be readily obtained from curves 1 and 2 in Fig. 8.1 by multiplying the ordinates by the ratio $-(p_{i,\,e}/p_{i,\,i})$.

It should be pointed out once again that the analysis described above, which leads to the results summarized in Figs. 8.1 and 8.2, is based not only on the maximum distortion energy yield theory, but also on assuming an ideal material exhibiting linear elastic-perfectly plastic behavior such as that shown in Fig. 7.1. For a real material, the pressure $p_{i,\,e}$ producing the fully plastic state will increase as a result of strain hardening. It should also be noted that this analysis can be used to predict the circular cylinder's bursting pressure if the yield-point stress in the above relations is substituted with the material's ultimate stress. The latter, in fact,

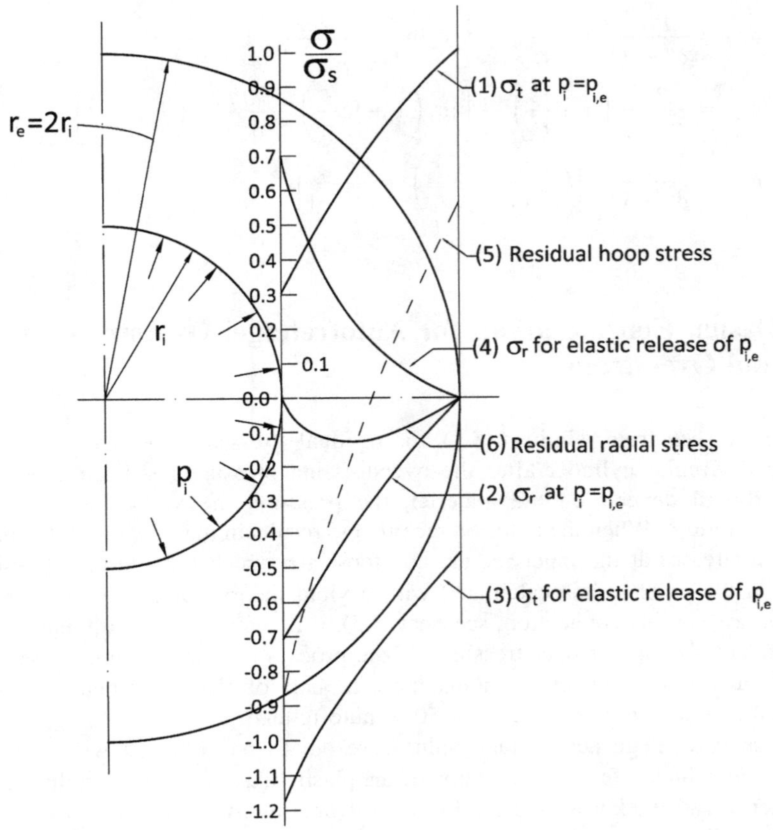

Fig. 8.2 Distribution *curves* of the dimensionless residual stresses, calculated with the maximum distortion energy yield theory, after full yielding of the cylinder's wall thickness due to an autofrettage pressure $p_{i,e}$ ($\delta = 2$)

will not be reached, as there will always be a certain modest amount of plastic flow. Likewise, it is advisable to remember that an autofrettaged circular cylinder's resistance to internal pressure increases at the expense of its plastic reserve capacity, and that the autofrettage produces no increase in bursting strength.

The autofrettage process discussed above makes it possible to load circular cylinders with working (or service) pressures $p_{i,l}$ well above that which would cause yielding at the inner radius if the structure consisted of virgin material. These working pressures must lie in the range $p_{i,i} \leq p_{i,l} \leq p_{i,e}$ if the stress state they produce is to be elastic (see Fig. 7.1). This stress state will be superposed on the residual stress state given by relations (8.1) and will result in an overall stress state in working conditions defined by the relations below, which apply to a structure whose whole wall thickness is in the fully plastic state (the second subscript l designates the working condition):

$$\begin{aligned}\sigma_{r,l} &= \frac{p_{i,l}}{\delta^2-1}\left(1-\frac{r_e^2}{r^2}\right)+\left[k\sigma_s \ln\frac{r}{r_e}-\frac{p_{i,e}}{\delta^2-1}\left(1-\frac{r_e^2}{r^2}\right)\right]\\ \sigma_{t,l} &= \frac{p_{i,l}}{\delta^2-1}\left(1+\frac{r_e^2}{r^2}\right)+\left[k\sigma_s\left(1+\ln\frac{r}{r_e}\right)-\frac{p_{i,e}}{\delta^2-1}\left(1+\frac{r_e^2}{r^2}\right)\right]\\ \sigma_{z,l} &= \frac{p_{i,l}}{\delta^2-1}+\left[k\sigma_s\left(\frac{1}{2}+\ln\frac{r}{r_e}\right)-\frac{p_{i,e}}{\delta^2-1}\right].\end{aligned} \tag{8.2}$$

8.3 Design Considerations for Autofrettage: Overstressing and Overstrain

As can be seen from relations (8.1), the residual stresses at the bore in an autofrettaged circular cylinder after the overstressing pressure has been completely released will depend on the material, the pressure, the yield theory and the diameter ratio δ. When the autofrettage process results in the maximum theoretical residual stresses at the inner radius, i.e., those for which the equivalent residual stress at the inner radius calculated with a yield theory equals the compressive yield stress (in this connection, see Sect. 8.5), it is referred to as full autofrettage (100 % autofrettage or overstressing). If the process produces residual stresses at the circular cylinder's inner radius that are half of the theoretical maximum residual stresses, it is referred to as 50 % autofrettage.

These percentage autofrettage values are not to be confused with those for *overstrain*, which refer to the location of the plastic-elastic interface in the circular cylinder's wall thickness, and are likewise expressed as a percentage of the wall thickness. Saying that there is complete overstrain (or full overstrain, or 100 % overstrain) thus means that the autofrettage process has been carried out up to a pressure that brought the plastic-elastic interface to the outer radius. As can be readily demonstrated, a 100 % autofrettage can in some cases be associated with a much lower percentage overstrain (only 10 %, for instance). Depending on the circular cylinder, various percentage values of overstrain are required to obtain various percentage values of autofrettage.

In constructing a circular cylinder to be subjected to autofrettage, it is often preferable to use a unalloyed steel which, though it has limited mechanical strength, is relatively inexpensive, since the process can produce the same results that would be achieved by using a much more costly high strength alloyed steel. It should also be recognized that, even without carrying out autofrettage as described above, it is frequently possible to increase the mechanical strength of a material with low tensile properties by means of appropriate heat treatments that produce a strain-aging effect. Depending on the material used, such treatments can increase yield strength by 5–50 % or more. Even more significant improvements in the material's mechanical strength properties can be achieved if autofrettage (a typically mechanical process) is employed synergistically together with heat treatment.

As mentioned earlier, when a circular cylinder consisting of virgin material is subjected to a sufficiently high internal pressure, the fibers at the bore will start to yield, i.e., plastic flow begins. If, on the other hand, there were residual stresses in the circular cylinder before the working pressure was applied, the pressure leading to the onset of plastic flow will be higher if the residual stresses are compressive and lower if they are tensile. The residual stresses, if present, must be considered in calculations together with the elastic stress state associated with the external loads and defined by relations (4.29) and by one of the three relations (4.30), (4.31) and (4.33).

For circular cylinders with free closed ends subjected only to internal pressure, Faupel and Furbeck (1953) demonstrated that if there is a residual stress state, the internal pressure $p_{i,i}$ which brings to fibers at the bore to the onset of yielding is given by the following relation:

$$\begin{aligned} &p_{i,i}^2\left(\frac{\delta^2}{\delta^2-1}\right)^2+p_{i,i}\left(\frac{\delta^2}{\delta^2-1}\right)\sigma_{t,res,r_i} \\ &+\frac{1}{3}\left(\sigma_{t,res,r_i}^2+\sigma_{z,res,r_i}^2+\sigma_{t,res,r_i}\sigma_{z,res,r_i}-\sigma_s^2\right)=0 \end{aligned} \tag{8.3}$$

where σ_{t,res,r_i} and σ_{z,res,r_i} are the residual hoop and axial stresses at the inner radius (where, as will be recalled, the residual radial stress is zero, i.e., $\sigma_{r,res,r_i}=0$). Where these residual stresses result from manufacturing processes or heat treatments, they are not generally known and must be determined experimentally, with the blind hole method or boring-out method, for example. If, however, they are associated with autofrettage or shrink-fit processes, they can be determined analytically as we have described.

8.4 Residual Stresses Resulting from Partial Overstrain, Without Reverse Yielding

The calculations to be performed will entail fairly significant analytical complications in cases where the autofrettage pressure is so high that it causes reverse yielding of the material once it has been completely released. This section will consider only normal cases involving ordinary elastic recovery, i.e., without reverse yielding. The latter will not occur if the diameter ratio δ is below the threshold that will be quantified below.

After releasing the pressure $p_{i,p}$ that brought the plastic-elastic interface to radius r_p, which is correlated to pressure $p_{i,p}$ by relation (7.57), there will be a residual stress and strain state in the circular cylinder's wall thickness because of the permanent radial expansion resulting from radial plastic flow. As indicated earlier, these residual stresses and strains are obtained by calculating the stresses and strains due to the load, i.e., to the internal pressure $p_{i,p}$ that brought the plastic-elastic interface to radius r_p, and subtracting the elastic stresses and strains that would have existed if the material had still exhibited elastic behavior at this autofrettage pressure.

For a circular cylinder with free closed ends subjected to partial overstrain, the residual stresses in the elastic region determined by subtracting relations (4.39) from relations (7.56) and recalling relation (7.57) are as follows:

$$\begin{aligned}
\sigma^e_{r,res} &= \frac{k\sigma_s}{2}\left(1-\frac{r_e^2}{r^2}\right)\left\{\frac{r_p^2}{r_e^2}-\frac{1}{\delta^2-1}\left[1-\frac{r_p^2}{r_e^2}+2\ln\frac{r_p}{r_i}\right]\right\}\\
\sigma^e_{t,res} &= \frac{k\sigma_s}{2}\left(1+\frac{r_e^2}{r^2}\right)\left\{\frac{r_p^2}{r_e^2}-\frac{1}{\delta^2-1}\left[1-\frac{r_p^2}{r_e^2}+2\ln\frac{r_p}{r_i}\right]\right\}\\
\sigma^e_{z,res} &= \frac{k\sigma_s}{2}\left\{\frac{r_p^2}{r_e^2}-\frac{1}{\delta^2-1}\left[1-\frac{r_p^2}{r_e^2}+2\ln\frac{r_p}{r_i}\right]\right\}.
\end{aligned} \tag{8.4}$$

If the circular cylinder has free open ends, relations (8.4) still apply, but there will obviously be zero longitudinal stress $\left(\sigma^e_{z,res}=0\right)$.

Again for a circular cylinder with free closed ends subjected to partial overstrain, the residual stresses in the plastic region determined by subtracting relations (4.39) from relations (7.61) and recalling relation (7.57) are as follows:

$$\begin{aligned}
\sigma^p_{r,res} &= \frac{k\sigma_s}{2}\left\{-1+\frac{r_p^2}{r_e^2}+2\ln\frac{r}{r_p}-\frac{1}{\delta^2-1}\left(1-\frac{r_e^2}{r^2}\right)\left[1-\frac{r_p^2}{r_e^2}+2\ln\frac{r_p}{r_i}\right]\right\}\\
\sigma^p_{t,res} &= \frac{k\sigma_s}{2}\left\{1+\frac{r_p^2}{r_e^2}+2\ln\frac{r}{r_p}-\frac{1}{\delta^2-1}\left(1+\frac{r_e^2}{r^2}\right)\left[1-\frac{r_p^2}{r_e^2}+2\ln\frac{r_p}{r_i}\right]\right\}\\
\sigma^p_{z,res} &= \frac{k\sigma_s}{2}\left\{\frac{r_p^2}{r_e^2}+2\ln\frac{r}{r_p}-\frac{1}{\delta^2-1}\left[1-\frac{r_p^2}{r_e^2}+2\ln\frac{r_p}{r_i}\right]\right\}.
\end{aligned} \tag{8.5}$$

Once again, if the circular cylinder has free open ends, relations (8.5) still apply, but in this case the residual longitudinal stress $\sigma^p_{z,res}$ given by the third of these relations after omitting the term in square brackets will not be zero. Figure 8.3 shows a typical distribution of residual stresses in a free closed-end circular cylinder with $r_e = 5r_i$ ($\delta = 5$) under an internal pressure which brings the plastic-elastic interface to radius $r_p = 1,75r_i$.

8.5 Limit Conditions for Reverse Yielding

As internal pressure increases from the value $p_{i,i}$ for the onset of plastic flow at the inner radius [given by relation (7.53)], the plastic-elastic interface will move from the bore towards the outer radius. Once pressure $p_{i,e}$ given by relation (7.54) has been reached, the entire wall thickness will be in the plastic state, i.e., full overstrain will have been reached. Pressure $p_{i,e}$ that brought the plastic-elastic interface to the outer radius is the maximum pressure that can be applied to the circular

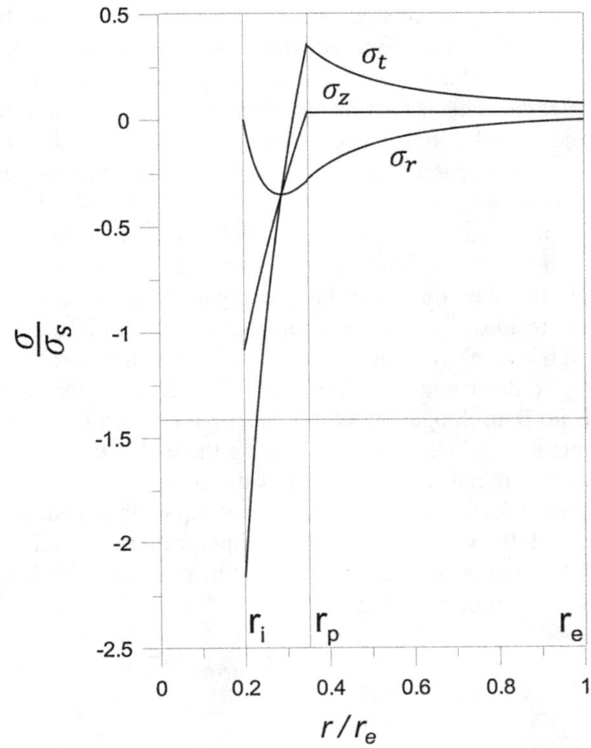

Fig. 8.3 Distribution *curves* of residual stresses in a free closed-end circular cylinder obtained by releasing the pressure that brought the plastic-elastic interface to radius $r_p = 1,75r_i$ ($k = 2/\sqrt{3}$)

cylinder without causing the structure to collapse, assuming that the material's behavior is linear elastic-perfectly plastic up to failure. In this fully plastic state, the residual stresses are given by relations (8.1), which can also be obtained from relations (8.5) by setting $r_p = r_e$ in the latter. It can be deduced that the local values of residual stresses at the radius r_i of a circular cylinder autofrettaged up to complete overstrain are given by the following relations (the superscript p has been omitted from these relations for the sake of simplicity):

$$\begin{aligned}
\sigma_{r,\,res,\,r_i} &= 0 \\
\sigma_{t,\,res,\,r_i} &= k\sigma_s\left(1 - \frac{2\delta^2}{\delta^2 - 1}\ln\delta\right) \\
\sigma_{z,\,res,\,r_i} &= \frac{k\sigma_s}{2}\left(1 - \frac{2\delta^2}{\delta^2 - 1}\ln\delta\right) = \frac{1}{2}\sigma_{t,\,res,\,r_i}.
\end{aligned} \tag{8.6}$$

As was indicated in Sect. 8.3, the condition of maximum theoretical residual stresses at the inner radius, which defines full autofrettage, is that at which the equivalent residual stress at the inner radius equals the material's compressive yield stress, i.e., $\sigma_{e,\,res,\,r_i} = -\sigma_s$. It should be specified that it is not always necessary to bring overstrain to a full 100 % in order to reach this condition. However, when the residual stress state at the inner radius is such that the equivalent residual stress reaches and exceeds the material's compressive yield stress, further yielding like that which took place during pressurization will occur when the pressure is released. This is the phenomenon known as *reverse yielding* or *reyielding* (Dawson and Siegel 1963, Faupel 1964).

As is apparent from relation (7.54), the pressure required to cause plastic flow throughout the circular cylinder's entire wall thickness increases along with diameter ratio δ. In turn, the higher the pressure that is released, the higher the compressive residual stresses at radius r_i will be. When the wall is in the fully plastic state, there will thus be a particular value of the diameter ratio δ at which the residual stresses remaining after the autofrettage pressure is released will be high enough to cause reverse yielding at the bore.

To determine this threshold value for diameter ratio δ, still assuming that there is no Bauschinger effect, the maximum distortion energy yield theory expressed by relation (7.3) is used, substituting the yield stress σ_s with the equivalent stress σ_e. As the radial residual stress at the inner radius given by the first relation (8.6) is zero, the maximum value of the equivalent residual stress at the inner radius correlated with this threshold value for diameter ratio δ will be given by one of the following relations, which apply to circular cylinders with free closed ends and free open ends respectively:

$$\sigma^2_{e,\,res,\,r_i} = \sigma^2_{t,\,res,\,r_i} + \sigma^2_{z,\,res,\,r_i} - \sigma_{t,\,res,\,r_i}\sigma_{z,\,res\,,r_i} \tag{8.7}$$

and

$$\sigma^2_{e,\,res,\,r_i} = \sigma^2_{t,\,res,\,r_i}. \tag{8.8}$$

We will first examine the circular cylinder with free closed ends, for which, as relations (8.6) confirm, $\sigma_{z,\,res,\,r_i} = (\sigma_{t,\,res,\,r_i} + \sigma_{r,\,res,\,r_i})/2$. Using the maximum distortion energy yield theory given by relation (8.7) with σ_s substituted for $\sigma_{e,\,res,\,r_i}$ and recalling relations (8.6), we find that the residual stresses remaining in this structure when the autofrettage pressure has been completely released will produce further yielding at the inner radius, this time in compression, when the hoop residual stress reaches $2\sigma_s/\sqrt{3}$, i.e., when $\sigma_{t,\,res,\,r_i} = (2/\sqrt{3})\sigma_s$. Equaling this value of $\sigma_{t,\,res,\,r_i}$ to $-k\sigma_s$ (with the maximum distortion energy yield theory, as will be recalled, $k = 2/\sqrt{3}$), we can conclude that the threshold value for diameter ratio δ at which reverse yielding will not occur is given by the relation

$$\frac{\delta^2}{\delta^2 - 1}\ln\delta = 1. \tag{8.9}$$

Solving this relation gives the limiting diameter ratio (or critical value of the diameter ratio or critical diameter ratio) $\delta \cong 2.22$.

Consequently, in a closed-end circular cylinder with diameter ratio $\delta = 2.22$, autofrettaged at a pressure $p_{i,e}$ that brought the plastic-elastic interface to the outer radius and whose wall thickness is thus in the fully plastic state, residual stresses at the inner radius which are high enough to cause reverse yielding in this area will be generated when the autofrettage pressure is fully released. For diameter ratios $\delta < 2.22$, the residual stresses that will be generated are lower than those required to produce reverse yielding for the circular cylinder in the fully plastic state, while for $\delta > 2.22$ they will be high enough to do so. The limiting condition for which reverse yielding will not occur in a fully plastic circular cylinder with free closed ends is thus that where diameter ratio δ does not exceed 2.22.

As diameter ratio δ increases beyond this threshold value, the value of ratio r_p/r_e which is strictly necessary to reach the compressive yield-point stress at the inner radius when the autofrettage pressure has been completely released. This means that there is also an autofrettage pressure threshold in cases where the limiting design condition is reverse yielding. Using the relations presented earlier, we can demonstrate that the limiting autofrettage pressure $p_{i,lsr}$ which, upon release, causes incipient reverse yielding at the bore of a circular cylinder with a diameter ratio $\delta > 2.22$ is given by the relation

$$p_{i,lsr} = k\sigma_s \frac{\delta^2 - 1}{\delta^2}. \tag{8.10}$$

This pressure is thus twice that, given by relation (7.53), which causes initial yielding in tension during pressurizing, i.e., $p_{i,lsr} = 2p_{i,i}$. Reverse yielding can thus be obtained by pressurizing the circular cylinder at a pressure greater than twice the elastic-breakdown pressure. This threshold value of the autofrettage pressure is known as the *critical pressure*.

We will now move on to an examination of the circular cylinder with free open ends, for which $\sigma_{r,res,r_i} = \sigma_{z,res,r_i} = 0$. Again using the maximum distortion energy yield theory, which in this case is given by relation (8.8) with σ_s substituted for σ_{e,res,r_i}, and recalling the second relation (8.6), we find that the residual stresses remaining in this structure when the autofrettage pressure has been completely released will produce further yielding at the inner radius, once again in compression, when the hoop residual stress reaches σ_s, or in other words when $\sigma_{t,res,r_i} = \sigma_s$. Equaling this value of σ_{t,res,r_i} to $-\sigma_s$, we can conclude that the threshold value for diameter ratio δ at which reverse yielding will not occur is given by the relation

$$\frac{\delta^2}{\delta^2 - 1} \ln\delta = \frac{1}{2}\left(1 + \frac{\sqrt{3}}{2}\right). \tag{8.11}$$

Solving this relation gives the limiting diameter ratio or critical diameter ratio $\delta \cong 2.03$.

With the necessary changes, the considerations that can be made for the circular cylinder with free open ends are similar to those for the closed-end circular cylinder. In this case, to avoid reverse yielding of an open-ended circular cylinder whose wall thickness is fully plastic, diameter ratio δ must not exceed 2.03. For diameter ratios δ above this threshold, reverse yielding will take place upon release of the autofrettage pressure $p_{i,e}$ given by relation (7.54) when the latter is greater than $1.85p_{i,i}$.

Accordingly, for any circular cylinder, whether it has free closed or free open ends, there are three limiting curves for the ratio $p_i/\sigma_s = f(\delta)$:

- The curve correlated to the pressure at which yielding initially occurs at the bore, given by relation (7.53).
- The curve correlated to the pressure required to bring the wall to the fully plastic state, which for $\delta < 2.22$, produces residual stresses that are low enough to prevent reverse yielding, and is given by relation (7.54).
- The curve correlated to the limiting autofrettage pressure given by relation (8.10) which, for $\delta > 2.22$, causes reverse yielding at inner radius r_i.

The results of these analyses thus demonstrate that the theoretically optimal autofrettage pressure for a circular cylinder with free closed or free open ends is around twice the pressure that causes initial yielding at the bore for circular cylinders with diameter ratios δ over 2.22 and approximately equal to the pressure making the wall thickness fully plastic for circular cylinders with ratios $\delta < 2.22$.

However, it is apparent that if a circular cylinder could always, i.e., regardless of its diameter ratio δ, be subjected to an autofrettage process involving pressures given by relation (7.54), the results that could be achieved in terms of residual stresses, and hence load capability, would be far superior to those given by relation (8.10). But in this case, as noted earlier, since the material's compressive yield-point stress is not unlimited, we would have reverse yielding at the bore upon releasing the autofrettage pressure if $\delta > 2.22$ or $\delta > 2.03$ for the circular cylinders with free closed and free open ends respectively.

8.6 Residual Stresses Resulting from Full Overstrain, with Reverse Yielding

Given that, as was shown in the preceding section, the results obtained for circular cylinders with free closed or free open ends are approximately identical, the analysis of reverse yielding carried out here covers practically all cases. It is performed using the procedure first advanced by Prager and Hodge (1951), and further developed by Dawson and Seigel (1963). Our analysis, in any case, applies to the closed-end circular cylinder.

The basic assumptions for the procedure are still those that were used earlier, viz., linear elastic-perfectly plastic, i.e., non-strain hardening material, zero Bauschinger effect, yield theory given by relation (7.41), but this time with the minus sign in the second member, and axial stress equal to the arithmetic average of the hoop stress and the radial stress.

We will consider a closed-end circular cylinder with diameter ratio $\delta > 2.22$ whose entire wall has been brought to the fully plastic state by applying an internal autofrettage pressure $p_{i,e}$ given by relation (7.54). At the end of pressurizing up to the maximum pressure $p_{i,e}$, our circular cylinder will have a stress state described by relations (7.64). As the autofrettage pressure is gradually released, the circular cylinder will deform elastically until the equivalent residual stress σ_{e,res,r_i} at the bore equals the material's compressive yield-point stress. Thereafter, each reduction in pressure will cause a new plastic flow progressing outward from the bore. When the autofrettage pressure has been completely released, and the internal pressure thus reaches zero, the circular cylinder will consist of two regions: an inner region or core which has undergone reverse yielding, i.e., a new yielding but in compression, and an outer region that has previously yielded in tension during autofrettage, which exhibits elastic behavior and thus acts as a sort of hooping sleeve surrounding the reyielded core.

The relations whereby the stress state in the reyielded inner core can be calculated are once again the equilibrium equation (4.7) and the equation expressing the yield theory, given by relation (7.41) with the minus sign in the second member. Through the same procedure described in Sect. 7.5.2, these two equations together with the boundary condition that the radial stress at inner radius r_i is zero lead to the following expression for this stress state (the last subscript n denotes the reyielded core):

$$\begin{aligned}\sigma_{r,res,n} &= -k\sigma_s \ln\frac{r}{r_i}\\ \sigma_{t,res,n} &= -k\sigma_s\left(1+\ln\frac{r}{r_i}\right)\\ \sigma_{z,res,n} &= -k\sigma_s\left(\frac{1}{2}+\ln\frac{r}{r_i}\right).\end{aligned} \tag{8.12}$$

These relations apply to $r_i \leq r \leq r_n$, where r_n denotes the outer radius of the core (note that the subscript n here denotes the re-yielded core, and thus has a different meaning from that used in the analysis of multilayer composite structure), i.e., the radius of the interface, also called reverse-yield boundary, separating the inner region where the material has undergone reverse yielding and the outer region where compressive reyielding did not take place. They represent the residual stresses in the reyielded region after pressure release. It should be noted that these relations are simply relations (7.48) for $p_i = 0$, with the sign changed.

Before the autofrettage pressure is released, the stress state in the outer hooping sleeve surrounding the reyielded core is given by relations (7.64). During autofrettage pressure release, this outer region is deformed elastically. Consequently, the stress state in it after autofrettage pressure has been fully released can be obtained by means of the method of superposition using the following general relation

$$\sigma' = \sigma + \Delta\sigma, \tag{8.13}$$

where $\sigma'$and σ are respectively the stresses after and before pressure release, and $\Delta\sigma$ is the change in stress due to the change in effective pressure Δp at the interface of radius r_n between the two regions, which is given by the difference in the effective pressures acting at the interface after and before the pressure has been released. Since the effective pressure at the interface between the two regions (core and hooping sleeve) is equal to the radial stress at the interface with the sign changed, we can calculate Δp as the difference, with the sign changed, between the radial stresses acting at the interface after and before pressure release, i.e., setting $(\Delta p)_{r_n} = -(\Delta\sigma_r)_{r_n}$.

Recalling the first relation (8.12) and the first relation (7.64), both written for $r = r_n$, we thus obtain the following expression for $(\Delta p)_{r_n}$:

$$(\Delta p)_{r_n} = k\sigma_s\left(\ln\frac{r_n}{r_i} - \ln\frac{r_e}{r_n}\right). \tag{8.14}$$

The elastic stresses acting in the outer region and correlated with $(\Delta p)_{r_n}$ are given by the relations:

$$\begin{aligned}
\sigma_r &= \frac{(\Delta p)_{r_n}}{\left(\frac{r_e}{r_n}\right)^2 - 1}\left(1 - \frac{r_e^2}{r^2}\right)\\
\sigma_t &= \frac{(\Delta p)_{r_n}}{\left(\frac{r_e}{r_n}\right)^2 - 1}\left(1 + \frac{r_e^2}{r^2}\right)\\
\sigma_z &= \frac{(\Delta p)_{r_n}}{\left(\frac{r_e}{r_n}\right)^2 - 1}.
\end{aligned} \tag{8.15}$$

Substituting relations (8.15) and the first two relations (7.64) into Eq. (8.13) and recalling relation (8.14) gives

$$\begin{aligned}
\sigma_{r,res,ze} &= -k\sigma_s\left\{\ln\frac{r_e}{r} + \left(\ln\frac{r_n}{r_i} - \ln\frac{r_e}{r_n}\right)\left[\frac{\left(\frac{r_e}{r}\right)^2 - 1}{\left(\frac{r_e}{r_n}\right)^2 - 1}\right]\right\}\\
\sigma_{t,res,ze} &= -k\sigma_s\left\{\ln\frac{r_e}{r} - 1 - \left(\ln\frac{r_n}{r_i} - \ln\frac{r_e}{r_n}\right)\left[\frac{\left(\frac{r_e}{r}\right)^2 + 1}{\left(\frac{r_e}{r_n}\right)^2 - 1}\right]\right\}\\
\sigma_{z,res,ze} &= -k\sigma_s\left\{\ln\frac{r_e}{r} - \frac{1}{2} - \frac{\ln\frac{r_n}{r_i} - \ln\frac{r_e}{r_n}}{\left(\frac{r_e}{r_n}\right)^2 - 1}\right\}.
\end{aligned} \tag{8.16}$$

These relations apply for $r_n \le r \le r_e$, i.e., for the outer region designated with the double subscript *ze*. They thus describe the distribution of residual radial, hoop and longitudinal stresses after the pressure has been released in the outer once yielded hooping sleeve, i.e., the outer region that yielded a single time in tension during pressurizing.

Since the hoop stress at radius $r = r_n$ must, because of continuity, be equal in the two regions inside and outside the reverse-yield boundary, equating the second relation (8.16) to the second relation (8.12) gives

$$\left(\ln\frac{r_n}{r_i} - \ln\frac{r_e}{r_n}\right) = \left(\frac{r_n}{r_e}\right)^2 - 1. \tag{8.17}$$

By solving this relation, we obtain the radius r_n which defines the radial extent of the core, i.e., the inner region that underwent reverse yielding. As it can be deduced from this relation that dimensionless ratio r_n/r_i is relatively small compared to diameter ratio $\delta = r_e/r_i$, the radial dimension of the reyielded plastic core is not very large compared to the wall thickness $s = (r_e - r_i)$. For $\delta = 3.0$ and $\delta = 5.0$, for example, $r_n/r_i = 1.13$ and $r_n/r_i = 1.41$, meaning that the radial thickness of the reverse yielded plastic core is respectively 6.50 % and 10.25 % of the circular cylinder's wall thickness. Figure 8.4 illustrates the variation of the dimensionless ratio r_n/r_i versus δ deriving from relation (8.17).

Recalling relation (8.17), (8.16) can be written in the following more compact form:

$$\begin{aligned}
\sigma_{r,\,res,\,ze} &= -\,k\sigma_s\left\{\ln\frac{r_e}{r} - \left(\frac{r_n}{r}\right)^2 + \left(\frac{r_n}{r_e}\right)^2\right\}\\
\sigma_{t,\,res,\,ze} &= -\,k\sigma_s\left\{\ln\frac{r_e}{r} - 1 + \left(\frac{r_n}{r}\right)^2 + \left(\frac{r_n}{r_e}\right)^2\right\}\\
\sigma_{z,\,res,\,ze} &= -\,k\sigma_s\left\{\ln\frac{r_e}{r} - \frac{1}{2} + \left(\frac{r_n}{r_e}\right)^2\right\}.
\end{aligned} \tag{8.18}$$

Relations (8.12) and (8.18), which apply respectively to the reyielded plastic core and the area surrounding it where the material behaves elastically, define the residual stresses developed in a fully autofrettaged thick-walled circular cylinder with $\delta > 2.22$ when the autofrettage pressure has been completely released. If the circular cylinder has free open ends, longitudinal stress $\sigma_{z,\,res,\,ze}$ must be regarded as zero in relations (8.18).

The pressure $p_{i,n}$ at the bore, which is correlated with radius r_n, can be calculated by means of the relation (7.57), applying to an ordinary load (that at which reverse yielding does not occur), by substituting r_n for r_p. We thus have:

$$p_{i,n} = k\sigma_s\left[\frac{1}{2}\left(1 - \frac{r_n^2}{r_e^2}\right) + \ln\frac{r_n}{r_i}\right]. \tag{8.19}$$

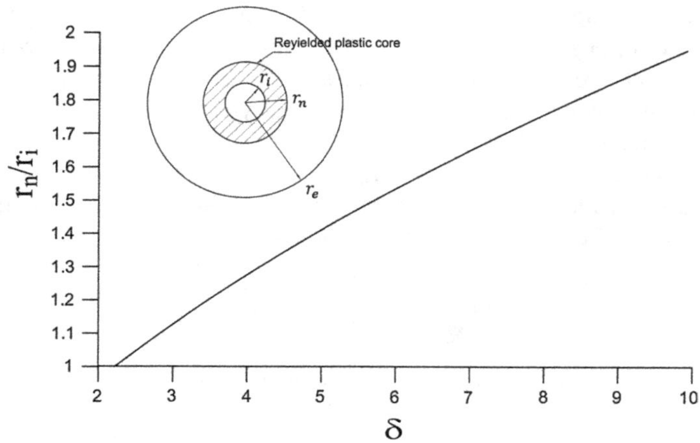

Fig. 8.4 Dimensionless ratio r_n/r_i versus δ in a circular cylinder with $\delta > 2.22$, after autofrettage pressure $p_{i,e} = k\sigma_s \ln\delta$ has been completely released, with $k = 2/\sqrt{3}$

To illustrate the effects of reverse yielding, we will consider a circular cylinder with diameter ratio $\delta = 5.0$ and consisting of a material having a tensile yield-point stress $\sigma_s = 240$ MPa. As this circular cylinder has a diameter ratio $\delta > 2.22$, reverse yielding can be induced by subjecting it to an internal pressure more than twice that causing initial yielding at the bore. As our structure's entire wall thickness is in the fully plastic state (100 % overstrain), the distribution curves of the three principal stresses through the wall are as shown in Fig. 8.5.

When the autofrettage pressure has been completely released, reverse yielding will begin at the radius r_n defined by relation (8.19). The final residual stresses in the inner reyielded core $(r_i \le r \le r_n)$ and in the outer region surrounding it $(r_n \le r \le r_e)$ can be calculated using relations (8.12) and (8.18) respectively. We thus obtain plots such as those shown in Fig. 8.6, which apply to the case with $k = 2/\sqrt{3}$. The solid-line curves in this figure show the distribution curves of residual stresses $\sigma_{r,res}$, $\sigma_{t,res}$ and $\sigma_{z,res}$ and of the correlated equivalent stress σ_e, calculated using the maximum distortion energy yield theory in a reyielded circular cylinder having a diameter ratio $\delta = 5.0$. The equivalent stress is calculated with the two relations

$$\sigma_e = -\frac{\sqrt{3}}{2} k\sigma_s \tag{8.20}$$

and

$$\sigma_e = -\frac{\sqrt{3}}{2} k\sigma_s \left[1 + \frac{4r_n^2}{r^2}\left(\frac{r_n^2}{r^2} - 1\right)\right]^{1/2} \tag{8.21}$$

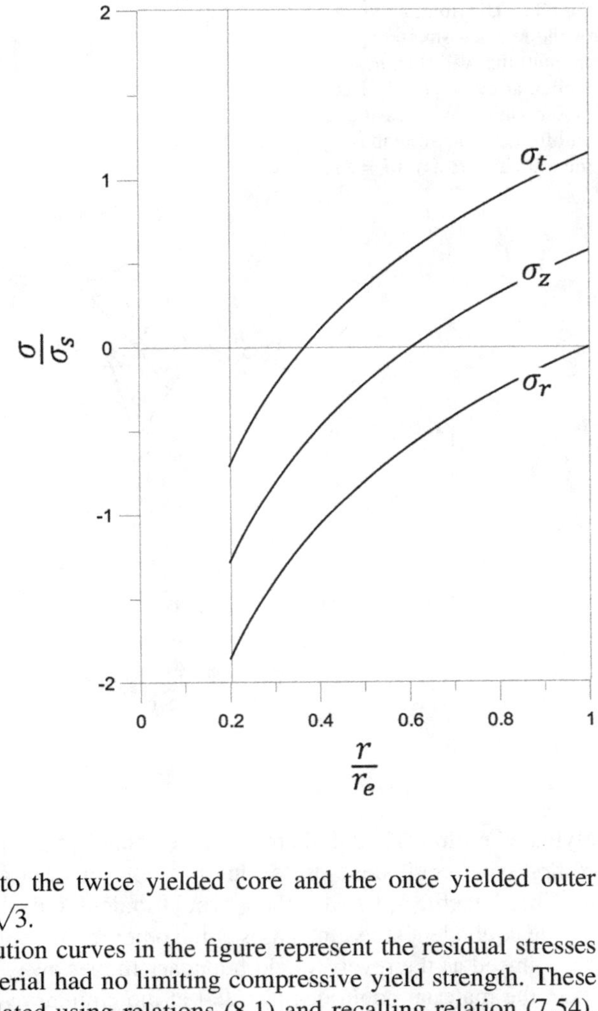

Fig. 8.5 Distribution *curves* of the dimensionless principal stresses in a circular cylinder with free closed ends, in the fully plastic state at an autofrettage pressure of 304 MPa ($\delta = 5$)

which apply respectively to the twice yielded core and the once yielded outer region, again with $k = 2/\sqrt{3}$.

The dashed-line distribution curves in the figure represent the residual stresses that would exist if the material had no limiting compressive yield strength. These residual stresses are calculated using relations (8.1) and recalling relation (7.54), still with $k = 2/\sqrt{3}$. As can be seen from a comparison of the two families of curves, the residual stresses are only slightly modified as a result of reverse yielding in the elastic region, whereas the differences are much greater in the reyielded core, particularly for hoop stress σ_t. The results thus confirm that a circular cylinder with $\delta > 2.22$, if autofrettaged to the fully plastic condition, will have a reyielded core after pressure release and that this plastic core has relatively small radial dimensions compared to the circular cylinder's outer radius.

Figure 8.6 also makes it clear that compressive stress increases along with the distance from the bore. This does not necessarily mean that the heightened benefit has been achieved by allowing reverse yielding to take place. To demonstrate this, it is sufficient to consider that the equivalent stress remains constant throughout the

Fig. 8.6 Distribution *curves* of the residual stresses through the wall thickness of a circular cylinder with free closed ends after releasing an autofrettage pressure above the critical pressure ($\delta = 5$)

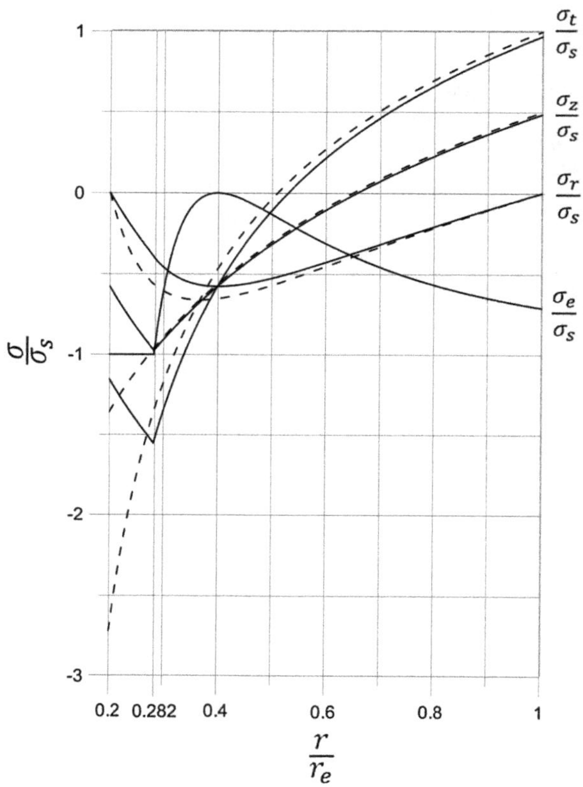

reyielded region. Though there could be some benefit in having a region with high compressive equivalent stress, this has not yet been confirmed. Along with this possible benefit, moreover, there would undoubtedly be a negative effect, as the tensile equivalent stress increases at the outer radius, and this increase will be more pronounced as the reverse-yield boundary moves away from the bore. Essentially, then, the material could begin to fail at the circular cylinder's outer radius, especially if stresses vary over time or there are non-negligible stress concentrations. It is likely that reverse yielding phenomenon can be very useful for autofrettaged circular cylinders whose bore must be machined after the autofrettage treatment.

8.7 Repressurizing a Circular Cylinder After Reverse Yielding

Suppose that we reapply pressure to a reverse yielded circular cylinder, gradually increasing the internal pressure. The core, which has already yielded twice in succession, first in tension and then in compression, will initially deform elastically. However, when the internal pressure becomes sufficiently high, the tensile stresses in the core will cause it to begin to yield a third time, in tension, at the bore. If the pressure is further increased beyond this point, the region of plastic deformation will extend outwards from the bore to radius r_b.

The relations whereby the stress state in this thrice yielded core can be calculated are, once again, the equilibrium equation (4.7) and equation expressing the yield theory, given by relation (7.41) with the plus sign in the second member. Again through the same procedure described in Sect. 7.5.2, these two equations together with the boundary condition $\sigma_{r,r_i} = -p_i$at inner radius r_i lead to the following expression for this stress state (the second subscript b denotes the thrice yielded core):

$$\begin{aligned}\sigma_{r,b} &= -p_i + k\sigma_s \ln\frac{r}{r_i}\\ \sigma_{t,b} &= -p_i + k\sigma_s\left(1 + \ln\frac{r}{r_i}\right)\\ \sigma_{z,b} &= -p_i + k\sigma_s\left(\frac{1}{2} + \ln\frac{r}{r_i}\right).\end{aligned} \tag{8.22}$$

These relations clearly coincide with relations (7.48), but apply in the range $r_i \le r \le r_b$. There are thus three regions in the circular cylinder: an outer region in the range $r_n \le r \le r_e$, once yielded in tension; an intermediate region in the range $r_b \le r \le r_n$, twice yielded, first in tension and then in compression; and the innermost region in the range $r_i \le r \le r_b$, thrice yielded (first in tension, then in compression, and again in tension).

As the strains in the circular cylinder outside the thrice yielded inner core are elastic, the stress state in the two outer regions showing elastic behavior can be obtained by the use of the method of superposition. Consequently, for the two regions of radius $r > r_b$, the principal stresses will be expressed as the sum of those existing before internal pressure application and those due to the change in effective pressure at the interface of radius $r = r_b$.

However, the stresses before reapplying internal pressure are obviously the residual stresses calculated earlier (see Sect. 8.6). The change in the effective pressure at the interface of radius $r = r_b$, on the other hand, is equal to the radial stress at this interface with the sign changed. Thus, the principal stresses after re-pressurizing can be expressed as the sum of the residual stresses before reapplying internal pressure and the stresses due to the change in the effective pressure at the interface. The latter is obtained by subtracting the first relation (8.12) from the first relation (8.22) and changing the sign of the relation so obtained. For a generic interface of radius r_f, for which $(\Delta p)_{r_f} = -(\Delta\sigma_r)_{r_f}$, from relations (8.15) we will have:

$$\sigma_r = \sigma_{r,res} + \frac{(\Delta\sigma_r)_{r_f}}{\left(\frac{r_e}{r_f}\right)^2 - 1}\left(1 - \frac{r_e^2}{r^2}\right)$$

$$\sigma_t = \sigma_{t,res} + \frac{(\Delta\sigma_r)_{r_f}}{\left(\frac{r_e}{r_f}\right)^2 - 1}\left(1 + \frac{r_e^2}{r^2}\right) \tag{8.23}$$

$$\sigma_z = \frac{1}{2}\left(\sigma_{r,res} + \sigma_{t,res}\right) + \frac{(\Delta\sigma_r)_{r_f}}{\left(\frac{r_e}{r_f}\right)^2 - 1}.$$

For the twice yielded region bounded by radii r_b and r_n, i.e., for $r_b \leq r \leq r_n$, where the residual stresses are given by relations (8.12), the above relations are specialized as follows:

$$\sigma_r = -k\sigma_s \ln\frac{r}{r_i} + \frac{p_i - 2k\sigma_s \ln\frac{r_b}{r_i}}{\left(\frac{r_e}{r_b}\right)^2 - 1}\left(1 - \frac{r_e^2}{r^2}\right)$$

$$\sigma_t = -k\sigma_s\left(1 + \ln\frac{r}{r_i}\right) + \frac{p_i - 2k\sigma_s \ln\frac{r_b}{r_i}}{\left(\frac{r_e}{r_b}\right)^2 - 1}\left(1 + \frac{r_e^2}{r^2}\right) \tag{8.24}$$

$$\sigma_z = -k\sigma_s\left(\frac{1}{2} + \ln\frac{r}{r_i}\right) + \frac{p_i - 2k\sigma_s \ln\frac{r_b}{r_i}}{\left(\frac{r_e}{r_b}\right)^2 - 1}.$$

Since the circumferential stress at radius $r = r_b$ must, because of continuity, be equal in the two regions inside and outside the interface, equating the second relation (8.24) to the second relation (8.22) gives the following expression linking internal pressure p_i to radius r_b:

$$\frac{p_i}{k\sigma_s} = 1 + 2\ln\frac{r_b}{r_i} - \frac{r_b^2}{r_e^2}. \tag{8.25}$$

If we set $r_b = r_i$ in this expression, we obtain (8.10). Consequently, our reverse yielded circular cylinder will begin to yield a third time at the bore surface when the reapplied internal pressure reaches the value given by relation (8.10). Essentially, the circular cylinder, upon reapplication of internal pressure, will behave elastically by withstanding the same internal pressure that it would have withstood had it been autofrettaged in such a way as to leave an equivalent residual stress at the bore equal to σ_s, i.e., the material's yield strength.

Substituting relation (8.25) into relations (8.24) then gives:

$$\begin{aligned}\sigma_r &= -k\sigma_s\left\{\ln\frac{r}{r_i} + \left(\frac{r_b}{r}\right)^2 - \left(\frac{r_b}{r_e}\right)^2\right\}\\ \sigma_t &= -k\sigma_s\left\{\ln\frac{r}{r_i} + 1 - \left(\frac{r_b}{r}\right)^2 - \left(\frac{r_b}{r_e}\right)^2\right\}\\ \sigma_z &= -k\sigma_s\left\{\ln\frac{r}{r_i} + \frac{1}{2} - \left(\frac{r_b}{r_e}\right)^2\right\}.\end{aligned} \tag{8.26}$$

These relations, as was indicated earlier, apply to the region for which $r_b \le r \le r_n$. In the region where $r_n \le r \le r_e$, relations (8.23) with $r_f = r_n$ become:

$$\begin{aligned}\sigma_r &= -k\sigma_s\left\{\ln\frac{r_e}{r} - \left(\frac{r_n}{r}\right)^2 + \left(\frac{r_n}{r_e}\right)^2\right\} + k\sigma_s\left(\frac{r_b}{r_e}\right)^2\left(1 - \frac{r_e^2}{r^2}\right)\\ &= -k\sigma_s\left\{\ln\frac{r_e}{r} - \left(\frac{r_n}{r}\right)^2 + \left(\frac{r_n}{r_e}\right)^2 + \left(\frac{r_b}{r}\right)^2 - \left(\frac{r_b}{r_e}\right)^2\right\}\\ \sigma_t &= -k\sigma_s\left\{\ln\frac{r_e}{r} - 1 + \left(\frac{r_n}{r}\right)^2 + \left(\frac{r_n}{r_e}\right)^2\right\} + k\sigma_s\left(\frac{r_b}{r_e}\right)^2\left(1 + \frac{r_e^2}{r^2}\right)\\ &= -k\sigma_s\left\{\ln\frac{r_e}{r} - 1 + \left(\frac{r_n}{r}\right)^2 + \left(\frac{r_n}{r_e}\right)^2 - \left(\frac{r_b}{r}\right)^2 - \left(\frac{r_b}{r_e}\right)^2\right\}\\ \sigma_z &= -k\sigma_s\left\{\ln\frac{r_e}{r} - \frac{1}{2} + \left(\frac{r_n}{r_e}\right)^2\right\} + k\sigma_s\left(\frac{r_b}{r_e}\right)^2\\ &= -k\sigma_s\left\{\ln\frac{r_e}{r} - \frac{1}{2} + \left(\frac{r_n}{r_e}\right)^2 - \left(\frac{r_b}{r_e}\right)^2\right\}.\end{aligned} \tag{8.27}$$

From these relations, we obtain:

$$\sigma_t - \sigma_r = -k\sigma_s\left\{2\left(\frac{r_n}{r}\right)^2 - 2\left(\frac{r_b}{r}\right)^2 - 1\right\}, \tag{8.28}$$

which for $r_n = r_b$ gives the equation of the yield theory with positive sign. This criterion is independent of r and states that the outermost region which was elastic in the circular cylinder re-yielded in compression as a result of reverse yielding becomes plastic instantaneously when the reapplied internal pressure reaches a level high enough to bring the third reyielding interface to radius $r = r_n$. From relations (8.25) and (8.17), with $r = r_n$, we obtain $p_{i,e} = k\sigma_s \ln\delta$, i.e., the relation (7.54) giving the internal pressure that brings the plastic boundary to outer radius r_e. It can thus be deduced that when the internal pressure is reapplied, the bore begins to yield for a third time, in tension, when the pressure reaches the value

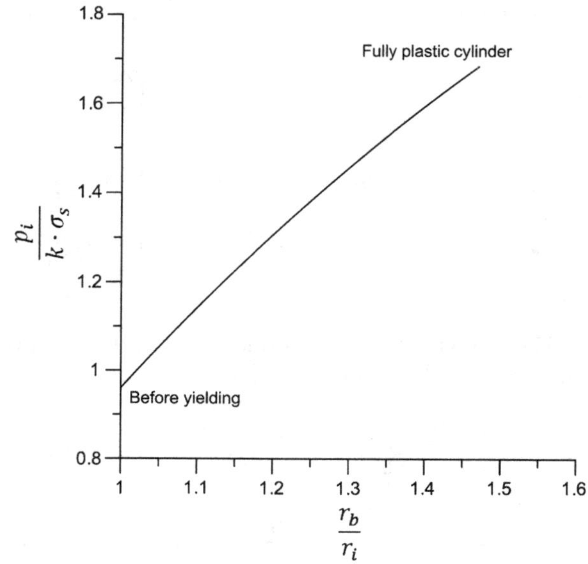

Fig. 8.7 Internal pressure required to extend the plastic zone to a given radius in a reyielded circular cylinder with diameter ratio $\delta = 5$

given by relation (8.10) and the new plastic boundary progresses outwards from the bore towards the outer surface as the pressure is further increased. When the pressure reaches the value given by relation (7.54), plastic flow will arrive at the radius $r = r_n$ and instantaneously at outer radius r_e, so that the entire wall thickness will again have become plastic.

Figure 8.7, which is a plot of relation (8.25), gives the internal pressure required to extend the plastic region to a given radius in a reyielded circular cylinder having an assigned diameter ratio δ.

8.8 Design Criteria for Autofrettage

As regards the design criteria for autofrettage, it should first be noted that, at least for steel—for which it has been confirmed experimentally—a circular cylinder's ability to undergo effective autofrettage depends on the material's yield strength as well as on the diameter ratio δ. This is clear from Fig. 8.8, which for various diameter ratios δ shows what the material's yield strength will have to be if a 100 % autofrettaged circular cylinder ($p = 2p_{i,i}$) is to exhibit linear elastic behavior when it is subjected during service to pressure cycles following autofrettage pressure application. For the material having the lowest strength, the minimum required diameter ratio δ is approximately 3.75, whereas for the highest-strength material it is around 2.50. Though this is a narrow range of variation for ratio δ, it includes practically all of the specifications for products that are required to operate from the elastic range to bursting pressure.

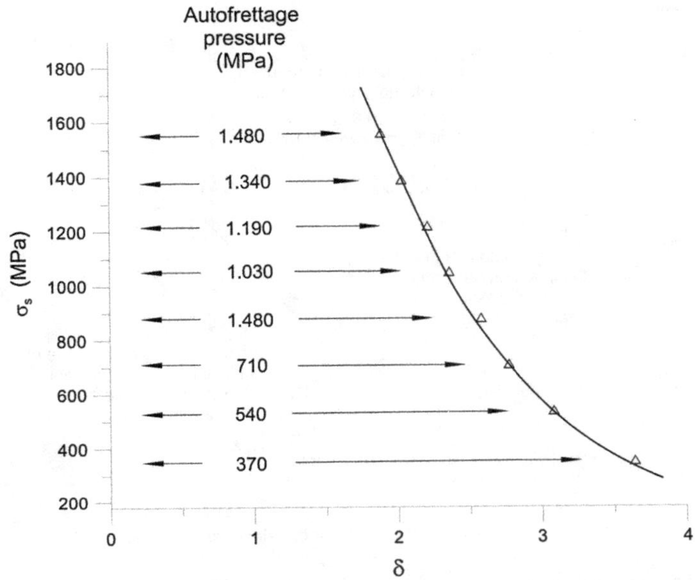

Fig. 8.8 Autofrettage design curve: yield strength of the material versus the minimum diameter ratio δ required for linear elastic behavior

The numerical values shown to the left of the curve in Fig. 8.8 represent the linear elastic range available for circular cylinders consisting of materials with different mechanical strength, having the required minimum diameter ratio δ and pressurized so as to achieve 100 % autofrettage (full autofrettage). For example, for a circular cylinder with $\delta = 3.75$ consisting of a material with a yield strength σ_s of 345 Mpa, the pressure $p_{i,\,i}$ that causes initial yielding at the bore is 185 Mpa. After the circular cylinder has undergone 100 % autofrettage, the range in which it shows linear elastic behavior is thus limited by a working pressure $p = 2p_i = 370$ Mpa. Through an appropriate choice of the material's elastic strength properties and of diameter ratios δ, circular cylinders can thus be designed for almost any pressure-strain specification.

Because of restrictions on the size of the circular cylinder and the material used in its manufacture, it is not always possible to achieve 100 % autofrettage. In order to formulate design specifications for the material's strength and the geometric proportions of the circular cylinder in such cases, it is important to know the range in which behavior will be linearly elastic during operation and how the circular cylinder must be autofrettaged to obtain this range. For this purpose, diagrams such as that shown in Fig. 8.9 can be used, which correlate maximum overstrain expressed as a percentage of wall thickness with the material's offset yield strength determined from a standard tensile test (here we use an offset yield strength corresponding to a strain offset or permanent set of 0.01 %).

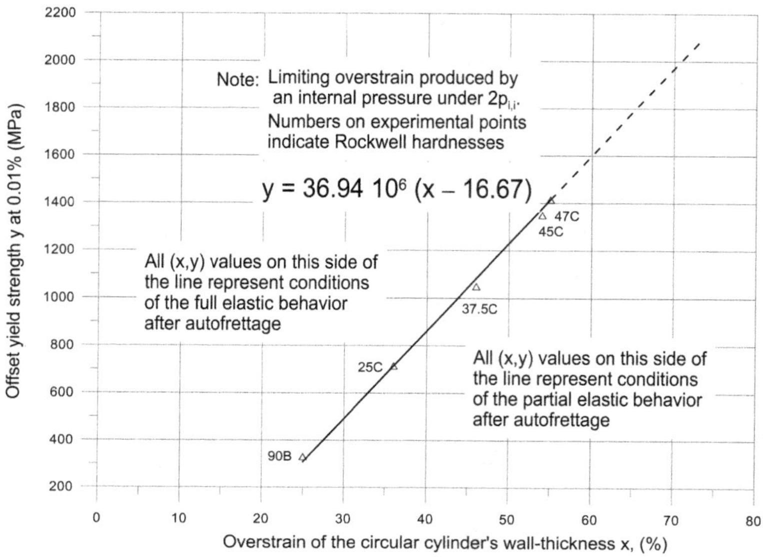

Fig. 8.9 Limiting design conditions for autofrettaged cylindrical bodies: maximum wall overstrain versus conventional yield strength

For example, it can be inferred from this figure that if the material's yield strength is around 345 MPa, the maximum overstrain that can be tolerated if the circular cylinder is to show linear elastic behavior upon being subjected to pressure cycles after autofrettage is approximately 25 %. The diagram in Fig. 8.9 can be used only in cases where the autofrettage pressure is not high enough to cause reverse yielding upon pressure release. In other words, and again as an example, if a material with $\sigma_s = 345$ Mpa is used, the limiting overstrain is 25 % and must be obtained with an internal pressure not exceeding that associated with 100 % autofrettage. For circular cylinders with diameter ratios $\delta < 2.22$, this means overstrain must be achieved with a pressure less than twice that which causes elastic-breakdown and hence initial yielding at the bore.

The straight line in Fig. 8.9, representing equation

$$y = 36.94 \cdot 10^6 (x - 16.67), \tag{8.29}$$

permits immediate calculation of the autofrettage parameters (overstrain as a percentage of wall thickness, autofrettage pressure and percentage autofrettage) required to ensure that a circular cylinder of given diameter ratio δ and consisting of a material having a given yield strength will operate elastically under working loads after the autofrettage process. For a circular cylinder with $\delta = 2.25$ made of a material having $\sigma_s = 860$ Mpa, for example, it can be deduced from this diagram that the maximum percentage overstrain will be 40 %. As the latter can be expressed as the percentage ratio $100\left[(r_p - r_i)/(r_e - r_i)\right]$, it corresponds to a ratio

$r_p/r_i = 1.50$. Relation (7.57) then gives $p_{i,p} = 680$ Mpa. The pressure $p_{i,i}$ at which yielding begins at the bore is 400 MPa. Achieving 100 % autofrettage would require an internal pressure of $2p_{i,\,i}$, i.e., equal to 800 MPa. Consequently, the autofrettage obtained will be $100\left[\left(p_{i,p} - p_{i,i}\right)/p_{i,i}\right] = 70$ %.

It should be borne in mind that autofrettage of steel circular cylinders is often followed by a stabilization heat treatment at temperatures between 300 and 370 °C. Exactly what this treatment does is not currently known, but its effects could be wide-ranging. First, any residual stresses exceeding the material's yield strength at operating temperature would be relieved. Second, in certain circumstances that depend chiefly on the steel's chemical composition, there could also be a strain-aging effect, thus increasing the material's strength as a result of strain hardening and compensating to some extent for residual stress relief. A number of tests presented in the scientific literature would appear to indicate that the beneficial effects of a heat treatment after autofrettage of metallurgical nature, and derive essentially from strain-aging.

From the design standpoint, it is interesting to compare three different structures with the same material and diameter ratio δ, and thus geometry: the single-walled or monobloc-type structure which has not been subjected to autofrettage, the double-walled composite structure assembled with a shrink-fit, and the autofrettaged monobloc-type structure. For these three structures, all other conditions remaining equal (i.e., with the same material and diameter ratio δ for the entire structure, including the double-walled composite structure), Fig. 8.10 shows the distribution curves of the elastic-breakdown pressure (the pressure causing initial yielding) at the bore, made dimensionless relative to $\sigma_s/\sqrt{3}$, versus diameter ratio δ.

Curve *A*, which refers to the monobloc-type structure with no initial stresses, i.e., which has not been subjected to treatment of any kind, for diameter ratios δ over 3.5 shows a modest gain in terms of pressure at the onset of yield. By

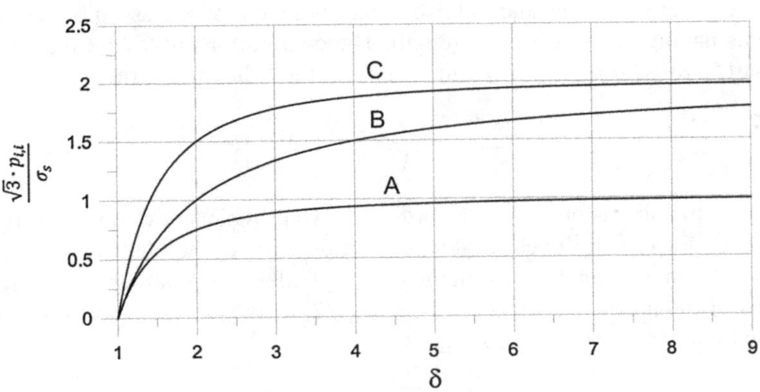

Fig. 8.10 Dimensionless elastic-breakdown pressure at the bore versus δ for three types of structure: *A*, non-autofrettaged monobloc-type structure; *B*, double-walled composite structure assembled with a shrink-fit; *C*, autofrettaged monobloc-type structure

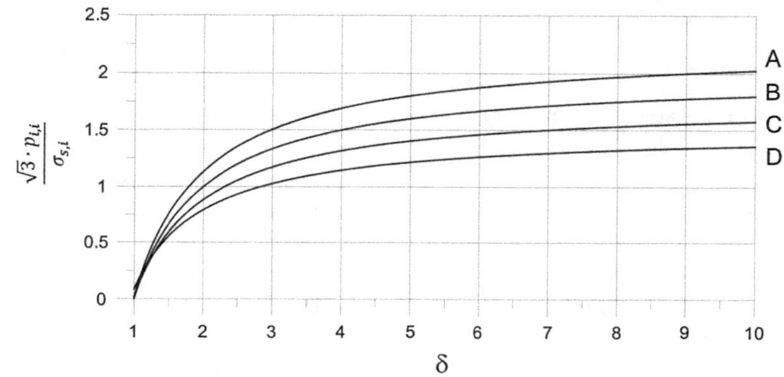

Fig. 8.11 Dimensionless elastic-breakdown pressure at the bore versus δ for a double-walled, shrink-fit composite structure, in which the yield strength of component members are different, i.e. $\sigma_{s,e} \neq \sigma_{s,i}$: *A*, $\sigma_{s,e}/\sigma_{s,i} = 1.25$; *B*, $\sigma_{s,e}/\sigma_{s,i} = 1.00$; *C*, $\sigma_{s,e}/\sigma_{s,i} = 0.75$; *D*, $\sigma_{s,e}/\sigma_{s,i} = 0.50$

contrast, the autofrettaged monobloc-type structure (curve *C*) indicates that the increase in pressure at the onset of yield is lower when δ exceeds values of (5–6). Lastly, for the double-walled composite structure (curve *B*), the optimal pressure at the onset of yield is reached only when $\delta > 10$. Consequently, for applications calling for high pressures at the onset of yield, it may be advantageous to use steels of higher mechanical strength rather than increase diameter ratio δ. It should also be noted that the autofrettaged structure, from the standpoint of pressure at the onset of yield, is superior to the shrink-fit composite structure in the entire range of δ except for $\delta \to \infty$.

Figure 8.11 shows four distribution curves correlating the elastic-breakdown pressure at the bore, also made dimensionless relative to $\sigma_s/\sqrt{3}$, with diameter ratio δ for double-walled composite structures assembled with a shrink-fit, in the case where the two inner and outer circular cylinders of the assembly consist of materials having different yield strength. The distribution curves are plotted using relation (5.38) with $p_{i,opt} = p_{i,i}$ and written in the following form

$$\frac{p_{i,i}\sqrt{3}}{\sigma_{s,i}} = 2\left(\frac{1+\phi}{2} - \frac{\sqrt{\phi}}{\delta}\right), \tag{8.30}$$

whose first member is plotted as the ordinates. To provide an example, this relation shows that, for $\delta = 4$, the elastic-breakdown pressure at the onset of yield can be increased by more than 33 %, with the outer cylinder's yield strength passing from half that of the inner cylinder, i.e., $\sigma_{s,e} = \sigma_{s,i}/2$, to a value 25 % above that of the inner cylinder, viz., $\sigma_{s,e} = 1.25\sigma_{s,i}$.

8.9 Autofrettage of Composite Multilayer Structures

As was demonstrated in Chap. 5, Sect. 5.7, the equivalent stress $\sigma_{e,i}$ at the inner radius of each layer of a composite multilayer structure in the optimized condition under internal pressure p_i and external pressure p_e is given by the following relation

$$\sigma_{e,i} = \frac{2k\delta^2(p_i - p_e)}{n(\delta^2 - 1)} \tag{8.31}$$

where n is the number of layers and δ is the ratio of the inner and outer radii of the entire structure, while $k = 1$ if the τ_{max} strength theory is used and $k = 2/\sqrt{3}$ if the maximum distortion energy theory is used.

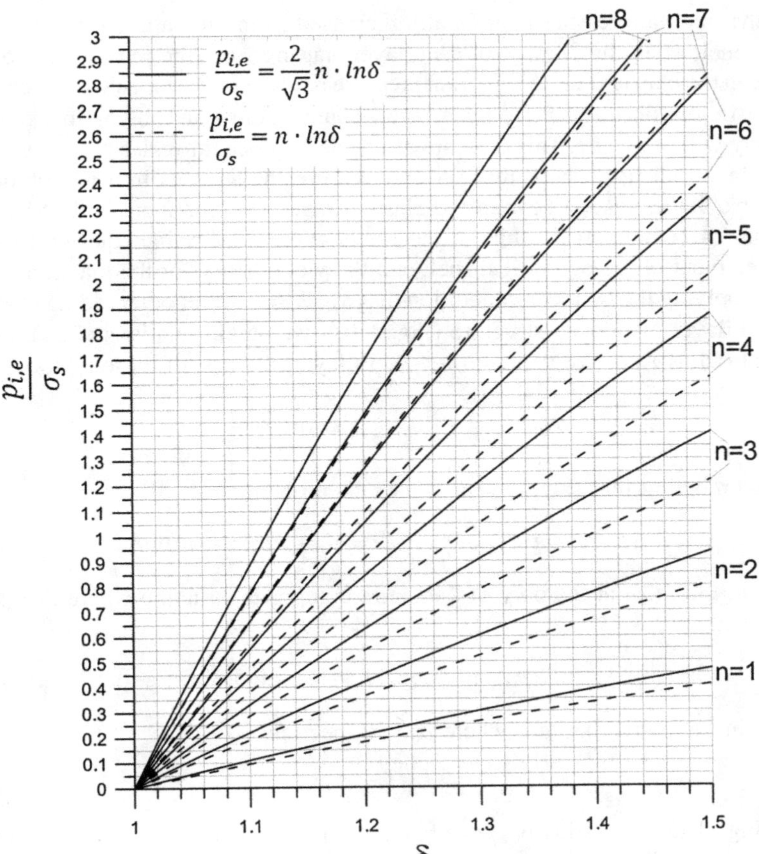

Fig. 8.12 Ratios $p_{i,e}/\sigma_s = f(\delta)$ for a multilayer structure with n varying from 1 to 8, calculated with the ED_{max} and τ_{max} yield theories

Once again, it is assumed that the proportional limit, the elastic limit and the yield point of the material used for the structure's layers coincide, and that the strength limit stress is the yield stress σ_s, as is done in current design practice. In accordance with the considerations made in the preceding sections, it can be deduced from relation (8.31) that the internal pressure $p_{i,i}$ that causes elastic-breakdown and hence initial yield at the inner radius of each layer of this structure is given by the relation:

$$p_{i,i} = p_e + \frac{n(\delta^2 - 1)\sigma_s}{2k\delta^2}. \tag{8.32}$$

Obviously, $p_e = 0$ in the particular case where the multilayer composite structure is under internal pressure but not external pressure. The distribution curves for ratio $p_{i,i}/\sigma_s = f(\delta)$ obtained in this case will be as shown in Fig. 5.11 after substituting $p_{i,i}/\sigma_s$ for $p_i/\sigma_{e,i}$.

If the internal pressure is gradually increased from the value given by relation (8.32), increasing thicknesses of each layer starting from the inner radius of each will be in the plastic state. An internal pressure $p_{i,e}$ will thus be reached which brings the plastic-elastic interface to each layer's outer radius, and the entire composite structure will be plastic. This pressure can be determined through a procedure which is similar in all respects to that used to analyze the stresses in a monobloc-type circular cylinder loaded beyond the elastic range. The treatment is thus based on relations (4.7) and (7.41), which respectively express the radial equilibrium of the small element of the cylindrical shell and the yield theory for the material.

By applying the boundary condition that $\sigma_{r,r=r_e} = -p_e$ at outer radius $r_e = r_n$ of the multilayer composite structure (Fig. 5.10), we obtain relations (7.31). For the internal radius r_m of the generic $(m+1)$ th layer, the first of these relations gives

$$\sigma_{r,r=r_m} = -(p_m)_{i,e} = k\sigma_s \ln\frac{r_m}{r_e} - p_e. \tag{8.33}$$

Remembering that $(r_e/r_m) = \delta^{n-m}$, we then obtain:

$$(p_m)_{i,e} = p_e + k\sigma_s \ln\delta^{n-m} = p_e + k\sigma_s(n-m)\ln\delta. \tag{8.34}$$

For $m = 0$, i.e., at the bore of the entire composite structure where $r_0 = r_i$, we will have:

$$(p_0)_{i,e} = p_{i,e} = p_e + k\sigma_s n\ln\delta, \tag{8.35}$$

which, in the particular case where $p_e = 0$, becomes:

$$p_{i,e} = k\sigma_s n\ln\delta. \tag{8.36}$$

In Fig. 8.12, the ratios $p_{i,e}/\sigma_s$ obtained with the maximum distortion energy ED_{max} and maximum shear stress τ_{max} yield theories are plotted versus δ for a multilayer structure under internal pressure only. Here again, the results with n varying from 1 to 8 are represented.

Chapter 9
Thick-Walled Circular Cylinders in the Elasto-Plastic State or Viscous State as a Result of Steady-State Creep

9.1 General

Chapters 7 and 8 addressed the problem of thick-walled circular cylinders under pressure in the partially plastic or fully plastic state, with the assumption that the material's behavior is linear elastic-perfectly plastic, i.e., that the stress–strain curve $\sigma = \sigma(\varepsilon)$ obtained from tensile tests on a standard specimen is as shown in Fig. 7.1. On the basis of this assumption, which sees the material as exhibiting linear elastic behavior up to yielding, and perfectly plastic, i.e., non-strain hardening, behavior thereafter, curves representing the three principal stresses, residual stresses and effective stresses under service conditions were plotted which show discontinuities in the stress gradient along the radius, and thus cusps, at the interface between the innermost region where the material's behavior is plastic, and the outermost region showing elastic behavior.

As a number of researchers have emphasized (see, for example, Manna 1967), the fact that the distribution curves for the stresses of interest here (the principal stresses due to autofrettage pressure, the residual stresses remaining when the autofrettage pressure is released, and the stresses in actual service conditions, given by the sum of the residual stresses and the stresses correlated with the working pressure) are cusped is clearly a physical absurdity: in accordance with Linnaeus' dictum that "natura non facit saltus", it has been demonstrated experimentally that the transition between these two regions is gradual, and takes place in a cylindrical band of annular cross section between the two regions (Steele 1952).

The treatment presented in the two preceding chapters is thus highly approximate, as it is based on simplifying assumptions and very rough evaluation criteria whose purpose is merely to permit an initial approach to the problem and, though they can provide information that is useful for design purposes, cannot claim to interpret the materials' actual behavior. To overcome the limitations of this treatment and the physical absurdity deriving from it, and to reflect experimental evidence more accurately, this chapter will tackle the problem in much more general terms by introducing a constitutive law that can describe the actual behavior of the material used to construct the circular cylinder as faithfully as possible. In fact, the

V. Vullo, *Circular Cylinders and Pressure Vessels*, Springer Series in Solid and Structural Mechanics 3, DOI: 10.1007/978-3-319-00690-1_9,

constitutive law that will be considered here is capable of simulating not only the elasto-plastic behavior of a strain-hardening material, but also the viscous behavior in steady state conditions associated with creep at high temperatures.

Creep is typical of materials that are subjected to high temperatures and loads for long periods. Accordingly, by contrast with what can be expected from classic plasticity theory (in this case, the stress–strain law is independent of the rate of deformation, but is dependent on the history of deformation), creep strains are defined as the time-dependent strains observed during a constant stress test at an elevated temperature. For the structures of interest to us here, a circular cylinder under high pressures and in service at elevated temperatures can fail as a result of creep. Failure will take place when the creep strains in the cylinder's material, which increase continuously over time, reach threshold values such that the structure is no longer able to fulfill its design functions. In cases where there is a danger that creep will occur under actual working conditions, it is necessary that the design process ensure stringent limits on this creep strain throughout the structure's expected service life.

9.2 Introductory Remarks on Creep

As the mechanisms underlying the creep behavior of structural materials are beyond our scope here, the reader is directed to specialized texts on the subject (see Penny and Marriott 1971, Christensen 1971, Hult 1973, Odqvist 1974, Gittus 1975, Bernasconi and Piatti 1979, Kraus 1980, Ponter and Hayhurst 1981, Shames and Cozzarelli 1997, Dill 2007). However, we must keep in mind that, on the basis of a simple tensile creep test at constant temperature and load, the total strain at any time t is given by the relation

$$\varepsilon(t) = \varepsilon_e + \varepsilon_p + \varepsilon_c(t) \tag{9.1}$$

where ε_e is the elastic strain that occurs immediately, i.e., in a very short time interval, upon loading at the test temperature, ε_p is the plastic strain that follows if the applied stress exceeds the yield point stress at the test temperature (ε_p occurs over a greater time interval than ε_e, but this plastic strain time interval is still short by comparison with the time scale of a creep test; for design purposes, both ε_e and ε_p can thus be considered as instantaneous responses) and $\varepsilon_c(t)$ is the total amount of creep strain at time t (the subscript c thus refers to magnitudes associated with creep).

For our purposes in analyzing the stress and strain state of circular cylinders subject to creep, it is necessary and sufficient to have a relation which links creep strain ε_c to stress as well as to the other magnitudes on which it depends (see the next section). Although a more precise characterization of the material's creep behavior with changing loads and temperatures would be necessary in many problems of practical interest, in this specific sector it is often found that the only experimental data that are available are those provided by uniaxial tensile creep tests (or simple tensile creep tests) at constant temperature and load.

A typical $\varepsilon_c = \varepsilon_c(t)$ curve from a simple tensile creep test at constant temperature and load is conventionally regarded as consisting of three stages: the first stage (primary creep stage) usually characterized by a decreasing creep rate $\dot{\varepsilon}_c = d\varepsilon_c/dt$, the second stage (secondary creep stage or steady-state creep), characterized by a steady-state creep rate, and the third stage (tertiary creep stage), where the creep rate increases until fracture occurs. Figure 9.1 shows the idealized shape of this type of creep curve obtained by means of a one-dimensional tensile creep test at constant temperature and load. It should be noted, however, that the degree to which these three stages are readily distinguishable depends on the temperature and applied stress. The portion of the creep curve most commonly used in structural design is the secondary creep stage, although the initial deformation, i.e., the deformation before creep actually starts correlated to instantaneous initial elastic and plastic response (see Fig. 9.1) is of great importance when dealing with short-life items like missiles.

Further data regarding the material's creep behavior can be obtained from systematic testing campaigns at different temperatures and loads. Results from creep tests can be cross-plotted in many ways, because the basic quantities involved are temperature, stress, strain, and time. Any two of these quantities can be selected as coordinates, and creep curves drawn with the other two held constant. Still more combinations result from combining strain and time and then using strain rate as a coordinate, or plotting curves of constant strain rate. Various types of curves (total strain versus time curves at constant stress, isochronous stress–strain curves, strain rate versus stress curves at constant temperature, stress versus temperature curves for constant strain rate, etc.) can thus be plotted which can play an important role at the design stage. Once again, the reader is directed to specialized texts for details. We will limit ourselves here to discussing only a few creep theories, focusing in particular on those that will be used in the following sections. These theories provide a description of the creep behavior of a material subjected to multiaxial loads on the basis of one-dimensional tensile creep test data.

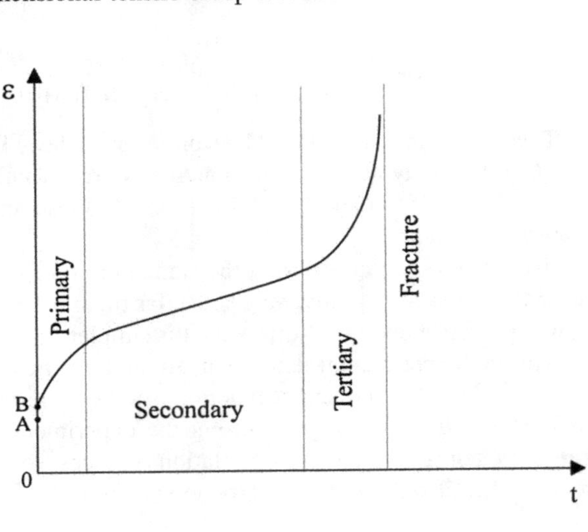

Fig. 9.1 Typical $\varepsilon_c = \varepsilon_c(t)$ curve from a uniaxial creep test at constant temperature and load (*OA*, elastic strain; *AB*, plastic strain)

9.2.1 Theories for Creep Under Constant Uniaxial Stress

During creep, a standard specimen under a constant uniaxial tensile load will deform continuously over time. This deformation depends mainly on the three following parameters: stress, time and temperature. In reality, creep behavior of materials is sensitive not only to these parameters, but also to atmosphere, physical properties of the material, past strain history of the material, etc. Here we will focus only on these three main parameters. From this point of view, the most general relation that can be written to link creep strain to these parameters is thus the following:

$$\varepsilon_c = f(\sigma, t, T) \tag{9.2}$$

As such a thorough correlation is extremely complex, in addressing the technical problems that the designer is normally required to solve, it is replaced with the following commutative relation that, though less general, can give results that provide a sufficiently close approximation, and are thus both reliable and have proven practical significance:

$$\varepsilon_c = f_1(\sigma) f_2(t) f_3(T). \tag{9.3}$$

Assuming the separation of functions $f_1(\sigma)$ and $f_2(t)$, as has been implicit in most studies of creep, is now generally accepted in the design analysis of structural components. The same cannot be said of the separation of function $f_3(T)$, as many researchers use a combined expression which takes the influence of both time and temperature into simultaneous account, and is thus not consistent with the separation criterion embodied in relation (9.3).

Among the many functions $f_1(\sigma)$ in the scientific literature (a wide summary is given by Kennedy 1963), the following must be mentioned:

$$f_1(\sigma) = B\sigma^q \tag{9.4}$$

$$f_1(\sigma) = Ce^{\sigma/\sigma_0} \tag{9.5}$$

$$f_1(\sigma) = A[\mathrm{sh}(\sigma/\sigma_0)]^q. \tag{9.6}$$

These relations, known as Norton's power law, Dorn's exponential function, and Garafalo's hyperbolic sine function, were formulated respectively by Norton (1929), Dorn (1955) and Garofalo (1965). In these relations, B, C, A, σ_0 and q are material constants.

Each of these relations (and the same is true of those that are not shown here) has advantages and disadvantages. By far the most commonly used function is the power law developed by Norton, as it is simpler than the others. It provides a good fit with the experimental data for medium–low stresses, but not for high stresses, for which Dorn's exponential function is to be preferred. Garofalo's more complex relation has the advantage of fitting the experimental data well for any level of stress: when $\sigma/\sigma_0 < 0.8$, the relation reduces to the power law, and when $\sigma/\sigma_0 > 1.2$, it reduces to the exponential form.

Of the equally numerous time functions $f_2(t)$ found in the scientific literature (see Kennedy 1963), the following should be mentioned:

$$f_2(t) = Dt^{\zeta} \tag{9.7}$$

$$f_2(t) = \sum a_i t^{n_i} \tag{9.8}$$

$$f_2(t) = \theta_1\left(1 - e^{-\theta_2 t}\right) + \dot{\varepsilon}_{c,min} t. \tag{9.9}$$

In these relations, which were formulated by Bailey (1935), Graham and Walles (1955) and Garofalo (1965) respectively, D, ζ (this exponent is normally in the range $1/3 \le \zeta \le 1/2$), a_i, n_i, θ_1 and θ_2 are material constants, while $\dot{\varepsilon}_{c,min}$ is the minimum creep rate, i.e., the steady-state creep rate in the secondary creep stage. It should be recognized that function $f_2(t)$ is heavily dependent on the context of the experimental data underlying its formulation. Consequently, the above relations must be chosen with greater caution than is the case for stress function $f_1(\sigma)$.

Lastly, determining function $f_3(T)$ poses a number of problems, as the temperature influences the material's microstructure as well as the physical constants indicated earlier. Without going into the details of the complex issues involved in this area, mention should be made of the following relation suggested by Dorn (1955)

$$f_3(T) = e^{-Q/RT} \tag{9.10}$$

where Q is the activation energy, R is the Boltzmann's constant and T is the absolute temperature. It should be noted that determining the activation energy is far from simple, both when relying on experimental measurements, and when using theoretical models based on the fact that it is equal to the activation energy for self-diffusion for pure metals.

On the basis of Dorn's observation that experimental creep data at various temperatures can all be plotted on a single curve if creep strain is plotted against parameter $\left(te^{-Q/RT}\right)$, Penny and Marriott (1971, 1995) proposed the following relation, which is of great interest for the analysis of structures subject to creep:

$$\varepsilon_c = f_1(\sigma)\left(te^{-\frac{Q}{RT}}\right)^n. \tag{9.11}$$

This is of the same type as relation (9.3), as it also separates the influence exercised by the three main parameters stress, time and temperature on creep strain.

9.2.2 Theories for Creep Under Variable Uniaxial Stress

In the case of variable uniaxial stress, the difficulty of developing reliable creep theories is even greater, both because the physical phenomenon is more complex, and because almost all of the available data are from uniaxial tensile tests with constant stress. A number of different creep theories have been put forward over the years to solve this crucial problem, but none have been able to account for all of the experimental data; many of these theories predict widely differing results under the same stress histories. Although the simplicity of some of these theories makes them extremely useful in arriving at a rough approximation for the magnitudes of interest, their limitations are clear from the experimental evidence. We will present a few of these theories below.

9.2.2.1 Time Hardening

The time hardening theory assumes that the creep strain rate is a function of stress, time and temperature, i.e., that

$$\dot{\varepsilon}_c = f(\sigma, t, T). \tag{9.12}$$

On the basis of the considerations advanced in the preceding section, this relation can be stated in the following form for isothermal conditions:

$$\dot{\varepsilon}_c = f_1(\sigma)\frac{df_2(t)}{dt}f_3(T). \tag{9.13}$$

9.2.2.2 Strain Hardening

The strain hardening theory assumes that the creep strain rate is a function of stress, temperature and accumulated creep strain, i.e., that:

$$\dot{\varepsilon}_c = f(\sigma, \varepsilon_c, T). \tag{9.14}$$

In the more confined commutative form that is usually accepted as useful first approximation, this relation is written as follows:

$$\dot{\varepsilon}_c = g_1(\sigma)g_2(\varepsilon_c)g_3(T). \tag{9.15}$$

A special case of the latter relation is when g_1 and g_2 are both expressed as power functions, so that:

$$\dot{\varepsilon}_c = \frac{\zeta h^{1/\zeta}\sigma^{q/\zeta}}{\varepsilon_c^{(1-\zeta)/\zeta}}. \tag{9.16}$$

This relation integrates for constant stress to give:

$$\varepsilon_c = h\sigma^q t^\zeta. \tag{9.17}$$

It should be noted that relation (9.17) can be considered as the resultant of relations (9.4) and (9.7), with h grouping the constants B and D that appear in them. The temperature plays an important role in this relation, which is implicit in the constant's variability with T.

9.2.2.3 Combined Theories

Finnie and Heller (1959), Manson (1961) and Johnson, Henderson and Kahn (1963) have suggested combined theories that are intermediate between the time hardening and strain hardening theories discussed above. For isothermal conditions, they propose the following relation:

$$\dot{\varepsilon}_c = C\sigma^{\gamma}\varepsilon_c^{\delta}t^{\eta} \tag{9.18}$$

where C, γ, δ and η are material constants.

9.2.2.4 Other Theories

For other and even more elaborate theories (Marin and Pao 1953; McVetty 1943; Graham and Walles 1955; Garofalo 1965; Maruyama et al. 1990; Evans and Wilshire 1985; Rabotnov 1953, 1969; Taira et al. 1960; Orowan and Becker 1953; Webster and Piercey 1967; Cockcroft 1969; Kachanov 1967; etc.), the reader is directed to specialized texts.

9.2.3 Effects of the Multiaxial Stress State

As has been indicated several times in the preceding chapters, the problems of interest to us here are characterized by triaxial stress and strain states. As was the case for the elastic problem and the plastic problem, it is thus necessary to generalize the concepts regarding creep strain to include the multiaxial stress state. To this end, all researchers agree that the assumptions at the basis of the classical plasticity theory or simply plasticity theory (i.e., the theory of rate-independent plasticity which assumes that plastic strains occur instantaneously), can be extended to creep (Larson and Miller 1952, Nagdi 1960, Bland 1960, Ohji and Marin 1964, Perzyna 1966, Marriott 1968, Sim and Penny 1971, Godono and De Iorio 1973, Katchanov 1975, Chaboche 1986, Miller 1987, Krempl 1987, Skrzypek 1993).

To obtain the general relations applying to creep, it is necessary to define the equivalent quantities which reduce the multiaxial problem to an equivalent uniaxial problem. In conformity with the theory of instantaneous plastic strain, it is accepted that these quantities are the equivalent stress σ_e and the equivalent strain increment $\delta\varepsilon_e$, produced by the combined action of the components of the multiaxial stress system involved in defining stress σ_e. Using, for example, relation (9.17) with $\zeta = 1$ would give an equivalent strain increment $\delta\varepsilon_e = h\sigma_e^q\delta t$.

On the basis of the maximum distortion energy equivalent theory (the maximum shear stress equivalent theory has also been used in this sector, but with less success), the equivalent stress σ_e will be given by relation (7.3), substituting σ_e for σ_s. Quantifying the equivalent strain increment entails first defining how the strain correlated with a multiaxial stress state is distributed in the three principal directions.

While referring the reader to specialized texts for details, it should be noted that according to the incremental plastic flow theory, which is assumed as the basis of the analogous viscous flow theory of creep, the constitutive laws or flow rules to be used in this case are the Prandtl-Reuss laws (Prandtl 1924, Reuss 1930), which apply for a homogeneous, isotropic and incompressible material (deformation in the material will thus take place at constant volume), assuming that the principal axes of stress and strain coincide. These laws are given by the following relations:

$$\begin{aligned}\delta\varepsilon_r &= l\frac{\delta\varepsilon_e}{2\sigma_e}[2\sigma_r - (\sigma_t + \sigma_z)]\\ \delta\varepsilon_t &= \frac{\delta\varepsilon_e}{2\sigma_e}[2\sigma_t - (\sigma_z + \sigma_r)]\\ \delta\varepsilon_z &= \frac{\delta\varepsilon_e}{2\sigma_e}[2\sigma_z - (\sigma_r + \sigma_t)].\end{aligned} \tag{9.19}$$

9.3 Stress Analysis in the Full Plastic or Steady-State Viscous State

As part of the generalization proposed in Sect. 9.1, here we will analyze the stress and strain states of a thick-walled circular cylinder, extending the analysis to the case in which the material exhibits strain hardening plastic behavior and/or steady-state viscous behavior due to creep. To this end, we will consider the circular cylinder with free closed ends under simultaneous internal pressure p_i and external pressure p_e. Our purpose here is thus to calculate the distributions of the three principal stresses σ_r, σ_t and σ_z and the three principal strains ε_r, ε_t and ε_z through the wall thickness, assuming that the circular cylinder is undergoing steady-state viscous creep as a result of the external loads deriving from these surface forces.

In this context, with reference to the problems of interest to us here, the equivalent strain ε_e can be expressed as a function of equivalent stress σ_e by means of the Bailey-Popov relation, proposed by Bailey (1935) and Popov (1949):

$$\varepsilon_e = h\sigma_e^q t \tag{9.20}$$

where t is time, while constants h and q depend on the material of which our circular cylinder consists, as well as on the operating temperature in service conditions. The constitutive law given by relation (9.20) derives from the even more general one expressed through Serensen's relation (see also Manna 1973, 1983):

$$\varepsilon_e = h\sigma_e^q t^\zeta \tag{9.21}$$

as formulated by Serensen et al. (1963), in which constants h, q and ζ depend on the material and temperature. For $\zeta = 1$, relation (9.21) brings us to the Bailey-Popov relation (9.20) used here, which is inherent to steady-state viscous creep. It should be pointed out that relation (9.21) is simply relation (9.17) rewritten in terms of equivalent strain ε_e and equivalent stress σ_e. For the triaxial stress state, the latter is given by relation (7.3), substituting σ_e for σ_s.

Solving for stress σ_e, relation (9.20) can be written in the following form:

$$\sigma_e = \left(\frac{1}{ht}\right)^{1/q} \varepsilon_e^{1/q} = A\varepsilon_e^{1/q}, \tag{9.22}$$

where $A = (1/ht)^{1/q}$. If there are no temperature gradients, as Serensen also demonstrated, both relations (9.20) and (9.22) make it possible to treat full yielding of our circular cylinder as a particular case of viscous creep.

As we are now considering only those problems involving steady-state plastic or viscous flow, relations (9.19) can be further simplified. In fact, the constitutive laws of interest to us here, which assuming small strains or small deformations correlate the principal strains with the principal stresses, are as follows:

$$\begin{aligned} \varepsilon_r &= \frac{\varepsilon_e}{2\sigma_e}[2\sigma_r - (\sigma_t + \sigma_z)] \\ \varepsilon_t &= \frac{\varepsilon_e}{2\sigma_e}[2\sigma_t - (\sigma_z + \sigma_r)] \\ \varepsilon_z &= \frac{\varepsilon_e}{2\sigma_e}[2\sigma_z - (\sigma_r + \sigma_t)]. \end{aligned} \tag{9.23}$$

In the plastic range and in the viscous range due to creep under steady-state conditions, these relations substitute the generalized Hooke's laws, which apply in the elastic range.

Thus, subtracting the first of relations (9.23) from the second gives:

$$\sigma_t - \sigma_r = \frac{2\sigma_e}{3\varepsilon_e}(\varepsilon_t - \varepsilon_r). \tag{9.24}$$

The equilibrium equations of interest are that for radial equilibrium given by relation (4.7), and that for longitudinal equilibrium given by relation (1.8); here, however, the latter should be expressed in the following form:

$$2\pi \int_{r_i}^{r_e} \sigma_z r dr = \pi\left(p_i r_i^2 - p_e r_e^2\right) = \pi \int_{-p_i r_i^2}^{-p_e r_e^2} d\left(\sigma_r r^2\right), \tag{9.25}$$

in as much as the boundary conditions given by relations (1.14) apply.

Expanding the differential operator in relation (9.25) yields:

$$d\left(\sigma_r r^2\right) = \left(2\sigma_r + r\frac{d\sigma_r}{dr}\right) r dr. \tag{9.26}$$

Remembering the equilibrium equation (4.7), relations (9.25) and (9.26) allow us to write:

$$2\int_{r_i}^{r_e} \sigma_z r dr = \int_{r_i}^{r_e} (\sigma_r + \sigma_t) r dr. \tag{9.27}$$

We thus have:

$$\sigma_z = \frac{1}{2}(\sigma_r + \sigma_t). \tag{9.28}$$

Accordingly, the second relation (9.23) yields:

$$\varepsilon_z = 0. \tag{9.29}$$

For the problem in question, then, the longitudinal stress σ_z must be the arithmetic average of the radial stress and the hoop stress (the assumption made in Chaps. 7 and 8 is thus proven) and the strain state must be plane, i.e., there is no axial translation of the circular cylinder's generic cross-section. Given the assumption that volume remains constant during deformation, meaning that cubical dilatation or volume expansion e will thus be zero, i.e., that $e = \varepsilon_r + \varepsilon_t + \varepsilon_z = 0$, it can be inferred that we must necessarily have:

$$\varepsilon_t = -\varepsilon_r. \tag{9.30}$$

Introducing the geometric relations (4.8) in the latter gives the differential equation

$$\frac{d}{dr}(ur) = 0 \tag{9.31}$$

which integrates to

$$u = \frac{C}{r} \tag{9.32}$$

with $C = const.$ Introducing the latter together with its first derivative in geometric relations (4.7) gives the equations:

$$\varepsilon_r = -\frac{C}{r^2} \quad \text{and} \quad \varepsilon_t = \frac{C}{r^2}. \tag{9.33}$$

Consequently, relation (9.24) becomes:

$$\sigma_t - \sigma_r = \frac{4\sigma_e}{3\varepsilon_e}\frac{C}{r^2}. \tag{9.34}$$

However, the equivalent stress in a circular cylinder with free closed ends is given by the relation

$$\sigma_e = \frac{1}{k}(\sigma_t - \sigma_r) \tag{9.35}$$

which is simply relation (7.41) with the plus sign after substituting σ_e for σ_s (it should be recalled that k is respectively $2/\sqrt{3}$ and 1 for the maximum distortion energy strength theory and for the maximum shear stress strength theory). By expanding the equilibrium equation (4.7) and substituting the difference $(\sigma_t - \sigma_r)$ given by relation (9.35) in it, we obtain the differential equation:

$$r\frac{d\sigma_r}{dr} = k\sigma_e. \tag{9.36}$$

Recalling the difference $(\sigma_t - \sigma_r)$ given by relation (9.35), relation (9.34) yields

$$\varepsilon_e = \frac{4}{3k}\frac{C}{r^2}. \tag{9.37}$$

Substituting the latter in relation (9.22) and the result in relation (9.36) gives the following differential equation:

$$\frac{d\sigma_r}{dr} = C_2\frac{1}{r^{\frac{2}{q}+1}} \tag{9.38}$$

where

$$C_2 = Ak\left(\frac{4C}{3k}\right)^{\frac{1}{q}}. \tag{9.39}$$

By integrating equation (9.38) and denoting the integration constant with C_1, we obtain:

$$\sigma_r = C_1 - \frac{q}{2}C_2\frac{1}{r^{2/q}}. \tag{9.40}$$

Expanding the equilibrium equation (4.7) and substituting relation (9.40) in it together with its first derivative given by relation (9.38) yields:

$$\sigma_t = C_1 - \frac{q-2}{2}C_2\frac{1}{r^{2/q}}. \tag{9.41}$$

Recalling relations (9.40) and (9.41), relation (9.28) then gives:

$$\sigma_z = C_1 - \frac{q-1}{2}C_2\frac{1}{r^{2/q}}. \tag{9.42}$$

By grouping relations (9.40)–(9.42), we obtain the following relations that define the stress state in the circular cylinder with free closed ends under internal and external pressure, non considering the integration constants C_1 and C_2:

$$\begin{aligned} \sigma_r &= C_1 - \frac{q}{2}C_2\frac{1}{r^{2/q}} \\ \sigma_t &= C_1 - \frac{q-2}{2}C_2\frac{1}{r^{2/q}} \\ \sigma_z &= C_1 - \frac{q-1}{2}C_2\frac{1}{r^{2/q}}. \end{aligned} \tag{9.43}$$

Constants C_1 and C_2 can be determined by using the first relation (9.43) and imposing the boundary conditions given by relation (1.14). Proceeding in this way and considering the product $(q/2)C_2$ as a single constant, we obtain:

$$\begin{aligned} C_1 &= \frac{p_i r_i^{2/q} - p_e r_e^{2/q}}{r_e^{2/q} - r_i^{2/q}} \\ \frac{q}{2} C_2 &= (p_i - p_e)\frac{r_e^{2/q} r_i^{2/q}}{r_e^{2/q} - r_i^{2/q}}, \end{aligned} \tag{9.44}$$

which differ from those given by relations (4.28) in that the exponent of radii r_i and r_e is $2/q$ rather than 2. Given the relation (9.39) and remembering that $A = (1/ht)^{1/q}$, the second of these relations can be written in the following form:

$$\frac{q}{2} C_2 = \frac{q}{2} k \left(\frac{4C}{3htk}\right)^{1/q}. \tag{9.45}$$

Once C_2 has been determined using the second relation (9.44), relation (9.45) can be used to obtain C and, by means of relation (9.32), the radial displacement u. Lastly, relations (9.33) are used to obtain strains ε_r and ε_t. Consequently, equaling the second members of relation (9.45) and the second expression (9.44) gives:

$$C = \frac{3}{2} ht \frac{(k/2)^{1-q}}{q^q} (p_i - p_e)^q \frac{r_e^2 r_i^2}{\left(r_e^{2/q} - r_i^{2/q}\right)^q}. \tag{9.46}$$

Substituting this relation in relations (9.32) and (9.33) makes it possible to define the radial displacement and the strain state of the circular cylinder in conditions of steady-state viscous flow univocally for both strength theories considered here. The stress state in the circular cylinder is determined from relations (9.43), recalling relations (9.44) and (9.46); we thus obtain:

$$\begin{aligned} \sigma_r &= \frac{p_i r_i^{2/q} - p_e r_e^{2/q}}{r_e^{2/q} - r_i^{2/q}} - (p_i - p_e)\frac{r_e^{2/q} r_i^{2/q}}{r_e^{2/q} - r_i^{2/q}} \frac{1}{r^{2/q}} \\ \sigma_t &= \frac{p_i r_i^{2/q} - p_e r_e^{2/q}}{r_e^{2/q} - r_i^{2/q}} - \frac{q-2}{q}(p_i - p_e)\frac{r_e^{2/q} r_i^{2/q}}{r_e^{2/q} - r_i^{2/q}} \frac{1}{r^{2/q}} \\ \sigma_t &= \frac{p_i r_i^{2/q} - p_e r_e^{2/q}}{r_e^{2/q} - r_i^{2/q}} - \frac{q-1}{q}(p_i - p_e)\frac{r_e^{2/q} r_i^{2/q}}{r_e^{2/q} - r_i^{2/q}} \frac{1}{r^{2/q}}. \end{aligned} \tag{9.47}$$

Note that since k does not appear in relations (9.47), the stress state does not change for the two strength theories considered. Relations (9.47) can then be specialized for the particular cases of the circular cylinder under internal pressure alone ($p_i \neq 0$ and $p_e = 0$) and the circular cylinder under external pressure alone ($p_i = 0$ e $p_e \neq 0$). Figure 9.2 shows the distribution curves (solid lines) for the three principal stresses, made dimensionless relative to internal pressure p_i, versus

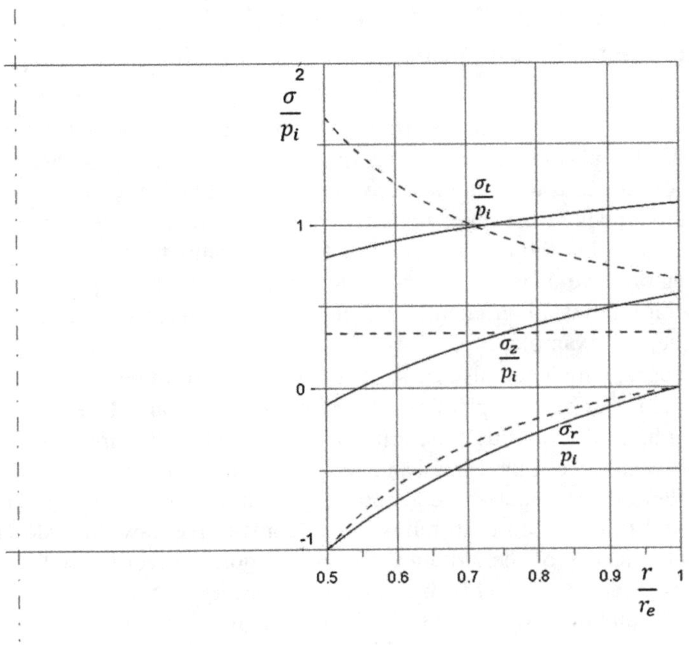

Fig. 9.2 Distribution *curves* for the dimensionless principal stresses in a circular cylinder with free closed ends with $\delta = 2$ in the fully plastic state or steady-state viscous creep (*solid lines*) and in the elastic range (*dotted lines*)

r/r_e in a circular cylinder with diameter ratio $\delta = 2$ under internal pressure (($p_e = 0$) alone and in the fully plastic state or steady-state viscous state due to creep in accordance with relation (9.22) with $q = 3$.

For $q = 1$, the first two relations (9.47) reduce to relations (4.29), while the third relation (9.47) reduces to relation (4.31). We thus obtain the elastic range relations given in Chap. 4. Again for $q = 1$, the constants C_1 and $(q/2)C_2$ given by relations (9.44) reduce to constants A and B given by relations (4.28). Figure 9.2 also shows the distribution curves for the three dimensionless principal stresses, again as a function of r/r_e, in the same circular cylinder considered above, but consisting of a material exhibiting elastic behavior (dotted lines).

The curves shown in Fig. 9.2 highlight the substantial difference between the distributions of the three principal stresses for plastic flow or steady-state viscous flow, and those for the elastic range. In the case of plastic or viscous flow, the distribution of hoop stress σ_t in particular (and thus of equivalent stress σ_e, though the latter is not shown to avoid over-complication) is much more uniform than its elastic counterpart. We can thus say that as q increases, i.e., as the plastic state progresses, the σ_t gradient tends to invert and there is a value of q at which hoop stress and equivalent stress will depart as little as possible from uniformity.

9.4 Partial Plasticization: Mixed Plastic and Elastic Strain State

As was seen in Chaps. 7 and 8, if we assume that the material exhibits linear elastic-perfectly plastic behavior, that the plastic boundary progresses outwards from the bore towards the outer surface as the internal pressure increases, resulting in two regions, where the inner region is in the plastic state and the outer one is still elastic, and that the constitutive law linking strains to stresses is different in the two regions, we arrive at distribution curves for the principal stresses which show discontinuities in gradient, i.e., cusps, at the interface between the two regions (see, for example, Fig. 7.14).

These cusps, which can also be seen in the stress–strain curve $\sigma = \sigma(\varepsilon)$ in Fig. 7.1, come close to being a physical absurdity and thus cannot be justified in any way. In the light of Lucretius's dictum that *natura non facit hiatus* (which does not differ in substance from the Linnaean assertion quoted above), the problem will be addressed here with a much more general approach which does not give rise to this physical absurdity because it relies on a constitutive law that describes the mechanical behavior of the structure's material more correctly.

Again with an eye to achieving greater generality, the problem will here be solved using relations (4.71) and (4.72) which, as was indicated in Chap. 4, apply to all four yield theories given in Fig. 7.2. These relations oblige us to focus attention on the circular cylinder with free closed ends, which is in any case that of greatest interest from the standpoint of the most advanced technologies employing the autofrettage process. A single comprehensive treatment can thus be used for both the partially plastic state and the fully plastic state, taking the predictions of the four yield theories into account simultaneously. For materials showing linear elastic-perfectly plastic behavior, relations (4.71) and (4.72) also apply for the ε_{max} theory, providing that we assume $\nu = 0.5$. In fact, setting $a = 1 - 2\nu = 0$ and $b = 1 + \nu = 1.5$ in relation (4.71) yields a relation identical to that obtained from the second relation (9.23) with $\varepsilon_e = \varepsilon_t$ and $\sigma_z = (\sigma_r + \sigma_t)/2$.

In the entire inner region showing plastic behavior, located in the domain $r_i \leq r \leq r_p$, we can assume $\sigma_e = \sigma_s$. Consequently, relation (4.71) yields:

$$\sigma_t = \frac{2\sigma_s - (a - b)\sigma_r}{(a + b)}. \tag{9.48}$$

Substituting the latter in the equilibrium equation (4.7) and setting $c = 2a/(a + b)$ gives the following differential equation

$$r\frac{d\sigma_r}{dr} + c\sigma_r - c\frac{\sigma_s}{a} = 0, \tag{9.49}$$

which integrates to:

$$\sigma_r = \frac{\sigma_s}{a} + \frac{C}{r^c} \tag{9.50}$$

where C is an integration constant to be determined by imposing the condition that $\sigma_r = -p_i$ at radius $r = r_i$. We thus obtain:

$$C = -r_i^c\left(p_i + \frac{\sigma_s}{a}\right). \tag{9.51}$$

Recalling relations (9.48) and (9.28), the plastic stress state in our circular cylinder is given by the following set of three relations:

$$\begin{aligned}
\sigma_r &= \frac{\sigma_s}{a}\left[1 - \left(\frac{r_i}{r}\right)^c\right] - p_i\left(\frac{r_i}{r}\right)^c \\
\sigma_t &= \frac{\sigma_s}{a}\left[1 + \frac{a-b}{a+b}\left(\frac{r_i}{r}\right)^c\right] + p_i\frac{a-b}{a+b}\left(\frac{r_i}{r}\right)^c \\
\sigma_z &= \frac{\sigma_s}{a}\left[1 - \frac{b}{a+b}\left(\frac{r_i}{r}\right)^c\right] - p_i\frac{b}{a+b}\left(\frac{r_i}{r}\right)^c.
\end{aligned} \tag{9.52}$$

In the outer region showing elastic behavior and located in domain $r_p \leq r \leq r_e$, the stress state is defined by Lamé's relations (4.21) as well as by relation (4.32). As $\sigma_r = 0$ for $r = r_e$, the first relation (4.21) gives $A = B/r_e^2$. Consequently, relations (4.21) and (4.32) yield:

$$\begin{aligned}
\sigma_r &= B\left(\frac{1}{r_e^2} - \frac{1}{r^2}\right) \\
\sigma_t &= B\left(\frac{1}{r_e^2} + \frac{1}{r^2}\right) \\
\sigma_z &= A = \frac{B}{r_e^2}.
\end{aligned} \tag{9.53}$$

On the other hand, relation (4.72) will assume the form

$$\sigma_e = B\left(\frac{a}{r_e^2} + \frac{b}{r^2}\right). \tag{9.54}$$

But since it is necessary that $\sigma_e = \sigma_s$ at radius $r = r_p$, we will have:

$$B = \sigma_s\frac{r_p^2 r_e^2}{ar_p^2 + br_e^2}. \tag{9.55}$$

By substituting the latter in relations (9.53), we obtain:

$$\begin{aligned}
\sigma_r &= \sigma_s\frac{r_p^2}{ar_p^2 + br_e^2}\left(1 - \frac{r_e^2}{r^2}\right) \\
\sigma_t &= \sigma_s\frac{r_p^2}{ar_p^2 + br_e^2}\left(1 + \frac{r_e^2}{r^2}\right) \\
\sigma_z &= \sigma_s\frac{r_p^2}{ar_p^2 + br_e^2}.
\end{aligned} \tag{9.56}$$

This set of three equations defining the stress state in the elastic region depends on the radius r_p of the boundary surface between the plastic and elastic regions. As it can be determined by setting $\sigma_r^p = \sigma_r^e$ (continuity of σ_r) for $r = r_p$, we can write:

$$\frac{\sigma_s}{a}\left[1 - \left(\frac{r_i}{r_p}\right)^c\right] - p_i\left(\frac{r_i}{r_p}\right)^c = \sigma_s\frac{r_p^2}{ar_p^2 + br_e^2}\left(1 - \frac{r_e^2}{r_p^2}\right). \tag{9.57}$$

Solving the latter relation for p_i gives:

$$p_i = \frac{\sigma_s}{a}\left[\frac{(a+b)r_e^2}{ar_p^2 + br_e^2}\left(\frac{r_p}{r_i}\right)^c - 1\right] \tag{9.58}$$

which links internal pressure p_i to the radius r_p of the plastic-elastic interface. For $r_p = r_i$, relation (9.58) gives the following value for the elastic-breakdown pressure $p_{i,i}$

$$p_{i,i} = \sigma_s\frac{r_e^2 - r_i^2}{ar_i^2 + br_e^2}, \tag{9.59}$$

while for $r_p = r_e$, we obtain the following value for pressure $p_{i,e}$ at which the circular cylinder is in the fully plastic state:

$$p_{i,e} = \frac{\sigma_s}{a}\left[\left(\frac{r_e}{r_i}\right)^c - 1\right]. \tag{9.60}$$

It should be noted that relation (9.59) can also be obtained from relation (4.73), setting $p_e = 0$, $\sigma_e = \sigma_s$ and $r = r_i$ in the latter.

If the σ_{max} yield theory is used, we will have $a = b = c = 2a/(a+b) = 1$ (see also Table 4.7), so that relations (9.52) and (9.56) will respectively become:

$$\begin{aligned} \sigma_r &= \sigma_s - (p_i + \sigma_s)\frac{r_i}{r} \\ \sigma_t &= \sigma_s \\ \sigma_z &= \sigma_s - \frac{1}{2}(p_i + \sigma_s)\frac{r_i}{r} \end{aligned} \tag{9.61}$$

and

$$\begin{aligned} \sigma_r &= \sigma_s\frac{r_p^2}{r_p^2 + r_e^2}\left(1 - \frac{r_e^2}{r^2}\right) \\ \sigma_t &= \sigma_s\frac{r_p^2}{r_p^2 + r_e^2}\left(1 + \frac{r_e^2}{r^2}\right) \\ \sigma_z &= \sigma_s\frac{r_p^2}{r_p^2 + r_e^2}. \end{aligned} \tag{9.62}$$

Relations (9.59) and (9.60) will respectively give

$$\frac{p_{i,i}}{\sigma_s} = \frac{r_e^2 - r_i^2}{r_e^2 + r_i^2} \quad \text{and} \quad \frac{p_{i,e}}{\sigma_s} = \frac{r_e}{r_i} - 1, \tag{9.63}$$

while relations (9.57) or (9.58) will yield the following second degree algebraic equation in terms of (r_p/r_e):

$$\left(\frac{r_p}{r_e}\right)^2 - 2f\left(\frac{r_p}{r_e}\right) + 1 = 0 \tag{9.64}$$

where

$$f = \frac{\sigma_s r_e}{(p_i + \sigma_s) r_i}. \tag{9.65}$$

By solving equation (9.64), we obtain:

$$r_p = r_e\left(f - \sqrt{f^2 - 1}\right) \tag{9.66}$$

as we must necessarily have $r_p \le r_e$; for $r_p = r_e$, we will have $f = 1$ and from Eq. (9.65) we will again obtain relation (9.60).

Given that $a = 0$ (it should be recalled that $v = 0.5$ in the plastic range), singularity problems will be encountered if the ε_{max}, τ_{max} and maximum distortion energy yield theories are used, since the term σ_s/a appears in the second member of the set of three equations (9.52). However, it can be readily demonstrated that the following limits exist:

$$\begin{aligned}
&\lim_{a\to 0}\frac{\sigma_s}{a}\left[1 - \left(\frac{r_i}{r}\right)^c\right] = \frac{\sigma_s}{b}\ln\left(\frac{r}{r_i}\right)^2 \\
&\lim_{a\to 0}\frac{\sigma_s}{a}\left[1 + \frac{a-b}{a+b}\left(\frac{r_i}{r}\right)^c\right] = \frac{\sigma_s}{b}\left[2 + \ln\left(\frac{r}{r_i}\right)^2\right] \\
&\lim_{a\to 0}\frac{\sigma_s}{a}\left[1 - \frac{b}{a+b}\left(\frac{r_i}{r}\right)^c\right] = \frac{\sigma_s}{b}\left[1 + \ln\left(\frac{r}{r_i}\right)^2\right].
\end{aligned} \tag{9.67}$$

The plastic stress state given by Eq. (9.52) can thus be expressed by means of the following set of three relations:

$$\begin{aligned}
\sigma_r &= -p_i + \frac{\sigma_s}{b}\ln\left(\frac{r}{r_i}\right)^2 \\
\sigma_t &= -p_i + \frac{\sigma_s}{b}\left[2 + \ln\left(\frac{r}{r_i}\right)^2\right] \\
\sigma_z &= -p_i + \frac{\sigma_s}{b}\left[1 + \ln\left(\frac{r}{r_i}\right)^2\right];
\end{aligned} \tag{9.68}$$

whereas the elastic stress state given by Eq. (9.56) is expressed by the set of three equations below:

$$\begin{aligned}\sigma_r &= \frac{\sigma_s}{b}\frac{r_p^2}{r_e^2}\left(1-\frac{r_e^2}{r^2}\right)\\ \sigma_t &= \frac{\sigma_s}{b}\frac{r_p^2}{r_e^2}\left(1+\frac{r_e^2}{r^2}\right)\\ \sigma_z &= \frac{\sigma_s}{b}\frac{r_p^2}{r_e^2}.\end{aligned} \tag{9.69}$$

From relations (9.59) and (9.60) we obtain the relations

$$p_{i,i}=\frac{\sigma_s}{b}\left(1-\frac{r_i^2}{r_e^2}\right) \quad \text{and} \quad p_{i,e}=\frac{\sigma_s}{b}\ln\frac{r_e^2}{r_i^2} \tag{9.70}$$

the second of which is obtained by recalling that

$$\lim_{a\to 0}\frac{\sigma_s}{a}\left[\left(\frac{r_e}{r_i}\right)^c-1\right]=\frac{\sigma_s}{b}\ln\left(\frac{r_e}{r_i}\right)^2. \tag{9.71}$$

Lastly, and again for the limit $a \to 0$, relation (9.57) or relation (9.58) yield the following equation

$$\ln\frac{r_p^2}{r_i^2}-\frac{r_p^2}{r_e^2}-\frac{bp_i}{\sigma_s}+1=0, \tag{9.72}$$

which is to be regarded as the most appropriate for calculating radius r_p.

9.5 Comparison Between the Four Yield Theories Considered

We will now compare the results obtained with the four yield theories considered above. To this end, we will take a circular cylinder with diameter ratio $\delta = 2$.

With the σ_{max} yield theory, relations (9.63) will give:

$$\frac{p_{i,i}}{\sigma_s}=\frac{\delta^2-1}{\delta^2+1}=\frac{3}{5} \quad \text{and} \quad \frac{p_{i,e}}{\sigma_s}=\delta-1=1. \tag{9.73}$$

Consequently, for a dimensionless load $p_i/\sigma_s = 0.65$ (midway between the extremes indicated above) and thus for $f = \delta/(1 + p_i/\sigma_s) = 1.212$, relation (9.66) gives $r_p/r_e = 0.527$, which corresponds to $r_p/r_i = 1.054$. These values are shown in Table 9.1.

Table 9.1 Values of comparison parameters obtained by applying the four yield theories to a circular cylinder with $\delta = 2$, in the plastic state ($\nu = 0.5$)

Yield theory	σ_{max}	ε_{max}	τ_{max}	ED_{max}
b	1.0	1.5	2.0	$\sqrt{3}$
$p_{i,i}/\sigma_s$	0.60	0.50	0.375	0.43
$p_{i,e}/\sigma_s$	1.0	1.08	0.69	0.80
p_i/σ_s	0.65	0.65	0.65	0.65
$(r_p/r_i)_{p_i/\sigma_s=0.65}$	1.054	1.173	1.600	1.328

Repeating the same calculations for the remaining three yield theories with all other conditions remaining equal, remembering that b is equal to 1.5, 2.0 and $\sqrt{3}$ for the ε_{max}, τ_{max} and maximum distortion energy theories respectively, this time using relations (9.70) and (9.72) and, once again, a load ratio $p_i/\sigma_s = 0.65$, which is still midway between the extreme values $p_{i,i}/\sigma_s$ and $p_{i,e}/\sigma_s$, though the latter differ for each yield theory, we obtain the values given in Table 9.1.

As can be seen from Table 9.1, at the value of the diameter ratio considered here ($\delta = 2$), the highest value of ratio $p_{i,e}/\sigma_s$ is that for the ε_{max} yield theory, while the lowest value is obtained using the τ_{max} yield theory. With any given ratio p_i/σ_s, plastic flow is more pronounced with the latter theory, as the ratio r_p/r_i in the last line of the table clearly shows. As regards the τ_{max} yield theory, it should be noted that the second relation (9.70), with $p_{i,e}/\sigma_s = p_i/\sigma_s$, is identical to the approximate relation (1.17) for $\sigma_{e.i} = \sigma_s$, which shows a high degree of reliability with theories that explicitly allow the material to show marked plastic adaptation. These results can in any case be generalized to any value of diameter ratio δ.

9.6 Partial Plasticization in More General Terms

In Sect. 9.3, we addressed the full plasticization of circular cylinders with free closed ends in the plastic or steady-state viscous state. In this section, the same theoretical approach will be extended to the partial plasticization of such cylinders in order to overcome the limitations of the treatment presented in Chaps. 7 and 8, which is based on the assumption that the material exhibits linear elastic-perfectly plastic behavior as represented by the stress–strain curve $\sigma = \sigma(\varepsilon)$ in Fig. 7.1. For the reasons explained in Sects. 7.2 and 7.3, we will here consider only the τ_{max} and maximum distortion energy yield theories.

We will begin at radius r_p, i.e., at the boundary separating the inner region showing plastic behavior and the outer region showing elastic behavior. At this radius, strain will be equal to that at the proportional limit, designated here as ε_P. Recalling that $k = 2/b$, relation (9.37) thus enables us to write:

$$\varepsilon_P = \frac{4C}{3kr_P^2} = \frac{2}{3}b\frac{C}{r_P^2}. \tag{9.74}$$

Figure 9.3a shows the stress–strain curve $\sigma = \sigma(\varepsilon)$ for a material in the virgin state, in which it is assumed that the proportional limit, the elastic limit and the yield-point stress coincide. In this curve, point P having coordinates $(\varepsilon_P, \sigma_P)$ and corresponding to the yield-point stress, divides the curve portion showing elastic behavior from that showing inelastic behavior. For the materials of interest here, the stress σ and its derivative $d\sigma/d\varepsilon$ are functions that monotonically increase and decrease respectively with ε, as these materials are strain-hardening. We will thus have $\sigma(d\sigma^2/d\varepsilon^2)<0$ for any ε, while it is assumed that the constitutive relation $\sigma = f(\varepsilon)$ applies to each cross-section of the specimen during the entire loading process. The curve in Fig. 9.3a highlights the fact that if we start from a generic point A having coordinates $(\varepsilon_A, \sigma_A)$, located past point P and the load is reduced, the representative point will move along path AB, which is essentially parallel to the linear elastic line segment in this curve. Consequently, when the load reaches zero, there will be a permanent plastic strain ε_e^p, while the recoverable elastic strain will be ε_e^e.

Obviously, we will have $\varepsilon_A = \varepsilon_e^p + \varepsilon_e^e = \varepsilon_e$, where ε_e is the equivalent strain (the other coordinate of the generic point A represents the equivalent stress σ_e). Conversely, if load is increased again starting from point B, the representative point will move along a path close to, but not coinciding with AB, thus resulting in a small hysteresis cycle (omitted from Fig. 9.3a to avoid over-complication, but shown schematically in Fig. 9.3b) correlated with the energy losses per unit volume occurring in the loading and unloading process. Once point A has been reached, further deformation will not occur unless load is increased: this is the strain hardening phenomenon. In the plastic range, the stress thus depends on the total loading operation and/or the material's overall load history. The figure also shows the elastic magnitudes E and E_T, i.e., the modulus of elasticity or Young's modulus and the tangent modulus, the first constant and the second variable point to point, i.e., a point function.

The relation expressing the constitutive law $\sigma = \sigma(\varepsilon)$,, which is similar to Swift's (1952) and thus a bit more elaborate than those formulated by Ludwik (1909) and by Ramberg and Osgood (1943), is as follows:

$$\sigma_e = \frac{\sigma_P}{\mu - 1}\left[\mu - \left(\frac{\varepsilon_P}{\varepsilon_e}\right)^{\mu-1}\right] \tag{9.75}$$

where σ_P is the proportional limit stress (obviously corresponding to ε_P), while μ is a constant parameter depending on the material. More complex constitutive laws, such as the polynomial correlation proposed and used by Vullo and Vivio (2013) for rotor stress analysis, would not lead to final equations written in closed form.

As is clear from Fig. 9.4, the correlation $\sigma = \sigma(\varepsilon)$ given by the constitutive law (9.75) is very general, since it provides a good fit with the experimental stress–strain curve for both materials having linear elastic-perfectly plastic behavior (for high values of μ) and for strain hardening materials (for lower values of μ). Moreover, this correlation has the twofold advantage of not giving rise to

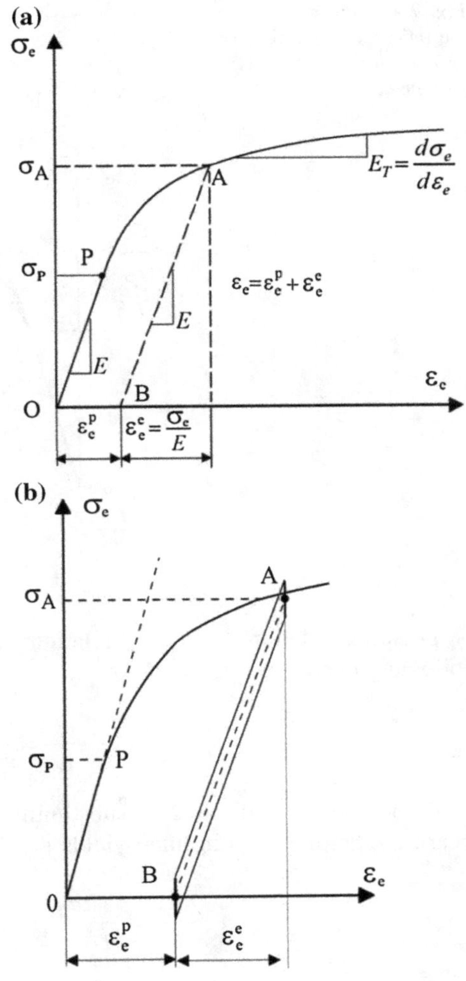

Fig. 9.3 **a** Stress–strain curve $\sigma = \sigma(\varepsilon)$ for a real material in the virgin state and characteristic magnitudes, and **b** idem, with hysteresis cycle in schematic form

discontinuities at the proportional limit, where $(d\sigma/d\varepsilon)_{\varepsilon_e=\varepsilon_P} = E$, and of applying not only for $\varepsilon_e > \varepsilon_P$, but also for $\varepsilon_e = \varepsilon_P$ as well as for $\varepsilon_e < \varepsilon_P$, i.e., in the proportional range, providing that we set $\mu = 0$ in it. For $\mu = 0$, in fact, relation (9.75) reduces to the relation $\sigma_e = (\sigma_P/\varepsilon_P)\varepsilon_e = E\varepsilon_e$, or in other words to Hooke's law for uniaxial stress. For $\mu = 1$, moreover, as can be verified immediately by means of De L'Hospital's theorem, relation (9.75) specializes in the following extremely simple relation

$$\sigma_e = \sigma_P\left(1 + \ln\frac{\varepsilon_e}{\varepsilon_P}\right). \tag{9.76}$$

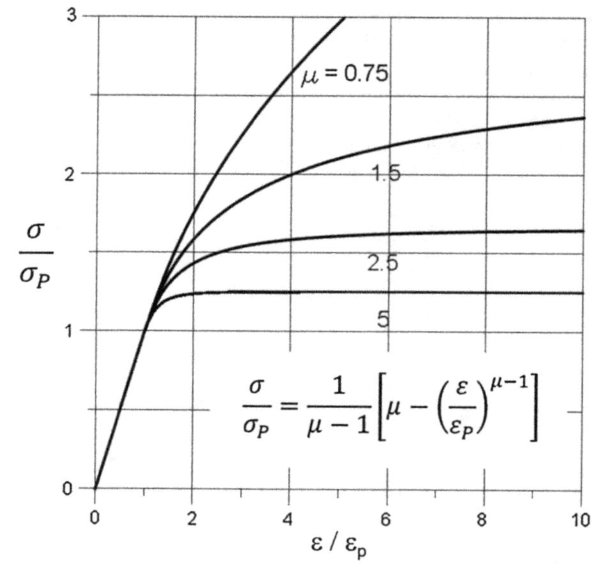

Fig. 9.4 Ratio σ/σ_P versus the ratio $\varepsilon/\varepsilon_P$ from relation (9.75), for different values of parameter μ

Dividing relation (9.74) by relation (9.37) member by member gives the following expression of the ratio

$$\frac{\varepsilon_P}{\varepsilon_e} = \frac{r^2}{r_p^2}. \tag{9.77}$$

Remembering that $k = 2/b$, substituting relation (9.75) in relation (9.36), and bearing relation (9.77) in mind yields the following differential equation

$$r\frac{d\sigma_r}{dr} = \frac{2\sigma_P}{b(\mu - 1)}\left[\mu - \left(\frac{r^2}{r_P^2}\right)^{\mu-1}\right] \tag{9.78}$$

which integrates to

$$\sigma_r = \frac{\sigma_P}{b(\mu - 1)^2}\left[\mu(\mu - 1)\ln r^2 - \left(\frac{r^2}{r_p^2}\right)^{\mu-1}\right] + C \tag{9.79}$$

where C is an integration constant to be determined by imposing the boundary condition at radius $r = r_i$, where $\sigma_r = -p_i$. We thus have:

$$C = -p_i - \frac{\sigma_P}{b(\mu - 1)^2}\left[\mu(\mu - 1)\ln r_i^2 - \left(\frac{r_i^2}{r_p^2}\right)^{\mu-1}\right]. \tag{9.80}$$

Substituting this in relation (9.79) then gives:

$$\sigma_r = -p_i + \frac{\sigma_P}{b(\mu-1)^2}\left[\mu(\mu-1)\ln\frac{r^2}{r_i^2} + \left(\frac{r_i^2}{r_p^2}\right)^{\mu-1} - \left(\frac{r^2}{r_p^2}\right)^{\mu-1}\right]. \tag{9.81}$$

Hoop stress σ_t can be determined by expanding the equilibrium equation (4.7) and substituting into the expanded relation the relations (9.81) and (9.78) appearing in it. We thus obtain:

$$\sigma_t = -p_i + \frac{\sigma_P}{b(\mu-1)^2}\left[\mu(\mu-1)\left(2+\ln\frac{r^2}{r_i^2}\right) + \left(\frac{r_i^2}{r_p^2}\right)^{\mu-1} - (2\mu-1)\left(\frac{r^2}{r_p^2}\right)^{\mu-1}\right]. \tag{9.82}$$

Lastly, longitudinal stress σ_z, which is the arithmetic average of the radial stress and the hoop stress, is given by:

$$\sigma_z = -p_i + \frac{\sigma_P}{b(\mu-1)^2}\left[\mu(\mu-1)\left(1+\ln\frac{r^2}{r_i^2}\right) + \left(\frac{r_i^2}{r_p^2}\right)^{\mu-1} - \mu\left(\frac{r^2}{r_p^2}\right)^{\mu-1}\right]. \tag{9.83}$$

By grouping relations (9.81)–(9.83), we obtain the following relations that define the stress state in the plastic region of the circular cylinder:

$$\begin{aligned}
\sigma_r &= -p_i + \frac{\sigma_P}{b(\mu-1)^2}\left[\mu(\mu-1)\ln\frac{r^2}{r_i^2} + \left(\frac{r_i^2}{r_p^2}\right)^{\mu-1} - \left(\frac{r^2}{r_p^2}\right)^{\mu-1}\right]\\
\sigma_t &= -p_i + \frac{\sigma_P}{b(\mu-1)^2}\left[\mu(\mu-1)\left(2+\ln\frac{r^2}{r_i^2}\right) + \left(\frac{r_i^2}{r_p^2}\right)^{\mu-1} - (2\mu-1)\left(\frac{r^2}{r_p^2}\right)^{\mu-1}\right]\\
\sigma_z &= -p_i + \frac{\sigma_P}{b(\mu-1)^2}\left[\mu(\mu-1)\left(1+\ln\frac{r^2}{r_i^2}\right) + \left(\frac{r_i^2}{r_p^2}\right)^{\mu-1} - \mu\left(\frac{r^2}{r_p^2}\right)^{\mu-1}\right].
\end{aligned} \tag{9.84}$$

For $\mu = 1$ the above equations particularize in the following relations

$$\begin{aligned}
\sigma_r &= -p_i + \frac{2\sigma_P}{b}\left[\ln\frac{r}{r_i} + \left(\ln\frac{r_i}{r_p}\right)^2 - \left(\ln\frac{r}{r_p}\right)^2\right]\\
\sigma_t &= -p_i + \frac{2\sigma_P}{b}\left[1+\ln\frac{r}{r_i} + \left(\ln\frac{r_i}{r_p}\right)^2 - \left(\ln\frac{r}{r_p}\right)\left(2+\ln\frac{r}{r_p}\right)\right]\\
\sigma_z &= -p_i + \frac{2\sigma_P}{b}\left[\frac{1}{2}+\ln\frac{r}{r_i} + \left(\ln\frac{r_i}{r_p}\right)^2 - \left(\ln\frac{r}{r_p}\right)\left(1+\ln\frac{r}{r_p}\right)\right].
\end{aligned} \tag{9.85}$$

The first relation (9.85) can either be obtained from relation (9.81) by applying De L'Hospital's theorem twice in succession, or directly considering the relation (9.76). The other two relations are obtained by using the same procedure employed for relations (9.82) and (9.83).

We must now calculate the stress state in the elastic region. The radial stress can be inferred from the first of a Lamé's equations (4.21), where the constant B can be expressed by means of relation (4.72), setting $a = 0$, $\sigma_e = \sigma_P$ and $r = r_p$. We thus obtain the following relation:

$$\sigma_r = A - \frac{\sigma_P}{b}\frac{r_P^2}{r^2}, \tag{9.86}$$

which can also be inferred from relation (9.81) for $\mu = 0$. Imposing that $\sigma_r = -p_e$ at radius $r = r_e$, we can use relation (9.86) to calculate constant A, which we then substitute into relation (9.86) to obtain:

$$\sigma_r = -p_e + \frac{\sigma_P}{b}\left(\frac{r_P^2}{r_e^2} - \frac{r_P^2}{r^2}\right). \tag{9.87}$$

This relations differs from the first relation (9.69) in that it contains the term in p_e, since the case examined here, which also considers the effect of external pressure, is more general than the one we dealt with earlier, which led to relation (9.69). Obviously p_e must be lower than p_i to ensure that the onset of plastic flow takes place at the bore.

Repeating the procedure that led from relation (9.81) to relations (9.82) and (9.83) brings us to the following relations

$$\sigma_t = -p_e + \frac{\sigma_P}{b}\left(\frac{r_P^2}{r_e^2} + \frac{r_P^2}{r^2}\right) \tag{9.88}$$

and

$$\sigma_z = -p_e + \frac{\sigma_P}{b}\frac{r_P^2}{r_e^2} \tag{9.89}$$

which also differ from the corresponding relations (9.69) in containing the term in p_e. Here as before, grouping relations (9.87)–(9.89) gives the following equations defining the stress state in the elastic region of our circular cylinder under internal and external pressure:

$$\begin{aligned}
\sigma_r &= -\,p_e + \frac{\sigma_P}{b}\left(\frac{r_P^2}{r_e^2} - \frac{r_P^2}{r^2}\right)\\
\sigma_t &= -\,p_e + \frac{\sigma_P}{b}\left(\frac{r_P^2}{r_e^2} + \frac{r_P^2}{r^2}\right)\\
\sigma_z &= -\,p_e + \frac{\sigma_P}{b}\frac{r_P^2}{r_e^2}.
\end{aligned} \tag{9.90}$$

The relations (9.84), or their equivalents (9.85) which apply to the particular case where $\mu = 1$, and relations (9.90) univocally define the stress state in the circular cylinder after partial yielding, once the radius r_p of the plastic boundary has been determined as a function of the applied loads and geometry. This stress state is continuous throughout the range of variation of r, i.e., for $r_i \leq r \leq r_e$, including the region of transition between plastic and elastic behavior. It can readily be shown, in fact, that:

$$\begin{aligned}
\left(\frac{d\sigma_r^p}{dr}\right)_{r=r_p} &= \left(\frac{d\sigma_r^e}{dr}\right)_{r=r_p} = \frac{2\,\sigma_P}{b\,r_p} \\
\left(\frac{d\sigma_t^p}{dr}\right)_{r=r_p} &= \left(\frac{d\sigma_t^e}{dr}\right)_{r=r_p} = \frac{2\,\sigma_P}{b\,r_p} \\
\left(\frac{d\sigma_z^p}{dr}\right)_{r=r_p} &= \left(\frac{d\sigma_z^e}{dr}\right)_{r=r_p} = 0.
\end{aligned} \tag{9.91}$$

This continuity thus applies not only to the principal stresses, but also to their derivatives with respect to r [for the sake of clarity, the indices p and e are used in relations (9.91) to denote relations for the plastic range and for the elastic range respectively].

Our next objective is to determine the radius r_p of the plastic boundary once the external load, the material used for our circular cylinder and its geometry are known. To this end, for $r = r_p$, we will equal the radial stress values obtained from the first relation (9.84) or the first relation (9.85), depending on whether we are dealing with the general case of any μ or the particular case with $\mu = 1$, and the radial stress value obtained from the first relation (9.90). In other words, we will impose that $\left(\sigma_r^p\right)_{r=r_p} = \left(\sigma_r^e\right)_{r=r_p}$. This procedure yields the following relation expressing the external load $(p_i - p_e)$ made dimensionless relative to the material's proportional limit stress σ_P:

$$\frac{p_i - p_e}{\sigma_P} = \eta + \vartheta \tag{9.92}$$

where in all cases we have

$$\eta = \frac{1}{b}\left(1 - \frac{r_P^2}{r_e^2}\right) \tag{9.93}$$

while we will have

$$\vartheta = \frac{1}{b(\mu-1)^2}\left[\mu(\mu-1)\ln\frac{r_P^2}{r_i^2} + \left(\frac{r_i^2}{r_P^2}\right)^{\mu-1} - 1\right] \tag{9.94}$$

or

$$\vartheta = \frac{2}{b}\ln\frac{r_P}{r_i}\left(1 + \ln\frac{r_P}{r_i}\right) \tag{9.95}$$

depending on whether $\mu \neq 1$ or $\mu = 1$.

The external load in the elasto-plastic state given by relation (9.92) can vary from a lower limit corresponding to elastic-breakdown at the bore, with the material stressed entirely in the elastic range, to an upper limit at which the circular cylinder's entire wall thickness is in the plastic state. By particularizing the pairs of relations (9.93) and (9.94) or (9.93) and (9.95), depending on whether $\mu \neq 1$ or $\mu = 1$, for $r = r_i$ and $r = r_e$, it can be inferred that these upper and lower limits for the dimensionless external load, denoted here with the subscripts *low* and *up*, correspond to equally particular values of the magnitudes η and ϑ, viz.:

$$\left(\frac{p_i - p_e}{\sigma_P}\right)_{low} \rightarrow \vartheta_{low} = 0 \tag{9.96}$$

and

$$\left(\frac{p_i - p_e}{\sigma_P}\right)_{up} \rightarrow \eta_{up} = 0. \tag{9.97}$$

For a given diameter ratio δ and a given material, each value of the external load $(p_i - p_e)$ will correspond to a specific value of ratio r_p/r_i defining the location of the elastic–plastic boundary radius. Figure 9.5 can be useful in calculating this radius quickly. Magnitudes η and ϑ are shown on the ordinate, while the abscissa consists of the ratio r_p/r_i. The line segment intercepted parallel to the ordinate axis by the ϑ curve for a given material, i.e., a given value of μ, and by the η curve for a given circular cylinder geometry defined by ratio δ, where it is read on the ordinate scale, represents the dimensionless external load, i.e., the ratio $(p_i - p_e)/\sigma_P$, while the corresponding abscissa represents the ratio r_p/r_i we wish to find.

The limit values of the dimensionless external load for predetermined values of μ and δ, i.e., for a given material and a given geometry, can be identified immediately from this figure. It should be noted that the diagram in Fig. 9.5 applies under the assumption that the yield theory is that of the maximum distortion energy, for which $b = \sqrt{3}$. A similar diagram can be constructed for the τ_{max} yield theory.

Once the elastic–plastic boundary radius r_p has been determined, the strain state in the plastic region can be calculated just as quickly. To do so, it is sufficient to determine the constant C as a function of ε_P from relation (9.74) and substitute it into relations (9.33). We thus obtain [it will be recalled that, on the basis of relation (9.29), $\varepsilon_z = 0$]:

$$\varepsilon_r = -\frac{3\varepsilon_P r_p^2}{2br^2} \quad \text{and} \quad \varepsilon_t = \frac{3\varepsilon_P r_p^2}{2br^2}. \tag{9.98}$$

It should be noted that the three principal stresses σ_r, σ_t and σ_z are heavily dependent on the exponent μ appearing in the constitutive Eq. (9.75). This is immediately apparent from Fig. 9.6, which applies to the particular case where $\delta = 3$, $p_e = 0$ and $p_i/\sigma_P = 1.5$. The curves representing the distributions of the

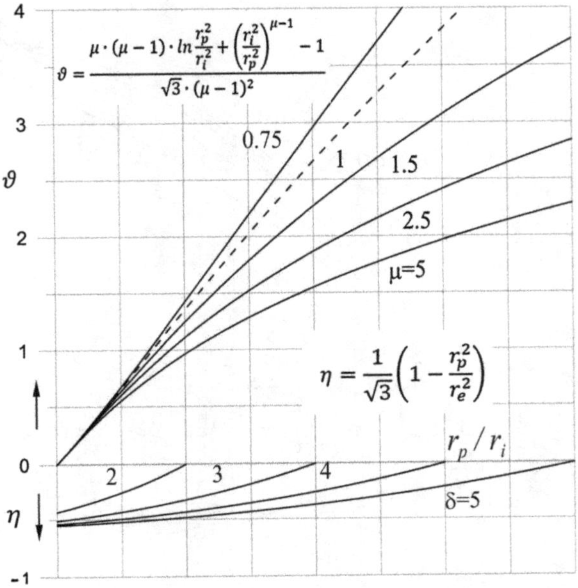

Fig. 9.5 Calculation diagrams of the elastic–plastic boundary radius r_p as a function of the dimensionless external load, for a given material and a given diameter ratio δ of the circular cylinder

three principal stresses σ_r, σ_t and σ_z through the wall thickness are clearly continuous and provide a better interpretation of the experimental data than the cusped curves obtained on the basis of the more restrictive simplifying assumptions used in Chaps. 7 and 8. A number of researchers starting from Steele (1952) have demonstrated experimentally that the two regions of a circular cylinder in the partially plastic state are not bounded by a well-defined cylindrical surface (such as that corresponding to the plastic-elastic interface identified earlier), but by a transitional region of annular cross-section, where highly deformed elastic material is mixed with spiral layers or wedge-shaped layers of plastic material.

9.7 Multilayer Circular Cylinders

We will now return to multilayer circular cylinders such as that examined in Sect. 5.7, consisting of n concentric components shrunk together with a process capable of producing the desired state of coaction in the composite structure. Once again, we will address the problem in completely general terms, using relations that can simultaneously interpret the four σ_{max}, ε_{max}, τ_{max} and maximum distortion energy strength theories.

For this purpose, we will again take relation (4.73) expressing the equivalent stress in a circular cylinder with free closed ends under internal and external pressure, and we will particularize it for $r = r_i$ and $r = r_e$. We thus obtain:

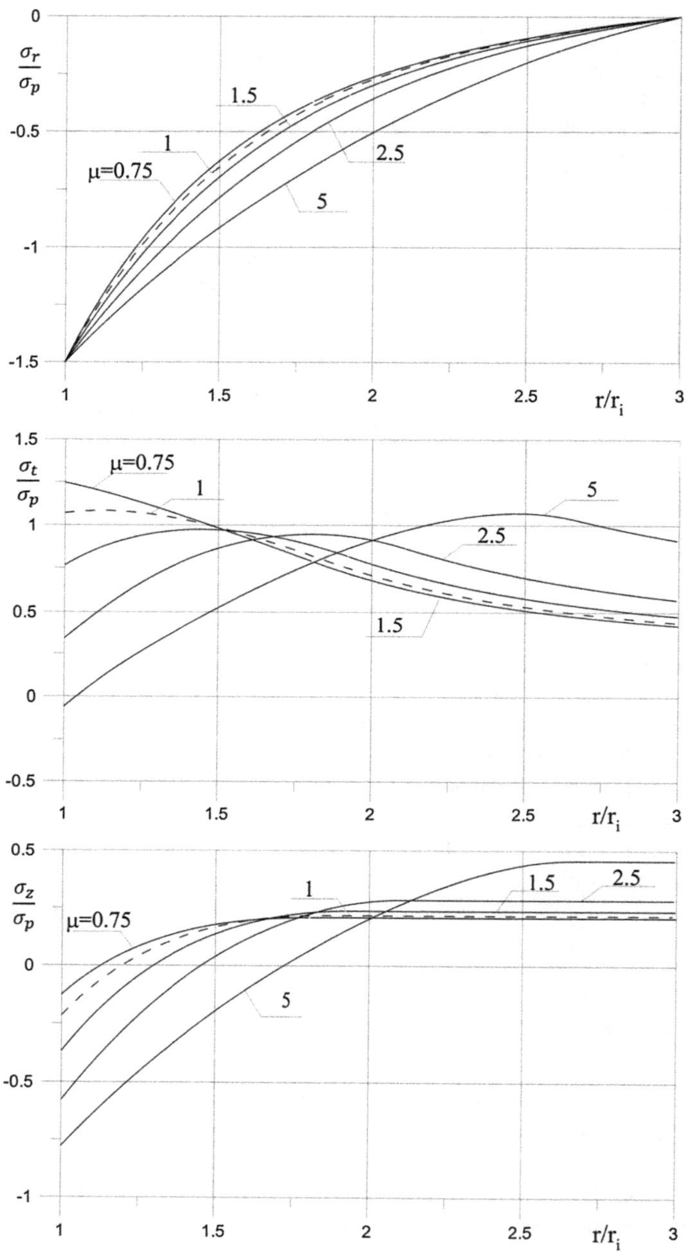

Fig. 9.6 Distribution curves for the three dimensionless principal stresses through the wall thickness of a circular cylinder in the plastic state ($\delta = 3$; $p_e = 0$; $p_i/\sigma_P = 1.5$)

$$\begin{aligned}\sigma_{e,i} &= \frac{p_i\left(ar_i^2 + br_e^2\right) - p_e(a+b)r_e^2}{r_e^2 - r_i^2} \\ \sigma_{e,e} &= \frac{p_i(a+b)r_i^2 - p_e\left(ar_e^2 + br_i^2\right)}{r_e^2 - r_i^2}.\end{aligned} \tag{9.99}$$

We will now impose that the equivalent stress $\sigma_{e,i}$ generated at the inner radius of each layer as a result of internal pressure p_i is the same for all of the layers making up the structure in question. According to the first relation (9.99), this condition, applied to two generic adjoining layers (the mth and $(m+1)$th layers as illustrated in Figs. 5.10 and 9.7), can be expressed through the following relation

$$a\frac{p_{i,m} - \alpha_m p_{e,m}}{\alpha_m - 1} = a\frac{p_{i,m+1} - \alpha_{m+1}p_{e,m+1}}{\alpha_{m+1} - 1} \tag{9.100}$$

where

$$\alpha_m = \frac{(a+b)r_{e,m}^2}{ar_{i,m}^2 + br_{e,m}^2} \quad \text{and} \quad \alpha_{m+1} = \frac{(a+b)r_{e,m+1}^2}{ar_{i,m+1}^2 + br_{e,m+1}^2}. \tag{9.101}$$

In the above relations, subscripts i and e denote geometrical magnitudes associated with the inner and outer radii of each layer, whereas subscripts m and $m+1$ refer to the generic layers m and $m+1$. As it is obvious that we must have $p_{e,m} = p_{i,m+1}$ at the interface between the two mth and $(m+1)$th generic adjoining layers, relation (9.100) will give:

$$p_{e,m} = \frac{p_{i,m}(\alpha_{m+1} - 1) + p_{e,m+1}\alpha_{m+1}(\alpha_m - 1)}{\alpha_m\alpha_{m+1} - 1}. \tag{9.102}$$

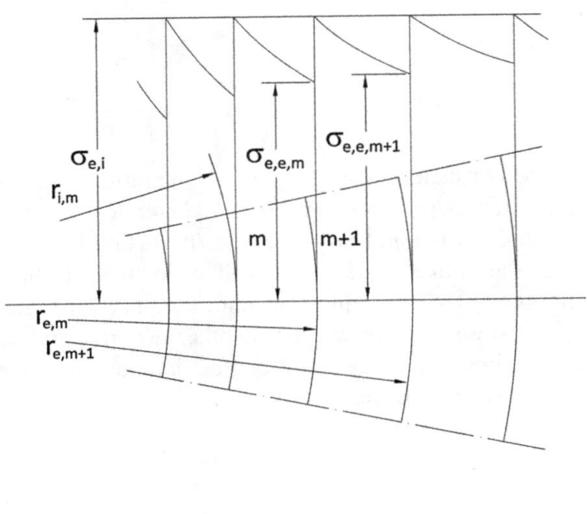

Fig. 9.7 Multilayer structure in the optimal configuration

Substituting this relation for $p_{e,m}$ into the first relation (9.100) results in the following expression for equivalent stress $\sigma_{e,i}$:

$$\sigma_{e,i} = a\frac{p_{i,m} - \alpha_m\alpha_{m+1}p_{e,m+1}}{\alpha_m\alpha_{m+1} - 1}, \tag{9.103}$$

which is assumed as the equivalent or ideal or comparison stress.

The value of radius $r_{e,m} = r_{i,m+1}$ at which this stress is minimum is that for which the product $\alpha_m\alpha_{m+1}$ is maximum. As $\partial(\alpha_m\alpha_{m+1})/\partial r_{e,m}$ drops to zero for $r_{e,m} = \sqrt{r_{i,m}r_{e,m+1}}$, i.e., when the radius of the interface between two adjoining layers is equal to the geometric mean of the inner radius of the mth layer and the outer radius of the $(m + 1)$th layer, for a multilayer structure with n layers and characterized by a ratio $\delta = r_e/r_i$ of the outer radius of the last layer to the inner radius of the first layer, this optimal condition will entail complying with the following relation:

$$\frac{r_{e,1}}{r_i} = \frac{r_{e,2}}{r_{i,2}} = \cdots = \frac{r_{e,m}}{r_{i,m}} = \cdots = \frac{r_e}{r_{i,n}} = \delta^{\frac{1}{n}}. \tag{9.104}$$

In other words, whatever strength theory is selected, all of the layers in a multilayer structure must be geometrically similar.

Since the constants given by relations (9.101) are equal under these conditions, eliminating the subscripts and recalling relations (9.104) will give:

$$\alpha = \frac{(a+b)\delta^{2/n}}{a + b\delta^{2/n}}. \tag{9.105}$$

with $\alpha = \alpha_m = \alpha_{m+1}$, relation (9.102) yields:

$$p_{e,m} = \frac{p_{i,m} + \alpha p_{e,m+1}}{\alpha + 1}. \tag{9.106}$$

This then gives

$$\alpha = \frac{p_{i,m} - p_{e,m}}{p_{e,m} - p_{e,m+1}}. \tag{9.107}$$

The constant α, which represents the ratio of the difference between the internal and external pressures of the mth layer to the difference between the external pressures of the mth layer and the $(m + 1)$th layer, has a clear physical meaning. With the values of factors a and b given in Table 4.2, it can be inferred from relation (9.105) that this constant is unitary for the τ_{max} and maximum distortion energy strength theories. For the other two σ_{max} and ε_{max} strength theories, constant α is greater than unity and it can be demonstrated that, for any value of Poisson's ratio ν, we will have:

$$\alpha_{\sigma_{max}} > \alpha_{\varepsilon_{max}} > 1. \tag{9.108}$$

Solving relation (9.107) for $p_{e,m+1}$ and substituting the result into relation (9.103) gives:

$$\sigma_{e,i} = a\frac{p_{i,m} - \alpha p_{e,m}}{\alpha - 1} \tag{9.109}$$

on the basis of which we can write the following for all layers, from the first to the nth:

$$\begin{aligned} p_{i,1} &= p_i = \frac{\alpha - 1}{a}\sigma_{e,i} + \alpha p_{e,1} \\ p_{i,2} &= \frac{\alpha - 1}{a}\sigma_{e,i} + \alpha p_{e,2} \\ &\dots \\ p_{i,m} &= \frac{\alpha - 1}{a}\sigma_{e,i} + \alpha p_{e,m}. \end{aligned} \tag{9.110}$$

If we now multiply the second of relations (9.110) by α, the third by α^2, … and the nth by α^{m-1}, adding member by member and recalling that

$$\sum_{i=0}^{m-1} \alpha^i = \frac{\alpha^m - 1}{a - 1}, \tag{9.111}$$

we obtain the equation

$$p_i = \frac{\alpha^m - 1}{a}\sigma_{e,i} + \alpha^m p_{e,m}. \tag{9.112}$$

Particularizing the latter for the outermost layer, i.e., writing it for $m = n$, gives the stress we wish to identify

$$\sigma_{e,i} = \frac{a p_i}{\alpha^n - 1}. \tag{9.113}$$

Substituting relation (9.113) into (9.112) yields the relation

$$p_{e,m} = \frac{\alpha^{n-m} - 1}{\alpha^n - 1} p_i, \tag{9.114}$$

while the pressure at the inner radius of the mth layer will be given by:

$$p_{i,m} = \frac{\alpha^{n-m+1} - 1}{\alpha^n - 1} p_i. \tag{9.115}$$

Relations (9.113), (9.114) and (9.115) do not give rise to singularity problems with the σ_{max} and ε_{max} strength theories. This is not the case with the τ_{max} and maximum distortion energy strength theories, with which there will be singularities because $a = 0$. For these theories, we must thus use the following relations instead of relations (9.113)–(9.115):

$$\begin{aligned}\sigma_{e,i} &= \frac{bp_i\delta^{\frac{2}{n}}}{n\left(\delta^{\frac{2}{n}}-1\right)}\\ p_{e,m} &= \left(1-\frac{m}{n}\right)p_i\\ p_{i,m} &= \left(1-\frac{m-1}{n}\right)p_i.\end{aligned}\tag{9.116}$$

These latter equations are obtained from relations (9.113)–(9.115) by passing to the limit for $a \rightarrow 0$ and noting that relation (9.105) gives:

$$\left(\frac{d\alpha}{da}\right)_{a=0} = \frac{b\delta^{\frac{2}{n}}}{\left(\delta^{\frac{2}{n}}-1\right)}.\tag{9.117}$$

The correlation between the ratio $\delta = r_e/r_i$ for the multilayer structure and the ratio $p_i/\sigma_{e,i}$ taken once again as a parameter for comparing the strength theories considered here, is determined by introducing the value of α obtained from relation (9.113) into relation (9.105) and solving the equation thus obtained for δ. We thus arrive at the following relation:

$$\delta = \left(1+\frac{ap_i}{\sigma_{e,i}}\right)^{1/2}\left\{1-\frac{b}{a}\left[\left(1+\frac{ap_i}{\sigma_{e,i}}\right)^{1/n}-1\right]\right\}^{-n/2}\tag{9.118}$$

which applies for the σ_{max} and ε_{max} strength theories. The two curves $p_i/\sigma_{e,i} = f(\delta)$ obtained in this way exhibit a horizontal asymptote given by the relation

$$\frac{p_i}{\sigma_{e,i}} = \frac{1}{a}\left[\left(1+\frac{a}{b}\right)^n-1\right].\tag{9.119}$$

For the other two strength theories, i.e., the τ_{max} and maximum distortion energy theories, it is necessary to use relations (9.118) and (9.119) because of the singularity deriving from $a = 0$, passing to the limit for $a \rightarrow 0$.We thus obtain the relations

$$\delta = \left(1-\frac{bp_i}{n\sigma_{e,i}}\right)^{-n/2}\tag{9.120}$$

and

$$\frac{p_i}{\sigma_{e,i}} = \frac{n}{b}.\tag{9.121}$$

Both relations (9.120) and (9.121) can be obtained directly from the first relation (9.116).

Figure 9.8 shows the $p_i/\sigma_{e,i} = f(\delta)$ curves obtained for a three-layer structure ($n = 3$) using the four σ_{max}, ε_{max}, τ_{max} and maximum distortion energy strength theories. These curves were obtained using relation (9.118) for the σ_{max} and ε_{max}

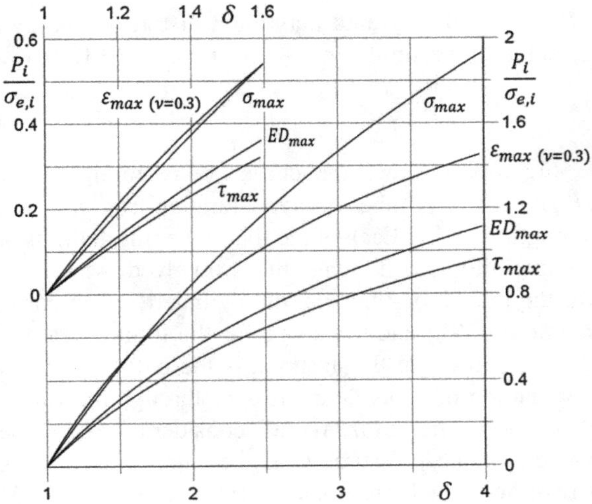

Fig. 9.8 Ratios $p_i/\sigma_{e,i} = f(\delta)$ for a multilayer structure with $n = 3$ calculated with the four σ_{max}, $\varepsilon_{max}(\nu = 0.3)$, τ_{max} and maximum distortion energy strength theories

strength theories (in this case with $\nu = 0.3$) and relation (9.120) for the τ_{max} and maximum distortion energy theories. Obviously, the values of factors a and b are those given in Table 4.2.

Figure 9.8 highlights the fact that, for the multilayer structure as for the monobloc-type circular cylinder, the most conservative strength theory is once again the τ_{max} theory. For this composite structure, it can be readily demonstrated on the basis of the relations given above that we have $\delta_{\tau_{max}} > \delta_{ED_{max}} > \delta_{\varepsilon_{max}}$ in all cases for any given ratio $p_i/\sigma_{e,i}$. The τ_{max} strength theory is thus that which, again for any given ratio $p_i/\sigma_{e,i}$, results in maximum thickness. It is equally easy to locate the points where the curve for the σ_{max} strength theory intersects the remaining three curves, and which delimit the area where prudence would dictate the use of one theory rather than another.

As regards the impact of the number of layers n, it should be pointed out that there is a limit of the ratio $p_i/\sigma_{e,i}$ for every diameter ratio δ or, equivalently, for every ratio $s/r_i = \delta - 1$. Thus, in calculating the limit for $n \to \infty$, relation (9.118) yields:

$$\delta_{n=\infty} = \left(1 + \frac{ap_i}{\sigma_{e,i}}\right)^{\frac{a+b}{2a}}. \tag{9.122}$$

For the σ_{max} strength theory, where $a = b = 1$, this relation particularizes to the following equation, which is the equation of a line

$$\delta_{n=\infty} = 1 + \frac{p_i}{\sigma_{e,i}}. \tag{9.123}$$

Where $a = 0$ (τ_{max} and maximum distortion energy strength theories), and again calculating the limit for $n \to \infty$, relation (9.120) yields:

$$\delta_{n=\infty} = e^{\frac{bp_i}{2\sigma_{e,i}}}. \tag{9.124}$$

Figure 9.9 shows the curves representing the function $p_i/\sigma_{e,i} = f(\delta - 1) = f(s/r_i)$ with a number of layers ranging from $n = 1$ to $n = \infty$ for each of the four strength theories. Each of the four diagrams delimits the domain of variation, and of compatibility, of the magnitudes involved, which is clearly bounded by the curve for the monobloc-type circular cylinder [which corresponds to that obtained from relation (4.40)] and the curve for the limiting case of a multilayer structure with $n = \infty$, which, in the abstract, is the only structure where an absolutely uniform distribution of equivalent stress σ_e through the wall thickness can be achieved.

It can be inferred from these considerations that the following purely theoretical values for the pressures that guarantee a uniform distribution of the equivalent stress, and which are obtained from relations (9.123) and (9.124) respectively

$$p_i = \sigma_{e,i}(\delta - 1) \quad \text{and} \quad p_i = \frac{\sigma_{e,i}}{b} ln\delta^2 \tag{9.125}$$

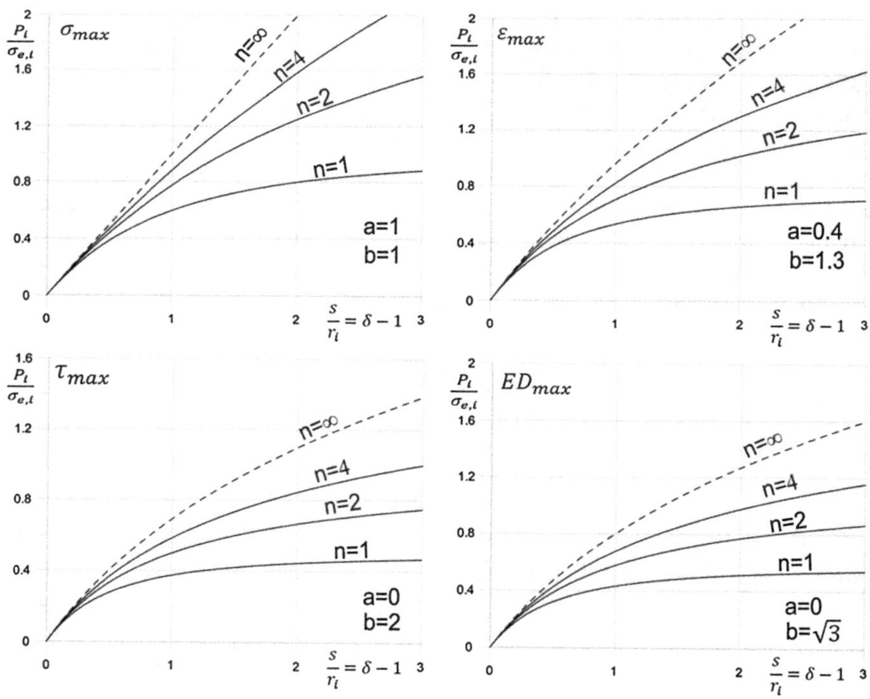

Fig. 9.9 Ratios $p_i/\sigma_{e,i} = f(\delta - 1) = f(s/r_i)$ for a multilayer structure calculated with the four strength theories considered, and comparison with the monobloc-type circular cylinder

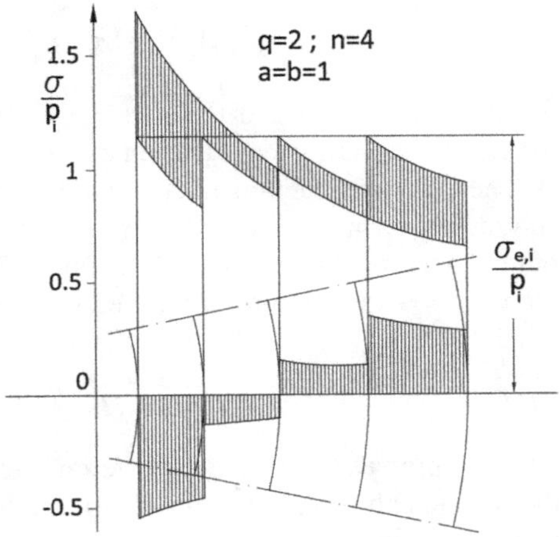

Fig. 9.10 Optimal distribution of equivalent stress and distributions resulting from shrink fitting and the pressure on the circular cylinder, here regarded as single-walled

are formally identical to those that bring the wall thickness of the circular cylinder to the fully plastic state, which are obtained from the second relations (9.63) and (9.70), setting $\sigma_{e,i} = \sigma_s$ in relations (9.125).

The state of coaction in the multilayer circular cylinder produced by shrink fitting must be such that it, together with that produced by the pressure p_i acting at the inner radius of the cylinder (which for this purpose is regarded as single-walled), results in the optimal distribution of equivalent stress shown in Fig. 9.10. It is thus necessary to determine the interferences i, or rather, the dimensionless ratios $i/r_{e,m}$, that make it possible to achieve this important goal.

For the pressures, we have the following relations:

$$\left(\sigma_{e,i}\right)_{m+1} = a\frac{p_{e,m} - \alpha p_{e,m+1}}{\alpha - 1} \tag{9.126}$$

and

$$\left(\sigma_{e,e}\right)_m = a\frac{p_{i,m}\left[1 - (\alpha - 1)\frac{b}{a}\right] - p_{e,m}\left[\alpha - (\alpha - 1)\frac{b}{a}\right]}{\alpha - 1}. \tag{9.127}$$

The first of these relations, which is fairly obvious, can be inferred from the first relation (9.99) recalling the second member of relation (9.100), while the second is obtained from the second relation (9.99). Once $p_{e,m+1}$ has been determined from relation (9.106), we then use relations (9.126) and (9.127) to arrive at the following after a few manipulations:

$$\left(\sigma_{e,i}\right)_{m+1} - \left(\sigma_{e,e}\right)_m = b\left(p_{i,m} - p_{e,m}\right). \tag{9.128}$$

Consequently, given relations (9.114) and (9.115), the stress difference given by the relation:

$$\left(\sigma_{e,i}\right)_{m+1} - \left(\sigma_{e,e}\right)_m = p_i \frac{b(\alpha - 1)}{\alpha^n - 1} \alpha^{n-m}, \tag{9.129}$$

which is invariable and equal to bp_i/n for $a = 0$, will be reduced as the interference radius is increased when $a \neq 0$.

The state of coaction is characterized by this stress difference, as a common subtractive term expresses the stress in the single-walled circular cylinder. On the other hand, on the basis of relation (4.71), which applies to all four strength theories considered here, the difference between the hoop stresses in the relation

$$\frac{i}{r_{e,m}} = \frac{1}{E}\left(\sigma_{t,r_i,m+1} - \sigma_{t,r_e,m}\right), \tag{9.130}$$

which expresses the specific interference to be produced between the mth layer and the $(m + 1)$th layer, is linked to the difference between the corresponding equivalent stresses by the relation

$$\sigma_{t,r_i,m+1} - \sigma_{t,r_e,m} = \frac{2}{a+b}\left[\left(\sigma_{e,i}\right)_{m+1} - \left(\sigma_{e,e}\right)_m\right]. \tag{9.131}$$

Substituting relation (9.129) into (9.131) and the relation thus obtained into relation (9.130) then yields

$$\frac{i}{r_{e,m}} = \frac{2bp_i}{E(a+b)} \frac{\alpha - 1}{\alpha^n - 1} \alpha^{n-m} \tag{9.132}$$

which applies for $m \leq n - 1$.

The specific interference $i/r_{e,m}$ is invariable and equal to $2p_i/En$ when the τ_{max} and maximum distortion energy strength theories are used. When the σ_{max} and ε_{max} strength theories are employed, this specific interference is found to decrease as m is increased. The extent of this decrease is such that $\left(i/r_{e,m}\right)_{\sigma_{max}} < \left(i/r_{e,m}\right)_{\varepsilon_{max}}$, which is in line with what we have seen regarding the physical meaning of constant α.

Chapter 10
Bending Theory of Circular Cylindrical Shells Under Axisymmetric Loads

10.1 General Considerations and Fundamental Relations

In addition to internal pressure, circular cylindrical shell structures are often subject to concentrated or localized bending moments and forces of varying nature (external loads, loads arising from the interaction between structural components of differing stiffness, loads due to constraint reactions, and so forth), distributed symmetrically around the rotational axis. This occurs in pressurized aeronautical and aerospace structures consisting of assembled axisymmetric substructures with different geometries, in cylindrical pressure vessels with closures consisting of flat-plate closures or formed heads, in boilers and heat exchangers, and in tubular shell structures with stiffening rings spaced at a predetermined distance along the axis, such as in the ribbed monocoque fuselages, structures and hulls used in the aeronautical, shipbuilding and submarine industries, for example, or the penstocks in hydroelectric powerplants, the cylindrical vessels under external pressure reinforced with circumferential stiffeners and distillation columns in oil refineries, the vessels used in the chemical and nuclear industries, missiles, etc (see Brownell and Young 1968, Jaward and Farr 1989).

If only the membrane theory discussed in Sect. 1.6 were to be used in analyzing such structures, discontinuity areas would be found where the compatibility of deflections (displacements and rotations) would not be satisfied. At locations where the deflections are restricted, or there is a change in geometry, the membrane theory is inadequate. Bending loads (forces and moments), also called discontinuity loads, are generated in these localized areas which, together with the membrane loads, restore the compatibility of displacements and rotations. Though these discontinuity loads are local, they are sufficiently intense to cause significant alterations in the stress distributions in the immediate vicinity of the area where they are generated. In addition, these stresses are localized over a small area of the shell, and dissipate rapidly along the structure (Houghton 1960, Meyer 1961).

To determine the relations governing the stress and strain states of these structures, we will assume that the intensity of the loads is such that the material works in the linear elastic range and that the displacements are small compared to

V. Vullo, *Circular Cylinders and Pressure Vessels*, Springer Series in Solid and Structural Mechanics 3, DOI: 10.1007/978-3-319-00690-1_10,

the radius. We will consider a circular cylinder under internal pressure p_i and having a geometry of revolution, small thickness s such that the cylinder can be regarded as thin-walled as defined in Chap. 1, and mean radius R (Fig. 10.1a and b). As the thickness is small, the inner radius r_i, outer radius r_e and mean radius $r_m = R$ will frequently be regarded as equivalent in the relations given below. The thin-walled circular cylinder is then a circular cylindrical shell.

With reference to the middle surface, the circular cylindrical shell considered here is generated by moving a straight line (also called the meridian line) along the circumference having radius $r_m = R$, while maintaining this meridian line parallel to its original direction. The infinite straight lines related to this motion of generation are called the generators. All planes normal to the generators intersect the middle surface in identical circumferences which are called directrices or profiles. The middle surface of a circular cylindrical shell is thus a developable surface, i.e., a surface with zero Gaussian curvature (see Gauss 1828).

As illustrated in Fig. 10.1a and b, let $O(r, t, z)$ be the three-dimensional rectangular Cartesian reference system having its origin at the left end of a cylinder generator, where the r axis is along a radial line whose positive direction is centripetal, the t axis is in the cross-sectional plane and directed tangentially to the directrix, and the z axis coincides with the generator and is positive if directed from left to right. Unlike the reference system used in Chap. 1, the z axis thus does not coincide with the rotational axis and the r axis is positive if directed inward. This reference system is used here in compliance with the usual practice employed with the theory of shells of revolution. When the formulas presented in the preceding chapters are used, care must thus be taken with the signs.

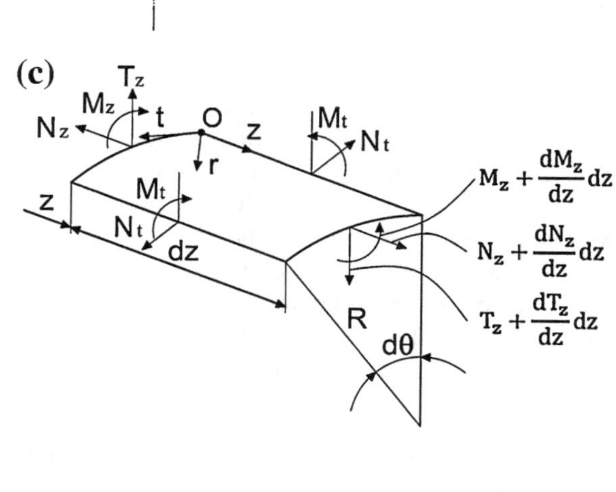

Fig. 10.1 Axisymmetric circular cylindrical shell and Cartesian reference system in cross-sectional (**a**) and side view (**b**); **c** small element of the cylindrical shell and non-zero stress resultants

We will then consider (Fig. 10.1c) a small element of the circular cylindrical shell bounded by two axial planes having angular coordinates θ and $\theta + d\theta$, which consequently form the angle $d\theta$ between them, and by two parallel planes perpendicular to the rotational axis, of abscissa z and $z + dz$ from the origin. This small element is thus bounded by two adjacent generators θ and $\theta + d\theta$ and by two adjacent directrices z and $z + dz$. As R is the mean radius of the cylindrical structure, the small element in question has axial length dz, while the dimension in the tangential direction (transverse direction) is $R\,d\theta$.

A circular cylindrical shell structure with free closed ends will undergo axisymmetric deformation if the external load, the constraint conditions and hence the boundary conditions, and the material's mechanical properties are symmetrical with respect to the axis. The material must thus be homogeneous and isotropic (this will be the only case examined below) or axisymmetrical orthotropic. In addition, the boundary conditions must be independent of the angular coordinate θ, which in a cylindrical coordinate system defines the position of a generic axial plane relative to another axial plane taken as a reference. In this condition, of the three displacements u, v and w in the direction of the coordinate axes r, t, z, which are assumed to be positive if they are in the same direction as the latter, the tangential displacement v will be zero ($v = 0$), while the other displacements, the rotations, the strains, the internal forces and the bending moments will be a function only of axial coordinate z.

As we know from the standard texts (Timoshenko and Woinowsky-Krieger 1959, Tsui 1968, Baker et al. 1968, Novozhilov 1970, Cicala 1971, Flügge 1973, Ventsel and Krauthammer 2001), in the most general non-axisymmetric conditions, the small element of the circular cylindrical shell is subject to ten stress resultants per unit length, four in-plane $(N_z, N_t, N_{zt}, N_{tz})$ and six out-of-plane $(T_z, T_t, M_z, M_t, M_{zt}, M_{tz})$. The in-plane stress resultants are also termed membrane stress resultants. Imposing equilibrium conditions, it can be inferred that, as a result of axisymmetry, the membrane shearing forces N_{zt} and N_{tz} must be zero $(N_{zt} = N_{tz} = 0)$, the transverse shearing forces T_t acting on the sides of length dz must be zero $(T_t = 0)$ and the twisting moments M_{zt} and M_{tz} must also be zero $(M_{zt} = M_{tz} = 0)$. As five out of the ten stress resultants are thus zero, viz., N_{zt}, N_{tz}, T_t, M_{zt}, M_{tz}, they are not shown in Fig. 10.1c.

Again because of the combined effect of axisymmetry and the equilibrium conditions, the in-plane circumferential forces N_t and the circumferential bending moments M_t are independent of the angular coordinate θ and thus are constant along the directrix circumference and vary only with axial coordinate z. In the symmetry conditions considered here, three of the six equilibrium conditions of the small element of our circular cylindrical shell are identically satisfied: that for the resultant in the tangential direction t and those for the moment about coordinate axes r and z. We must thus consider, and write, the three remaining equilibrium conditions, which are those obtained by projecting the forces on the direction of the r and z axes and by taking the moments of the forces about the t axis.

In the case where the only external forces are those due to internal pressure p_i, these three equilibrium equations are:

$$
\begin{aligned}
&\frac{dN_z}{dz} R\,dz\,d\theta = 0 \\
&\frac{dT_z}{dz} R\,dz\,d\theta + N_t\,dz\,d\theta - p_i\,R\,dz\,d\theta = 0 \\
&\frac{dM_z}{dz} R\,dz\,d\theta - T_z\,R\,dz\,d\theta = 0,
\end{aligned}
\tag{10.1}
$$

where N_z is the axial membrane force, T_z is the transverse shearing force, and M_z is the bending moment acting on an axial plane.

Integrating the first Eq. (10.1) gives $N_z = const$. The in-plane meridian stress resultant, which coincides with the axial stress resultant, is thus constant. Here, it is convenient to assume that this constant is zero, i.e., that $N_z = 0$. Obviously, if $N_z = const \neq 0$ (this occurs, for example, because of the internal pressure acting on the closures), the stress and strain states correlated with this non-zero axial force will be calculated separately and added to those due to the radial load acting on our shell structure. As we have assumed that we are in the linear elastic range, in fact, the method of superposition applies.

Dividing all members by $Rdzd\theta$ (which represents the area of a differential element of the cylindrical shell's middle surface), the other two Eq. (10.1) can be written in the following form:

$$
\begin{aligned}
&\frac{dT_z}{dz} + \frac{N_t}{R} = p_i \\
&\frac{dM_z}{dz} - T_z = 0.
\end{aligned}
\tag{10.2}
$$

These equations contain three unknown quantities: N_t, T_z and M_z. To solve the problem, which is clearly hyperstatic, it is necessary to consider the kinematic relations that link the strains ε_t and ε_z in the circumferential and axial directions and the changes in curvatures χ_t and χ_z of the middle surface in the same directions to displacements u, v and w. As was mentioned earlier, $v = 0$ because of symmetry, i.e., the tangential displacement is zero. And as a result of what we saw in Sect. 1.1, radial strain ε_r is zero, while hoop strain ε_t is $-u/R$, given that here the r axis is positive in the centripetal direction. Axial strain ε_z is readily calculated since, as can be seen from Fig. 10.1c, the left end of the small element of the shell considered here moves by w, while the right end moves by $w + dw$. Consequently, the element's initial length dz expands by an amount equal to dw; we thus have $\varepsilon_z = dw/dz$. The kinematic relations that link ε_t and ε_z to displacements u and w are thus:

$$
\varepsilon_t = -\frac{u}{R} \qquad \varepsilon_z = \frac{dw}{dz}.
\tag{10.3}
$$

Noting that σ_r is negligible and that $\sigma_t = N_t/s$ and $\sigma_z = N_z/s$, Hooke's law for the biaxial stress state, with $N_z = 0$, leads to the following relations:

$$\begin{aligned} N_z &= \frac{E\,s}{1-\nu^2}(\varepsilon_z + \nu\varepsilon_t) = \frac{E\,s}{1-\nu^2}\left(\frac{dw}{dz} - \nu\frac{u}{R}\right) = 0 \\ N_t &= \frac{E\,s}{1-\nu^2}(\varepsilon_t + \nu\varepsilon_z) = \frac{E\,s}{1-\nu^2}\left(-\frac{u}{R} + \nu\frac{dw}{dz}\right). \end{aligned} \tag{10.4}$$

From the first of these relations, we can infer

$$\frac{dw}{dz} = \nu\frac{u}{R}. \tag{10.5}$$

Substituting the latter into the second relation (10.4) gives:

$$N_t = -\frac{E\,s\,u}{R}. \tag{10.6}$$

If we accept the Donnell (1933), Mushtari (1938), Vlasov (1964) theory of thin shells of any geometric shape—the so-called DMV theory, confirmed by Gol'denweizer (1961) and Novozhilov (1964)—which is based on the Kirchhoff (1850) and Love (1944) hypotheses for thin plates but adds a static assumption (the effects of the transverse shearing forces T_z and T_t in the in-plane equilibrium equations are negligible) and a geometric assumption (the influence of the out-of-plane radial displacement u will predominate over the influence of the in-plane displacements v and w in the bending response of the shell), the bending moments M_z and M_t are linked to the changes in curvatures of the middle surface χ_z and χ_t in the directions of the z- and t-coordinate lines respectively in the meridian or axial plane (rz plane) and in the cross-sectional plane (rt plane) by the following relations:

$$\begin{aligned} M_z &= -D(\chi_z + \nu\,\chi_t) \\ M_t &= -D(\chi_t + \nu\,\chi_z), \end{aligned} \tag{10.7}$$

where $D = Es^3/12(1-\nu^2)$ is the shell's flexural rigidity. It should be noted that, in this case, the twist χ_{zt} of a differential element $Rdzd\theta$ of the middle surface due to shell bending is zero.

However, for the symmetrically loaded circular cylindrical shell considered here under the assumption of small displacements, there are no changes of curvature in the cross-sectional plane, so that $\chi_t = 0$. With the usual approximation of neglecting the square of the first derivative dw/dz relative to unity and keeping the sign conventions in mind, we can infer that the curvature that any generator will assume in the rz plane after deformation will be equal to d^2u/dz^2. Consequently, the geometric relations linking the change in curvatures to displacements are as follows:

$$\chi_t = 0 \qquad \chi_z = \frac{d^2u}{dz^2}. \tag{10.8}$$

Substituting these relations into relations (10.7) yields:

$$M_z = -D\frac{d^2u}{dz^2}$$
$$M_t = \nu M_z. \tag{10.9}$$

Eliminating T_z, relations (10.2) give

$$\frac{d^2M_z}{dz^2} + \frac{N_t}{R} = p_i; \tag{10.10}$$

this equation, recalling the first relation (10.9) and relation (10.6), yields:

$$\frac{d^2}{dz^2}\left(D\frac{d^2u}{dz^2}\right) + \frac{Es}{R^2}u = -p_i. \tag{10.11}$$

All problems of symmetrical deformation of circular cylindrical shells reduce to the integration of differential equation (10.11). If the shell in question is of constant thickness, we have $D = const$, and Eq. (10.11) becomes:

$$D\frac{d^4u}{dz^4} + \frac{Es}{R^2}u = -p_i. \tag{10.12}$$

By introducing the notation

$$\beta^4 = \frac{Es}{4R^2D} = \frac{3(1-\nu^2)}{R^2s^2}, \tag{10.13}$$

where β is a geometric parameter, also depending on the Poisson's ratio of the material, having the dimension of the reciprocal of a length (in the SI system, m^{-1}), Eq. (10.12) can be transformed into:

$$\frac{d^4u}{dz^4} + 4\beta^4u = -\frac{p_i}{D}. \tag{10.14}$$

This non-homogeneous fourth order differential equation, like its equivalent given by relations (10.12), is the governing differential equation for a symmetrically loaded circular cylindrical shell. The associated homogeneous equation, on the other hand, is the governing differential equation for the same shell subjected to concentrated bending moments and forces distributed symmetrically relative to the rotational axis.

Lastly, by taking the first relation (10.9) into account, we see from the second relation (10.2) that shear stress resultant T_z can be expressed by the following:

$$T_z = \frac{dM_z}{dz} = -D\frac{d^3u}{dz^3}. \tag{10.15}$$

It is readily demonstrated that differential equation (10.14) formally coincides with the differential equation governing the deflection of an elemental longitudinal strip-beam (a Winkler beam, Winkler 1867) bounded on the cylindrical structure

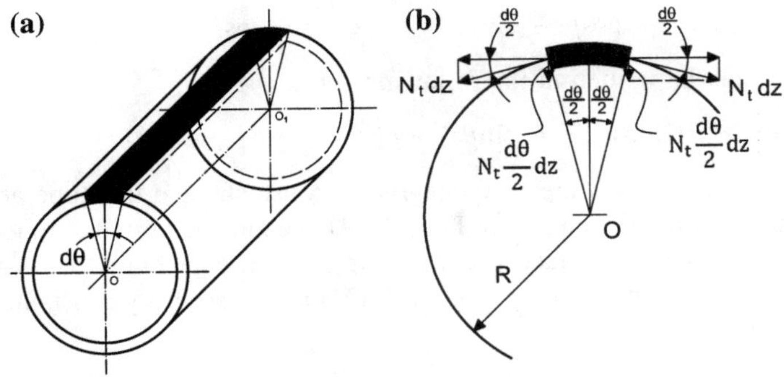

Fig. 10.2 **a** Elemental longitudinal strip-beam of the cylindrical shell; **b** Circumferential membrane forces $N_t\,dz$ exerted by the strip-beam on the remaining part of the circular cylindrical shell

by two axial planes having angular coordinates θ and $\theta + d\theta$ (Fig. 10.2a) and whose long sides rest continuously on equally continuous supports which act as a continuous elastic Winkler foundation. This foundation is thus the remaining part of the circular cylindrical shell, which resists the beam's deflection thanks to the reactive circumferential membrane forces N_t (Fig. 10.2b). In the direction of the normal to the elemental strip-beam, which coincides with the r axis, the resultant of these forces is equal to $2N_t\,dz\sin(d\theta/2) = N_t\,dz\,d\theta$, which can be interpreted as the reactive force exerted by the elastic foundation on the elemental strip-beam.

The general solution of Eq. (10.14), which is a non-homogeneous fourth order differential equation with constant coefficients, can be found by adding the complementary solution u_h of the associated homogeneous equation and a particular integral u_p of the complete equation, i.e., the Eq. (10.14) with the right-hand side. The complementary solution of the associated homogeneous equation is found from Eq. (10.14), omitting the term in the right-hand side and setting $u = Ce^{az}$, where C and a are constants. Substituting this expression for u into Eq. (10.14), together with its fourth derivative, leads to equation $a^4 + 4\beta^4 = 0$, which admits four roots given by the relation $a_{1,2,3,4} = \sqrt[4]{-4\beta^4} = \sqrt{2}\beta\sqrt[4]{-1}$. Since $\sqrt[4]{-1}$ has four roots given by $(\pm 1 \pm i)/\sqrt{2}$, the four roots of the equation $a^4 + 4\beta^4 = 0$ can be expressed as: $a_{1,2} = (1 \pm i)\beta$ and $a_{3,4} = -(1 \pm i)\beta$. Recalling that $i = \sqrt{-1}$, the complementary solution of the homogeneous equation associated with Eq. (10.14) can be expressed in the following form:

$$u_h = e^{\beta z}\left(A_1 e^{i\beta z} + A_2 e^{-i\beta z}\right) + e^{-\beta z}\left(A_3 e^{i\beta z} + A_4 e^{-i\beta z}\right) \tag{10.16}$$

where A_1, A_2, A_3 and A_4 are complex integration constants. Remembering, moreover, that on the basis of Euler's relations, $\cos\beta z = \left(e^{i\beta z} + e^{-i\beta z}\right)/2$ and $\sin\beta z = \left(e^{i\beta z} - e^{-i\beta z}\right)/2$ (Abramowitz and Stegun 1972, Kreyszig 1993), the exponentials appearing in brackets in Eq. (10.16) can be expressed in terms of trigonometric functions, i.e., that

$$e^{i\beta z} = \cos\beta z + \sin\beta z \qquad e^{-i\beta z} = \cos\beta z - \sin\beta z, \tag{10.17}$$

Eq. (10.16) can be written in the following form:

$$u_h = e^{\beta z}(C_1\cos\beta z + C_2\sin\beta z) + e^{-\beta z}(C_3\cos\beta z + C_4\sin\beta z), \tag{10.18}$$

where C_1, C_2, C_3 and C_4 are new integration constants, and this time are real.

To find the particular integral of Eq. (10.14), we must identify the function $p_i = p_i(z)$ expressing how the surface force p_i acting on the circular cylindrical shell varies with z. If p_i is constant, setting $u = C$, with C being an arbitrary constant, gives:

$$u_p = -\frac{p_i}{4\beta^4 D} = -\frac{p_i R^2}{E s}. \tag{10.19}$$

We reach the same conclusion if the surface force $p_i$varies with z according to a power function whose exponent is not greater than three. In cases where p_i varies with z according to more complex functions, the particular solution of Eq. (10.14) can be found by using Lagrange's method of variation of arbitrary constants (Lagrange 1811, see also Tricomi 1967, Smirnov 1970 and 1972, Amerio 1977, Forsyth 1996), also called method of variation of parameters. It should be noted that the particular integral represents, in terms of radial displacement, the solution that can be obtained by assuming membrane loading for the shell, i.e., by considering the structure as a freely deformable membrane shell, without constraints. Obviously, this solution often fails to satisfy the boundary conditions specified for the shell. The complementary solution of the homogeneous equation is thus the correction to be made to the membrane solution in order to ensure that the specified boundary conditions are satisfied by the sum $u_h + u_p$.

To fully define the general solution of the differential equation (10.14), i.e., to calculate displacement u as the sum

$$u = u_h + u_p, \tag{10.20}$$

it is necessary to determine the integration constants C_1, C_2, C_3 and C_4 that appear in relation (10.18). To this end, we must impose the boundary conditions at the two edges (the two end faces) bounding the circular cylindrical shell longitudinally. In general, as two boundary conditions can be imposed for each edge, the problem of determining the integration constants is perfectly defined. A few of the cases that are most frequently encountered in designing structures of this kind will be analyzed below.

10.2 Long Circular Cylindrical Shells Subjected to Concentrated Loads on One Edge

We will consider a long circular cylindrical shell (a long circular pipe) loaded by bending moments M_0 and shear forces T_0 distributed uniformly at one of its edges, that having axial coordinate $z = 0$ (Fig. 10.3). We will also assume that these are

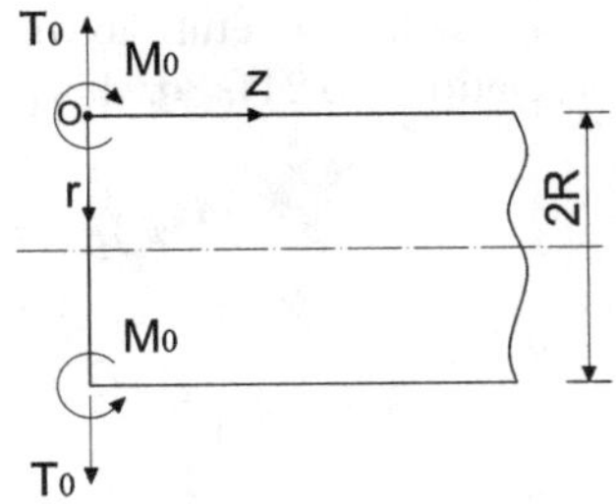

Fig. 10.3 Long circular cylindrical shell subject to M_0 and T_0 at one edge

the only loads acting on the structure, and that the loaded edge is free of all constraints.

As in this case $p_i = 0$, the particular integral of equation (10.14) will be zero, i.e., $u_p = 0$. Regarding the complementary solution of the homogeneous equation associated with Eq. (10.14), given by Eq. (10.18), it is clear that the second function at the right-hand side (the exponentially damped trigonometric function), will die out rapidly as βz increases, while the first function at the right-hand side, i.e., the exponentially amplified trigonometric function, will show an equally rapid rise along with βz. As the loads applied at the edge having axial coordinate $z = 0$ cause local bending that is rapidly damped out as βz increases, i.e., as the distance z from the loaded edge increases, the first function in the second member of Eq. (10.18) must vanish. In other words, the first two integration constants must be zero, and we will thus have $C_1 = C_2 = 0$. Consequently, the complementary solution of the homogeneous equation is given by the relation:

$$u = u_h = e^{-\beta z}(C_3 \cos\beta z + C_4 \sin\beta z). \tag{10.21}$$

Before clarifying what we mean by a *long* thin-walled circular cylinder, we will begin by asking what influence the axial length of the structure has, and what problems it poses in calculating the integration constants. Let L be the axial length of our circular cylindrical shell. Equation (10.21) describes the displacement field $u = u_h$ in the immediate vicinity of the loaded edge, i.e., for small values of z. As will be recalled, we have assumed $p_i = 0$ and thus $u_p = 0$, but the following considerations would still apply if $p_i \neq 0$ and thus $u_p \neq 0$.

To determine the displacement field at the other non-loaded edge of the circular cylindrical shell in question, that having coordinate $z = L$, and in its immediate vicinity, we must use relation (10.18) rather than (10.21), as the latter holds only in the immediate neighborhood of $z = 0$. When relation (10.18) is used, however, the exponentially amplified trigonometric function cannot be neglected given that the displacement values associated with it are very high. To continue to use relation (10.21), we must change reference system, i.e., we must set the origin of the z axis at the edge opposite to that which we considered in calculating the integration constants C_3 and C_4 correlated with the boundary conditions at the edge having abscissa $z = 0$. In this way, we calculate two new values of the integration constants C_3 and C_4, in this case correlated with the boundary conditions at the edge having abscissa $z = L$.

We will now return to calculating the integration constants C_3 and C_4 appearing in relation (10.21). For this purpose, we impose the following boundary conditions:

$$\begin{aligned}
(M_z)_{z=0} &= -D\left(\frac{d^2u}{dz^2}\right)_{z=0} = M_0 \\
(T_z)_{z=0} &= \left(\frac{dM_z}{dz}\right)_{z=0} = -D\left(\frac{d^3u}{dz^3}\right)_{z=0} = T_0.
\end{aligned} \tag{10.22}$$

Deriving relation (10.21) three times, substituting the second derivative and the third derivative into the first and second relations (10.22) respectively and setting $z = 0$ in them gives:

$$C_3 = -\frac{1}{2\beta^3 D}[T_0 + \beta M_0] \qquad C_4 = \frac{M_0}{2\beta^2 D}. \tag{10.23}$$

Consequently, relation (10.21) becomes:

$$u = -\frac{e^{-\beta z}}{2\beta^3 D}[\beta M_0(\cos\beta z - \sin\beta z) + T_0\cos\beta z]. \tag{10.24}$$

Deriving relation (10.24), again three times, and taking relations (10.9), (10.15) and (10.6) into account, we obtain the following relations which express the rotation φ of the normal to the cylinder's generator as a function of z, as well as the stress resultants acting on the structure in question:

$$\begin{aligned}
\varphi &= \frac{du}{dz} = \frac{e^{-\beta z}}{2\beta^2 D}[2\beta M_0\cos\beta z + T_0(\cos\beta z + \sin\beta z)] \\
M_z &= -D\frac{d^2u}{dz^2} = \frac{e^{-\beta z}}{\beta}[\beta M_0(\cos\beta z + \sin\beta z) + T_0\sin\beta z] \\
M_t &= \nu M_z = \frac{\nu e^{-\beta z}}{\beta}[\beta M_0(\cos\beta z + \sin\beta z) + T_0\sin\beta z] \\
T_z &= -D\frac{d^3u}{dz^3} = -e^{-\beta z}[2\beta M_0\sin\beta z - T_0(\cos\beta z - \sin\beta z)] \\
N_t &= -\frac{E\,s\,u}{R} = 2\beta R e^{-\beta z}[\beta M_0(\cos\beta z - \sin\beta z) + T_0\cos\beta z].
\end{aligned} \tag{10.25}$$

The normal stresses are given by the relations

$$\begin{aligned}
\sigma_t &= \frac{N_t}{s} + \frac{12M_t}{s^3}r \\
\sigma_r &= \frac{12M_z}{s^3}r
\end{aligned} \tag{10.26}$$

and will be maximum at the inner and outer radius, i.e., for $r = \pm s/2$. Obviously, if the structure is also under internal pressure p_i, the membrane stresses given by the first relations (1.7) and relation (1.13) must be added to the stresses given by relation (10.26).

Relations (10.24) and (10.25) univocally define the displacement field and stress resultants in a long circular cylindrical shell subjected to concentrated loads (bending moment and shear force) applied at one edge. All six quantities defined by these relations are exponentially damped trigonometric functions, since the following functions of βz are exponentially damped

$$\begin{aligned} \phi_1 &= e^{-\beta z}(\cos\beta z + \sin\beta z) \\ \phi_2 &= e^{-\beta z}(\cos\beta z - \sin\beta z) \\ \phi_3 &= e^{-\beta z}\cos\beta z \\ \phi_4 &= e^{-\beta z}\sin\beta z \end{aligned} \quad (10.27)$$

These functions are shown versus βz in Fig. 10.4. Their wavelength λ, i.e., the distance between two points having the same phase in two consecutive cycles of the periodic function, along the direction of the z-axis, is given by the period of functions $\sin\beta z$ and $\cos\beta z$ and is thus:

$$\begin{aligned} \lambda &= \frac{2\pi}{\beta} = 2\pi\sqrt[4]{\frac{4R^2D}{Es}} = 2\pi\sqrt[4]{\frac{R^2s^2}{3(1-\nu^2)}} \\ \lambda &= \frac{2\pi}{\beta} \cong 4.88\sqrt{Rs}\,(\text{for } \nu = 0.3). \end{aligned} \quad (10.28)$$

As can be seen from Fig. 10.4, all four of these functions are practically damped out in one half-wavelength, i.e., for $\beta z = \pi$. The maximum values of

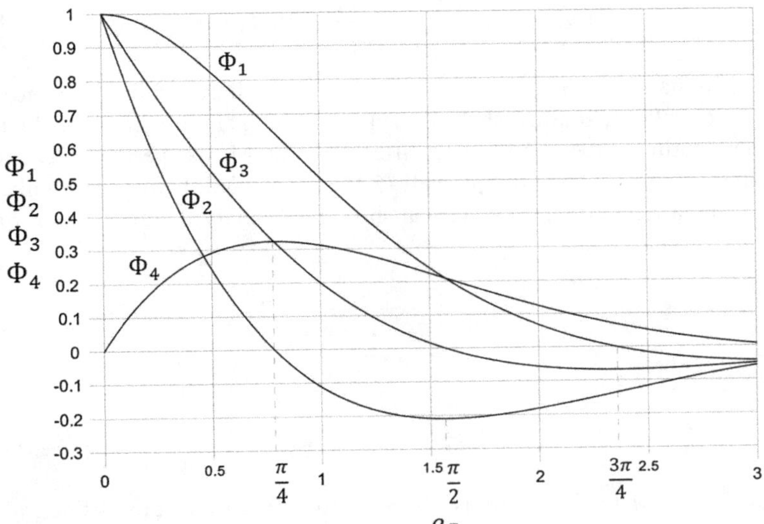

Fig. 10.4 Functions ϕ_1, ϕ_2, ϕ_3 and ϕ_4 versus βz in the domain $0 \leq \beta z < \pi$

functions ϕ_1, ϕ_2 and ϕ_3 are unitary and are found for $\beta z = 0$, while $\phi_4 = 0$, again for $\beta z = 0$. The maximum value of ϕ_4, on the other hand, occurs with $\beta z = \pi/4$ and is approximately 0.32. The zeros of these functions, like the zeros of their first derivatives, are spaced at intervals π of variable βz. The successive maxima and minima of these functions are within each of these intervals and are out of phase with their respective zeros by $\pi/4$. For example, function $\phi_3 = e^{-\beta z}\cos\beta z$ has a first zero for $\beta z = \pi/2$, a first minimum for $\beta z = 3\pi/4$, a subsequent zero for $\beta z = 3\pi/2$, a subsequent maximum for $\beta z = 7\pi/4$, and so forth. It can be readily demonstrated that the ratio of the absolute values of a maximum and the subsequent minimum is equal to $e^{-\pi} \cong 0.043$.

The spacing between two successive zeros of these functions on the axis of the abscissa, like that between a maximum and the subsequent minimum, is one half-wavelength, i.e., for steels, ($\nu = 0.3$), 2.44 $\sqrt{Rs}$. For any given radius R, it thus decreases along with thickness. Consequently, the smaller the structure's thickness, the smaller the region affected by localized loads T_0 and M_0 will be.

As can be seen from relation (10.24), the maximum radial displacement takes place with $\beta z = 0$, i.e., at the loaded edge, and is

$$(u)_{z=0} = -\frac{1}{2\beta^3 D}[\beta M_0 + T_0] \tag{10.29}$$

where the minus sign indicates that it is in the direction opposite to the positive direction established for the radial coordinate r. The slope of the deflection curve, which is equal to the rotation of the normal to the generator, is maximum for $z = 0$ and, as can be inferred from the first relation (10.25), is:

$$(\varphi)_{z=0} = \left(\frac{du}{dz}\right)_{z=0} = \frac{1}{2\beta^2 D}[2\beta M_0 + T_0]. \tag{10.30}$$

On the basis of the foregoing considerations, we can now clarify what we mean by a *long* circular cylindrical shell. In fact, once the quantities calculated with the procedure outlined above have been used to establish the permissible error for calculating the stress state in the shell (5 %, for example), we can readily demonstrate that a long circular cylindrical shell must satisfy the disequation $\beta L \geq \pi$, i.e.:

$$L \geq \frac{\pi}{\beta} = \frac{\pi}{\sqrt[4]{3(1-\nu^2)}}\sqrt{Rs}, \tag{10.31}$$

and, for steels ($\nu = 0.3$),

$$L \geq \frac{\pi}{\beta} \cong 2.44\sqrt{Rs}. \tag{10.32}$$

Shells that satisfy disequations (10.31) and (10.32) are thus defined as *long circular cylindrical shells* or *semi-infinite cylindrical shells.* They can be treated on the basis of approximate relation (10.21) with an error of not more than 5 %.

10.3 Long Circular Cylindrical Shells with Constrained Edges Under Uniform Internal Pressure

We will now consider a circular cylindrical shell with built-in (Fig. 10.5a) or simply supported edges (Fig. 10.5b) under uniform internal pressure p_i. We will assume that these constraints are idealized constraints capable of absorbing the axial load due to internal pressure p_i, and thus of relieving the structure of the longitudinal stress σ_z correlated with this pressure.

If the edges of the structure in question were free, at each cross-section and regardless of the distance from the edges, there would be a hoop stress given by the first relation (1.7) and an increase in radius u_R, i.e., a radial displacement of the middle generator that, on the basis of relation (1.20) with $\sigma_e = \sigma_t$, or of relation (1.21), setting $r_i \cong R$ and ignoring the contribution of axial stress σ_z, can be expressed in absolute value as:

$$u_R = \frac{R\,\sigma_t}{E} = \frac{p_i\,R^2}{E\,s}. \tag{10.33}$$

This free radial expansion at the edges cannot take place, however, since the edges are constrained. If they are built-in (Fig. 10.5a), the constraint reaction at the absolutely rigid built-in edge (or clamped edge) is associated with a localized bending moment M_0 and shearing force T_0, which can be determined by considering the radial displacement u and the slope φ of the deflection curve to be zero at the built-in edge, i.e., for $z = 0$. Taking relations (10.29) and (10.30) into account, and equaling their second members to u_R (the algebraic sum of the displacements correlated with the internal pressure and with the hyperstatic unknowns M_0 and T_0 must be zero) and to zero respectively yields a system of two equations in two unknowns whose solution gives:

$$M_0 = 2\beta^2 D u_R = \frac{p_i}{2\beta^2};\; T_0 = -4\beta^3 D u_R = -\frac{p_i}{\beta}. \tag{10.34}$$

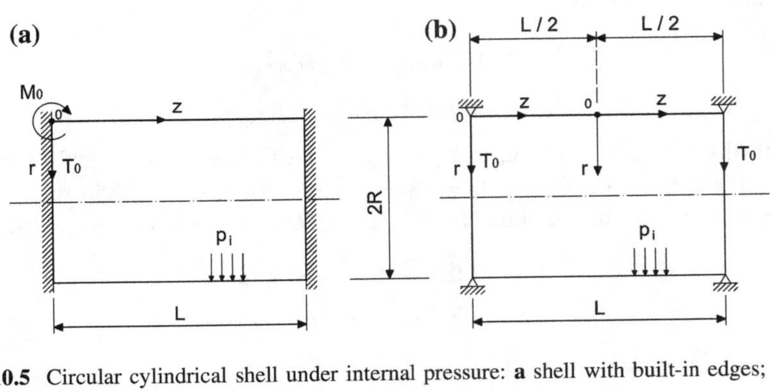

Fig. 10.5 Circular cylindrical shell under internal pressure: **a** shell with built-in edges; **b** shell with simply supported edges

The constraint reaction thus consists of a positive bending moment M_0 and a negative shearing force T_0. Lastly, substituting the values thus found for M_0 and T_0 into relations (10.24) and (10.25) yields all of the complete relations expressing the displacement field and stress resultants correlated with the edge effects, i.e., with M_0 and T_0, in the circular cylindrical shell with built-in edges under internal pressure.

If the shell has simply supported edges (Fig. 10.5b), the constraint reaction can be reduced to the localized shearing force T_0 alone, while the boundary conditions to be imposed for $z = 0$ regard the radial displacement and the bending moment, which must be zero. In this case, relation (10.29) with the second member equal to u_R and setting $M_0 = 0$ gives:

$$T_0 = -2\beta^3 D u_R. \tag{10.35}$$

Once again, substituting this value of T_0 as well as $M_0 = 0$ in relations (10.24) and (10.25) yields all of the complete relations which univocally define the displacement field and stress resultants correlated with the edge effects in the circular cylindrical shell with simply supported edges under internal pressure.

10.3.1 Calculation Example

Consider a circular cylindrical vessel under internal pressure p_i and having thickness $s_1 = 0.01$ m and mean radius $R = 1.00$ m. We will suppose that the vessel has closed ends, with the closures consisting of two circular flat plates which we will assume to be infinitely rigid for loads acting both in the middle plane and transversely to it. The end closures can thus be assumed to be built-in with the shell edges (absolutely rigid built-in edges or idealized clamped edges), as they do not permit either radial and axial displacements or rotations of the cylinder's end faces. The boundary conditions to be imposed are thus the following (the subscript 1 denotes the vessel, while the subscript 2 which will be introduced later denotes the closure):

$$\begin{aligned} (u_1)_{z=0} &= \left(u_{1,T_0,M_0}\right)_{z=0} + \left(u_{1,p_i}\right)_{z=0} = 0 \\ (\varphi_1)_{z=0} &= \left(\varphi_{1,T_0,M_0}\right)_{z=0} + \left(\varphi_{1,p_i}\right)_{z=0} = 0. \end{aligned} \tag{10.36}$$

With the data given above, and assuming that the cylinder is steel ($E = 200$ GPa and $\nu = 0.3$), we have: $D_1 = 18.315 \times 10^3$ Nm; $\beta_1 = 12.854\ \text{m}^{-1}$. Introducing these values in relations (10.34), setting $D = D_1$ and $\beta = \beta_1$, gives:

$$\begin{aligned} M_0 &= 3.026 \times 10^{-3} p_i\, \text{Nm/m} \\ T_0 &= -7.780 \times 10^{-2} p_i\, \text{N/m}. \end{aligned}$$

As a result of the internal pressure and the closure's stiffness, the circular cylindrical shell is deflected inward in the area adjacent to the built-in edge. The bending moment M_z assumes its maximum value, equal to M_0, at the built-in edge and is rapidly damped out as z increases, in accordance with the second relation (10.25). The stresses at the edge of the cylindrical shell are as follows:

$$(\sigma_z)_{z=0} = \frac{6M_0}{s_1^2}$$
$$(\sigma_t)_{z=0} = \frac{p_i R}{s_1} + \frac{6\nu M_0}{s_1^2} \tag{10.37}$$

and, in our case, will be:

$$(\sigma_z)_{z=0} = 181.560\, p_i$$
$$(\sigma_t)_{z=0} = 154.468\, p_i.$$

In the areas of the cylindrical shell far from the clamped edge, the axial and hoop stresses are respectively:

$$\sigma_z = 50.00\, p_i \qquad \sigma_t = 100.00\, p_i.$$

We will now suppose that the two end constraints are not absolutely rigid, as they permit axial displacements while still preventing radial displacements and rotations of the cylinder's end faces. In this case, the radial displacement (u_{i,p_i}) due to the internal pressure will be that given by relations (1.21), while $(\varphi_{i,p_i}) = 0$. Consequently, recalling relation (10.24) and the first relation (10.25), both calculated for $z = 0$, the above boundary conditions lead to the following system of equations in the two unknowns M_0 and T_0:

$$-\frac{1}{2\beta_1^3 D_1}[\beta_1 M_0 + T_0] - \frac{p_i R^2}{E s_1}\left(1 - \frac{\nu}{2}\right) = 0$$
$$\frac{1}{2\beta_1^2 D_1}[2\beta_1 M_0 + T_0] = 0. \tag{10.38}$$

Solving this system gives the following relations defining the hyperstatic unknowns M_0 and T_0:

$$M_0 = \frac{p_i}{2\beta_1^2}\left(1 - \frac{\nu}{2}\right)$$
$$T_0 = -\frac{p_i}{\beta_1}\left(1 - \frac{\nu}{2}\right). \tag{10.39}$$

With the values of the quantities found earlier, we obtain :

$$M_0 = 2.572 \times 10^{-3}\, p_i\, \mathrm{Nm/m}$$
$$T_0 = 6.613 \times 10^{-2}\, p_i\, \mathrm{N/m}.$$

The constraint's axial deformation (or axial compliance) thus reduces the local stress resultants due to edge effect by a factor of $(1 - \nu/2)$ and thus, for steels, by 15 %. The stresses at the clamped edge are given by the following relations:

$$\begin{aligned}(\sigma_z)_{z=0} &= \frac{p_i R}{2s_1} + \frac{6M_0}{s_1^2}\\ (\sigma_t)_{z=0} &= \frac{p_i R}{s_1} + \frac{6\nu M_0}{s_1^2}\end{aligned} \qquad (10.40)$$

and in our case will be:

$$\begin{aligned}(\sigma_z)_{z=0} &= 204.326\,p_i\\ (\sigma_t)_{z=0} &= 146.298\,p_i.\end{aligned}$$

The axial stress, however, will increase, since despite the reduction of the bending moment M_0, the effect of the pressure on the closures will come into play. In fact, the closures now behave as constraints that are deformable in the axial direction, and the pressure acting on them stresses the cylindrical shell longitudinally. In the previous case of absolutely rigid or idealized clamped edges, the latter absorbed the entire axial load resulting from axial pressure.

10.4 Long Circular Cylindrical Shells Subjected to Concentrated Line Load Distributed Uniformly Along a Circumference

We will now consider a long circular cylindrical shell subjected to a concentrated line load P distributed uniformly over the outer circumference delimiting a cross-section located at a long distance from the ends (Fig. 10.6a). Considering the free-body diagram of a small element of our shell (Fig. 10.6b) of axial length dz straddling the circumference to which line load P is applied, we see that the problem can be regarded as involving two half shells symmetrical with respect to the loading plane, each of which is a semi-infinite shell loaded at the interface by a distributed shearing force T_0 of known value equal to $P/2$, and by a bending moment M_0, the latter being unknown. We can thus refer once again to the diagram shown in Fig. 10.3, remembering to invert the sign of T_0, as in this case the shearing force is directed towards the interior of the shell rather than the exterior.

Thus, using (10.24) and setting $T_0 = -P/2$ in it gives:

$$u = \frac{e^{-\beta z}}{2\beta^3 D}\left[\beta M_0(\sin\beta z - \cos\beta z) + \frac{P}{2}\cos\beta z\right]. \qquad (10.41)$$

As Fig. 10.6a shows, the z coordinate is measured in this case starting from the loading plane. As the problem examined here is axially symmetric relative to this loading plane, it is clear that in any axial plane the slope of the deflection curve for

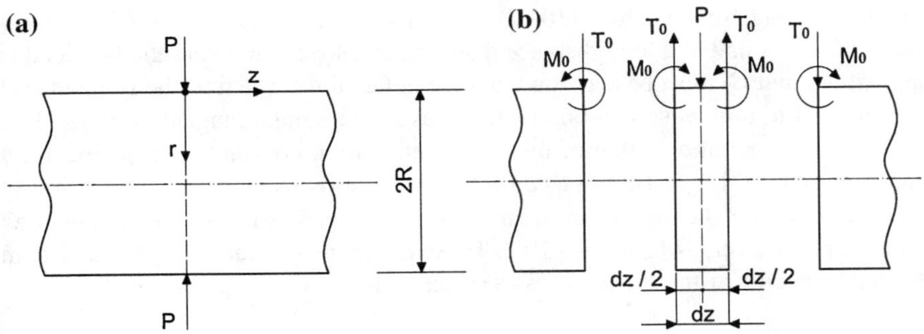

Fig. 10.6 **a** Circular cylindrical shell subjected to a concentrated line load P; **b** Free-body diagram of a small element dz straddling the loaded section

$z = 0$ must be zero. In other words, we must have $\varphi_{z=0} = (du/dz)_{z=0} = 0$. Consequently, using the first relation (10.25) and setting $T_0 = -P/2$ in it, for $z = 0$, gives us the following value of the hyperstatic unknown M_0:

$$M_0 = \frac{P}{4\beta}. \tag{10.42}$$

Substituting this value of M_0 into Eq. (10.41) yields the following expression for radial displacement:

$$u = \frac{P}{8\beta^3 D} e^{-\beta z}(\cos\beta z + \sin\beta z). \tag{10.43}$$

Here again, deriving relation (10.43) three times and taking relations (10.9), (10.15) and (10.6) into account gives the following relations expressing the other quantities of interest as a function of z:

$$\begin{aligned}
\varphi &= \frac{du}{dz} = -\frac{P}{4\beta^2 D} e^{-\beta z}\sin\beta z \\
M_z &= -D\frac{d^2u}{dz^2} = \frac{P}{4\beta} e^{-\beta z}(\cos\beta z - \sin\beta z) \\
M_t &= \nu M_z = \frac{\nu P}{4\beta} e^{-\beta z}(\cos\beta z - \sin\beta z) \\
T_z &= -D\frac{d^3u}{dz^3} = -\frac{P}{2} e^{-\beta z}\cos\beta z \\
N_t &= -\frac{Esu}{R} = -\frac{P\beta}{2} e^{-\beta z}(\cos\beta z + \sin\beta z).
\end{aligned} \tag{10.44}$$

Here as before, relations (10.43) and (10.44) univocally define the displacement field and stress resultants in a long circular cylindrical shell subjected to a concentrated line load applied to the outer circumference of an annular cross-section located at a long distance from the ends. The same relations can be obtained directly starting from relation (10.18), considering that since the radial displacement must be small at a long distance from the cross-section where the line load is applied, C_1 and C_2 must be zero, and that the deflection curve must be symmetrical with respect to this cross-section and thus have a horizontal tangent, thus yielding $C_3 = C_4$. The common value of these two integration constants is then found by imposing that $(T_z)_{z=0} = T_0 = -P/2$ in the loaded cross-section.

The curves for the significant quantities characterizing the problem addressed here are plotted versus βz in Fig. 10.7. These diagrams indicate that the maximum deflection occurs at the loaded cross-section and its value is given by:

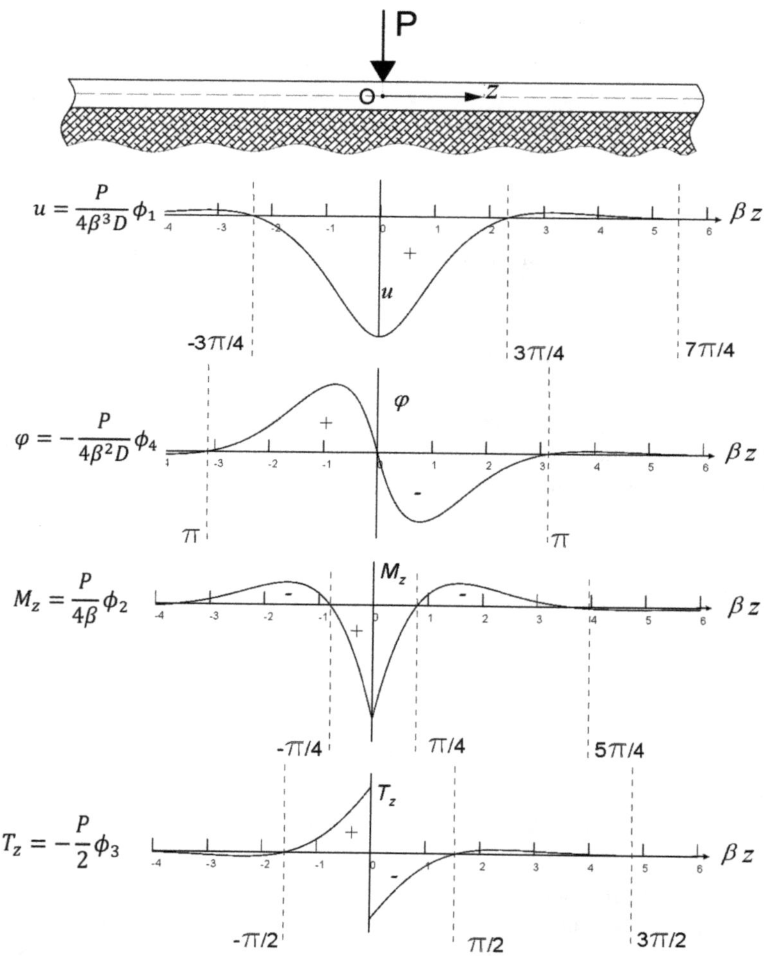

Fig. 10.7 Distribution curves for the significant quantities u, φ, M_z and T_z versus βz

$$u_{max} = \frac{P}{8\beta^3 D} = \frac{PR^2\beta}{2Es}. \quad (10.45)$$

The maximum bending moment given by relation (10.42) is also at this cross-section, as is the maximum shearing force, which is obviously equal, in absolute value, to $P/2$. As the distance from the loaded cross-section increases, the values of all these quantities drop rapidly and, at distance $z > \pi/\beta$ from this cross-section, are small enough to be ignored. In other words, the bending of the circular cylindrical shell considered here is of a local character, so that a cylindrical shell of length $L = 2\pi/\beta$ loaded at the middle behaves as if it were very long. The figure does not show the curves for M_t and N_t, as they are entirely similar to those for M_z and for u. For some of these quantities, the zeros following the first are indicated in order to highlight the fact that at a long distance from the origin, all of the quantities considered above oscillate from positive values to negative values around zero, though the values are highly attenuated.

10.5 Long Circular Cylindrical Shells Subjected to Load Distributed Through a Portion of Finite Length

We will now consider a long circular cylindrical shell subjected to an axisymmetric radial load distributed over a portion of the structure of length L, according to an arbitrary function (Fig. 10.8). Let this function be $p = p(\xi)$, with ξ being the axial coordinate with origin at a point A that can be assumed to be outside or inside the loaded region, depending on whether we wish to analyze the displacement field and the stress state occurring outside the region or in it. We will also assume that the loaded region is at a considerable distance from the ends of the circular cylindrical shell, so that the relations determined in the preceding section apply.

The relation to be used in this case is (10.43), which here is written in more compact form employing the first relation (10.27) and thus becomes:

$$u = \frac{P}{8\beta^3 D}\phi_1(\beta z). \quad (10.46)$$

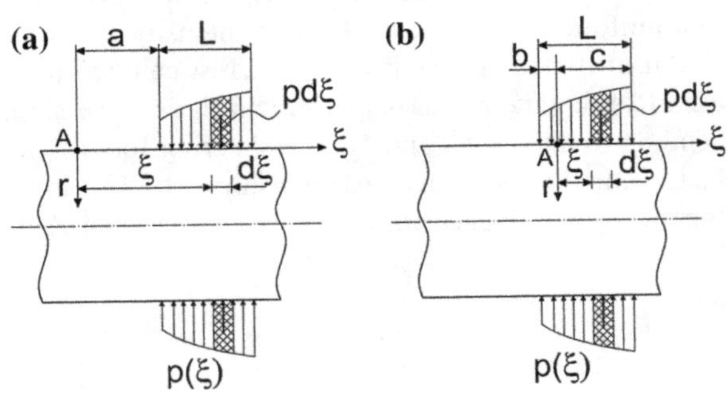

Fig. 10.8 Axisymmetric distribution of load $p = p(\xi)$ over a portion of length L: **a** A, point outside the loaded region; **b** A, point inside the loaded region

We will first consider a point A outside the loaded region (Fig. 10.8a), which we will take as the origin of axial coordinate ξ. The elementary deflection or elementary radial displacement du_A at point A caused by the elementary radial load $pd\xi$ acting between the cross-section of abscissa ξ and the cross-section of abscissa $\xi + d\xi$ can be expressed by means of relation (10.46) where we substitute $pd\xi$ for P and the coordinate ξ for z. The deflection or radial displacement u_A caused at A by the total load distributed over the portion of length L will thus be given by the relation:

$$u_A = \frac{1}{8\beta^3 D} \int_a^{a+L} p\phi_1(\beta\xi)d\xi, \tag{10.47}$$

which can be calculated once the load distribution function $p = p(\xi)$ is known. Where $p = const$, relation (10.47) integrates to:

$$u_A = \frac{pR^2}{2Es} e^{-\beta a}\left[\cos\beta a - e^{-\beta L}\cos\beta(a + L)\right]. \tag{10.48}$$

If point A is located inside the loaded region (Fig. 10.8b), the same procedure leads to the following relation expressing the deflection u_A caused at A by the generic load $p = p(\xi)$:

$$u_A = \frac{1}{8\beta^3 D}\left[\int_0^b p\phi_1(\beta\xi)d\xi + \int_0^c p\phi_1(\beta\xi)d\xi\right]. \tag{10.49}$$

For $p = const$, this relation integrates to:

$$u_A = \frac{pR^2}{2Es}\left[2 - e^{-\beta b}\cos\beta b - e^{-\beta c}\cos\beta c\right]. \tag{10.50}$$

It can be inferred from relation (10.48) that where L is of finite value and a is large, i.e., in a point A located far from the loaded area, deflection u_A assumes a very small value, at the limit tending to zero. This is entirely in line with the local character of the deformation due to bending, which dies out rapidly as the distance from the loaded region increases. It can then be inferred from relation (10.50) that for high values of b and c, the deflection assumes a value close to pR^2/Es, i.e., $u_A \cong pR^2/Es$. This is the radial displacement of a long cylindrical shell subjected to a uniform axisymmetric load of intensity $p = const$.

All of the other quantities of interest can be obtained with the same procedure used to calculate radial displacement u, viz., the slope of the deflection curve, the bending moments M_z and M_t, the shearing force T_z and the membrane force N_t. In fact, with this approach, relationships (10.43) and (10.44) are used as Green's functions or influence functions (see Green and Zerna 1963, Melnikov 1995).

10.6 Analysis of Short Circular Cylindrical Shells

In Sect. 10.2, we demonstrated that the radial displacement u, the rotation φ and the stress resultants M_z, M_t, T_z and N_t in a long circular cylindrical shell subject to bending moments and shearing forces applied at the ends die out rapidly as z increases and, for $\nu = 0.3$, become negligible for $z > 2.44\sqrt{Rs}$. This rapid reduction of the values of the significant quantities as z increases simplifies the problem of solving Eq. (10.18), given that we must have $C_1 = C_2 = 0$. If the circular cylindrical shell is short, however, i.e., when its length L is less than the value that can be obtained from the relation (10.31) and, for steels ($\nu = 0.3$), for L less than $2.44\sqrt{Rs}$, the constants C_1 and C_2 in Eq. (10.18) can no longer be considered zero. All four constants C_1, C_2, C_3 and C_4 must thus be taken into consideration, as the bending at one end cannot be considered as independent of the conditions at the other end, and the stress and strain states correlated with the two ends of the circular cylinder shell influence each other. Circular cylindrical shells whose length does not satisfy the inequalities (10.31) or (10.32) for steels, are called short circular cylindrical shells.

Suppose that the circular cylindrical shell is under constant internal pressure p_i. In this case, the general solution of the governing differential equation (10.14) is obtained by adding the complementary solution of the associated homogeneous equation given by (10.18), considered with all its terms, and the particular integral given by Eq. (10.19). Recalling, however, that on the basis of Euler's relations $\cosh\beta z = \left(e^{\beta z} + e^{-\beta z}\right)/2$ and $\sinh\beta z = \left(e^{\beta z} - e^{-\beta z}\right)/2$ (Abramowitz and Stegun 1972), the exponential functions appearing in Eq. (10.18) can be expressed in terms of hyperbolic functions, i.e., that

$$e^{\beta z} = \cosh\beta z + \sinh\beta z \qquad e^{-\beta z} = \cosh\beta z - \sinh\beta z, \tag{10.51}$$

the general solution of Eq. (10.14) can be written in the following form:

$$\begin{aligned} u = {} & C_1\cos\beta z\cosh\beta z + C_2\cos\beta z\sinh\beta z + C_3\sin\beta z\cosh\beta z + C_4\sin\beta z\sinh\beta z \\ & - \frac{p_i R^2}{E s}. \end{aligned} \tag{10.52}$$

To evaluate the four integration constants C_1, C_2, C_3 and C_4, we can introduce the so-called Krylov's functions, which are combinations of trigonometric and hyperbolic functions (see, e. g., Ventsel and Krauthammer 2001). With this procedure, the relation (10.52) changes form and becomes:

$$u = C_1V_1(\beta z) + C_2V_2(\beta z) + C_3V_3(\beta z) + C_4V_4(\beta z) - \frac{p_iR^2}{Es}, \tag{10.53}$$

where $V_1(\beta z)$, $V_2(\beta z)$, $V_3(\beta z)$ and $V_4(\beta z)$ are the four Krylov's functions (Krylov 1902).

But here we will follow the procedure used by Timoshenko and Woinowsky-Krieger (1959). To address the problem with this approach, it is convenient to shift the origin of the coordinates to the middle cross-section of the cylindrical shell, as shown in Fig. 10.5b. In this way, since relation (10.52) must in this case be an even function of the variable z, constants C_2 and C_3 will be zero, i.e., we will have $C_2 = C_3 = 0$. Constants C_1 and C_4 must thus be calculated by imposing boundary conditions, which will also be symmetrical with respect to this middle cross-section.

If we assume that the structure is simply supported at the ends as illustrated in Fig. 10.5b, both the radial displacement or deflection u and the bending moment M_z and, consequently, the second derivative d^2u/dz^2 will be zero at the structure's ends, i.e., for $z = \pm L/2$. Considering the end having coordinate $z = L/2$ and imposing these two boundary conditions, relation (10.52) and its second derivative, both with the terms in C_2 and C_3 omitted, gives the following system of equations in C_1 and C_4:

$$\begin{aligned} &C_1 \cos\alpha \cosh\alpha + C_4 \sin\alpha \sinh\alpha - p_i R^2/E s = 0 \\ &- C_1 \sin\alpha \sinh\alpha + C_4 \cos\alpha \cosh\alpha = 0, \end{aligned} \tag{10.54}$$

where for the sake of simplicity we set $\alpha = \beta L/2$, the latter being a dimensionless quantity that depends on the structure's dimensions as well as on Poisson's ratio ν. Solving this system gives:

$$\begin{aligned} C_1 &= \frac{p_i R^2}{Es} \frac{\cos\alpha \cosh\alpha}{\cos^2\alpha \cosh^2\alpha + \sin^2\alpha \sinh^2\alpha} = \frac{p_i R^2}{E s} \frac{2\cos\alpha \cosh\alpha}{\cos 2\alpha + \cosh 2\alpha} \\ C_4 &= \frac{p_i R^2}{E s} \frac{\sin\alpha \sinh\alpha}{\cos^2\alpha \cosh^2\alpha + \sin^2\alpha \sinh^2\alpha} = \frac{p_i R^2}{E s} \frac{2\sin\alpha \sinh\alpha}{\cos 2\alpha + \cosh 2\alpha}. \end{aligned} \tag{10.55}$$

By substituting the constants thus obtained into Eq. (10.52) without the terms in C_2 and C_3 and observing from Eq. (10.13), with $\beta = 2\alpha/L$, that $R^2/Es = 1/4D\beta^4 = L^4/64D\alpha^4$, we obtain the following expression for radial displacement u:

$$\begin{aligned} u = -\frac{p_i L^4}{64 D\alpha^4}\bigg(1 &- \frac{2\cos\alpha \cosh\alpha}{\cos 2\alpha + \cosh 2\alpha} \cos\beta z \cosh\beta z \\ &- \frac{2\sin\alpha \sinh\alpha}{\cos 2\alpha + \cosh 2\alpha} \sin\beta z \sinh\beta z \bigg). \end{aligned} \tag{10.56}$$

Deriving Eq. (10.56) twice and substituting the derivative thus found into the first relation (10.9) gives the following general expression of bending moment M_z:

$$\begin{aligned} M_z = -D\frac{d^2u}{dz^2} = -\frac{p_i L^2}{4\alpha^2}\bigg(&\frac{\sin\alpha \sinh\alpha}{\cos 2\alpha + \cosh 2\alpha} \cos\beta z \cosh\beta z \\ &- \frac{\cos\alpha \cosh\alpha}{\cos 2\alpha + \cosh 2\alpha} \sin\beta z \sinh\beta z \bigg). \end{aligned} \tag{10.57}$$

Starting from Eq. (10.56), we can obtain the other quantities of interest, i.e., φ, T_z, M_t and N_t in the same way.

In the middle cross-section, i.e., that having abscissa $z = 0$, the local values of radial displacement u and bending moment M_z are given by the relations:

$$(u)_{z=0} = -\frac{p_i L^4}{64 D \alpha^4}\left(1 - \frac{2\cos\alpha\cosh\alpha}{\cos 2\alpha + \cosh 2\alpha}\right)$$
$$(M_z)_{z=0} = -\frac{p_i L^2}{4\alpha^2}\frac{\sin\alpha\sinh\alpha}{\cos 2\alpha + \cosh 2\alpha}. \qquad (10.58)$$

Where the shell is of considerable length, α becomes large, and the second term in the parenthesis in the first of these relations is thus small. Consequently, the deflection approaches that which we would have if the ends were free, given by relation (10.33). This confirms that for very long shells the effect of end supports on the deflection at the middle cross-section between the two end faces is negligible. Again in the case of high values for α, i.e., of very long shells, the second relation (10.58) indicates that the bending moment at the middle of cylinder is negligibly small. It follows that the portion of the shell spanning the middle cross-section behaves as if it were a membrane under internal pressure alone, with a hoop stress $\sigma_t = p_i R/s$.

To return to the deflection, if α were very small, i.e., if the shell were of very short axial length, performing a power series expansion of the trigonometric and hyperbolic functions appearing in the first relation (10.58) would show that the expression in parentheses in this relation approaches the value $5\alpha^4/6$; in other words, the deflection would approach that for a uniformly loaded and simply supported beam of length L and flexural rigidity D.

If, on the other hand, the circular cylindrical shell has built-in ends as shown in Fig. 10.5a and we again shift the origin of the coordinates to the middle cross-section of the structure, both the radial displacement u and the rotation of the tangent to the deflection curve, i.e., the first derivative du/dz, will be zero at the structure's ends, where $z = \pm L/2$. Again considering the end having coordinate $z = L/2$ and imposing these boundary conditions, relation (10.52) and its first derivative, both with the terms in C_2 and C_3 omitted, gives the following system of equations in C_1 and C_4:

$$C_1\cos\alpha\cosh\alpha + C_4\sin\alpha\sinh\alpha - p_i R^2/E s = 0$$
$$C_1(-\sin\alpha\cosh\alpha + \cos\alpha\sinh\alpha) + C_4(\cos\alpha\sinh\alpha + \sin\alpha\cosh\alpha) = 0, \qquad (10.59)$$

with $\alpha = \beta L/2$.

Solving this system gives:

$$C_1 = \frac{p_i R^2}{E s}\frac{\sin\alpha\cosh\alpha + \cos\alpha\sinh\alpha}{\sinh\alpha\cosh\alpha + \sin\alpha\cos\alpha}$$
$$C_4 = \frac{p_i R^2}{E s}\frac{\sin\alpha\cosh\alpha - \cos\alpha\sinh\alpha}{\sinh\alpha\cosh\alpha + \sin\alpha\cos\alpha}. \qquad (10.60)$$

By substituting the constants thus obtained into Eq. (10.52) without the terms in C_2 and C_3, we obtain the following expression for radial displacement u:

$$u = -\frac{p_i L^4}{64\,D\,\alpha^4}\left[1 - \frac{2(\sin\alpha\cosh\alpha + \cos\alpha\sinh\alpha)}{\sinh 2\alpha + \sin 2\alpha}\cos\beta z\cosh\beta z - \frac{2(\sin\alpha\cosh\alpha - \cos\alpha\sinh\alpha)}{\sinh 2\alpha + \sin 2\alpha}\sin\beta z\sinh\beta z\right]. \tag{10.61}$$

Deriving Eq. (10.61) twice and substituting the derivative thus found into the first relation (10.9) gives the following general expression of bending moment M_z:

$$M_z = -D\frac{d^2u}{dz^2} = -\frac{p_i L^2}{4\alpha^2}\left(\frac{\sin\alpha\cosh\alpha - \cos\alpha\sinh\alpha}{\sinh 2\alpha + \sin 2\alpha}\cos\beta z\cosh\beta z - \frac{\sin\alpha\cosh\alpha + \cos\alpha\sinh\alpha}{\sinh 2\alpha + \sin 2\alpha}\sin\beta z\sinh\beta z\right). \tag{10.62}$$

Starting from Eq. (10.61), we can obtain the other quantities of interest, i.e., φ, T_z, M_t and N_t, in the same way.

We find from relation (10.62) that the bending moment acting in the built-in sections at the ends is:

$$(M_z)_{z=\pm L/2} = \frac{p_i}{2\beta^2}\frac{\sinh 2\alpha - \sin 2\alpha}{\sinh 2\alpha + \sin 2\alpha}. \tag{10.63}$$

For very large values of α, i.e., for long circular cylindrical shells, the fraction in the second member of relation (10.63) containing the trigonometric and hyperbolic functions approaches unity. Consequently, the bending moment at the built-in sections will equal that given by the first relation (10.34). For shorter circular cylindrical shells, this bending moment is calculated using relation (10.63).

We can proceed in the same way for a short circular cylindrical shell of thickness s subjected at both ends, which are assumed to be free, to a bending moment M_0 (Fig. 10.9a) or a shearing force T_0 (Fig. 10.9b). Only the results for deflection and the rotation of the generator at these ends will be presented below.

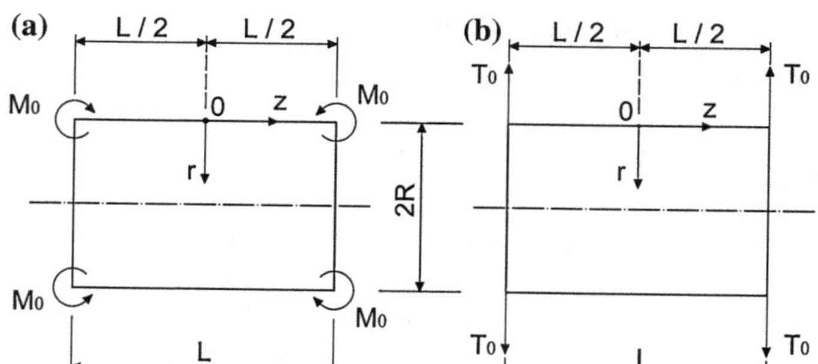

Fig. 10.9 Short circular cylindrical shell subjected at the two free ends to: **a** bending moments M_0; **b** shearing forces T_0

If the structure is loaded only by a bending moment M_0 (Fig. 10.9a), we will have:

$$(u)_{z=\pm L/2} = -\frac{2M_0\beta^2R^2}{Es}\frac{\sinh 2\alpha - \sin 2\alpha}{\sinh 2\alpha + \sin 2\alpha} = -\frac{2M_0\beta^2R^2}{Es}\psi_1$$
$$(\varphi)_{z=\pm L/2} = \left(\frac{du}{dz}\right)_{z=\pm L/2} = \pm\frac{4M_0\beta^3R^2}{Es}\frac{\cosh 2\alpha - \cos 2\alpha}{\sinh 2\alpha + \sin 2\alpha} = \pm\frac{4M_0\beta^3R^2}{Es}\psi_2 \quad (10.64)$$

where for the sake of brevity $\psi_1 = \psi_1(2\alpha)$ and $\psi_2 = \psi_2(2\alpha)$ denote the two fractions with hyperbolic and trigonometric functions.

If the structure is subjected only to shear force T_0 (Fig. 10.9b), we will have:

$$(u)_{z=\pm L/2} = -\frac{2T_0\beta R^2}{Es}\frac{\cosh 2\alpha + \cos 2\alpha}{\sinh 2\alpha + \sin 2\alpha} = -\frac{2T_0\beta R^2}{Es}\psi_3$$
$$(\varphi)_{z=\pm L/2} = \left(\frac{du}{dz}\right)_{z=\pm L/2} = \pm\frac{2T_0\beta^2R^2}{Es}\frac{\sinh 2\alpha - \sin 2\alpha}{\sinh 2\alpha + \sin 2\alpha} = \pm\frac{2T_0\beta^2R^2}{Es}\psi_1 \quad (10.65)$$

where $\psi_3 = \psi_3(2\alpha)$ represents the fraction with hyperbolic and trigonometric functions characterizing the radial displacement u, while ψ_1 coincides with that indicated above.

For very large values of α, i.e., for long circular cylindrical shells, the fraction in the second member of relations (10.64) and (10.65) containing the trigonometric and hyperbolic functions are also close to unity, so the results will coincide with those given by relations (10.29) and (10.30).

10.7 Analysis of Circular Cylindrical Shells Reinforced by Equidistant Stiffening Rings

Many practical applications involve structures similar to long tubes, i.e., long circular cylindrical shells under constant internal or external pressure and reinforced with stiffening rings spaced uniformly in the direction of the longitudinal axis (Fig. 10.10a). Typical examples include the fuselage shell structures with ring frames used in the aeronautical industry, the penstocks in hydroelectric powerplants, the ribbed hulls of submarines and bathyscaphes, reinforced double-walled vessels with heating or cooling fluid circulating in the space between the two walls, and so on.

The stress resultants and consequent stress state in these structures can be analyzed on the basis of relations (10.64) and (10.65). Let R, s and p_i be respectively the radius of the tube's middle surface, its thickness, and the internal pressure loading, which is assumed to be constant (attention should be paid to the

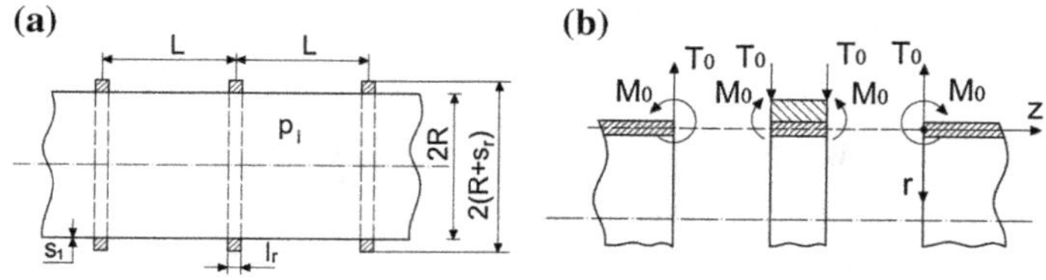

Fig. 10.10 **a** Circular cylindrical shell reinforced by equidistant stiffening rings; **b** disassembled structure at a stiffening ring and stress resultants

fact that, when the thicknesses of other structural components come into play, we will use the symbol s_1 for the thickness of the circular cylindrical shell, as indicated in Fig. 10.10a). In addition, let L be the center-to-center spacing, i.e., the axial distance between the middle planes of two successive stiffening rings, which we will initially assume to be of very high, theoretically infinite, stiffness. Without these absolutely rigid rings, the structure under the action of internal pressure p_i would be subjected to a uniform hoop stress $\sigma_t = p_i R/s$, obviously coinciding with the first relation (1.7), and its mean radius $R \cong d_i/2$ would expand uniformly, with a radial displacement u_R which, assuming zero axial stress σ_z (circular cylinder with free open ends), is given by relation (10.33).

The actual structure, however, is reinforced by the uniformly spaced stiffening rings, at which reactive forces per unit length of the circumference of the tube are generated. These forces denoted by P are distributed along the directrix circumference of each ring. To determine P, it is sufficient to impose that the deflection due to these forces in the cross-section of the structure lying on the ring's middle plane is equal to the radial displacement u_R caused by the internal pressure. For this purpose, the portion of circular cylindrical shell between the middle planes of two successive stiffening rings can be assimilated to the structures shown schematically in Fig. 10.9a and b, with $T_0 = -P/2$ and M_0 determined from the condition that $du/dz = 0$ at the stiffener. Taking the second relation (10.64) and the second relation (10.65) into account, this condition leads to the relation:

$$\frac{\beta^2 R^2}{Es}(4\,M_0\,\beta\psi_2 - P\psi_1) = 0, \tag{10.66}$$

which gives:

$$M_0 = \frac{P\psi_1}{4\,\beta\psi_2}. \tag{10.67}$$

We must now determine P. To do so, it is sufficient to impose that the sum of the radial displacements correlated with M_0 and with $T_0 = -P/2$, given by the first relation (10.64) and the first relation (10.65) respectively, equals u_R. We thus have:

$$\frac{\beta R^2}{Es}(-2M_0\beta\psi_1 + P\psi_3) = u_R = \frac{p_i\,R^2}{E\,s}. \tag{10.68}$$

Substituting the expression for M_0 given by relation (10.67) into relation (10.68) and solving the relation thus obtained for P then gives:

$$P = \frac{u_R E s}{\beta R^2 \left[\psi_3 - \frac{1}{2} \frac{\psi_1^2}{\psi_2} \right]} = \frac{p_i}{\beta \left[\psi_3 - \frac{1}{2} \frac{\psi_1^2}{\psi_2} \right]}. \tag{10.69}$$

For very large values of α, i.e., for very long distances L between two successive stiffening rings (in generally, as should be recalled, $\alpha = \beta L/2 = L\sqrt[4]{3(1-\nu^2)}/2\sqrt{Rs}$, while for $\nu = 0.3$ we have $\alpha \cong 0.643L/\sqrt{Rs}$), the functions ψ_1 and ψ_2 appearing in relation (10.67) approach unity, so that M_0 approaches the value given by relation (10.42). For the same reason, given that function ψ_3 also approaches unity for high values of α, relation (10.69) reduces to relation (10.45) with $u_{max} = u_R$. For small values of α, P is calculated with relation (10.69). Substituting this value into relation (10.67) gives the following final expression for M_0:

$$M_0 = \frac{p_i}{2\beta^2} \psi_1, \tag{10.70}$$

which coincides with relation (10.63) obtained for the circular cylinder shell with built-in edges.

In reality, however, the rigidity of the stiffening rings cannot be said to be infinite. As their cross-sectional dimensions are usually not very large, they are not absolutely rigid, and are thus deformable and subject to radial expansion (deformable stiffening rings). If their actual behavior is to be taken into account, we must necessarily consider the fact that the forces P produce a tensile force PR in the cross-section of any of the stiffening rings with an axial plane, and that this force produces an increase of the inner radius of the ring (given the small thicknesses involved, it is assumed that this radius is equal to R) given by the relation:

$$u_{r=r_i=R} = PR^2/AE \tag{10.71}$$

where A is the cross-sectional area of the ring. To take this radial expansion into account, it is sufficient to substitute the difference $(u_R - u_{r=r_i=R})$ for u_R into relation (10.69), so that we obtain:

$$P = \frac{p_i}{\left\{ \beta \left[\psi_3 - \frac{1}{2} \frac{\psi_1^2}{\psi_2} \right] + \frac{s}{A} \right\}}. \tag{10.72}$$

The bending moment M_0 is obtained from relation (10.70) by substituting p_i with the difference $(p_i - Ps/A)$.

If the circular cylindrical shell in question has free closed ends, a longitudinal force per unit length $N_z = p_i R/2$ will also be produced. In this case, the radial displacement of the middle generator of the cylindrical shell will no longer be given by relation (10.33), but by the following:

$$u_R^{'} = \frac{p_i R^2}{E s}\left(1 - \frac{\nu}{2}\right). \tag{10.73}$$

Accordingly, $p_i(1 - \nu/2)$ must be substituted for p_i in relations (10.70) and (10.72).

It should be pointed out that relations (10.69) and (10.72) are entirely general and can thus also be used when the circular cylindrical shell in question is under uniform external pressure. In this case, it is obviously necessary to verify that the compressive stresses generated in the circular cylindrical shell and in the stiffening rings remain below the critical stresses at which buckling may occur. This is the case of the double-walled vessels with fluid circulating between the walls and the ribbed submarine and bathyscaphe hulls mentioned at the beginning of this section.

10.8 Stress Analysis in a Reinforced Circular Cylindrical Shell

We will now return to the structure represented in Fig. 10.10a, which we will suppose has free closed ends and is under internal pressure p_i. Let L be the center-to-center spacing between two successive stiffening rings located at a uniform distance from each other along the axis. We will assume that the spacing L between the rings is large enough satisfy the inequalities given by relations (10.31) or (10.32). Let R, s_1, s_r and l_r be respectively the radius of the middle surface of the circular cylindrical shell, its wall thickness, and the thickness and width of the stiffening rings. We will determine the stress state in the areas of the structure around the rings, i.e., at ring attachments, assuming that both the structure and the stiffeners are made of the same material, that the rings' cross-sectional dimensions are small by comparison with R and that the rings permit the rotation of the circular cylindrical shell's generators.

To frame the problem, we will consider the free-body diagram of any of the stiffening rings and the two portions of the circular cylindrical shell adjacent to it (Fig. 10.10b). Let T_0 and M_0 be the uniformly distributed shearing forces and bending moments per unit length that load the three substructures into which we have divided the structure in question, which is also under internal pressure p_i. If we consider the symmetry of the problem with respect to the middle plane of any stiffening ring, the two conditions of displacement and rotation compatibility to be imposed at the interface having axial coordinate $z = 0$, where the substructures are connected, are as follows (the cylindrical shell and stiffening ring are denoted by 1 and 2 respectively):

$$\begin{aligned} (u_1)_{z=0} &= (u_2)_{z=0} \\ (\varphi_1)_{z=0} &= 0. \end{aligned} \tag{10.74}$$

The first of these conditions imposes equality of radial displacements in the two substructures at this interface. The second, deriving from the conditions of symmetry to the right and left of the middle plane bisecting the length of the ring in question, imposes that the rotation of the circular cylindrical shell at the interface with the ring and due to the combined effect of internal pressure p_i and local loads M_0 and T_0 be zero.

With the sign conventions employed in Fig. 10.10b, however, the radial displacement of the circular cylindrical shell's generator at the interface having axial coordinate $z = 0$ is given by the following relation:

$$(u_1)_{z=0} = -\frac{1}{2\beta_1^3 D_1}[\beta_1 M_0 + T_0] - \frac{p_i R^2}{E s_1}\left(1 - \frac{\nu}{2}\right), \tag{10.75}$$

which is obtained by adding the contribution of loads M_0 and T_0 due to local bending, as inferred from relation (10.24) for $z = 0$, and that correlated with internal pressure p_i, as given by relation (10.72).

The ring's radial displacement at the interface is given by the relation:

$$(u_2)_{z=0} = \frac{2T_0 R^2}{AE} \tag{10.76}$$

where $A = l_r s_r$ is the ring's effective cross-sectional area. Bearing in mind that, as shown in Fig. 10.10b, the ring is subjected to a centripetal radial load per unit length equal to $2T_0$, the latter relation can be obtained from Eq. (1.20), where we set $r_i = R$ and $\sigma_e = 2T_0 R/A$. As can be seen from Fig. 1.1b, in fact, relation (1.3), with $\sigma_t = \sigma_e$, $dF_i = 0$, $dF_e = 2T_0 R d\theta$ and $sdz = A$, becomes

$$2\sigma_e A = \int_0^{\pi} 2T_0 R \sin\theta \, d\theta \tag{10.77}$$

so that we obtain $\sigma_e = 2T_0 R/A$. It should be pointed out that displacement $(u_2)_{z=0}$ is positive, given that it is oriented along the positive direction of the r axis.

To calculate the rotation angle of the circular cylindrical shell's generator at the interface with the stiffening ring having coordinate $z = 0$, it is also necessary to add the contribution of internal pressure p_i and that correlated with loads M_0 and T_0 due to local bending. But the internal pressure does not contribute to this rotation, as it only causes the generator to move in translation parallel to itself. The contribution correlated with M_0 and T_0 is determined using the first relation (10.25) for $z = 0$, which gives:

$$(\varphi_1)_{z=0} = \frac{1}{2\beta_1^2 D_1}[2\beta_1 M_0 + T_0]. \tag{10.78}$$

Note that the contributions of M_0 and T_0 are in the same direction.

By introducing relations (10.75), (10.76) and (10.78) into relations (10.74), we obtain the following system of two equations in the two hyperstatic unknowns M_0 and T_0:

$$
\begin{aligned}
&-\frac{1}{2\beta_1^3 D_1}[\beta_1 M_0 + T_0] - \frac{p_i R^2}{E s_1}\left(1 - \frac{\nu}{2}\right) = \frac{2 T_0 R^2}{AE} \\
&\frac{1}{2\beta_1^2 D_1}[2\beta_1 M_0 + T_0] = 0,
\end{aligned}
\tag{10.79}
$$

whose solution gives:

$$
\begin{aligned}
M_0 &= \frac{p_i}{4\beta_1 s_1}\frac{2-\nu}{(\beta_1/s_1 + 2/A)} \\
T_0 &= -\frac{p_i}{2 s_1}\frac{2-\nu}{(\beta_1/s_1 + 2/A)}.
\end{aligned}
\tag{10.80}
$$

In view of the considerations demonstrated above, the hoop stress in the stiffening ring is given by the relation:

$$
(\sigma_t)_2 = \sigma_e = \frac{2 T_0 R}{A} = -\frac{p_i R(2-\nu)}{s_1(\beta_1 A/s_1 + 2)}. \tag{10.81}
$$

The maximum longitudinal stress in the circular cylindrical shell arises at the cross-section interfacing with the ring ($z = 0$), and is:

$$
\begin{aligned}
(\sigma_z)_{1,max} &= \frac{p_i R}{2 s_1} + \frac{6 M_0}{s_1^2} = \frac{p_i R}{2 s_1} + \frac{3 p_i (2-\nu)}{2 s_1^3 \beta_1 (\beta_1/s_1 + 2/A)} \\
&= \frac{p_i}{2 s_1}\left[R + \frac{3(2-\nu)}{s_1^2 \beta_1 (\beta_1/s_1 + 2/A)}\right].
\end{aligned}
\tag{10.82}
$$

The circular cylindrical shell's radial displacement, which is obtained by introducing relations (10.80) into (10.75), is maximum in the interface cross-section, and is:

$$
(u_1)_{z=0} = -\frac{p_i R^2 (2-\nu)}{E(2 s_1 + A\beta_1)}. \tag{10.83}
$$

The hoop stress resultant in the circular cylindrical shell, again at the cross-section having axial coordinate $z = 0$, is calculated as the algebraic sum of the membrane contribution due to internal pressure, which can be found from the first relation (1.7) and is $(N_t)_{p_i} = p_i R$, and of the contribution due to local stress resultants M_0 and T_0, given by the last relation (10.25) by setting $z = 0$ and substituting relations (10.80). We thus obtain:

$$
N_t = (N_t)_{p_i} + (N_t)_{M_0,T_0} = p_i R\left[1 - \frac{\beta_1 (2-\nu)}{2 s_1 (\beta_1/s_1 + 2/A)}\right]. \tag{10.84}
$$

The maximum hoop stress in the circular cylindrical shell (which also arises at the cross-section connected to the ring, having axial coordinate $z = 0$) is the sum of the contribution of the hoop stress resultant N_t and of the circumferential bending moment νM_0. We thus have:

$$(\sigma_t)_{1,max} = \frac{N_t}{s_1} + \frac{6\nu M_0}{s_1^2} = \frac{p_i}{s_1}\left[R + \frac{(2-\nu)}{2s_1(\beta_1/s_1 + 2/A)}\left(\frac{3\nu}{\beta_1 s_1} - \beta_1 R\right)\right]. \quad (10.85)$$

10.8.1 Calculation Example

As a calculation example, we will determine the stresses in the reinforced circular cylindrical vessel under constant internal pressure p_i shown in Fig. 10.10 using the following data: $s_1 = 0.01$ m; $s_r = 0.04$ m; $l_r = 0.05$ m; $R = 1.00$ m; $L = 1.00$ m; $E = 200$ GPa; $\nu = 0.3$. These data give us the values of the following quantities: $D_1 = 18.315 \times 10^3$ Nm; $A = 2 \times 10^{-3}$ m^2; $\beta_1 = 12.854$ m^{-1}. Introducing these values in relations (10.80) yields:

$$M_0 = 14.467 \times 10^{-4} p_i \text{ Nm/m}$$
$$T_0 = -3.719 \times 10^{-2} p_i \text{ N/m}.$$

The hoop stress in the stiffening ring is given by relation (10.81), and will be:

$$(\sigma_t)_2 = 37.193\, p_i.$$

The maximum longitudinal stress in the circular cylindrical shell is given by relation (10.82), and will be:

$$(\sigma_z)_{1,max} = 136{,}804\, p_i.$$

Lastly, the maximum hoop stress in the circular cylindrical shell is given by relation (10.85), and will be:

$$(\sigma_t)_{1,max} = 78.234\, p_i.$$

Consequently, the absolute maximum value in the structure's stress state is that reached by the maximum longitudinal stress in the circular cylindrical shell at the cross-section interfacing with the reinforcing ring.

Chapter 11
Cylindrical Pressure Vessels

11.1 General

Pressure vessels consist of shells of various shapes, assembled together using suitable methods. Circular cylindrical vessels used as pressure equipment are closed at the ends by closures, also termed heads, having different geometrical configurations (flat-plate and conical closures or formed heads, e.g. torispherical, ellipsoidal, super ellipsoidal, hemispherical, toriconical etc., heads). These closures are shells of revolution belonging to a highly general class of shells frequently used in engineering. Flat plates or cones (the latter have zero Gaussian curvature since they are developable surfaces) may be used as closures because they are easily formed, but their use is usually limited to low working pressure and small-diameter vessels. Flat plates are also used as closures or cover plates for hand holes, manholes, and so on (Cicala 1966, Brownell and Young 1968, Megyesy 1984, Jaward and Farr 1989, Moss 2004).

For higher working pressure, a wide variety of formed closures, termed *heads*, have been developed, standardized, and extensively used in the fabrication of process pressure vessels. Generally these closures have a non-developable surface, and are subclasses of shells of revolution with non-zero Gaussian curvature. They are stiffer, stronger, and more stable than shells with zero Gaussian curvature (in this regard, we should remember that we define *Gaussian curvature* and *mean curvature* of a surface as the product and semi-sum of the principal curvatures $1/r_I$ and $1/r_{II}$ respectively.

A non-developable surface is defined completely and uniquely by simultaneously using the first and second quadratic form of surfaces. In a system of orthogonal curvilinear coordinates (α, β), the necessary conditions for the existence of a surface (the middle surface of the shell) consist of the fact that the two conditions of Codazzi-Mainardi-Peterson (Codazzi 1868-1869, Mainardi 1857, Peterson 1853; see also Kline 1972, Ivanov 2001) and the condition of Gauss are met (Gauss 1828). These three conditions are related to one another by four functions of the two curvilinear coordinates α and β, given by the two Lamé parameters appearing in the equation expressing the first quadratic form of the surface, generally indicated with A and B, and the two principal radii of curvature r_I and r_{II}.

V. Vullo, *Circular Cylinders and Pressure Vessels*, Springer Series in Solid and Structural Mechanics 3, DOI: 10.1007/978-3-319-00690-1_11,

The majority of the vessels used in practical applications have either semielliptical or torispherical dished heads, both because they are significantly stronger than flat-plate and conical closures, and because the stress concentrations arising from the discontinuity in shape at the welded junction with the circular cylindrical shell are considerably lower. Structurally, the best closures are hemispherical, but the fact that they are much more costly to fabricate than elliptical and torispherical dished heads limits their use to niche applications.

From a general standpoint, the discontinuity stress resultants can be reduced to a bending moment M_0 and a shearing force T_0 uniformly distributed around the mean circumference represented by the welded seam joining the closure and the shell. As was indicated in Chap. 10, these are stress resultant per unit length of this circumference. The shearing force T_0, in fact, is directed in such a way that it acts to restrain the extension of the radius of the middle surface of the cylindrical shell caused by internal pressure. This gives rise to a bending moment M_0 that in the junction and its immediate neighborhood deflects the cylindrical shell inward. This local bending, however, is resisted by the compressive circumferential forces per unit length of the cylindrical shell's generator, which oppose the decrease in radius caused by local stress resultants M_0 and T_0.

In these pressure vessels, the displacement field and stress and strain states in the portions of the cylindrical shell and formed heads far from the junction connecting the component substructures making up the structure as a whole are analyzed by means of the membrane theory of axisymmetrically loaded shells of revolution. However, this theory is known to fail when applied to the portions adjacent to the junction between the cylindrical shell and the flat-plate or formed closures, since the edge conditions usually cannot be completely satisfied by considering only membrane stresses. Thus, in these portions of the component substructures, it is not sufficient to consider only the membrane stresses, as it is also necessary to allow for the so-called *discontinuity stresses* correlated with the local stress resultants arising as a result of the differing stiffness response of the substructures to the external loads applied to the entire structure (Timoshenko and Woinowsky-Krieger 1959, Tsui 1968, Baker et al. 1968, Novozhilov 1970, Cicala 1971, Flügge 1973, Ventsel and Krauthammer 2001).

The intensity of these discontinuity stresses depends on the vessel's geometry, the variations in thickness and curvature of the component substructures at the junctions between them, and the mechanical properties of the materials used. These discontinuities produce local shearing forces and bending moments which are superposed on the membrane stress resultants and satisfy compatibility of displacements and rotations between adjoining substructures, which membrane theory alone cannot do. The discontinuity stresses are calculated using the bending theory for axisymmetrically loaded shells of revolution, which if rigorously applied is extremely complex.

However, it is also possible to arrive at approximate solutions that are as reliable as they are of undoubted practical utility for the designer. Though they introduce a few simplifying assumptions, these approximate methods yield results that differ by only a few percent from those obtained either through the rigorous application of bending theory or from experimental procedures such as strain gage measurements, and are essentially based on various hypotheses-models of elastic foundation, such as that suggested by Winkler (1867), termed Winkler elastic foundation theory. In this context, we can use the relations presented in Chap. 10, with the additions needed to address the three most interesting cases dealt with in this chapter, viz., circular cylindrical pressure vessels with flat-plate closures, and those with formed heads whose generating meridian is hemispherical or semi-elliptical.

To determine the discontinuity stress resultants, we will assume the substructures connected by welded seams have constant thickness on both sides of the junction plane, and that their middle surfaces are continuations of each other. On the basis of this assumption, the additional bending moments that would be generated by the eccentric meridian forces acting on the substructures are zero. If this condition were not respected, it would obviously be necessary to complete the analysis by considering the influence that these additional bending moments, which arise from the lack of continuity between the middle surfaces of adjoining structural components, exert on the structure's stress and strain states. Lastly, it should be emphasized that the following analysis, like those presented earlier, applies to thin-walled structures consisting of a homogeneous and isotropic material stressed in the linear elastic range, under the assumption that displacements and deformations are small.

11.2 Cylindrical Pressure Vessels with Flat-Plate Closures: Closure Absolutely Rigid in Its Own Plane

We will consider a cylindrical pressure vessel with flat-plate closures at both ends, as shown in Fig. 11.1a. In actual construction practice, for the reasons that will be described below, flat-plate closures are thicker, and often significantly thicker, than the cylindrical shell, which is almost always to be considered as thin-walled. Let 1 (cylinder) and 2 (flat-plate closure) be the two component substructures, and let s_1 and s_2 be their thicknesses. Let R be the outer radius of the two solid circular plates serving as the two flat-plate closures. The radius R_m of the cylinder's middle surface will be $(R + s_1)$; however, as s_1 is much smaller than R, we can treat R_m and R as roughly equivalent, i.e., we can assume $R_m \cong R$.

As was indicated for discontinuity stresses in the preceding chapter, at a sufficient distance from the plane in which the local stress resultants M_0 and T_0 act, which we will call the junction plane, the stress resultants in the cylindrical shell and in the two flat-plate closures can be calculated with a fair degree of accuracy using membrane theory for the shell and Kirchhoff's plate theory (Kirchhoff 1850) for the two closures. First, as we did for the problems considered above, we will idealize the two component substructures as being separate (Fig. 11.1b and c). As these substructures' displacements and rotations in the junction plane calculated with these two theories are not equal, it is necessary to introduce uniformly distributed shearing forces T_0 and bending moments M_0 in this plane as our

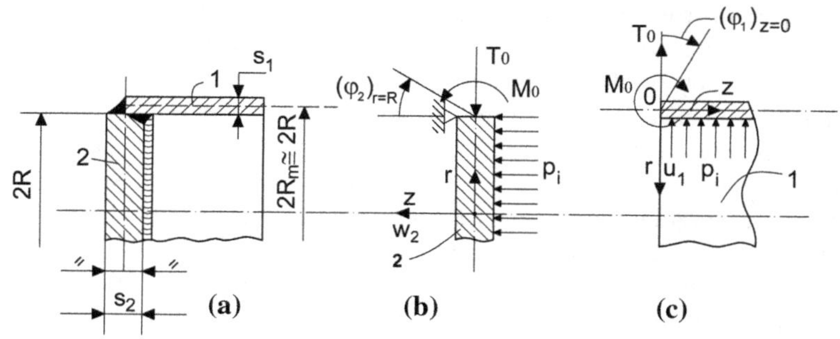

Fig. 11.1 **a** Cylindrical pressure vessel with flat-plate closures. **b** Flat-plate closure and local stress resultants M_0 and T_0 in the middle plane; **c** cylindrical shell and local stress resultants M_0 and T_0 at the end

hyperstatic unknowns. These forces and moments can be determined by imposing conditions of compatibility of displacements and rotations in the junction plane between the cylindrical shell and the flat-plate closure.

We will first assume that the closure is thicker than the shell and its thickness is such that it can be considered non-deformable in its own plane, i.e., that it is infinitely rigid or absolutely rigid in that plane. The deflected surface of its middle plane resulting from the so-called out-of-plane loads (the transverse load due to pressure p_i and the bending moment M_0) will thus be inextensional in the radial direction. In these conditions, the shearing load T_0 due to edge effect will not cause radial displacements. Moreover, since the flat-plate closures are absolutely rigid, the shell edges may be assumed to be built-in with the closures. Consequently, the radial displacement of the outer edge of the solid circular plate will be zero, i.e., we will have $(u_2)_{r=R}= 0$. The two conditions of compatibility of displacements and rotations in the junction plane will be as follows:

$$\begin{aligned} (u_2)_{r=R} &= (u_1)_{z=0} = 0 \\ (\varphi_2)_{r=R} &= (\varphi_1)_{z=0}. \end{aligned} \tag{11.1}$$

The values of displacements and rotations to be introduced in these compatibility relations must use the sign convention established in this connection. In this section and in those that follow, centripetal displacements (which are thus oriented along the positive direction of the displacement in the direction of the normal introduced in shell theory) and clockwise rotations will be considered positive, as shown in Fig. 11.1c. Conversely, centrifugal displacements and counterclockwise rotations will be considered negative.

The rotation angle of the cylinder's edge (equal to the slope of the deflection curve of the cylinder's middle generator at the edge) can be determined from the first relation (10.25) for $z = 0$, while its radial displacement is determined using relation (10.24), also calculated for $z = 0$, adding the contribution of internal pressure p_i given by relation (10.73) to the latter with the sign changed. We thus have:

$$(\varphi_1)_{z=0} = \frac{1}{2\beta_1^2 D_1}[2\beta_1 M_0 + T_0]$$
$$(u_1)_{z=0} = -\frac{1}{2\beta_1^3 D_1}[\beta_1 M_0 + T_0] - \frac{p_i R^2}{E s_1}\left(1 - \frac{\nu}{2}\right). \quad (11.2)$$

It should be noted that pressure p_i, as it causes the cylinder's generator to move in translation parallel to itself, does not contribute to rotation.

To calculate the displacements, the rotations and the other quantities of interest for the flat-plate closure and deriving both from discontinuity loads and from internal pressure using the assumptions of the classical plate theory (Kirchhoff's plate theory), it is convenient to consider the plate's middle plane, which in the case examined here (Fig. 11.1a), coincides with the junction plane (see Timoshenko and Woinowsky-Krieger 1959, Vivio and Vullo 2010). But the radial displacement at the edge of this solid circular plate can be determined without calculation, as we have assumed that it is zero, i.e., we have: $(u_2)_{r=R} = 0$. At the edge of the plate, the angle of rotation of the normal to the middle plane (which is obviously equal to the angle of rotation of the tangent to the deflected surface in any axial plane) can be determined by adding the contributions of bending moment M_0 and of internal pressure p_i. As Fig. 11.1b shows, the rotations produced by M_0 and by p_i differ in sign, the former being negative and the latter positive.

As a result of bending moment M_0, which is uniformly distributed along the outer edge of the plate, the latter deflects along a spherical surface whose principal curvatures are obviously identical and are given by the relation:

$$\frac{1}{r_I} = \frac{1}{r_{II}} = \frac{M_0}{D_2(1+\nu)}, \quad (11.3)$$

where r_I and r_{II} are the two principal radii of curvature and D_2 is the plate's flexural rigidity, given by the relation:

$$D_2 = \frac{E s_2^3}{1-\nu^2}. \quad (11.4)$$

Under the small displacements assumption made in the Kirchhoff's plate theory, however, the plate's second principal radius of curvature r_{II} (that in the second principal plane that contains the normal in a generic point of the deflected surface and is perpendicular to the first principal plane, with the latter coinciding with any axial plane, i.e., passing through the axis of symmetry) is correlated with the coordinate r (Fig. 11.1b) and the rotation angle φ_2 of the tangent, in any axial plane, to the deflected surface in a generic point of the latter by the relation:

$$\frac{1}{r_{II}} = \frac{\varphi_2}{r} = -\frac{dw_2}{rdr} \quad (11.5)$$

where w_2 is the displacement in the direction of the axis of symmetry z (Fig. 11.1b).

Substituting relation (11.3) into relation (11.5) gives the following first order differential equation with separable variables:

$$-\frac{dw_2}{rdr} = \frac{M_0}{D_2(1+\nu)}. \tag{11.6}$$

Integrating this equation, and imposing the boundary condition that $w_2 = 0$ for $r = R$, we obtain the relation:

$$w_{2,M_0} = \frac{M_0}{2D_2(1+\nu)}\left(R^2 - r^2\right). \tag{11.7}$$

The solid circular plate in question, however, is also subject to the uniformly distributed load due to internal pressure p_i. As Fig. 11.1b shows, the end constraint can be regarded as a hinge that permits the outer edge of the plate to rotate but prevents it from expanding in its middle plane. Recalling the following third order differential equation, which is the governing equation for circular plates subject to axisymmetrically distributed transverse load

$$\frac{d}{dr}\left[\frac{1}{r}\frac{d}{dr}\left(r\frac{dw}{dr}\right)\right] = \frac{T_r}{D} \tag{11.8}$$

where $w = w_2$, $D = D_2$ and T_r is the transverse shearing force at the generic radius r (in the case examined here, we have $T_r = p_i r/2$), and integrating it three times relative to r and, to calculate the integration constants resulting from it, imposing the following boundary conditions

$$\begin{aligned} w_2 &= 0 \quad \text{for} \quad r = R \\ M_r &= 0 \quad \text{for} \quad r = R \\ \varphi_2 &= 0 \quad \text{for} \quad r = 0 \end{aligned} \tag{11.9}$$

where M_r is the radial bending moment acting on the plate, gives the following relation:

$$w_{2,p_i} = \frac{p_i(R^2 - r^2)}{64D_2}\left(\frac{5+\nu}{1+\nu}R^2 - r^2\right). \tag{11.10}$$

Consequently, the total transverse displacement w_2 deriving from the sum of the contributions of internal pressure p_i and the bending moment M_0 is given by the relation:

$$w_2 = w_{2,p_i} - w_{2,M_0} = \frac{p_i(R^2 - r^2)}{64D_2}\left(\frac{5+\nu}{1+\nu}R^2 - r^2\right) - \frac{M_0}{2D_2(1+\nu)}\left(R^2 - r^2\right). \tag{11.11}$$

It should be noted that the transverse displacements (those in the direction of the z axis) correlated with M_0 and p_i in the plate have different signs, the former being negative and the latter positive. The rotation angle φ_2 of the tangent, in any axial plane, to the deflected surface in a generic point, which is equal to the difference of the rotation φ_{2,p_i} correlated with the internal pressure p_i and the rotation φ_{2,M_0} due to the bending moment M_0, is given by the relation:

$$\varphi_2 = -\frac{dw_2}{dr} = \varphi_{2,p_i} - \varphi_{2,M_0} = \frac{r}{D_2}\left[\frac{p_i}{16}\left(\frac{3+\nu}{1+\nu}R^2 - r^2\right) - \frac{M_0}{1+\nu}\right]. \quad (11.12)$$

At the outer edge of the plate, i.e., for $r = R$, this rotation will be:

$$(\varphi_2)_{r=R} = \left(\varphi_{2,p_i}\right)_{r=R} - \left(\varphi_{2,M_0}\right)_{r=R} = \frac{R}{D_2(1+\nu)}\left(\frac{p_i R^2}{8} - M_0\right). \quad (11.13)$$

Substituting relations (11.2) and (11.13) considering the sign conventions indicated above into relations (11.1) gives the following system of two equations in the two hyperstatic unknowns M_0 and T_0:

$$\begin{aligned} &\frac{M_0}{2\beta_1^2 D_1} + \frac{T_0}{2\beta_1^3 D_1} + \frac{p_i R^2}{E s_1}\left(1 - \frac{\nu}{2}\right) = 0 \\ &\frac{M_0}{\beta_1 D_1} + \frac{T_0}{2\beta_1^2 D_1} = -\frac{R M_0}{D_2(1+\nu)} + \frac{p_i R^3}{8 D_2(1+\nu)}. \end{aligned} \quad (11.14)$$

Solving the latter system then gives

$$\begin{aligned} M_0 &= \frac{p_i R^2 \beta_1 D_1 [E R s_1 + 4(1+\nu)(2-\nu)\beta_1 D_2]}{4 E s_1 [D_2(1+\nu) + 2R\beta_1 D_1]} \\ T_0 &= -\frac{p_i R^2 \beta_1^2 D_1 \{E R s_1 + 8(2-\nu)\beta_1 [D_2(1+\nu)] + R\beta_1 D_1\}}{4 E s_1 [D_2(1+\nu) + 2R\beta_1 D_1]}. \end{aligned} \quad (11.15)$$

11.2.1 Calculation Example

As a calculation example, we will determine the stresses in the cylindrical vessel under constant internal pressure p_i shown in Fig. 11.1, using the following data: $s_1 = 0.01$ m; $s_2 = 0.04$ m; $R = 1.00$ m; $E = 200$ GPa; $\nu = 0.3$. These data give us the values of the following quantities: $D_1 = 18.315 \times 10^3$ Nm; $D_2 = 1172 \times 10^3$ Nm; $\beta_1 = 12.854\ \mathrm{m}^{-1}$. Introducing these values in relations (11.15) yields:

$$\begin{aligned} M_0 &= 3.147 \times 10^{-2} p_i\ \mathrm{Nm/m} \\ T_0 &= -4.376 \times 10^{-1} p_i\ \mathrm{N/m}. \end{aligned}$$

Through the combined effect of the internal pressure and the infinite radial stiffness of the flat-plate closure, the cylindrical shell is deflected inward in the area adjacent to the junction. The bending moment M_z assumes its maximum value, equal to M_0 in the junction plane and rapidly dies out as z increases, in accordance with the second relation (10.25). The stresses at the edge of the cylindrical shell, which coincides with the junction plane, where radial displacement is zero, are as follows:

$$(\sigma_z)_{z=0} = \frac{p_i R}{2s_1} + \frac{6M_0}{s_1^2} = 1{,}938\, p_i$$

$$(\sigma_t)_{z=0} = \frac{p_i R}{s_1} + \frac{6\nu M_0}{s_1^2} = 666\, p_i.$$

These normal stresses are both tensile at the outer radius, where the membrane contribution of the internal pressure—with tensile stress uniformly distributed through the wall thickness—is added to the bending contribution due to edge effects, with a stress distribution varying linearly through the thickness which is tensile at the outer radius and compressive at the inner radius. The shear stress correlated with T_0 is known to vary through the thickness according to a parabolic function, and is zero at the outer and inner radii and maximum at the middle surface, where its value is one and a half times the mean shear stress. The latter is in any case much smaller than the normal stresses and is thus usually ignored. Accordingly, the maximum stress in the cylindrical shell is the axial stress at the outer radius in the junction plane, due to the two contributions we have just mentioned.

It should be borne in mind, however, that the following axial and hoop membrane stresses arise as a result of internal pressure in the areas far from the junction plane, which are not affected by the discontinuity problem:

$$\sigma_z = 50\, p_i \quad \sigma_t = 100\, p_i.$$

The maximum stress in the flat-plate closure occurs at the axis ($r = 0$), where the radial and hoop stresses assume the same value, viz.:

$$(\sigma_r)_{r=0} = (\sigma_t)_{r=0} = \frac{6}{s_2^2}\left(M_{r,p_i} + M_0\right) = \frac{6}{s_2^2}\left[\frac{(3+\nu)R^2 p_i}{16} + M_0\right] = 891\, p_i.$$

The stress state's absolute maximum stress is thus the axial stress at the outer radius of the cylindrical vessel on the junction plane.

11.3 Cylindrical Pressure Vessels with Flat-Plate Closures: Closure Deformable in Its Own Plane

We will now consider the same cylindrical pressure vessel shown in Fig. 11.1, but in this case will assume that the flat-plate closures are also deformable in their own plane. Here, by contrast with the case examined in Sect. 11.2, the shearing force T_0 due to the local discontinuity will cause a radial displacement of the plate's outer edge, so we will have $(u_2)_{r=R} \neq 0$. For this displacement to take place, the hinge in

Fig. 11.1b must be substituted with a simple support, i.e., a constraint that can simultaneously permit radial translation and rotation of the normal in the junction plane. The two conditions of compatibility of displacements and rotations will thus be as follows:

$$\begin{aligned}(u_2)_{r=R} &= (u_1)_{z=0}\\ (\varphi_2)_{r=R} &= (\varphi_1)_{z=0}.\end{aligned} \tag{11.16}$$

The angle of rotation and the radial displacement of the cylinder's edge will still be those given by relations (11.2). For the flat-plate closure, the angle of rotation of the normal to the plate's edge will once again be that given by relation (11.13). The displacement $(u_2)_{r=R}$ caused by shearing force T_0 is readily calculated, as the plate so loaded is nothing more than a non-rotating solid disk of constant thickness subjected at the outer radius to a uniform distribution of surface force $\sigma_{re} = -T_0/s_2$. The stress state in this non-rotating disk (Vullo and Vivio 2013) is defined by Lamé's relations (4–9), in which A and B are integration constants that can be determined by imposing boundary conditions. To avoid singularity problems at the axis, however, we must have $B = 0$, and since the radial stress at the outer radius must equal the surface force σ_{re}, we will have $A = \sigma_{re}$. Consequently, the stress state in our solid circular plate subjected to external radial load T_0, acting in its middle plane is given by the relation:

$$\sigma_r = \sigma_t = A = \sigma_{re} = -T_0/s_2. \tag{11.17}$$

On the basis of Hooke's law given by the second relation (1.17) for $\sigma_z = 0$ (biaxial stress state) and the second geometric relation (4–3), we can write:

$$u = r\varepsilon_t = \frac{r}{E}(\sigma_t - \nu\sigma_r). \tag{11.18}$$

Thus, considering the outer radius R and taking relation (11.17) into account, we find:

$$(u_2)_{r=R} = -\frac{RT_0}{Es_2}(1-\nu). \tag{11.19}$$

Substituting relations (11.2), (11.13) and (11.19) considering the sign conventions indicated above into relations (11.16) gives the following system of two equations in the two hyperstatic unknowns M_0 and T_0:

$$\begin{aligned}\frac{RT_0}{Es_2}(1-\nu) &= \frac{M_0}{2\beta_1^2 D_1} + \frac{T_0}{2\beta_1^3 D_1} + \frac{p_i R^2}{Es_1}\left(1-\frac{\nu}{2}\right)\\ \frac{M_0}{\beta_1 D_1} + \frac{T_0}{2\beta_1^2 D_1} &= -\frac{RM_0}{D_2(1+\nu)} + \frac{p_i R^3}{8D_2(1+\nu)}.\end{aligned} \tag{11.20}$$

Solving this system gives:

$$M_0 = \frac{p_i R^2 \beta_1 D_1 \{Rs_1 [Es_2 - 2(1-\nu)R\beta_1^3 D_1] + 4(1+\nu)(2-\nu)s_2 \beta_1 D_2\}}{4s_1\{Es_2[D_2(1+\nu) + 2R\beta_1 D_1] - 4(1-\nu)R\beta_1^3 D_1 [D_2(1+\nu) + R\beta_1 D_1]\}}$$

$$T_0 = -\frac{p_i R^2 \beta_1^2 D_1 s_2 \{ERs_1 + 8(2-\nu)\beta_1 D_1 [R\beta_1 + (1-\nu)]\}}{4s_1\{Es_2[D_2(1+\nu) + 2R\beta_1 D_1] - 4(1-\nu)R\beta_1^3 D_1 [D_2(1+\nu) + R\beta_1 D_1]\}}. \tag{11.21}$$

11.3.1 Calculation Example

As a calculation example, we will again consider the cylindrical pressure vessel with flat-plate closures examined in the preceding section, but will now assume that the closures are also deformable in their own plane. All other conditions remaining equal, we will compare the results thus obtained with those for the same vessel with closures assumed to be rigid in their own plane, analyzing and discussing the substantial differences.

Introducing the quantities found in the previous calculation example (used for purposes of comparison) into relations (11.21) gives:

$$M_0 = 3.165 \times 10^{-2} p_i \,\text{Nm/m}$$
$$T_0 = -4.430 \times 10^{-1} p_i \,\text{N/m} \,.$$

Ceteris paribus, a comparison with the results obtained from the calculation example in the preceding section, where the closure was assumed to be infinitely rigid in its own plane, shows that the effect of the closure's in-plane deformability will entail changes in the values of the hyperstatic unknowns that are entirely negligible, as they will at most be in the order of a few percent. In this specific case, we find that the value of the bending moment M_0 passes from $3.147 \times 10^{-2} p_i$ Nm/m to $3.165 \times 10^{-2} p_i$ Nm/m, increasing only by 5.72 ‰, while the value of the shearing force T_0, which passes from $-4.376 \times 10^{-1} p_i$ N/m to $-4.430 \times 10^{-1} p_i$ N/m, increases by 1.23 %.

As regards the stress state, repeating the procedures used in the calculation example in the preceding section lead us to infer that the maximum values of axial stress σ_z and hoop stress σ_t in the cylindrical shell occur in the junction plane and are respectively:

$$(\sigma_z)_{z=0} = \frac{p_i R}{2s_1} + \frac{6M_0}{s_1^2} = 1,949\, p_i$$
$$(\sigma_t)_{z=0} = \frac{p_i R}{s_1} + \frac{6\nu M_0}{s_1^2} = 670\, p_i.$$

Maximum stress in the closure also occurs at the axis ($r = 0$), where the radial and hoop stresses assume the same value, viz.:

$$(\sigma_r)_{r=0} = (\sigma_t)_{r=0} = \frac{T_0}{s_2} + \frac{6}{s_2^2}\left(M_{r,p_i} + M_0\right) = \frac{T_0}{s_2} + \frac{6}{s_2^2}\left[\frac{(3+\nu)R^2 p_i}{16} + M_0\right] = 903\,p_i.$$

The differences in the maximum stress values also amount to a few percent at most. More specifically, the maximum value of the axial stress in the cylindrical shell, which passes from 1938 p_i to 1949 p_i, increases by 5.64 ‰, while the maximum value reached by both radial and hoop stresses at the center of the plate passes from 891 p_i to 903 p_i, increasing by 1.35 %. It is confirmed that the absolute maximum value of the stress state is that reached by the axial stress in the cylindrical shell's junction plane. This tensile stress arises at the outer radius.

11.4 Cylindrical Pressure Vessels with Flat-Plate Closures: Closure Deformable in Its Own Plane and Middle Plane Which does not Coincide with the Junction Plane

We will now consider a cylindrical pressure vessel with flat-plate closures welded to both ends of the shell as shown in Fig. 11.2a. Here as in the case presented in Sect. 11.3, we will assume that the flat-plate closures are deformable in their own plane as well as transversely. As shown in Fig. 11.2b, the shearing force T_0 due to the local discontinuity is applied in the junction plane, and thus has eccentricity $s_2/2$ with respect to the plate's middle plane. The eccentric load T_0 is statically equivalent to a force acting in this middle plane and a bending moment given by $(T_0 s_2/2)$. Due to the deformability, this statically equivalent load system causes a

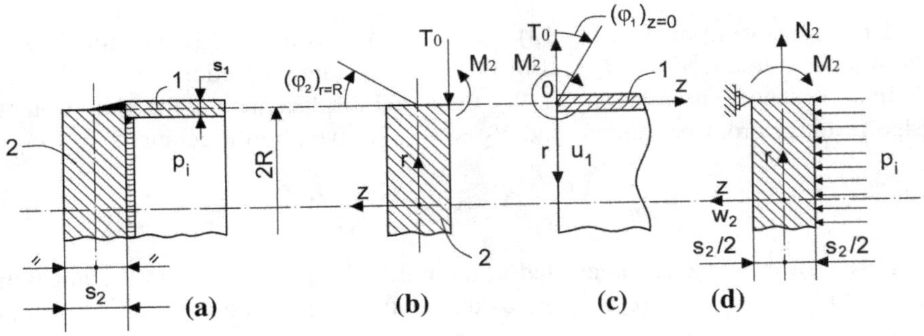

Fig. 11.2 **a** Cylindrical pressure vessel with flat-plate closures. **b** Flat-plate closure and local stress resultants M_0 and T_0 in the *junction* plane. **c** Cylindrical shell and local stress resultants M_0 and T_0 at the edge. **d** Flat-plate closure and reduced local stress resultants in its *middle* plane

radial displacement of the plate's outer edge as well as a rotation of the normal to the deflected surface at the edge of the plate.

To address the problem, we will assume that the outer radius of the two solid circular plates used as closures is equal to the radius R of the cylindrical shell's middle surface. All other conditions are the same as in the case examined in the preceding section, including the constraint at the outer edge of the plate which, as it must simultaneously permit rotation of the normal and radial translation of the plate's outer edge, must necessarily be treated as a simple support (Fig. 11.2d).

To determine the two hyperstatic unknowns M_0 and T_0, the two conditions of compatibility of displacements and rotations will again be those given by relations (11.16), noting however that some of the terms appearing in them will have different expressions correlated with the specific case examined here. As regards the cylindrical shell, the structure will be in the conditions shown in Fig. 11.1c, i.e., subjected to concentrated loads M_0 and T_0, for which relations (10.24) and (10.25) apply with $\beta = \beta_1$ and $D = D_1$ and we denote the quantities involving substructure 1 with subscript 1.

To calculate the quantities for the flat-plate closure (displacements, rotations, stress resultants, etc.), as specified in Sect. 11.2, it is convenient to consider the plate's middle plane, which is parallel to the junction plane and located at a distance $s_2/2$ from it (Fig. 11.2 b and d). To this end, we will use the new system of discontinuity loads N_2 and M_2, which from the static standpoint is equivalent to the T_0 and M_0 system and is defined by the relations:

$$N_2 = -T_0 \qquad\qquad M_2 = \frac{T_0 s_2}{2} - M_0. \tag{11.22}$$

With a procedure similar to that which leads to relation (11.13), we can demonstrate that the rotation of the normal to the deflected surface at the edge of the plate ($r = R$), which is equal to the sum of the contributions of the internal pressure p_i and the bending moment M_2, is given by the relation:

$$(\varphi_2)_{r=R} = (\varphi_{2,p_i})_{r=R} + (\varphi_{2,M_2})_{r=R} = \frac{R}{D_2(1+\nu)}\left(\frac{p_i R^2}{8} + M_2\right). \tag{11.23}$$

This differs from relation (11.13) with $M_0 = M_2$, as in the case examined here, the rotations produced by M_2 and by p_i are in the same direction (Fig. 11.2d).

In accordance with relation (11.19), the radial displacement at the plate's outer edge resulting from peripheral load $N_2 = -T_0$ is given by the relation:

$$(u_2)_{r=R} = \frac{RN_2}{Es_2}(1-\nu) = -\frac{RT_0}{Es_2}(1-\nu). \tag{11.24}$$

Here, however, we are interested in the radial displacement δ_R in the junction plane, which is equal to $(u_2)_{r=R}$ minus the contribution of rotation $(\varphi_2)_{r=R}$ and is thus:

$$\delta_R = (u_2)_{r=R} - \frac{s_2}{2}(\varphi_2)_{r=R}. \tag{11.25}$$

In the case examined here, relations (11.16) for compatibility of displacements in the junction plane become:

$$\begin{aligned}(u_1)_{z=0} &= \delta_R \\ (\varphi_1)_{z=0} &= (\varphi_2)_{r=R}.\end{aligned} \tag{11.26}$$

Introducing relations (11.2), (11.25) and (11.23) into these equalities and taking relations (11.22) and (11.24) into account together with the sign conventions indicated above, we arrive at the following system of two equations in the two hyperstatic unknowns M_0 and T_0:

$$\begin{aligned}\frac{(\beta_1 M_0 + T_0)}{2\beta_1^3 D_1} + \frac{p_i R^2}{E s_1}\left(1 - \frac{\nu}{2}\right) &= \frac{R T_0}{E s_2}(1-\nu) - \frac{R s_2}{2D_2(1+\nu)}\left(\frac{p_i R^2}{8} + M_2\right) \\ &= \frac{R T_0}{E s_2}(1-\nu) - \frac{R s_2}{2D_2(1+\nu)}\left(\frac{p_i R^2}{8} + \frac{T_0 s_2}{2} - M_0\right) \\ \frac{(2\beta_1 M_0 + T_0)}{2\beta_1^2 D_1} &= \frac{R}{D_2(1+\nu)}\left(\frac{p_i R^2}{8} + M_2\right) \\ &= \frac{R}{D_2(1+\nu)}\left(\frac{p_i R^2}{8} + \frac{T_0 s_2}{2} - M_0\right).\end{aligned} \tag{11.27}$$

Solving this system gives:

$$\begin{aligned}M_0 &= \frac{p_i R^2 \beta_1 D_1\left\{E R s_1 s_2(2+\beta_1 s_2) - 4R\beta_1^3 D_1\left[2(2-\nu)s_2^2 + (1-\nu)R s_1\right] + 8(2-\nu)(1+\nu)\beta_1 D_2 s_2\right\}}{8 s_1\left\{2ER\beta_1 D_1 s_2(1+\beta_1 s_2) - 4(1-\nu)R^2\beta_1^4 D_1^2 + (1+\nu)E D_2 s_2 + R\beta_1^3 D_1\left[E s_2^3 - 4(1-\nu^2)D_2\right]\right\}} \\ T_0 &= -\frac{p_i R^2 \beta_1^2 D_1 s_2\left\{E R s_1(1+\beta_1 s_2) + 8(2-\nu)\beta_1\left[R\beta_1 D_1 + (1+\nu)D_2\right]\right\}}{4 s_1\left\{2ER\beta_1 D_1 s_2(1+\beta_1 s_2) - 4(1-\nu)R^2\beta_1^4 D_1^2 + (1+\nu)E D_2 s_2 + R\beta_1^3 D_1\left[E s_2^3 - 4(1-\nu^2)D_2\right]\right\}}\end{aligned} \tag{11.28}$$

11.4.1 Calculation Example

As a calculation example, we will once again consider the cylindrical pressure vessel with flat-plate closures examined in the two preceding sections, and with all other conditions remaining equal will analyze the effects of the different type of junction between the two closures and the cylindrical shell shown in Fig. 11.2. We will consider the solid circular plates used as closures to be deformable both transversely and in their own plane.

By introducing the quantities for the closures found above into relations (11.28), we obtain:

$$\begin{aligned}M_0 &= 3.393 \times 10^{-2}\, p_i\, \mathrm{Nm/m} \\ T_0 &= -5.547 \times 10^{-1}\, p_i\, \mathrm{N/m}.\end{aligned}$$

Comparison with the results obtained from the calculation example in the preceding section, all other conditions remaining unchanged, shows that the effect of the displacement of the closure's middle plane relative to the junction plane caused by the different position of the welded seams (as Figs. 11.1 and 11.2 show, in the first case the displacement is zero and the two planes coincide, whereas in the second case the displacement is $s_2/2$) entails non-negligible changes in the values of the hyperstatic unknowns. In this specific case, we find that the value of the bending moment M_0 passes from $3.165 \times 10^{-2} p_i$ Nm/m to $3.393 \times 10^{-2} p_i$ Nm/m, with an increase of 7.20 %, while the value of the shearing force T_0, which passes from $-4.430 \times 10^{-1} p_i$ N/m to $-5.547 \times 10^{-1} p_i$ N/m, increases by a full 25.21 %.

Here again, maximum values of axial stress σ_z and hoop stress σ_t in the cylindrical shell occur in the junction plane and are respectively:

$$(\sigma_z)_{z=0} = \frac{p_i R}{2s_1} + \frac{6M_0}{s_1^2} = 2{,}086\, p_i$$

$$(\sigma_t)_{z=0} = \frac{p_i R}{s_1} + \frac{6\nu M_0}{s_1^2} = 711\, p_i.$$

In this case as before, the maximum stress in the closure occurs at the axis ($r = 0$), where the radial and hoop stresses assume the same value, viz.:

$$(\sigma_r)_{r=0} = (\sigma_t)_{r=0} = \frac{T_0}{s_2} + \frac{6}{s_2^2}\left(M_{r,p_i} - M_2\right)$$

$$= \frac{T_0}{s_2} + \frac{6}{s_2^2}\left[\frac{(3+\nu)R^2 p_i}{16} + M_0 - \frac{T_0 s_2}{2}\right] = 873\, p_i.$$

The latter normal stress is tensile and arises at the closure's outer face, on the axis. It is clear from a comparison of the stresses determined here and those obtained in the previous section that the differences in the maximum stress values are also significant. In particular, the maximum value of the axial stress in the cylindrical shell, which passes from 1949 p_i to 2,086 p_i, increases by 7.03 %, while the maximum value reached by both radial and hoop stresses at the plate's outer face on the axis ($r = 0$) passes from 903 p_i to 873 p_i, dropping by 3.32 %.

As the flat-plate closure is always thicker than the cylindrical shell, however, it is on the latter that the structural designer's attention must be focused. The absolute maximum value of the stress state of interest to the designer here is that reached by the axial stress in the shell at the junction. This maximum value is heavily influenced by the ratio s_2/s_1 of the two substructures' thicknesses, and, above all, by the construction solution used in welding the substructures together. The designer must concentrate on these factors in order to achieve maximum performance and efficiency.

As indicated by the scientific literature in this area (e.g., Watts and Lang 1952), to which the reader is directed for further details, it should be borne in mind that the stress concentration factor (Peterson 1974), defined as the ratio of the maximum stress to the reference stress given by the first relation (1.7) for any given ratio $d_i/s_1 \cong 2R/s_1$, rises rapidly as thickness ratio s_2/s_1 decreases. It is thus advisable that the flat-plate closure be as thick as possible by comparison with the cylindrical shell, i.e., that we have $s_2 \gg s_1$. For low values of thickness ratio s_2/s_1, the maximum stress (and, with it, the stress concentration factor) will be so high that it exceeds the material's yield-point stress. It is clear that the theory of elasticity on which the treatment used here comes up against its limits in this case, and the problem must be investigated further using theory of plasticity. It is equally clear that exceeding the yield-point stress is beneficial in redistributing the local stresses and thus bringing down their high peak values.

Lastly, it should be noted that for thickness ratios $s_2/s_1 > 1$, as is the case in almost all practical applications, where this ratio is rarely less than 4, the maximum stress is the axial stress in the cylindrical shell at the junction. If $s_2/s_1 \leq 1$, it can be demonstrated using the relations given above that the maximum stress is the radial stress in the flat-plate closure at the junction (Brownell and Young 1968). In this case, we will be able to optimize the design by selecting an appropriate optimal thickness ratio s_2/s_1, which in any case depends on ratio $2R/s_2$, which can ensure that these two stresses are equal. This possibility, however, is purely theoretical, given that in order to reduce the stress concentration factor and avoid the problems that would arise if there are large transverse displacements and the assumptions on which Kirchhoff's plate theory is based are thus not satisfied, ratios s_2/s_1 must be as high as possible, and in any case above those in the range indicated above.

11.5 Cylindrical Pressure Vessels with Flat-Plate Closures that are Deformable in Their Own Plane: Relations Obtained by Other Researchers

For purposes of comparison, the relations obtained by Watts and Lang (1952) and by Johns and Orange (1961) for the cylindrical pressure vessel with flat-plate closures discussed in the two preceding sections are given below.

On the basis of specific assumptions and considerations that can be found in their original paper, Watts and Lang arrived at the following two relations expressing the compatibility of displacements and rotations in the junction plane given by relations (11.26). These expressions are in dimensionless form and thus independent of the system of units employed, with all of the advantages that this entails.

$$\begin{aligned} a_1\frac{M_0}{p_i d_i^2} + a_2\frac{T_0}{p_i d_i} + a_3 &= a_4\frac{M_0}{p_i d_i^2} + a_5\frac{T_0}{p_i d_i} + a_6 \\ b_1\frac{M_0}{p_i d_i^2} + b_2\frac{T_0}{p_i d_i} + b_3 &= b_4\frac{M_0}{p_i d_i^2} + b_5\frac{T_0}{p_i d_i} + b_6. \end{aligned} \tag{11.29}$$

Solving this system gives the following expressions of the hyperstatic unknowns M_0 and T_0:

$$
\begin{aligned}
M_0 &= p_i d_i^2 \left[\frac{(a_3 - a_6)(b_5 - b_2) - (a_5 - a_2)b_3}{(a_4 - a_1)(b_5 - b_2) - (a_5 - a_2)(b_4 - b_1)} \right] \\
T_0 &= p_i d_i \left[\frac{(a_4 - a_1)b_3 - (a_3 - a_6)(b_4 - b_1)}{(a_4 - a_1)(b_5 - b_2) - (a_5 - a_2)(b_4 - b_1)} \right].
\end{aligned}
\tag{11.30}
$$

In these relations, we have $d_i \cong 2R$. The twelve factors appearing in them are given by the following expressions, which are clearly dependent on the elastic quantities (Poisson's ratio) and geometrical quantities that influence the substructure interaction we are dealing with here (note that in all cases $b_6 = 0$):

$$
\begin{aligned}
a_1 &= -3(1 - \nu)\tfrac{d_i}{s_2} & b_1 &= \tfrac{6(1-\nu)}{\beta_1^2 s_1 s_2} \\
a_2 &= 2(1 - \nu) & b_2 &= -\tfrac{3(1-\nu)}{\beta_1^2 d_i s_1} \\
a_3 &= \tfrac{3(1-\nu)d_i}{32 s_2} & b_3 &= -\tfrac{3(1-\nu)}{16\beta_1^2 s_1 s_2} \\
a_4 &= -\tfrac{\beta_1^2 d_i^2 s_2}{2 s_1} & b_4 &= -\beta_1 d_i \left(\tfrac{s_2}{s_1}\right)^2 \\
a_5 &= -\tfrac{\beta_1 d_i s_2}{2 s_1} & b_5 &= -\tfrac{1}{2}\left(\tfrac{s_2}{s_1}\right)^2 \\
a_6 &= -\tfrac{(2-\nu)s_2}{8 s_1} & b_6 &= 0.
\end{aligned}
\tag{11.31}
$$

Johns and Orange, on the other hand, use a structure diagram such as that shown in Fig. 11.1 to arrive at the following relations

$$
\begin{aligned}
M_0 &= \frac{p_i R^2}{8\lambda_1} \left\{ \frac{2c^3\lambda_1^3 + (1 - \nu)c^4\lambda_1^2 + 2(1 + \nu)(2 - \nu)}{2c^3\lambda_1^2 + \lambda_1[(1 - \nu)c^4 + (1 + \nu)] + (1 - \nu^2)c} \right\} \\
T_0 &= -\frac{p_i R}{4} \left\{ \frac{c^3\lambda_1^3 + 2(2 - \nu)c^3\lambda_1 + 2(1 + \nu)(2 - \nu)}{2c^3\lambda_1^2 + \lambda_1[(1 - \nu)c^4 + (1 + \nu)] + (1 - \nu^2)c} \right\},
\end{aligned}
\tag{11.32}
$$

where

$$
c = \frac{s_1}{s_2} \qquad \lambda_1 = \sqrt[4]{\frac{3(1 - \nu^2)R^2}{s_1^2}} = \beta_1 R. \tag{11.33}
$$

11.5.1 Calculation Example

As a calculation example, we will once again take the cylindrical pressure vessel with flat closures examined in the three preceding sections, comparing the results obtained using the relations proposed by Watts and Lang and by Johns and Orange with those obtained earlier.

In accordance with the procedure used by Watts and Lang, we introduce the problem data into relations (11.31) to obtain the following values of the dimensionless factors: $a_1 = -105$; $a_2 = 1.4$; $a_3 = 3.281$; $a_4 = -1321.802$; $a_5 = -51.416$; $a_6 = -0.85$; $b_1 = 63.550$; $b_2 = -0.635$; $b_3 = -1.986$; $b_4 = -411.328$; $b_5 = -8$; $b_6 = 0$. Substituting these factors into relations (11.30) then gives:

$$M_0 = 3.358 \times 10^{-2}\, p_i \,\mathrm{Nm/m}$$
$$T_0 = -5.432 \times 10^{-1}\, p_i \,\mathrm{N/m}\ .$$

Using the procedure proposed by Johns and Orange, on the other hand, calculating c and λ_1 with relations (11.33) and introducing the values thus found into relations (11.32) gives:

$$M_0 = 3.180 \times 10^{-2}\, p_i \,\mathrm{Nm/m}$$
$$T_0 = -4.324 \times 10^{-1}\, p_i \,\mathrm{N/m}.$$

A comparison of these results with those obtained assuming that the flat closure is also deformable in its own plane (see Sect. 11.3) indicates that:

- Using the Watts and Lang relations results in non-negligible differences in terms of bending moment M_0 (which passes from $3.165 \times 10^{-2}\, p_i$ Nm/m to $3.358 \times 10^{-2}\, p_i$ Nm/m and thus increases by 6.10 %) and very significant differences in terms of shearing force T_0 (which passes from $-4.430 \times 10^{-1}\, \mathrm{p_i}$ N/m to $-5.432 \times 10^{-1}\, \mathrm{p_i}$ N/m and thus increases by a full 22.62 %).
- Using the Johns and Orange relations results in entirely negligible differences both in terms of bending moment M_0 (which passes from $3.165 \times 10^{-2}\, p_i$ Nm/m to $3.180 \times 10^{-2}\, p_i$ Nm/m and thus increases by 4.74 ‰) and in terms of shearing force T_0 (which passes from $-4.430 \times 10^{-1}\, p_i$ Nm/m to $-4.324 \times 10^{-1}\, p_i$ N/m and thus drops by 2.39 %).

Comparing the same results with those obtained in Sect. 11.4, where we considered that the connection between the cylindrical shell and flat-plate closure is such that the junction plane is eccentric with respect to the closure's middle plane, shows that:

- Using the Watts and Lang relations results in entirely negligible differences both in terms of bending moment M_0 (which passes from $3.393 \times 10^{-2}\, p_i$ Nm/m to $3.358 \times 10^{-2}\, p_i$ Nm/m and thus drops by 1.03 %) and in terms of shearing force T_0 (which passes from $-5.547 \times 10^{-1}\, \mathrm{p_i}$ N/m to $-5.432 \times 10^{-1}\, \mathrm{p_i}$ N/m and thus drops by 2.07 %).
- Using the Johns and Orange relations results in non-negligible differences in terms of bending moment M_0 (which passes from $3.393 \times 10^{-2}\, p_i$ Nm/m to $3.180 \times 10^{-2}\, p_i$ Nm/m and thus drops by 6.28 %) and very significant differences in terms of shearing force T_0 (which passes from $-5.547 \times 10^{-1}\, \mathrm{p_i}$ N/m to $-4.324 \times 10^{-1}\mathrm{p_i}$ N/m and thus drops by a full 22.05 %).

These differences—some, some significant, some less so—depend in the different configuration of the connection between the flat-plate closure and the cylindrical shell, which can be such that the junction plane is eccentric with respect to the closure's middle plane. At the design stage, this fact must be taken into account by introducing calculation models that reflect the actual welded structural connection as closely as possible.

11.6 Cylindrical Pressure Vessels with Formed Closures Having Hemispherical Meridian Curve (Hemispherical Dished Heads)

We will consider a cylindrical pressure vessel having two end closures consisting of formed hemispherical heads (Fig. 11.3) whose middle surface has radius R equal to that of the middle surface of the cylindrical shell, which is of very long length; total length is thus equal to $(L + 2R)$, where L is very large. Let s be the thickness of all three component substructures making up the structure in question, which is under internal pressure p_i; as usual, we will assume that thickness s is small compared to R. As we did in the previous cases, in breaking down the structure into its component substructures, we will denote the quantities relating to the cylindrical shell and the hemispherical heads with the subscripts 1 and 2 respectively.

At a sufficient distance from the junction planes, the in-plane stress resultants acting on the three substructures can be accurately calculated with membrane theory, using relations (1.32) and (1.33). For the cylindrical shell, the meridian stress resultant, which coincides with the axial stress resultant (or axial membrane

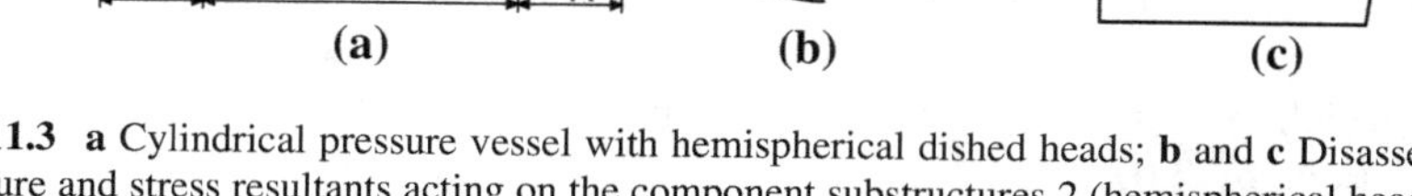

Fig. 11.3 **a** Cylindrical pressure vessel with hemispherical dished heads; **b** and **c** Disassembled structure and stress resultants acting on the component substructures 2 (hemispherical heads) and 1 (cylinder)

force), and the hoop stress resultant (or hoop membrane force) will thus be given by the relations:

$$N_{z,1} = \frac{p_i R}{2} \qquad N_{t,1} = p_i R \tag{11.34}$$

which, symbols apart, correspond to relations (1.37) and (1.36) respectively, while for the two hemispherical heads we will have:

$$N_2 = N_{z,2} = N_{t,2} = \frac{p_i R}{2}. \tag{11.35}$$

In the reference system shown in Fig. 4.1 (the z axis coinciding with the rotational axis and the r axis positive in the outward direction), the radial displacement of the cylindrical shell's middle generators, i.e., the extension of radius R (as the thickness is small, we assume $r_i \cong r_m = R$) as a result of the system of forces given by relations (11.34), is expressed by means of relation (1.40), which is written here in terms of inner radius $r_i \cong R$:

$$u_{R,1} = \frac{p_i R^2}{Es}\left(1 - \frac{\nu}{2}\right). \tag{11.36}$$

On the basis of relations (11.35) and the second relation (1.38) with $\varepsilon_\theta = \varepsilon_t = u/r$, the analogous radial displacement of the hemispherical head's middle meridian curve, i.e., the extension of the two hemispherical heads' equatorial radius, can be expressed by the following relation:

$$u_{R,2} = \frac{p_i R^2}{2Es}(1 - \nu). \tag{11.37}$$

A comparison of relations (11.36) and (11.37), in which the second subscripts denote the cylindrical shell 1 and the hemispherical head 2 respectively, shows that the cylindrical shell's radial displacement in the junction plane is greater than that of the hemispherical head. The ratio $u_{R,1}/u_{R,2}$ of the cylinder's radial displacement to that of the head is $(2 - \nu)/(1 - \nu)$, and is thus equal to around 2.43 for steel structures. The difference between the two radial displacements in the junction plane is $p_i R^2/2Es$. This discontinuity between the two displacements is shown in Fig. 11.3.

With membrane theory, a discontinuity in radial displacement is thus predicted in the junction plane. To overcome the membrane theory's limits and thus satisfy continuity between the mating edges of the two component substructures in this plane, it is necessary to apply bending theory (also called general theory) of shells of revolution. To do so, we introduce localized bending moments M_0 and shearing forces T_0 per unit length uniformly distributed along the circumference of radius R at the interface connecting the two component substructures.

To calculate the out-of-plane stress resultants M_0 and T_0 as well as the correlated discontinuity stresses, we first assume that they have a local character and that they rapidly die out as we move away from the junction plane. The cylindrical shell can be analyzed with more than sufficient accuracy using the relations developed and discussed in Sect. 10.2. Rigorous analysis of the deflection of the hemispherical head is a much more complex problem, as it can be performed only on the basis of general bending theory for shells of revolution under loads that are symmetrical with respect to the axis.

Here we will use an approximate solution based on the assumption that the bending is of importance only in the region of the hemispherical shell adjacent to the equatorial junction plane, and that this region can be treated locally as if it were cylindrical, i.e., as if it were an extension of the cylindrical shell. Meissner (1925), in fact, demonstrated that the error resulting from this approximation in calculating the stresses associated with local bending is small for thin hemispherical shells (less than 1 % for $R/s > 30$). Under this assumption, the cylindrical shell's mean generator and the mean meridian curve of the hemispherical head are continuations of each other, and the head can be regarded as locally cylindrical in a fairly wide neighborhood starting from the junction plane.

As the thicknesses of the cylindrical shell and the hemispherical head have been assumed to be equal, the uniformly distributed shearing forces T_0 on the two interface circumferences give rise to equal rotations of their common local tangent in the meridian plane, which coincides with the cylindrical shell's generator. This equal rotation of the two component substructures in the junction plane is indicated in Fig. 11.3. This fact means that the local value M_0 of the bending moment drops to zero, and the shearing force T_0 will be sufficient to eliminate the discontinuity mentioned above.

As under the above assumption, the hemispherical shell is considered to be the continuation of the circular cylindrical shell at the junction and the thicknesses of the two substructures are equal, the stress resultants arising from the discontinuity will cause the same radial displacement in both. In this context, the local value of the shearing force T_0 can be calculated simply by imposing that the two substructures' common radial displacement at their interface as calculated with bending theory equals the difference $\left(u_{R,1} - u_{R,2}\right)$ between the radial displacements obtained with membrane relations (11.36) and (11.37). Consequently, with regard to relation (10.24), where we set $M_0 = 0$ and $\beta z = 0$, this equality yields the relation:

$$-\frac{T_0}{\beta^3 D} = u_{R,1} - u_{R,2} = \frac{p_i R^2}{2Es} \tag{11.38}$$

from which, with the notations (10.13), we obtain:

$$T_0 = -\frac{p_i R^2 \beta^3 D}{2Es} = -\frac{p_i}{8\beta}. \tag{11.39}$$

In the latter relation, the minus sign indicates that the actual direction of T_0 is opposite to that shown in Fig. 11.3. It should be noted that relations (11.38) and (11.39), like the relations deriving from them, do not have subscripts identifying the quantities associated with edge effects in the two substructures, given that in this case we have $\beta_1 = \beta_2 = \beta$ and $D_1 = D_2 = D$.

The hyperstatic unknowns for the problem examined here are thus given by the relations:

$$M_0 = 0 \qquad T_0 = -\frac{p_i}{8\beta}. \tag{11.40}$$

Once hyperstatic unknowns M_0 and T_0 have been calculated, we need only substitute them into relations (10.24) and (10.25) to have a complete picture of the displacement field and all stress resultants. In particular, and also recalling relations (10.27), we obtain the following general relations expressing the deflection u, the rotation of the normal φ, the bending moments M_z and M_t, the shearing force T_z and the hoop stress resultant N_t as a function of βz:

$$\begin{aligned}
u &= -\frac{T_0}{2\beta^3 D} e^{-\beta z}\cos\beta z = -\frac{T_0}{2\beta^3 D}\phi_3 \\
\varphi &= \frac{du}{dz} = \frac{T_0}{2\beta^2 D} e^{-\beta z}(\cos\beta z + \sin\beta z) = \frac{T_0}{2\beta^2 D}\phi_1 \\
M_z &= -D\frac{d^2u}{dz^2} = -\frac{T_0}{\beta} e^{-\beta z}\sin\beta z = -\frac{T_0}{\beta}\phi_4 \\
M_t &= \nu M_z = \frac{\nu T_0}{\beta} e^{-\beta z}\sin\beta z = -\frac{\nu T_0}{\beta}\phi_4 \\
T_z &= -D\frac{d^3u}{dz^3} = -T_0 e^{-\beta z}(\cos\beta z - \sin\beta z) = -T_0\phi_2 \\
N_t &= -\frac{Esu}{R} = 2\beta R T_0 e^{-\beta z}\cos\beta z = 2\beta R T_0 \phi_3.
\end{aligned} \tag{11.41}$$

Taking relations (11.39) and (10.13) into account, the third of these relations gives the bending moment M_z:

$$M_z = -\frac{p_i R s}{8\sqrt{(1-\nu^2)}} e^{-\beta z}\sin\beta z = -\frac{p_i R s}{8\sqrt{(1-\nu^2)}}\phi_4. \tag{11.42}$$

By equating to zero the first derivative of M_z with respect to βz, we find that the maximum value of the bending moment occurs for $\beta z = \pi/4$, i.e., at a distance $z = \pi/4\beta$ from the junction interface. For the cylindrical shell, using the method of superposition and thus adding the tensile stress due to N_z, given by the first relation (11.34), and the tensile stress at the outer radius arising from M_z, which is equal to $6M_z/s^2$, with M_z given by relation (11.42) and with ϕ_4 calculated for $\beta z = \pi/4$, gives the following value for maximum axial stress:

$$
\begin{aligned}
(\sigma_z)_{max} &= \frac{p_i R}{2s} + \frac{3p_i R}{4s\sqrt{(1-\nu^2)}}(\phi_4)_{\beta z=\pi/4} \\
(\sigma_z)_{max} &= 1.293\frac{p_i R}{2s} \quad (\nu = 0.3).
\end{aligned}
\tag{11.43}
$$

For steels, then, the normal stress at the outer radius, to which both the axial load due to internal pressure and the local bending in the junction area contribute, is around 30 % higher than the axial stress correlated with membrane stress resultant N_z.

The total hoop stress in the cylindrical shell can also be calculated using the method of superposition. It is the sum of the stress correlated with the membrane stress resultant $N_\theta = N_t$, which is thus equal to $\sigma_t = N_\theta/s = p_i R/s$, of the hoop stress generated by radial displacement $u = u(z)$, and the stress produced by the bending moment $M_t = \nu M_z$. On the cylindrical shell's outer surface, we will thus have:

$$
\sigma_t = \frac{p_i R}{s} - \frac{Eu}{R} - \frac{6\nu M_z}{s^2} = \frac{p_i R}{2}\left[1 - \frac{\phi_3}{4} + \frac{3\nu}{4\sqrt{3(1-\nu^2)}}\phi_4\right].
\tag{11.44}
$$

It can be readily demonstrated that this stress reaches its maximum for $\beta z = 1.85$ and that, for $\nu = 0.3$, it is:

$$
(\sigma_t)_{max} = 1.032\frac{p_i R}{s}.
\tag{11.45}
$$

As the maximum stresses in the hemispherical heads are lower than those found in the cylindrical shell, the stresses to be considered in design analysis are those given by relations (11.43) and (11.44) or (11.45). In particular, the hoop stress given by the latter two relations is the stress state's absolute maximum stress, and must thus be closely monitored by the designer.

For the more general case of a cylindrical pressure vessel with heads whose form is that of a portion of a sphere and is thus generated by a meridian curve corresponding to a circular sector of central half-angle φ and radius R_2 which differs from the cylindrical shell's radius $R_1 = R$ (Fig. 11.4), and whose component substructures are of different thickness ($s_1 \neq s_2$), Johns and Orange (1961) arrive at the following relations:

$$
\begin{aligned}
M_0 &= \frac{p_i R_1 R_2}{2}\left\{\frac{\left[\frac{(1-\nu)}{\sin\varphi} - (2-\nu)\bar{c}\right]\left(\frac{1-\bar{c}^2}{2\lambda_2^2}\right) + \cos\varphi\left[\frac{\bar{c}^2}{\lambda_2}\left(1 + \frac{1}{\sqrt{\bar{c}\sin\varphi}}\right)\right]}{(1+\bar{c}^2)^2 + \frac{2\bar{c}^2}{\sqrt{\bar{c}\sin\varphi}}(1+\bar{c}\sin\varphi)}\right\} \\
T_0 &= \frac{p_i R_1}{2}\left\{\frac{\left[\frac{(1-\nu)}{\sin\varphi} - (2-\nu)\bar{c}\right]\left(\frac{1+\bar{c}^2\sqrt{\bar{c}\sin\varphi}}{\lambda_2}\right) - \cos\varphi(1+\bar{c}^2+2\bar{c}^2\sqrt{\bar{c}\sin\varphi})}{(1+\bar{c}^2)^2 + \frac{2\bar{c}^2}{\sqrt{\bar{c}\sin\varphi}}(1+\bar{c}\sin\varphi)}\right\}.
\end{aligned}
\tag{11.46}
$$

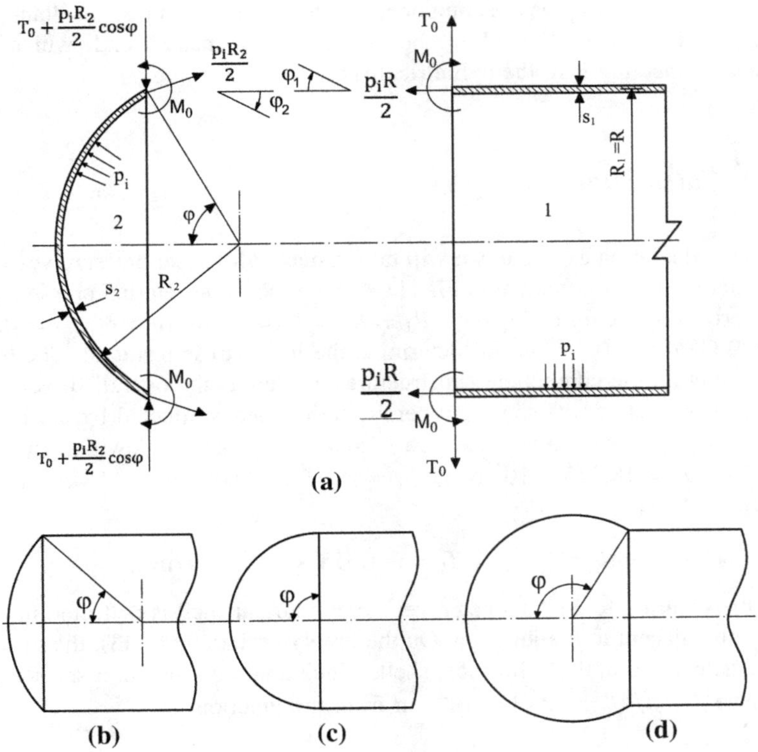

Fig. 11.4 Cylindrical pressure vessel with dished heads consisting of portions of a sphere: **a** Disassembled structure and stress resultants acting on component substructures 1 (cylinder) and 2 (head). **b** *Meridian curve* with $\varphi < \pi/2$. **c** *Meridian curve* with $\varphi = \pi/2$. **d** *Meridian curve* with $\varphi > \pi/2$

As usual, subscripts 1 and 2 denote the cylindrical shell and the head, while the new quantities appearing in the above relations are given by:

$$\bar{c} = \frac{s_2}{s_1} \quad \lambda_2 = \sqrt[4]{\frac{3(1-\nu^2)R_2^2}{s_2^2}} = \beta_2 R_2. \tag{11.47}$$

Relations (11.46) are entirely general and hold for $\varphi \lesseqgtr \pi/2$. For $\varphi = \pi/2$ (which entails that $R_1 = R_2 = R$), we obtain the solution for the hemispherical head. For $\varphi = \pi/2$ and $\bar{c} = 1$, from which $\lambda_2 = \beta_2 R_2 = \beta R$, relations (11.46) yield relations (11.40), which are thus a particular case of the hemispherical head.

It should be noted that when we are dealing with the general case of a spherical head generated by a meridian curve with $\varphi \neq \pi/2$, calculating the stresses and displacements to which it is subject will require that a load $T_0 + (p_i R_2/2)\cos\varphi$, i.e., a load equal to T_0 plus the component of the membrane stress resultant $p_i R_2/2$ in the junction plane (Fig. 11.4a) be applied to the head's end, which is the interface connecting it to the cylindrical shell.

11.6.1 Calculation Examples

As a *first* calculation example, we will take a steel cylindrical pressure vessel with hemispherical heads as shown in Fig. 11.3, under constant internal pressure p_i and characterized by the following data: $R_1 = R_2 = R = 2.0$ m; $s_1 = s_2 = s = 0.01$ m; $E = 200$ GPa; $\nu = 0.3$. We will determine the local stress resultants due to edge effect acting on the component substructures as well as the overall stresses in the regions of the cylindrical shell and hemispherical heads affected by the junction.

The data above give us the values of the following quantities: $D_1 = D_2 = D = 18.315 \times 10^3$ Nm; $\beta_1 = \beta_2 = \beta = 9.089$ m^{-1}. Using relations (11.40), we have:

$$M_0 = 0 \qquad T_0 = -1.375 \times 10^{-2} p_i \, \mathrm{N/m} \, .$$

As the shearing force T_0 is negative, the cylindrical shell is deflected inward in the region adjacent to the junction. On the basis of relation (11.43), the maximum axial tensile stress in the cylindrical shell, which arises on the outer surface and at a distance $z = \pi/4\beta = 8.641 \times 10^{-2}$ m from the junction, is:

$$(\sigma_z)_{max} = 129.3 \, p_i.$$

On the basis of relation (11.45), the maximum hoop stress in the cylindrical shell, which also arises on the outer surface and at a distance $z = 1.85/\beta = 0.204$ m from the junction, is:

$$(\sigma_t)_{max} = 206.4 \, p_i.$$

*****

As a *second* calculation example, we will take a steel cylindrical pressure vessel with heads whose meridian curve corresponds to a circular sector as shown in Fig. 11.4a, under constant internal pressure p_i and characterized by the following data: $R_1 = R = 2.0$ m; $R_2 = 2R_1 = 2R = 4.0$ m; $\varphi = 30°$; $s_1 = s_2 = s = 0.01$ m; $E = 200$ GPa; $\nu = 0.3$. In this case, however, we will determine only local stress resultants due to edge effect acting on the component substructures, and compare them with those obtained in the previous calculation example.

The data above give us the values of the following quantities: $D_1 = D_2 = D = 18.315 \times 10^3$ Nm; $\beta_1 = 9.089$ m^{-1}; $\beta_2 = 6.427$ m^{-1}; $c = 1$; $\lambda_2 = 25.708$ m/m. Introducing these quantities in relations (11.46) yields:

$$M_0 = 3.947 \times 10^{-2}\, p_i \text{ Nm/m}$$
$$T_0 = -3.611 \times 10^{-1}\, p_i \text{ N/m}.$$

By comparison with the results from the previous calculation example, the values of the local stress resultants increase significantly. First, it should be noted that the bending moment M_0 is no longer zero, and assumes a value that, though not quantifiable in percentage terms, is nevertheless appreciable. It should also be noted that there is an enormous increase in the shearing force T_0, which passes from $-1.375 \times 10^{-2} p_i$ N/m to $-3.611 \times 10^{-1} p_i$ N/m, a rise of 252.62 %. We can thus say that any departure from the hemispherical shape entails high local concentrations, with repercussions on the stress state.

11.7 Cylindrical Pressure Vessels with Formed Closures Having Semielliptical Meridian Curve (Elliptical Dished Heads)

We will now consider a cylindrical pressure vessel whose two end closures consist of heads having the form of semiellipsoids of revolution, and are thus generated by a semielliptical meridian curve with major semi-axis $a = R$, i.e., equal to the radius of the cylindrical shell, and minor semi-axis b (Fig. 11.5a). Here too, we consider the middle surfaces of the three component substructures. Let $K = a/b$ be the ellipse's flattening factor and s be the thickness of all three component substructures making up the structure in question, which is to be considered very long (total length $L + 2b$, where L is very large) and under internal pressure p_i. Here as for the cylindrical pressure vessel with hemispherical dished heads, at a sufficient distance from the junction planes, the stress resultants acting on the three substructures can be calculated with a sufficient degree of accuracy using membrane theory of shells, and thus, once again, with relations (1.32) and (1.33).

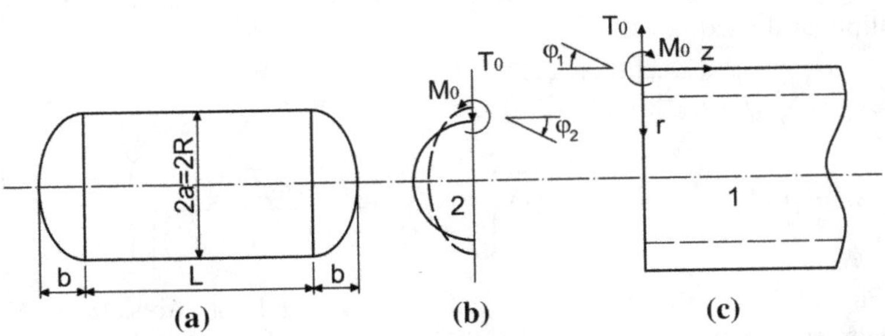

Fig. 11.5 **a** Cylindrical pressure vessel with semielliptical dished heads; **b** and **c** Disassembled structure and stress resultants acting on component substructures 2 (elliptical dished heads) and 1 (cylinder)

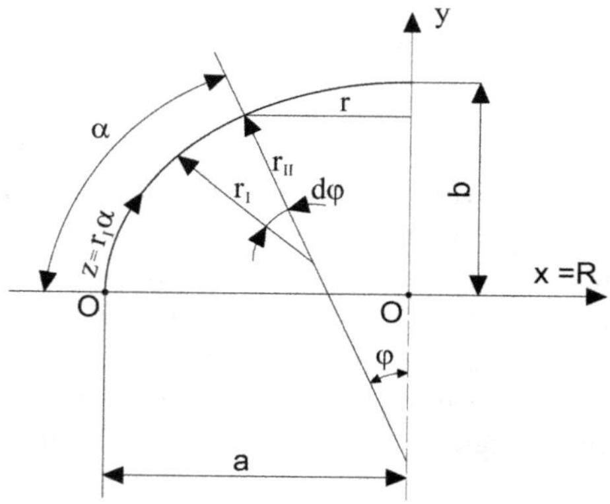

Fig. 11.6 Dished head having semielliptical *meridian curve*

For the cylindrical shell, the membrane stress resultants N_z and N_t are still those described by relations (11.34). For the two dished heads with semielliptical meridian curve, on the basis of relations (1.35) and (1.34) with the notations shown in Fig. 11.6 and recalling that $Q = \pi r^2 p_i$ and $Z = p_i$ (see Sect. 1.6), we will have (see also Huggaenberger 1925, 1927):

$$N_\varphi = \frac{p_i r_{II}}{2} \qquad N_\theta = p_i r_{II}\left(1 - \frac{r_{II}}{2r_I}\right). \tag{11.48}$$

Simple geometrical considerations show that the two principal radii of curvature r_I and r_{II} are given by the relations:

$$\begin{aligned} r_I &= \frac{\left[(aK)^2 + r^2(1-K^2)\right]^{3/2}}{(aK)^2} \\ r_{II} &= \left[(aK)^2 + r^2(1-K^2)\right]^{1/2}. \end{aligned} \tag{11.49}$$

Consequently, substituting relations (11.49) into relations (11.48) gives the following general expressions for the membrane stress resultants N_φ and N_θ in the semielliptical dished head:

$$\begin{aligned} N_\varphi &= \frac{p_i}{2}\sqrt{(aK)^2 + r^2(1-K^2)} \\ N_\theta &= p_i\sqrt{(aK)^2 + r^2(1-K^2)}\left\{1 - \frac{(aK)^2}{2\left[(aK)^2 + r^2(1-K^2)\right]}\right\}. \end{aligned} \tag{11.50}$$

In Fig. 11.7, the meridian stress $\sigma_\varphi = N_\varphi/s$ and hoop stress $\sigma_\theta = N_\theta/s$ in dimensionless form relative to the hoop stress $\sigma_{t,c} = p_i d_i/2s = p_i R/s$ in the cylindrical shell of inner radius $r_i = R = a$ under the same internal pressure p_i as the semielliptical dished head are plotted versus the radius r of the generic parallel

Fig. 11.7 Dimensionless membrane stresses in a semielliptical dished head: **a** Meridian stresses. **b** Hoop stresses

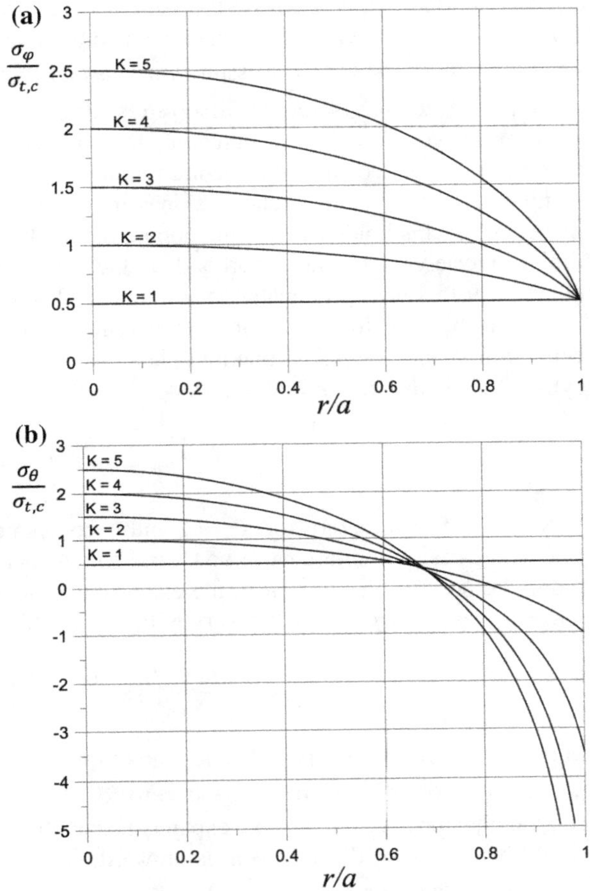

circle, likewise made dimensionless relative to the ellipse's major semi-axis $a = R$. The figure thus shows the dimensionless stresses $\sigma_\varphi/\sigma_{t,c}$ and $\sigma_\theta = \sigma_{t,c}$ as a function of ratio r/a. The curves in the figure refer to values of factor K varying from $K = 1$ (hemispherical head) to $K = 5$ (highly flattened semielliptical head). Relations (11.50) and the diagrams in Fig. 11.7 show:

- For K between 1 and 2, the maximum stresses occur at $r/a = 0$, i.e., on the rotational axis, where $(\sigma_\varphi)_{r/a=0} = (\sigma_\theta)_{r/a=0} = aKp_i/2s$ (that the two stresses are equal on the axis can also be seen from the fact that the head's apex is the tangent point of the osculating sphere).
- For $K = 2$, the meridian and hoop stresses on the axis are positive and equal in absolute value to the hoop stress in the equatorial plane, which is negative, i.e., compressive.
- For $K > 2$, the absolute maximum stress is the hoop stress in the equatorial plane, which is obviously negative.

The meridian stress is tensile throughout the entire range of variation for ratio r/a. From the axis, where it reaches its maximum value, it decreases towards the equatorial plane, at which it assumes a constant value of $(\sigma_\varphi)_{r/a=1} = ap_i/2s$ for any ellipse flattening factor. The hoop stress decreases from its maximum positive value at the axis as ratio r/a increases, drops to zero for values of r/a which can be readily calculated from the second relation (11.50), changes sign to become a compressive stress and reaches its maximum negative value in the equatorial plane. All of this holds true for membrane theory. The latter, however, reveals its limits precisely in the neighborhood of the junction plane, where the bending effects that this theory is unable to account for become predominant.

As can be seen from relations (11.50), the meridian and hoop stresses in the equatorial plane, i.e., the junction plane between the semi ellipsoid and the cylindrical shell where $r = R = a$, have the following values:

$$\sigma_\varphi = \frac{p_i a}{2s} \qquad \sigma_\theta = \frac{p_i a}{s}\left(1 - \frac{K^2}{2}\right). \tag{11.51}$$

Stress σ_θ drops to zero for $K^2 = 2$ and becomes negative for $K^2 > 2$.

On the basis of relations (11.51) and the second relation (11.38), where it should be recalled that $\varepsilon_\theta = u/r$, the extension of the equatorial radius of the two semielliptical dished heads is given by the relation:

$$u_{R,2} = \frac{p_i R^2}{Es}\left(1 - \frac{K^2}{2} - \frac{\nu}{2}\right), \tag{11.52}$$

where the second subscript 2 denotes the elliptical head shown in Fig. 10.5b. As we can see from this relation, $u_{R,2}$ is zero for $K = \sqrt{2-\nu}$ (for steels, $K \cong 1.3$), i.e., the equatorial radius neither expands nor contracts. In the junction plane, the dished head's radial displacement is outward, causing the major semi-axis of the ellipse to expand for $K < \sqrt{2-\nu}$. For $K > \sqrt{2-\nu}$, on the other hand, the radial displacement is inward, and the major semi-axis will obviously contract (as the minor semi-axis expands, the head approaches the hemispherical shape). In actual engineering practice, however, the flattening factor used for semielliptical dished heads is $K = 2$, as is strongly recommended by the ASME standards. Here, then, we must consider a deformation mechanics oriented towards the minimum elastic potential energy configuration (the spherical shape), resulting in contraction of the ellipse's major semi-axis and expansion of its minor semi-axis.

In the junction plane, the ratio $u_{R,1}/u_{R,2}$ of the cylindrical shell's radial displacement given by relation (11.36) to the semielliptical dished head's radial displacement in the equatorial plane given by relation (11.52) is $(2-\nu)/(2-\nu-K^2)$, while the difference between these quantities is given by relation:

$$u_{R,1} - u_{R,2} = \frac{p_i R^2 K^2}{2Es}. \tag{11.53}$$

Here again, as the thicknesses of the cylindrical shell and the semielliptical dished heads are equal, the uniformly distributed shearing forces T_0 in the junction interface cause equal rotations (Fig. 11.5b) of the common local tangent in the meridian plane, which coincides with the cylinder's generator. Once again, then, we will have $M_0 = 0$, and the shearing force T_0 will be sufficient to eliminate the discontinuity. As was done in the preceding section, we will calculate the local value of this shearing force by imposing that the radial displacements at the two structures' interface as calculated with bending theory equals the difference given by relation (11.53). Consequently, with regard to relation (10.24) where we set $M_0 = 0$ and $\beta z = 0$, this equality yields:

$$-\frac{T_0}{\beta^3 D} = u_{R,1} - u_{R,2} = \frac{p_i R^2}{2Es} K^2 \tag{11.54}$$

from which we obtain

$$T_0 = -\frac{p_i R^2 \beta^3 D}{2Es} K^2 = -\frac{p_i}{8\beta} K^2. \tag{11.55}$$

We thus see that the shearing force T_0 in the case of a semielliptical dished head is higher than that for a hemispherical dished head by a factor K^2. The discontinuity stresses obviously increase by the same factor. For $K = 2$ and $v = 0.3$, we will have the following in place of relations (11.43) and (11.45) respectively:

$$(\sigma_z)_{max} = \frac{p_i R}{2s} + \frac{3 p_i R}{s\sqrt{3(1-v^2)}} (\phi_4)_{\beta z = \pi/4} = 2.172 \frac{p_i R}{2s}. \tag{11.56}$$

$$(\sigma_t)_{max} = 1.128 \frac{p_i R}{s}. \tag{11.57}$$

In this case as before, the highest stress, and thus that which must be considered in the design conditions, is hoop stress σ_t.

11.8 The Problem of Discontinuity in Cylindrical Pressure Vessels with Formed Closures Having Semielliptical Meridian Curve

We will now return to the cylindrical pressure vessel with dished heads whose generating curve is a semielliptical meridian curve (see Fig. 11.5), under the most general assumption that the component substructures 1 (cylindrical shell) and 2 (semielliptical dished heads) have different thicknesses, i.e., that $s_1 \neq s_2$. All of the quantities that come into play in formulating and solving the problem of elastic interaction between these substructures are denoted by subscripts 1 and 2.

If we consider the cylindrical shell as separate from the dished heads, we are dealing with a long cylindrical shell subjected to concentrated loads on edges, such as that shown in Fig. 10.3 and treated in Sect. 10.2. It is obvious that the two edges can be treated in the same way by changing reference system from one end to the other end. All of the relations of interest in these conditions have already been obtained, and are shown below with the notations for the specific case examined here. We have:

$$
\begin{aligned}
u_1 &= -\frac{e^{-\beta_1 z}}{2\beta_1^3 D_1}[\beta_1 M_0(\cos\beta_1 z - \sin\beta_1 z) + T_0\cos\beta_1 z].\\
\varphi_1 &= \frac{e^{-\beta_1 z}}{2\beta_1^2 D_1}[2\beta_1 M_0\cos\beta_1 z + T_0(\cos\beta_1 z + \sin\beta_1 z)]\\
M_{z,1} &= \frac{e^{-\beta_1 z}}{\beta_1}[\beta_1 M_0(\cos\beta_1 z + \sin\beta_1 z) + T_0\sin\beta_1 z]\\
M_{t,1} &= \frac{\nu e^{-\beta_1 z}}{\beta_1}[\beta_1 M_0(\cos\beta_1 z + \sin\beta_1 z) + T_0\sin\beta_1 z]\\
T_{z,1} &= -e^{-\beta_1 z}[2\beta_1 M_0\sin\beta_1 z - T_0(\cos\beta_1 z - \sin\beta_1 z)]\\
N_{t,1} &= 2\beta_1 R e^{-\beta_1 z}[\beta_1 M_0(\cos\beta_1 z - \sin\beta_1 z) + T_0\cos\beta_1 z].
\end{aligned}
\tag{11.58}
$$

In the above relations, β_1 and D_1 are given by the relations:

$$
\beta_1^4 = \frac{Es_1}{4R^2 D_1} = \frac{3(1-\nu^2)}{R^2 s_1^2} \qquad D_1 = \frac{Es_1^3}{12(1-\nu^2)}. \tag{11.59}
$$

The stress resultants in the dished head having a semielliptical meridian curve must now be calculated. The general bending theory of axisymmetrically loaded and constrained shells of revolution should be used for this purpose. This bending theory, however, leads to governing differential equations whose solution is extremely complex, even for shells of constant thickness. Consequently, here we will employ an approximate method based on the assumptions first introduced by Reissner (1912) and Geckeler (1915) for cylindrical shells and generalized by Meissner (1913, 1915, 1925) to shells of revolution of an arbitrary shape of generating meridian curve. Under these assumptions, the two principal radii of curvature r_I and r_{II} are constant near the junction plane with the cylindrical shell (the closer the head's shape approaches the spherical, the more precise this assumption will be) and that all functions characterizing the stress and strain state in the shell near the junction plane, as well as their first derivatives, are small compared with their higher derivatives.

Using this approximate method, the complex governing equations can be reduced to a single fourth order homogeneous differential equation of the same type as the associated homogeneous equation of the non-homogeneous equation (10.14). This is known as the *edge effect differential equation* and will be used here in the formulation given to it by Coates (1929, 1930), who expressed these

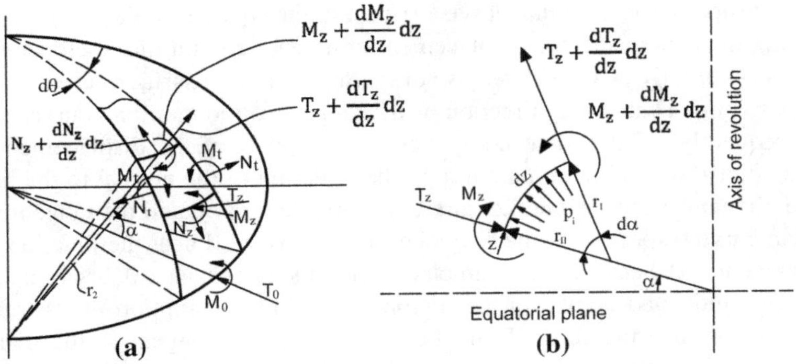

Fig. 11.8 **a** Elementary slice and differential element of an elliptical shell and stress resultants. **b** Small arc of the *meridian curve* and stress resultants acting on it

assumptions more precisely and adapted them to the treatment of shells having middle surface generated by a semielliptical meridian curve.

Coates considers a small element (differential element) of the elliptical shell (Fig. 11.8) bounded by two adjacent meridian planes forming the elementary angle $d\theta$ and by two adjacent parallel planes having angular coordinates α and $\alpha + d\alpha$ measured from the equatorial plane. The angular coordinate α is thus the complement of the spherical coordinate φ (Fig. 11.6); thus α and θ are latitude angle and colatitude angle, respectively. Stress resultants M_t and N_t act on the meridian sides of this differential element, and are constant because of axisymmetry. For the same reason, the transverse shearing force on these sides is zero. Stress resultants M_z, T_z and N_z act on the sides of the small element lying on the two parallels having angular coordinates α and $\alpha + d\alpha$. All five of these stress resultants are functions of curvilinear coordinate $z = r_I\alpha$, measured from the equatorial plane. Therefore, compared to the cylindrical shell where z is a linear coordinate, z here is a curvilinear coordinate.

The bending theory of shells of revolution having non-developable middle surface (or double curvature surface), and thus non-zero Gaussian curvature, unlike that for cylindrical shells, which on the contrary have zero Gaussian curvature, contemplates stress resultants that, though known to be very small, contribute to complicating the mathematical treatment of the problem to a very significant extent. As will be discussed in greater detail below, Coates omits these stress resultants in arriving at a solution that, though approximate, provides reliable results of proven practical significance, as has been demonstrated by the experimental data obtained by a number of researchers (see Brownell and Young 1968).

Coates bases this approximate treatment on the assumption that, for very small values of α (and thus for equally small values of the curvilinear coordinate z on the meridian curve, measured from the equatorial plane) and for sufficiently high values of ratio r_I/a (i.e., for values of the first principal radius of curvature at and near the junction plane which are high by comparison with the ellipse's major semi-axis, so that the portion of the meridian curve adjacent to the equatorial plane can regarded as a cylindrical segment), the displacement of a generic point of the ellipse in this zone can essentially be considered as normal to the middle surface, as we can thus ignore its component in the direction of the tangent to the meridian curve.

Consequently, of the three components u, v and w of the displacement of a generic point P on the meridian curve in the direction of the normal to the middle surface, the tangent to the parallel circle (or latitude circle), and the tangent to the meridian curve respectively, the only one to be considered near the junction is the first, since $v = 0$ because axisymmetry is assumed, and $w = 0$ because of the assumption indicated above. In the region where this assumption holds true, the component of shearing force T_0 in the direction of the tangent to the meridian curve can also be ignored, and, even admitting that it is transmitted entirely in that direction, the in-plane meridian stress resultant N_z and the correlated normal stress σ_z will thus be negligible.

In this case, moreover, the resistant force opposing the radial displacement of the small element and the elementary slice of shell between a pair of meridian planes indefinitely close together to which this differential element belongs is directly proportional to the same displacement and inversely proportional to the square of the second principal radius of curvature r_{II}. As this radius increases along with α, however, the ratio $1/r_{II}^2$ decreases with α, and thus with z. By analogy with the cylindrical shell, Coates assumes that the elementary slice of the elliptical shell shown in Fig. (11.8) behaves as a beam on a continuous elastic foundation with decreasing foundation modulus and subjected at the ends to a shearing force T_0 and a bending moment M_0.

The displacement u_2 in the direction of the normal to a generic point P on the generating meridian curve located at a distance from the junction plane defined by curvilinear coordinate z, and taken to be positive if directed inward, is governed by the following fourth order differential equation:

$$\frac{d^4 u_2}{dz^4} + 4\beta_2^4 u_2 = 0 \qquad (11.60)$$

where

$$\beta_2^4 = \frac{Es_2}{4r_{II}^2 D_2} = \frac{3(1-\nu^2)}{r_{II}^2 s_2^2} \qquad D_2 = \frac{Es_2^3}{12(1-\nu^2)}. \qquad (11.61)$$

The first of these last two relations contains the second principal radius of curvature r_{II} which, in shells generated by a semielliptical meridian curve as in most shells having non-zero Gaussian curvature, is variable. Consequently, β_2 is also variable. Differential equation (11.60) is thus the generalization of the homogeneous equation associated with Eq. (10.14), which is characterized by $r_{II} = a = R = const.$ The latter homogeneous equation, however, is a particular case of Eq. (11.60). As will be recalled, another particular case is that of the spherical shell, for which $r_I = r_{II} = R$; this can also be obtained from the elliptical shell for $K = 1$. Given these circumstances, Eq. (11.60) can be integrated by following the same procedure used to integrate the homogeneous equation associated with Eq. (10.14), where the only variation involves determining the integration constants.

Given that r_{II} varies with z and hence with β_2, Eq. (11.60) can be integrated numerically. If, however, we limit ourselves to a portion of the meridian curve whose curvilinear coordinate z is small enough that the second principal radius of curvature r_{II} can be regarded as constant in it, we can consider this elementary arc to be cylindrical, i.e., as if it were a continuation of the generator of the cylinder to which the semielliptical dished head is connected. This quasi-cylindrical path of the portion of the meridian curve adjacent to the junction is even more pronounced if the dished head is hemispherical or superelliptical, with appropriate exponents of the parameter governing its geometry (Gola and Vullo 1976). This quasi-cylindrical portion of the semielliptical dished head can be treated like the adjoining cylinder, with T_0 and M_0 changing signs with respect to those indicated in Fig. 10.3 and thus oriented as in Fig. 11.9, and with β_2 varying in accordance with ratio $1/\sqrt{r_{II}}$. It should be noted that we will have $\beta_2 = \beta_1$ in the equatorial plane, i.e., for $z = 0$, if and only if $s_1 = s_2$. Lastly, Coates assumes that the first derivative of β_2 with respect to z is negligible.

Considering the elementary arc of the meridian curve between the curvilinear coordinates z and $z + dz$ (Fig. 11.8b) to be under internal pressure p_i, recalling the correlation between M_z and radial displacement u_2 in the direction of the normal to the middle surface, and writing two equilibrium equations, the first for the

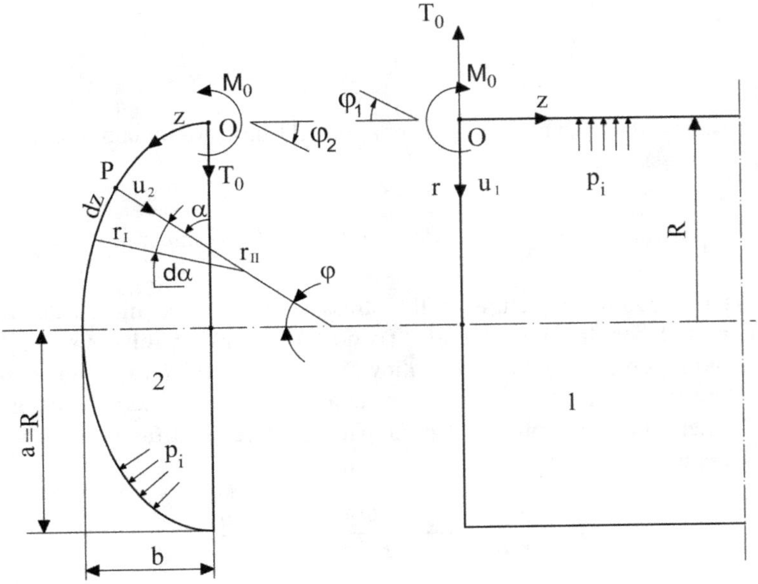

Fig. 11.9 Disassembled substructures, geometrical quantities and stress resultants in the equatorial plane

moments around, say, point O and the second for the components of the forces acting on it in the direction of the normal, gives the following relations:

$$\begin{aligned} M_{z,2} &= -D_2 \frac{d^2 u_2}{dz^2} \\ T_{z,2} &= -\frac{dM_{z,2}}{dz} = D_2 \frac{d^3 u_2}{dz^3} \\ p_i &= -\frac{dT_{z,2}}{dz} = -D_2 \frac{d^4 u_2}{dz^4}. \end{aligned} \tag{11.62}$$

Integrating relation (11.60) and noting that here again, for the same reasons described in Sect. 10.1, we must have $C_1 = C_2 = 0$, we arrive at the following relation:

$$u_2 = e^{-\beta_2 z}(C_3 \cos\beta_2 z + C_4 \sin\beta_2 z) \tag{11.63}$$

which gives displacement u_2 in the direction of the normal. Integration constants C_3 and C_4 are determined by imposing the following boundary conditions:

$$\begin{aligned} \left(M_{z,2}\right)_{z=0} &= -D_2 \left(\frac{d^2 u_2}{dz^2}\right)_{z=0} = M_0 \\ \left(T_{z,2}\right)_{z=0} &= D_2 \left(\frac{d^3 u_2}{dz^3}\right)_{z=0} = T_0. \end{aligned} \tag{11.64}$$

These conditions give:

$$C_3 = \frac{1}{2\beta_2^3 D_2}[T_0 - \beta_2 M_0] \qquad C_4 = \frac{M_0}{2\beta_2^2 D_2}. \tag{11.65}$$

Substituting these constants into relation (11.63), we obtain the following relation expressing displacement u_2;

$$u_2 = \frac{e^{-\beta_2 z}}{2\beta_2^3 D_2}[T_0 \cos\beta_2 z - \beta_2 M_0(\cos\beta_2 z - \sin\beta_2 z)]. \tag{11.66}$$

To have a complete picture of the stress resultants, we must calculate the circumferential bending moment M_t. To do so, by taking relations (10.7) into account and introducing in them the following relations which express the changes of curvature χ_z and χ_t in the meridian plane and in the plane perpendicular to the meridian, i.e., in the two planes of principal curvature, as a function of non-zero displacements u_2 and w_2:

$$\begin{aligned} \chi_z &= \frac{1}{r_I}\frac{d}{d\varphi}\left(\frac{w_2}{r_I} + \frac{du_2}{r_I d\varphi}\right) = \frac{1}{r_I}\frac{d\Theta_2}{d\varphi} \\ \chi_t &= \left(\frac{w_2}{r_I} + \frac{du_2}{r_I d\varphi}\right)\frac{\text{cotg}\varphi}{r_{II}} = \Theta_2 \frac{\text{cotg}\varphi}{r_{II}} \end{aligned} \tag{11.67}$$

where

$$\Theta_2 = \left(\frac{w_2}{r_I} + \frac{du_2}{r_I d\varphi}\right) \tag{11.68}$$

is the total rotation of the normal to the middle surface of the double curvature shell in the meridian plane (equal to the angle of rotation of the tangent to the shell meridian), taken as positive if it is in the positive direction of angle α, we obtain the following relations:

$$\begin{aligned} M_z &= -D_2\left[\frac{1}{r_I}\frac{d\Theta_2}{d\varphi} + \nu\Theta_2\frac{\text{cotg}\varphi}{r_{II}}\right] \\ M_t &= -D_2\left[\Theta_2\frac{\text{cotg}\varphi}{r_{II}} + \frac{\nu}{r_I}\frac{d\Theta_2}{d\varphi}\right]. \end{aligned} \tag{11.69}$$

Determining $d\Theta_2/r_I d\varphi$ from the first relation (11.69), substituting the expression thus found in the second relation (11.69), passing from colatitude angle φ to its complementary angle α (latitude angle) and recalling relation (11.68) with $w_2 \cong 0$ and $r_I d\varphi = r_I d\alpha = dz$, we obtain the following relation for the circumferential bending moment:

$$M_t = \nu M_z - D_2\left(1 - \nu^2\right)\frac{\text{tg}\alpha}{r_{II}}\frac{du_2}{dz}. \tag{11.70}$$

In these relations, it should be noted that the principal radii of curvature of the semielliptical dished head are indicated without the second subscript 2, which denotes substructure 2 (Fig. 11.9), as there is no danger of confusing them with the principal radii of curvature of the cylindrical shell. And in view of the foregoing considerations, the quantities of interest to us here are u_2, which is given by relation (11.66), and the following:

$$\begin{aligned} \Theta_2 &= -\frac{e^{-\beta_2 z}}{2\beta_2^2 D_2}[T_0(\cos\beta_2 z + \sin\beta_2 z) - 2\beta_2 M_0\cos\beta_2 z] \\ M_{z,2} &= -D_2\frac{d^2u_2}{dz^2} = -\frac{e^{-\beta_2 z}}{\beta_2}[T_0\sin\beta_2 z - \beta_2 M_0(\cos\beta_2 z + \sin\beta_2 z)] \\ M_{t,2} &= -\frac{\nu e^{-\beta_2 z}}{\beta_2}[T_0\sin\beta_2 z - \beta_2 M_0(\cos\beta_2 z + \sin\beta_2 z)] \\ &\quad + \frac{(1-\nu^2)\text{tg}\alpha}{r_{II}}\frac{e^{-\beta_2 z}}{2\beta_2^2}[T_0(\cos\beta_2 z + \sin\beta_2 z) - 2\beta_2 M_0\cos\beta_2 z] \\ T_{z,2} &= D_2\frac{d^3u_2}{dz^3} = e^{-\beta_2 z}[T_0(\cos\beta_2 z - \sin\beta_2 z) + 2\beta_2 M_0\sin\beta_2 z] \\ N_{t,2} &= -\frac{Es_2}{r_{II}}u_2 = -\frac{Es_2 e^{-\beta_2 z}}{2\beta_2^3 D_2 r_{II}}[T_0\cos\beta_2 z - \beta_2 M_0(\cos\beta_2 z - \sin\beta_2 z)]. \end{aligned} \tag{11.71}$$

Where $s_1 = s_2$, as is often the case in actual practice, noting that in the equatorial plane, which is the junction plane between the two substructures, we have $\beta_2 = \beta_1$ and $D_2 = D_1$, so that the integration constants given by relations (11.65) can be expressed in terms of β_1 and D_1 rather than quantities β_2 and D_2, and that these four quantities are expressed by relations (11.59) and (11.61), relations (11.66) and (11.71) are specialized as follows

$$
\begin{aligned}
u_2 &= \frac{e^{-\beta_2 z}}{2\beta_1^3 D_1}\left[T_0\cos\beta_2 z - \beta_1 M_0(\cos\beta_2 z - \sin\beta_2 z)\right] \\
\Theta_2 &= -\frac{\beta_2 e^{-\beta_2 z}}{2\beta_1^3 D_1}\left[T_0(\cos\beta_2 z + \sin\beta_2 z) - 2\beta_1 M_0\cos\beta_2 z\right] \\
M_{z,2} &= -\frac{\beta_2^2 e^{-\beta_2 z}}{\beta_1^3}\left[T_0\sin\beta_2 z - \beta_1 M_0(\cos\beta_2 z + \sin\beta_2 z)\right] \\
&= -\frac{Re^{-\beta_2 z}}{r_{II}\beta_1}\left[T_0\sin\beta_2 z - \beta_1 M_0(\cos\beta_2 z + \sin\beta_2 z)\right] \\
M_t &= \nu M_z - D_2\left(1 - \nu^2\right)\frac{\mathrm{tg}\alpha}{r_{II}}\Theta_2 \\
T_{z,2} &= \frac{\beta_2^3 e^{-\beta_2 z}}{\beta_1^3}\left[T_0\cos\beta_2 z - \sin\beta_2 z + 2\beta_1 M_0\sin\beta_2 z +\right] \\
&= -\frac{R}{r_{II}}\sqrt{\frac{R}{r_{II}}}\left[T_0\cos\beta_2 z - \sin\beta_2 z + 2\beta_1 M_0\sin\beta_2 z\right] \\
N_{t,2} &= -2\beta_1\frac{R^2 e^{-\beta_2 z}}{r_{II}}\left[T_0\cos\beta_2 z - \beta_1 M_0(\cos\beta_2 z - \sin\beta_2 z)\right].
\end{aligned}
\tag{11.72}
$$

It should be pointed out that the third and fifth of the above relations were written by expressing the ratios β_2^2/β_1^2 and β_2^3/β_1^3 by means of the relations given by the first relation (11.61) and (11.59).

To calculate the hyperstatic unknowns M_0 and T_0, it is necessary to impose boundary conditions in the junction plane. In this plane, which coincides with the equatorial plane of the semielliptical dished head, the two following conditions must be satisfied because of structural continuity:

- The total radial displacements at the edge of the cylindrical shell and at the edge of the semielliptical dished head, which are the algebraic sum of the displacements resulting from membrane loads correlated with internal pressure and the displacements due to discontinuity loads, i.e., the local bending moments and shearing forces, must have the same value.
- The two local deflection curves in any axial plane calculated taking both membrane loads and bending loads due to edge effects into account must have a common tangent at these two edges.

These two conditions are expressed through the following respective relations:

$$\begin{aligned} \left|\left(u_{2,m}\right)_0\right| - \left|\left(u_{1,m}\right)_0\right| &= \left|\left(u_{1,b}\right)_0\right| + \left|\left(u_{2,b}\right)_0\right| \\ (\varphi_1)_0 &= \left|(\Theta_2)_0\right| \end{aligned} \tag{11.73}$$

where $\left(u_{1,m}\right)_0$ and $\left(u_{2,m}\right)_0$, given by relations (11.36) and (11.52) respectively, are the radial displacements of the two substructures in the junction plane calculated with membrane theory, while $\left(u_{1,b}\right)_0$ and $\left(u_{2,m}\right)_0$, which are obtained respectively from the first relation (11.58) and from relation (11.66) both with $z = 0$, are the substructures' radial displacements in the junction plane calculated with bending theory. In relation (11.73), the first subscripts 1 and 2 denote the two substructures, the second subscripts m and b denote the membrane and bending theory respectively, the last subscript denotes the junction plane having coordinate $z = 0$, and the vertical bars outside the parenthesis denote the quantities' absolute values.

The first of these equations, that for continuity of displacements, requires that the difference between the absolute values of the radial displacements due to the membrane load arising as a result of internal pressure be compensated by the sum of the absolute values of the displacements due to discontinuity loads. The second equation requires that the rotations of the normals in the direction indicated in Fig. 11.9 (where $\varphi_2 = \Theta_2$) be equal.

To determine the hyperstatic unknowns M_0 and T_0, it is convenient to express the quantities for substructure 2 as a function of the corresponding quantities for substructure 1. To this end, introducing the ratio $c = s_1/s_2 = 1/\bar{c}$ of the two thicknesses and using relations (11.59) and (11.61), gives: $\beta_2 = \beta_1 c^{1/2}$; $\beta_2^2 = \beta_2^2 c$; $\beta_2^3 = \beta_1^3 c^{3/2}$; $D_2 = D_1 c^{-3}$. Taking the sign conventions for cylindrical shells introduced in Sect. 10.1 (see Fig. 10.1b) and those indicated in Fig. 11.9 into account, the second relation (11.73) is expressed as follows on the basis of the second relation (11.58) and the first relation (11.71), both written for $z = 0$:

$$\frac{2\beta_1 M_0 + T_0}{\beta_1^2 D_1} = \frac{T_0 - 2\beta_2 M_0}{\beta_2^2 D_2}. \tag{11.74}$$

With the quantities for substructure 2 expressed as a function of those for substructure 1 and solving for T_0, this relation gives:

$$T_0 = 2\beta_1 M_0 \frac{c^{5/2} + 1}{c^2 - 1}. \tag{11.75}$$

Again taking into account the sign conventions indicated above, relations (11.36) and (11.52) as well as the first relation (11.58) and relation (11.66), with the latter two written for $z = 0$, the first relation (11.73) becomes:

$$\frac{p_i R^2}{E s_2}\left(2 - K^2 - \nu\right) - \frac{p_i R^2}{E s_1}(2 - \nu) = \frac{\beta_1 M_0 + T_0}{\beta_1^3 D_1} + \frac{T_0 - \beta_2 M_0}{\beta_2^3 D_2}. \tag{11.76}$$

From this relation, again expressing the quantities for substructure 2 as a function of those for substructure 1, introducing the expression for T_0 as a function of M_0 given by relation (11.75) and solving for M_0, we obtain:

$$M_0 = \frac{p_i}{4\beta_1^2}\left\{\frac{[c(2-K^2-\nu)-(2-\nu)](c^2-1)}{(c^2+1)^2+2c^{3/2}(c+1)}\right\}. \tag{11.77}$$

Lastly, substituting the expression for M_0 thus found into relation (11.75) gives:

$$T_0 = \frac{p_i}{2\beta_1}\left\{\frac{[c(2-K^2-\nu)-(2-\nu)](c^{5/2}+1)}{(c^2+1)^2+2c^{3/2}(c+1)}\right\}. \tag{11.78}$$

In the particular case where the two substructures have the same thickness, so that $s_1 = s_2$, $c = 1$, $D_1 = D_2$ and $\beta_1 = \beta_2$, these two relations are specialized in the following:

$$\begin{aligned} M_0 &= 0 \\ T_0 &= -\frac{p_i K^2}{2\beta_1}. \end{aligned} \tag{11.79}$$

The second of these relations coincides with relation (11.55) with $\beta = \beta_1$. The results obtained in Sect. 11.7 are thus confirmed.

Once the hyperstatic unknowns have thus been found in both the general formulation (11.77) and (11.78) for different thicknesses and the particular formulation (11.79) for equal thicknesses, it is sufficient to substitute them into relations (11.58) to have all of the quantities of interest for the cylindrical shell, and into relations (11.66) and (11.71) to have those for the semielliptical dished head. Obviously, in the particular case of substructures of equal thickness, relations (11.72) can be used for the semielliptical dished head instead of relations (11.66) and (11.71).

As we are in the linear elastic range, the two substructures' stress state can be obtained by using the method of superposition, thus summing the effects of the membrane loads due to internal pressure p_i and the edge effect loads generated as a result of discontinuity. Recalling equations (11.34), we obtain the following membrane stresses resulting from the pressure load (the second subscript m denotes membrane stresses) for the cylinder:

$$\sigma_{z,m} = \frac{N_z}{s_1} = \frac{p_i R}{2s_1} \qquad \sigma_{t,m} = \frac{N_t}{s_1} = \frac{p_i R}{s_1}. \tag{11.80}$$

These stresses are uniformly distributed through the wall thickness. For the semielliptical dished head, recalling relations (11.50) the membrane stresses due to internal pressure are as follows:

$$\sigma_{\varphi,m} = \frac{N_\varphi}{s_2} = \frac{p_i}{2s_2}\sqrt{(aK)^2 + r^2(1-K^2)}$$
$$\sigma_{\theta,m} = \frac{N_\theta}{s_2} = \frac{p_i}{s_2}\sqrt{(aK)^2 + r^2(1-K^2)}\left\{1 - \frac{(aK)^2}{2\left[(aK)^2 + r^2(1-K^2)\right]}\right\}. \qquad (11.81)$$

These stresses are also uniformly distributed through the wall thickness.

As a result of the discontinuity due to edge effects, we will have the following stress state in the cylinder (note that the second subscript 1 denoting the cylinder is neglected here, as no confusion is possible):

$$\sigma_{z,b} = \pm\frac{6M_z}{s_1^2} \quad \sigma'_{t,b} = \pm\frac{6M_t}{s_1^2} \quad \tau_{z,b} = \frac{T_z}{s_1} \quad \sigma''_{t,b} = \frac{N_t}{s_1} \qquad (11.82)$$

where M_z, M_t, T_z and N_t are given by the last four relations (11.58) into which relations (11.77) and (11.78) are substituted. Whereas stresses $\sigma_{z,b}$ and $\sigma'_{t,b}$ correlated with bending moments M_z and M_t are distributed linearly through the wall thickness and reach their maximum values, one positive and the other negative, at the inner and outer radius in relation to the sign of M_0 and T_0, the stresses correlated with T_z and N_t are uniformly distributed through the wall thickness. It should also be noted that $\tau_{z,b}$ represents the average shear stress; assuming that it is distributed, as in beams, according to a parabolic function through the wall thickness, it will be zero at the outer radius and inner radius, and will assume its maximum value of $(3/2)\tau_{z,b}$ on the middle fiber (neutral fiber), where the stresses arising from bending moments drop to zero. Lastly, it should be pointed out that relations (11.82) are entirely general and apply for any value of the coordinate z, i.e., at any distance z from the junction plane.

Again a result of the discontinuity due to edge effects, we will have the following stress state in the semielliptical dished head (here as before, no confusion is possible, so the second subscript 2 denoting the head is omitted):

$$\sigma_{z,b} = \pm\frac{6M_z}{s_2^2} \quad \sigma'_{t,b} = \pm\frac{6M_t}{s_2^2} \quad \tau_{z,b} = \frac{T_z}{s_2} \quad \sigma''_{t,b} = \frac{N_t}{s_2} \qquad (11.83)$$

where M_z, M_t, T_z and N_t are given here by the last four relations (11.71) into which relations (11.77) and (11.78) are substituted. The observations made above for the cylindrical shell also apply here, but the following important notations should be added:

- As stress resultants M_0 and T_0 deriving from discontinuity have opposite directions in the cylinder and in the semielliptical dished head, the stresses correlated with bending moments M_z, M_t and circumferential forces N_t will have opposite signs in the two substructures. More specifically, the stress correlated with N_t in the cylinder will be compressive, and the outer radius and the inner radius will be respectively stretched and compressed as a result of the bending moments, while in the semielliptical dished head the stress deriving from the circumferential stress resultant N_t will be tensile, and the bending moments will cause the outer radius to be compressed and the inner radius to be stretched.

- Relations (11.83) do not have general validity and can be used only in a circumscribed region adjacent to the junction plane (i.e., for very small values of curvilinear coordinate z) where the second principal radius of curvature r_{II} can be considered constant with sufficient accuracy.

In the particular case where the thicknesses of the two substructures are equal (here, we will have $s_1 = s_2 = s$, $D_1 = D_2 = D_0$ and $\beta_1 = \beta_2 = \beta_0$), we will obtain the following relations for the cylinder on the basis of relations (11.58), (11.79) and (11.82):

$$\begin{aligned}
&u_1 = -\frac{p_i R^2 K^2}{4Es} e^{-\beta_0 z} \cos\beta_0 z && u_2 = -u_1\\
&\left(\sigma_{z,b}\right)_e = \frac{3p_i K^2}{4s^2\beta_0^2} e^{-\beta_0 z} \sin\beta_0 z && \left(\sigma'_{t,b}\right)_e = \nu\left(\sigma_{z,b}\right)_e\\
&\left(\sigma_{z,b}\right)_i = -\left(\sigma_{z,b}\right)_e && \left(\sigma'_{t,b}\right)_i = -\left(\sigma'_{t,b}\right)_e\\
&\tau_{z,b} = \frac{p_i K^2}{8s\beta_0} e^{-\beta_0 z}\left[\cos\beta_0 z - \sin\beta_0 z\right] && \sigma''_{t,b} = -\frac{p_i R K^2}{4s} e^{-\beta_0 z} \cos\beta_0 z
\end{aligned} \tag{11.84}$$

where the new subscripts e and i denote the substructure's outer and inner radii respectively. Under the same conditions, the same relations (11.84) but with the signs changed are obtained for the semielliptical dished head on the basis of relations (11.72), (11.79) and (11.83).

All of the quantities defined by relations (11.84) are exponentially damped trigonometric functions and, taking the constant factors appearing in them into account, vary according to the functions ϕ_1, ϕ_3 and ϕ_4 represented in Fig. 10.4. The two most important stresses due to edge effects in the cylinder are stresses $\left(\sigma_{z,b}\right)_e = -\left(\sigma_{z,b}\right)_i$ and the stress $\sigma'_{t,b}$, as their values are such as to make the contributions of the membrane stresses in the region adjacent to the junction plane almost entirely insignificant. Stresses $\left(\sigma'_{t,b}\right)_e = -\left(\sigma'_{t,b}\right)_i$ and $\tau_{z,b}$, on the other hand, are relatively lower. As mentioned elsewhere, all of the local stresses deriving from the discontinuity die out very quickly and their values are low enough to be ignored for distances $z > 2.44\sqrt{Rs}$ from the junction plane.

For a general evaluation of the stress state in the cylinder, the membrane stresses deriving from internal pressure must be superposed on the local stresses due to edge effects. For this purpose, it must be emphasized that some of these stresses are distributed uniformly through the wall thickness, while others vary linearly along it and assume their maximum positive value at the outer radius and their maximum negative value at the inner radius. It is thus advisable to calculate the stresses at the outer and inner radius and compare the resulting stresses with those allowable for the material used for the substructure in question.

For the cylinder, the highest stress value, viz., $-(p_i RK^2/4s)$, is that reached by stress $\sigma''_{t,b}$. It is negative, i.e., compressive, and occurs at the junction plane, where $e^{-\beta_0 z}\cos\beta_0 z$ is unity. The stresses correlated with the bending moments are zero in the junction plane, as $M_0 = 0$. The first maximum for $\sigma_{z,b}$, equal to

$\pm 0.32\left(3p_iK^2/4s^2\beta_0^2\right)$, occurs at $\beta_0 z = \pi/4$, i.e., at a distance $z = 0.61\sqrt{Rs}$ from the junction plane. Given that, for $\beta_0 z = \pi/4$, we have $e^{-\beta_0 z}\sin\beta_0 z = e^{-\beta_0 z}\cos\beta_0 z = 0.32$, we also have a compressive stress $\sigma''_{t,b} = -0.32(p_iRK^2/4s)$ in the same plane where the stresses deriving from the bending moments are maximum.

As was indicated earlier, the stresses in the semielliptical dished head are the same as in the cylindrical shell with the sign changed, providing that the assumptions at the basis of this treatment are satisfied. However, the portion of the semielliptical dished head adjacent to the junction plane, where the second principal radius of curvature r_{II} can be regarded as constant, may be too small by comparison with the region where the local stresses die out. In this case, a sufficient number of curvilinear segments where radius r_{II} can be regarded as constant must be identified on the meridian curve starting from the junction plane, so as to cover the entire region where local stresses are significant. Apart from in the first portion adjacent to the junction, which for the reasons indicated above may be too short, relations (11.84) with the sign changed no longer hold and must be substituted by the following:

$$
\begin{aligned}
u_2 &= \frac{p_iR^2K^2}{4Es}\left(e^{-\int_0^z \beta_2 dz}\right)\cos\left(\int_0^z \beta_2 dz\right)\\
\left(\sigma_{z,b}\right)_e &= -\frac{3p_iK^2R^2}{4sr_{II}\sqrt{3(1-\nu^2)}}\left(e^{-\int_0^z \beta_2 dz}\right)\sin\left(\int_0^z \beta_2 dz\right)\\
\left(\sigma'_{t,b}\right)_e &= \nu\left(\sigma_{z,b}\right)_e\\
\left(\sigma_{z,b}\right)_i &= -\left(\sigma_{z,b}\right)_e\\
\left(\sigma'_{t,b}\right)_i &= -\left(\sigma'_{t,b}\right)_e\\
\tau_{z,b} &= -\frac{\sqrt{2}p_iK^2R^2}{8s\beta_0 r_{II}^2}\left(e^{-\int_0^z \beta_2 dz}\right)\cos\left(\frac{\pi}{4}+\int_0^z \beta_2 dz\right)\\
\sigma''_{t,b} &= -\frac{p_iR^2K^2}{4sr_{II}}\left(e^{-\int_0^z \beta_2 dz}\right)\cos\left(\int_0^z \beta_2 dz\right).
\end{aligned}
\tag{11.85}
$$

To calculate the values of the quantities in whose expression the integral $\int_0^z \beta_2 dz$ occurs, it is necessary to proceed numerically.

For the two most significant stresses, the following stress intensification factors are often introduced where relations (11.84) hold. These factors are obtained by relating the absolute values of stresses $\left(\sigma_{z,b}\right)_e$ and $\sigma''_{t,b}$ to the hoop stress $\sigma_t = p_iR/s_1 = p_iR/s$ in the cylindrical shell due to internal pressure as determined using membrane theory and taken as the reference stress:

$$I_z = \frac{3K^2}{4sR\beta_0^2} e^{-\beta_0 z} \sin\beta_0 z$$
$$I_t = \frac{K^2}{4} e^{-\beta_0 z} \cos\beta_0 z. \tag{11.86}$$

Where relations (11.85) must be used, these stress intensification factors are expressed by the following relations:

$$I_z = \frac{3RK^2}{4r_{II}\sqrt{3(1-\nu^2)}} \left(e^{-\int_0^z \beta_2 dz} \right) \sin\left(\int_0^z \beta_2 dz \right)$$
$$I_t = \frac{RK^2}{4r_{II}} \left(e^{-\int_0^z \beta_2 dz} \right) \cos\left(\int_0^z \beta_2 dz \right). \tag{11.87}$$

11.8.1 Calculation Example

As a calculation example, we will take a steel cylindrical pressure vessel with dished heads having middle surface generated by a semielliptical meridian curve as shown in Fig. 11.5, under constant internal pressure p_i and characterized by the following data: $a = R = 1.0$ m; $K = a/b = 2$; $s_1 = s_2 = s = 0.01$ m; $E = 200$ GPa; $\nu = 0.3$. We will determine the local stress resultants due to edge effect acting on the component substructures as well as the overall stresses in the regions of the cylindrical shell and semielliptical dished heads affected by the junction.

The data above give us the values of the following quantities: $D_1 = D_2 = D = 18.315 \times 10^3$ Nm; $(\beta_1)_{z=0} = (\beta_2)_{z=0} = (\beta)_{z=0} = 12.854$ m^{-1}; $c = s_1/s_2 = 1$. Using relations (11.77) and (11.78) or relations (11.79), we have:

$$M_0 = 0; \quad T_0 = -3.890 \times 10^{-2} p_i \,\text{N/m} .$$

As the shearing force T_0 is negative, the cylindrical shell is deflected inward in the region adjacent to the junction plane. On the basis of relation (11.56), the maximum axial tensile stress in the cylindrical shell, which arises on the outer surface and at a distance $z = \pi/4\beta = 6.110 \times 10^{-2}$ m from the junction plane, is:

$$(\sigma_z)_{max} = 108.6\, p_i.$$

On the basis of relation (11.57), the maximum hoop stress in the cylindrical shell, which also arises on the outer surface and at a distance $z = 1.85/\beta = 0.144$ m from the junction plane, is:

$$(\sigma_t)_{max} = 112.8\, p_i.$$

Using the relations presented in this and in the previous section, it can be readily demonstrated that the maximum stresses in the dished heads with a semielliptical meridian curve are lower than those found in the cylindrical shell. Consequently, the stresses to be considered in design analysis for the structure in question are those given by the above relations. Clearly, the designer must devote particular attention to the maximum hoop stress in the cylindrical shell, as it is the stress state's absolute maximum stress.

11.9 Thickness Discontinuities Between Cylindrical Shells

All of the cases presented in the preceding sections involved cylindrical pressure vessels having constant thickness along their entire length. In some cases, however, we may be dealing with cylindrical pressure vessels consisting of two or more cylindrical shells of equal mean radius R but different thickness, welded together with circumferential seams.

Here as in the previous cases, local stress resultants due to the thickness discontinuity are generated at the interface plane between the two contiguous substructures. Idealizing the two disassembled substructures as shown in Fig. 11.10, we see that as a result of internal pressure p_i, the radial displacements of the two cylinders' common generator (once again, it is assumed that thicknesses s_1 and s_2, with $s_1 \neq s_2$, are small by comparison with the mean radius R and that the latter thus also represents the cylinders' inner and outer radii), calculated with membrane relation (11.36), will differ because the two thicknesses are different. In particular, the radial displacement of the thinner cylinder's generator will be greater than the that of the thicker cylinder.

The local discontinuity loads M_0 and T_0 arising as a result of the two substructures' different response to internal pressure p_i should guarantee structural continuity in the interface plane between them, in terms of both radial displacements and rotation of the normals. With reference to Fig. 11.10 and the sign conventions indicated therein, we can write the following compatibility conditions in the junction plane:

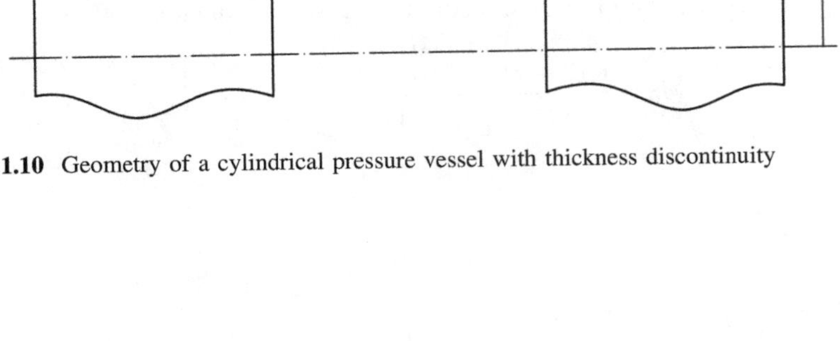

Fig. 11.10 Geometry of a cylindrical pressure vessel with thickness discontinuity

$$\begin{aligned} &\left|(u_{R,2})_0\right| - \left|(u_{R,1})_0\right| = \left|(u_{1,b})_0\right| + \left|(u_{2,b})_0\right| \\ &(\varphi_1)_0 = \left|(\varphi_2)_0\right|. \end{aligned} \tag{11.88}$$

The first of these equations, that for continuity of displacements, requires that the difference between the absolute values of the radial displacements due to the membrane load arising as a result of internal pressure be compensated by the sum of the absolute values of the displacements due to discontinuity loads. The second equation requires that the rotations of the normals in the direction indicated in Fig. 11.10 be equal. The meaning of the subscripts is as indicated in the preceding sections.

Here again, to determine the hyperstatic unknowns M_0 and T_0, it is convenient to express the quantities for substructure 2 as a function of the corresponding quantities for substructure 1. As in the preceding section, introducing the ratio $c = s_1/s_2 = 1/\bar{c}$ of the two thicknesses and using relations (11.59) and (11.61) gives: $\beta_2 = \beta_1 c^{1/2}$; $\beta_2^2 = \beta_2^2 c$; $\beta_2^3 = \beta_1^3 c^{3/2}$; $D_2 = D_1 c^{-3}$. Taking the sign conventions for cylindrical shells introduced in Sect. 10.1 (see Fig. 10.1b) and those indicated in Fig. 11.10 into account, the second relation (11.88) is expressed as follows on the basis of the second relation (11.58) written for $z = 0$ and specialized for both substructures 1 and 2:

$$\frac{2\beta_1 M_0 + T_0}{\beta_1^2 D_1} = \frac{T_0 - 2\beta_2 M_0}{\beta_2^2 D_2}. \tag{11.89}$$

With the quantities for substructure 2 expressed as a function of those for substructure 1 and solving for T_0, this relation gives:

$$T_0 = 2\beta_1 M_0 \frac{c^{5/2} + 1}{c^2 - 1}. \tag{11.90}$$

On the basis of relation (11.36) and the corresponding obtained from it for substructure 2, as well as on the basis of the first relation (11.58) written for $z = 0$ and specialized for both substructures, and once again taking into account the sign conventions for cylindrical shells as compared with those indicated in Fig. 11.10, the radial displacements compatibility equation given by the first relation (11.88) leads to:

$$\frac{p_i R^2 (2 - \nu)(c - 1)}{E s_1} = \frac{\beta_1 M_0 + T_0}{\beta_1^3 D_1} + \frac{T_0 - \beta_2 M_0}{\beta_2^3 D_2}. \tag{11.91}$$

From this relation, again expressing the quantities for substructure 2 as a function of those for substructure 1, introducing the expression for T_0 as a function of M_0 given by relation (11.90) and solving for M_0, we obtain:

$$M_0 = \frac{p_i(2 - \nu)}{4\beta_1^2} \left[\frac{(c - 1)(c^2 - 1)}{(c^2 + 1)^2 + 2c^{3/2}(c + 1)} \right]. \tag{11.92}$$

Lastly, substituting the expression for M_0 thus found into relation (11.90) gives:

$$T_0 = \frac{p_i(2-\nu)}{2\beta_1}\left[\frac{(c-1)(c^{5/2}+1)}{(c^2+1)^2+2c^{3/2}(c+1)}\right]. \tag{11.93}$$

When examining substructure 2, it is obvious that the sign of the shearing force T_0 must be changed because of the convention indicated in Fig. 11.10.

11.9.1 Calculation Example

To have an idea of the local stress resultants due to edge effects arising, with all other conditions remaining equal, from the thickness discontinuity, we will consider the cylindrical pressure vessel shown in Fig. 11.3. We will assume that the cylindrical shell is fabricated from roll formed tubular segments, each having axial length $L/2$, equal mean radius R and thicknesses such that $s_1 = 2s_2$, joined circumferentially by means of a butt welded seam located in the plane perpendicular to the axis bisecting the length of the structure. The vessel is under constant internal pressure p_i and characterized by the following data: $R = 1.0$ m; $s_1 = 0.02$ m; $s_2 = 0.01$ m; $E = 200$ GPa; $\nu = 0.3$. We will determine the hyperstatic unknowns that would be absent if the structure were to be considered as having uniform thickness.

The data above give us the values of the following quantities: $D_1 = 18.315\times 10^3$ Nm; $D_2 = 146.520 \times 10^3$ Nm; $\beta_1 = 12.854$ m^{-1}; $\beta_2 = 9.089$ m^{-1} ; $c = s_1/s_2 = 2$. Using relations (11.92) and (11.93), we have:

$$M_0 = 2.661 \times 10^{-4}\, p_i \,\mathrm{Nm/m}$$
$$T_0 = 4.561 \times 10^{-3}\, p_i \,\mathrm{N/m}\ .$$

The shearing force T_0 is positive for the thicker substructure and negative for the thinner substructure.

References

Abramowitz, M., & Stegun, I. H. (1972). *Handbook of mathematical functions*. New York: Dover Publications, Inc.

Alexandrov, A. V., & Potapov, V. D. (1990). *The fundamentals of the theory of elasticity and plasticity*. Moscow: Izd-vo Vyshaya Shkola (in russo).

Almen, J. O., & Black, P. H. (1963). *Residual stresses and fatigue in matals*. New York: McGraw-Hill Book Company, Inc.

Almroth, B. O.(1966) Influence of edge conditions on the stability of axially compressed cylindrical shells. *Journal of the American Institute of Aeronautics and Astronautics, 4*(1).

Amerio, L. (1977). *Analisi matematica con elementi di analisi funzionale* (Vol. 2). Torino: UTET.

Annaratone, A. (1998). *Generatori di vapore*. Milano: Edizioni Libreria CLUP.

Annaratone, A. (1998). *Recipienti in pressione*. Milano: Edizioni Libreria CLUP.

Aron, H. (1874). Das Gleichgewicht und die Bewegung einer unendlich dünnen, beliebig gekrümmten, elastischen Schale. *Journal Für reine und angewandte Mathematik, 78*.

ASM Handbook. (1990). *Proprierties and selection: Iron, steel and high performance alloys*. Materials Park: ASM International.

Bach, C. (1909). Versuche über die tatsächliche Widerstandsfähigkeit von Balkan mit [-förmigem Kuerschnitt. *Zeitschrift des Vereines Deutscher Ingenieure, 53*(44), 1790–1795.

Bach, C. (1910). Versuche über die tatsächliche Widerstandsfähigkeit von Trägern mit [-förmigem Kuerschnitt. *Zeitschrift des Vereines Deutscher Ingenieure, 54*(10), 382–387.

Bailey, R. W. (1935). The utilisation of creep test data in engineering design. *Proceedings of Institutions of Mechanical Engineers, 131*, 209–284.

Baker, E. H., Cappelli, A. P., Kovalevsky, L., Rish, F. L., & Verette, R. M. (1968). *Shell analysis manual*. North American Aviation, Inc., NASA CR-912, Washington, DC.

Banerjee, P. K., & Butterfield, R. (1981). *Boundary element methods in engineering science*. London: McGraw-Hill.

Barber, J. R. (1999). *Elasticity*. Dordrecht: Kluwer Academic Publishers.

Bassett, A. (1890). On the extension and flexure of cylindrical and spherical thin elastic shells. *Philosophical Transactions of the Royal Society, Series A, 181*(6), 433–480.

Bassett, A. (1890). On the extension and flexure of thin elastic shells. *Philosophical Transactions of the Royal Society, Series A, 179*.

Batdorf, S. B. (1947). *A simplifield method of elastic stability analysis for thin cylindrical shells*. Technical Report, No. 874, Washington, DC: NACA.

Batdorf, S. B., & Budiansky, B. (1949). *A mathematical theory of plasticity based on the concept of slip*. Technical Note, no. 1871, Washington, DC: NACA.

Bell, J. F. (1984). *Mechanics of solids: The experimental foundation of solid mechanics*. New York: Springer.

Belloni, G., & Bernasconi, G. (1984). *Sforzi, deformazioni e loro legami*. Milano: Edizioni Spiegel.

V. Vullo, *Circular Cylinders and Pressure Vessels*, Springer Series in Solid and Structural Mechanics 3, DOI: 10.1007/978-3-319-00690-1,

Belloni, G., & Lo Conte, A. (2002). *Costruzione di macchine: Resistenza dei materiali e sicurezza*. Milano: Hoepli.

Belluzzi, O. (1971). *Scienza delle costruzioni* (Vols. 1–4). Bologna: Zanichelli.

Bernardini, C., Ragnisco, O., & Santini, P. M. (2002). *Metodi Matematici della Fisica*. Roma: Carocci Editore.

Bernasconi, G., & Piatti, G. (1979). *Creep of engineering materials and structures*. London: Applied Science Publication Ltd.

Bickell, M. B., & Ruiz, C. (1967). *Pressure vessel design and analysis*. London: Macmillan.

Biezeno, C. B., & Grammel, R. (1953). *Technische Dynamik, Erster Band, Grundlagen und Einzelne Maschinenteile*. Berlin: Springer.

Bland, D. R. (1960). *Theory of linear viscoelasticity*. Oxford: Pergamon Press.

Boley, B. A., & Weiner, J. H. (1997). *Theory of thermal stresses*. Mineola: Dover Publications Inc.

Boller, C., & Seeger, T. (1987). *Materials science monographs, 42 B, materials data for cyclic loading, part B: Low-alloy steels*. Amsterdam: Elsevier.

Boresi, A. P., & Chong, K. P. (2000). *Elasticity in engineering mechanics*. New York: Wiley.

Brownell, L. E., & Young, E. H. (1968). *Process equipment design*. New York: Wiley.

Bruhn, E. F. (1973). *Analysis and design of flight vehicle structures*. NP: Jacobs Publishing Inc.

Brush, D. O., & Almroth, B. O. (1975). *Buckling of bars, plates, and shells*. New York: McGraw-Hill.

Budiansky, B., & Sanders, J. L. (1963). *On the "best" first-order linear shell theory. Progress in applied mechancs* (pp. 129–140). New York: McMillan.

Burr, A. H. (1982). *Mechanical analysis and design*. New York: Elsevier.

Byrne, R. (1944). *Theory of small deformations in a thin elastic shell*. Seminar Report in Mathematics, University of California, Publication in Mathematics (Vol. 2(1), pp. 103–152).

Caboni, F. (1971). *Costruzione di macchine*. Bologna: Pitagora Editrice.

Calderale, P. M., Vullo, V., Bracco, G., & Uslenghi, P. (1981). Analisi delle tensioni in un soffietto soggetto a carico non assialsimmetrico e verifica sperimentale. *Progetto, Rivista di Progettazione di Macchine, 3*(8/9), 45–50.

Calladine, C. R. (2001). A shell-buckling paradox resolved. In D. Durban, D. Givoli, & J. G. Simmonds (Eds.), *Advances in the mechanics of plates and shells* (pp. 119–134). Dordrecht: Kluwer Academic Publishers.

Calladine, C. R. (1985). *Plasticity for engineers*. Chichester: Ellis Horwood Limited Publishers.

Capurso, M. (1971). Analisi limite: Strutture Bidimensionali. In R. Baldacci, G. Cerandini & E. Giangreco (Eds.), *Plasticità* (Vol. 2). Collana Tecnico-Scientifica per la Progettazione di Strutture in Acciaio. Genova: Italsider.

Carslaw, H. S., & Jaeger, J. C. (1959). *Conduction of heat in solids* (2nd ed.). London: Oxford University Press.

Cauchy, A. L. (1828). Sur l'equilibre et le mouvement d'une plaque solide. *Exercise de Mathématique, 3*, 328.

Chaboche, J. L. (1986). Time-independent constitutive theories for cyclic plasticity. *International Journal of Plasticity, 2*(2), 149–188.

Chajes, A. (1974). *Principles of structural stability theory*. Englewood Cliffs: Prentice-Hall Inc.

Chen, F. Y. (1976). *Mohr's circle and its applications to engineering design*. Paper 76—DET-99, ASME, September 1976.

Christensen, R. M. (1971). *Theory of viscoelasticity: An introduction*. New York: Academic Press.

Cicala, P. (1997). *An engineering approach to the calculus of variations, with complete bibliography of the author* (2nd ed.). Torino: Libreria Editrice Universitaria Levrotto & Bella.

Cicala, P. (1977). *Asymptotic approach to linear shell theory* (p. 129). AIMETA Research Report, No. 6.

Cicala, P. (1940). La stabilità elastica del cilindro in parete sottile. *L'Aerotecnica, 20*(5), 355–373.

Cicala, P. (1940). La stabilità elastica del cilindro in parete sottile, Parte I, Parte II, Parte III. *L'Aerotecnica*, *20*(6), 460–479, (7), 595–612 and (10), 735–746.
Cicala, P. (1981). *La teoria non lineare d'elasticità*. Quaderno n. 1, Torino: Istituto di Scienza delle Costruzioni, Politecnico di Torino.
Cicala, P. (1971). *Le strutture a guscio, parte prima, la teoria lineare*. Torino: Libreria Editrice Universitaria Levrotto & Bella.
Cicala, P. (1978). *Linear shell, theories, an asymptotic approach*. Torino: Libreria Editrice Universitaria Levrotto & Bella.
Cicala, P. (1963). *Scienza delle Costruzioni, Parte Prima, Travature e Parte seconda, Teoria delle Travi*. Torino: Libreria Editrice Universitaria Levrotto & Bella.
Cicala, P. (1940). Sulla stabilità dell'equilibrio elastico. *Atti dell'Accademia delle Scienze di Torino*, *75*, 185–222.
Cicala, P. (1973). Sulla teoria non lineare del guscio elastico. *L'Aerotecnica, Missili e Spazio*, *5*, 325–331.
Cicala, P. (1950). Sulle deformazioni plastiche. *Rendiconti dell'Accademia delle Scienze di Torino*, *8*(6), 583–586.
Cicala, P. (1974). *Sulle tensioni critiche nelle strutture a guscio*. Torino: Istituto di Scienza delle Costruzioni, Politecnico di Torino.
Cicala, P. (1965). *Systematic approximation approach to linear shell theory*. Torino: Libreria Editrice Universitaria Levrotto & Bella.
Cicala, P. (1965). *Teoria Lineare della Parete Sottile*. Torino: Libreria Editrice Universitaria Levrotto & Bella.
Cicala, P. (1966). The calculation of axially symmetric vessels. *Meccanica*, *1–2*, 58–60.
Coates, W. M. (1930). The state of stress in full heads of pressure vessels. *Transactions of the American Society Mechanical Engineers*, *52*, 117–131 (APM 52-12).
Coates, W. M. (1929). The state of stress in thin-walled pressure vessels. *Transactions of the American Society Mechanical Engineers*, *51*, 829.
Cockcroft, R. D. H. (1969). *Creep in materials subjected to variable loading*. In Bi-axial stress PhD Dissertation, University of Cambridge.
Codazzi, D. (1868–1869). Sulle coordinate curvilinee d'una superficie dello spazio. *Annali di Matematica Pura ed Applicata*, 2, 101–119.
Colladine, C. R. (2001). A shell-buckling paradox resolved. In D. Durban, D. Givoli, & J. G. Simmonds (Eds.), *Advances in the mechanics of plates and shells* (pp. 119–134). Dordrecht: Kluwer Academic Publishers.
Collins, J. A. (1981). *Failure of materials in mechanical design, analysis, prediction, prevention*. New York: Wiley.
Como, M. (1973). Sull'imbozzamento elasto-plastico dei contenitori cilindrici. *Quaderni de "La Ricerca Scientifica"* (Vol. 78, pp. 42–56). Roma: Consiglio Nazionale delle Ricerche.
Cook, R. D. (1981). *Concepts and applications of finite element analysis* (2nd ed.). New York: Wiley.
Cook, R. D., & Young, W. C. (1999). *Advanced mechanics of materials* (2nd ed.). Englewood Cliffs: Prentice-Hall Inc.
Corradi Dell'Acqua, L. (1992). *Meccanica delle strutture*. Milano: McGraw-Hill Libri Italia.
Coulomb, C. A. (1773). *Essai sur une application des règles des maximis et minimis à quelques problèmes de statique relatifs à l'architecture* (Vol. 7, pp. 343–382). Paris: Mémoire présenté à l'Académie royale des sciences.
Crossland, B., & Bones, J. A. (1955). The ultimate strength of thick-walled cylinders subjected to internal pressure. *Engineering*, 80–83 and 114–117.
Crossland, B., & Bones, J. A. (1958). The ultimate strength of thick-walled cylinders. *Transactions of the ASME*, *80*.
Crossland, B., Jorgensen, S. M., & Bones, J. A. (1959). The strength of thick-walled cylinders. *Journal of Engineering for Industry*, *81*.

Dawson, V. C. D. (1960). *Elastic and plastic stress equations for hollow cylinders*. Navord Report No. 6786.

Dawson, V. C. D., & Seigel, A. E. (1963). *Reverse yielding of a fully autofrettaged tube of large wall ratio*. Meryland: United States Naval Ordnance Laboratory. Unclassified report NOLTR 63-123.

Den Hartog, J. P. (1952). *Advanced strength of materials* (1st ed.). New York: McGraw Hill.

Desai, C. S., & Sirilwardane, H. J. (1984). *Constitutive laws for engineering materials*. Englewood Cliffs: Prentice-Hall Inc.

Deutschman, A. D., Michels, W. J., & Wilson, C. E. (1975). *Machine design theory and practice*. New York: Macmillan Publishing Co. Inc.

Dieter, G. E. (1988). *Mechanical metallurgy, metric editions materials science and metallurgy*. London: McGraw-Hill Book Company.

Dill, E. H. (2007). *Continuum mechanics: Elasticity, plasticity, viscoelasticity*. Boca Raton: CRC Press, Taylor & Francis Group.

Donnell, L. H. (1934). A new theory for the buckling of thin cylinders under axial compression and bending. *Transactions of the ASME, 56*, 795–806.

Donnell, L. H. (1933). *Stability of thin walled tubes under torsion*. Report no. 479. Washington, DC: NACA.

Donnell, L. H., & Wan, C. C. (1950). Effect of imperfections on buckling of thin cylinders. *Journal of Applied Mechanics*, pp. 75–83.

Dorn, J. E. (1955). Some fundamental experiments on high temperature creep. *Journal of Mechanics, Physics and Solids, 3*.

Dowling, N. E. (1999). *Mechanical behavior of materials* (2nd ed.). Upper Saddle River: Pearson Education Inc.

Drucker, D., & Shield, R. (1958). Limit analysis of symmetrically loaded thin shells of revolution. *Journal of Applied Mechanics, 25*.

Dubbel, H. (1983). *Taschenbuch für den Maschinenbau* (Vol. 15). Berlin Heidelberg: Springer.

Evans, R. W., & Wilshire, B. (1985). *Creep of metals*. London: Institute of Metals.

Faupel, J. H. (1964). *Engineering design: A synthesis of stress analysis and materials engineering*. New York: Wiley.

Faupel, J. H., & Furbeck, A. R. (1953). Influence of residual stresses on behaviour of thick-walled closed-end cylinders. *Transactions of the ASME, 75*, 345–354.

Faupel, J. H. (1955). *Residual stresses in heavy-wall cylinders* (Vol. 259, pp. 405–419). Int. Franklin Institute.

Faupel, J. H. (1956). Yield and bursting characteristics of heavy-walled cylinders. *Transactions of the ASME, 78*(5), 1031–1064.

Favretti, G. (1967). *Corpi cilindrici e serbatoi in parete spessa*. Quaderno n. 36 de "La Ricerca Scientifica: Stato delle conoscenze sul dimensionamento dei recipienti in pressione", Roma.

Fazekas, G. A. (1967). A note on the bending of Euler beams. *Journal of Engineering Education, 57*(5).

Feinberg, S. (1939). On the problem of the construction of an approximate theory of thin shells of arbitrary shape. *Studies in the Theory of Structures, Gosstroiizdat*.

Fenner, D. N. (1987). *Engineering stress analysis: A finite element approach with FORTRAN 77 software*. Chichester: Ellis Horwood Limited Publishers.

Fenner, R. T. (1986). *Engineering elasticity: Application of numerical and analytical techniques*. Chichester: Ellis Horwood Limited.

Fenner, R. T. (1975). *Finite element methods for engineers*. London: The Macmillan Press Ltd.

Feodosyev, V. (1973). *Strength of materials* (2nd ed.). Moscow: MIR Publishers.

Filonenko-Borodich, M. (1968). *Theory of elasticity*. Moscow: MIR Publishers.

Findley, W. N., Lai, J. S., & Onaran, K. (1976). *Creep and relaxation of nonlinear viscoelastic materials*. Amsterdam: North-Holland Publishing Co.

Finnie, I., & Heller, W. R. (1959). *Creep of engineering materials*. New York: McGraw-Hill.

Flügge, W. (1932). Die stabilität der Kreiszylinderschale. *Ing-Arch, 3*, 463–506.

Flügge, W. (1960). *Stresses in shells* (2nd ed.). Berlin: Springer (reprint of 1973).
Ford, H., & Alexander, J. M. (1977). *Advanced mechanics of materials* (2nd ed.). Chichester: Ellis Horwood Limited.
Forsyth, A. R. (1996). *A treatise on differential equations* (6th ed.). Mineola, New York: Dover Pubblications Inc.
Franciosi, V. (1979). *Calcolo a Rottura: Lo Stato Limite Ultimo da Meccanismo*. Napoli: Liguori Editore.
Franciosi, V. (1981). *Fondamenti di Scienza delle Costruzioni, Teoria dell'Elasticità* (Vol. I). Napoli: Liguori Editore.
François, D., Pineau, A., & Zaoui, A. (1998). *Mechanical behaviour of materials*. Dordrect: Kluwer Academic Publishers.
Frost, H. J., & Ashby, M. F. (1982). *Deformation mechanism maps: The plasticity and creep of metals and ceramics*. Oxford: Pergamon Press.
Fung, Y. C. (1969). *A first course in continuum mechanics*. Englewood Cliffs, NJ: Prentice-Hall Inc.
Galerkin, B. G. (1934). On the theory of elastic cylindrical shells. *DAN SSSR*, *4*(5–6).
Garofalo, F. (1965). *Fundamentals of creep and creep-rupture in metals*. New York: McMillan Publishing Company.
Garro, A. (1992). *Progettazione Strutturale del Motore*. Torino: Levrotto & Bella.
Gauss, K. F. (1965). Disquisitiones Generales circa Superficies Curvas (in Latin), *Communication Society Science Gottingen*, *4*, 1828. English translation: *General Discussions about Curved Surfaces* (A. Hiltebeitel & J. Morehead, Trans.). New York: Raven Press.
Geckeler, J. (1915). *Uber die Festigkeit Achsensymmetrischer schalen* (p. 276). Forschungsarbeiten auf dem Gebiete des Ingenieurwesens, Berlin.
Geckeler, J. (1930). Die festigkeit achsensymmetrischer schalen. *Ing-Arch*, *1*, 22–71.
Genta, G. (1996). *Calcolo di Resistenza degli Organi Rotanti e dei Recipienti Cilindrici*. Torino: Levrotto & Bella.
Gere, J. M., & Timoshenko, S. P. (1997). *Mechanics of materials* (4th ed.). Boston: PWS Publishing Company.
Giacosa, D. (2000). *Motori Endotermici* (15th Edizione) aggiornata da A. Garro (Ed.). Milano: Editore Ulrico HOEPLI.
Gibbs, C. W. (Ed.). (1971). *New compressed air and gas data* (2nd ed.). Phillipsburg: Ingersoll-Rand Co.
Giovannozzi, R. (1965). *Costruzione di Macchine* (4th ed., Vols. 1–2). Bologna: Pàtron.
Gittus, J. (1975). *Creep, viscoelasticity and creep fracture in solids*. London: Applied Science Publishers Ltd.
Godono, G., & De Iorio, A. (1973). *Sull'impiego dei criteri energetici per la definizione dello stato viscoplastico dei corpi cilindrici di forte spessore*. Quaderni de "La Ricerca Scientifica" (Vol. 78, pp. 57–70). Roma: Consiglio Nazionale delle Ricerche.
Gol'denweizer, A. L. (1963). Derivation of an approximate theory of shells by means of asymptotic integration of the equations of theory of elasticity. *PriklMat Mekh, Akademiya Nauk. SSSR*, *30*, 593–608.
Gol'denweizer, A. L. (1940). The equations of the theory of shells. *Prikl Mat Mekh, Akademiya Nauk SSSR*, *4*(2).
Gol'denweizer, A. L. (1961). *Theory of elastic shells*. New York: Pergamon Press.
Gola, M. M., & Vullo, V. (1976). *Analisi Teorica di Fondi Curvi di Forma Superellittica*. In Atti del III Congresso Nazionale AIMETA, Associazione Italiana di Meccanica Teorica ed Applicata, Cagliari, 13–16 ottobre 1976.
Goodier, J. N., & Hodge, P. G. (1958). *Elasticity and plasticity*. New York: Wiley.
Gormann, D. J. (1967). Plastic-elastic stress distributions in partially yielded thick-wall cylinders. *Journal of Applied Mechanics*, 509.
Gradshteyn, I. S., & Ryzhik, I. M. (2000). *Tables of integrals, series and products*. San Diego: Academic Press.

Graham, A., & Walles, K. F. A. (1955). Relations between long and short time properties of a commercial alloy. *Journal of the Iron and Steel Institute, 179.*

Green, A. E., & Zerna, W. (1963). *Theoretical elasticity.* Oxford: Oxford University Press.

Haddad, Y. M. (2000). *Mechanical behaviour of engineering materials* (Vol. 1). Dordrecht (NE): Kluwer Academic Publishers. (Static and Quasi-Static Loading).

Hartmann, L. (1896). *Distribution des deformations dans les métaux soumis à des efforts*, Paris.

Haupt, P. (2000). *Continuum mechanics and theory of materials.* Berlino: Springer.

Hill, R. (1950). *Mathematical theory of plasticity.* Oxford: Clarendon Press.

Hodge, P. G. (1963). *Limit analysis of rotationally symmetric plates and shells.* Englewood Cliffs: Prentice-Hall.

Hoff, N. J., & Rehfield, L. W. (1964). *Buckling of axially compressed circular cylindrical shells at stresses smaller than the classical critical value.* Stanford University, Department of Aeronautics and Astronautics, Report SUDAER, No. 191, May 1964.

Hoff, N. J. (1958). Stress distribution in the presence of creep. In N. J. Hoff (Ed.), *High temperature effects in aircraft structures.* Oxford: Pergamon.

Hoffmann, O., & Sachs, G. (1953). *Introduction to the theory of plasticity for engineers.* New York: McGraw-Hill Co.

Houghton, D. S. (1960). Discontinuity problems in shell structures, in Nuclear reactor cointainment buildings and pressure vessels. Glasgow, Scotland, Butterworths, London: The Royal College of Science and Technology.

Huggaenberger, A. (1925). Calculation of the plate thickness of dished heads with knuckles according tot the hamburg rules. *Z Ver deal Ing, 69*(6), 159–162.

Huggaenberger, A. (1927). Calculation of the plate thickness of dished heads with knuckles according tot the hamburg rules. *Mech Eng, 49*(6), 629.

Hult, J. A. H. (1973). *Creep strength of structures.* Udine: CISM.

Iurzolla, E. (1981). *Il Calcolo dei Recienti a Pressione* (2nd ed.). Padova: Edizioni Libreria Cortina.

Ivanov, A. B. (2001). Peterson-codazzi equations. In M. Hazewinkel (Ed.), *Encyclopedia of mathematics.* Berlin: Springer.

Jaward, M. H., & Farr, J. R. (1989). *Structural analysis and design of process equipment* (2nd ed.). New York: Wiley.

Johns, R. H., & Orange, T. W. (1961). *Theoretical elastic stress distributions arising from discontinuities and edge loads in several shell-type structures.* Technical Report R-103. Washington, DC: NASA.

Johnson, A. E., Henderson, J., & Kahn, B. (1963). Multiaxial creep strain/complex stress/time relations for metallic alloys with some applications to structures. In *ASME/ASTM/I Mech E, Procedings Conference on Creep.* New York, London: Institute of Mechanical Engineering.

Jorgensen, S. M. (1958). Overstrain and bursting strength of thick-walled cylinders. *Transactions of the ASME, 80.*

Juvinall, R. C. (1967). *Engineering consideration of stress, strain and strength.* New York: McGraw Hill.

Juvinall, R. C. (1983). *Fundamentals of machine components design.* Wiley, New York.

Kalnins, A., & Updike, D. P. (2001). Limit pressures of cylindrical and spherical shells. *Journal of Pressure Vessel Technology, 123*, 288–292.

Katchanov, L. M. (1975). *Éléments de la Théorie de la Plasticité.* Moscou: Éditions MIR.

Katchanov, L. M. (1967). In A. J. Kennedy (Ed.), *The theory of creep.* Boston: National Lending Library.

Katchanov, L. M. (1986). *Introduction to continuum damage mechanics.* Amsterdam: Martinus Nijhoff Publishers.

Kennedy, A. J. (1963). *Process of creep and fatigue in metals.* New York: Wiley.

Kil'chevskii, N. A. (1939). A generalization of contemporary shell theory. Prikl Mat Mekh, Akademiya Nauk SSSR, *2*(4).

Kirchhoff, G. R. (1850). Uber das gleichgewichi und die bewegung einer elastishem scheibe. *J Fuer die Reine und Angewandte Mathematik, 40*, 51–88.

Kline, M. (1972). *Mathematical thought from ancient to modern times*. UK: Oxford University Press.

Koiter, W. T. (1970). *The stability of elastic equilibrium* (Chap. 5: *Shell theory for finite deflections*). Dissertation, Technical Report, AFFDL-TR-70-25, Air Force System Command, Wright-Patterson AFB.

Kraus, H. (1980). *Creep analysis*. New York: Wiley.

Krauss, F. (1929). Über die Grundgleichungen der Elastizitätstheorie schwachdeformierter Schalen. *Mathematische Annalen, 101*(1), 61–92.

Krempl, E. (1987). Models of viscoplasticity some comments on equilibrium (back) stress and drag stress. *Acta Mechanica, 69*, 25–42.

Kreyszig, E. (1993). *Advanced engineering mathematics* (7th ed.). New York: Wiley.

Krylov, A. N. (1902). *On stresses experienced by a ship in a sea way* (p. 44). London: Trans Institute Naval Architects.

Laforgia, A. (2001). *Equazioni Differenziali Ordinarie*. Roma: Accademica S.r.l.

Laforgia, A. (1999). *Successioni e Serie di Funzioni*. Roma: Accademica S.r.l.

Lagrange, J. L. (1811). *Mécanique analytique*. Paris: Tome Premier, MME VE Courcier.

Lamé, G., & Clapeyron, B. P. E. (1833). Mémoire sur l'équilibre intérieur des solides homogènes. *Mémoires de l'Académie des Sciences, 7*, 465–562.

Lamé, G. (1852). *Leçons sur la théorie mathématique de l'élasticité des corps solides*. Parigi: Bachelier.

Larson, F. R., & Miller, J. (1952). A Time-temperature relationship for rupture and creep stress. *Transactions of the ASME, 174*.

Lévy, M. (1870). Mémoire sur les équations générales des mouvements intérieurs des corps solides ductiles au-delà des limites où l'élasticité pourrait les ramener à leur premier état. *Comptes Rendus de l' Academie des Sciences Paris,70*, 1323–1325.

Lévy, M. (1871). Mémoire sur les équations générales des mouvements intérieurs des corps solides ductiles au-delà des limites où l'élasticité pourrait les ramener à leur premier état. *Journal de Mathématiques Pures et Appliquées, Séries II,16*, 369–372.

Lévy, M. (1871). Sur l'intégration des équations aux différences partielles, relatives aux mouvements intérieur des corps solides ductiles, lorsque ces mouvements ont lieu par plans parallèles. *Comptes Rendus de l' Academie des Sciences, Séries A,73*(16), 1098–1103.

Lidonnici, F. (2010). *Confronto tra le varie normative applicabili nella costruzione degli apparecchi a pressione non soggeti a fiamma*. Milano: Sant'Ambrogio Servizi Industriali SRL.

Liguori, A. (1983). *Sul Calcolo dei Recipienti Cilindrici in Parete Spessa Parzialmente Plasticizzati*. Napoli: Tipografia Pipola.

Lipson, C., & Juvinall, R. C. (1963). *Handbook of stress and strength*. New York: The MacMillan Company.

Lockett, F. J. (1972). *Nonlinear viscoelastic solids*. New York: Academic Press.

Love, A. E. H. (1888). On the small free vibrations and deformations in the elastic shells. *Philosophical Transactions of the Royal Society, 179A*.

Love, A. E. H. (1944). *A treatise on the mathematical theory of elasticity* (4th ed.). New York: Dover Publication Inc.

Lubahn, J., & Felgar, R. (1961). *Plasticity and creep of metals*. New York: Wiley.

Lubliner, J. (2006). *Plasticity theory*. Berkeley: Pearson Education Inc., University of California.

Lüders, W. (1860). Über die Äusserung der elasticität an stahlartigen Eisenstäben und Stahlstäben, und über eine beim Biegen solcher Stäbe beobachtete Molecularbewegung. *Dingler's Polytechnisches Journal,155*, 18–22.

Ludwik, P. (1909). *Elemente der Technologischen Mechanik*. Berlin: Julius Springer.

Lur'e, A. I. (1940). The general theory of thin elastic shells. *Prikl Mat Mekh, Akademiya Nauk SSSR, 4*(2).

Maddox, S. J. (2003). Assessment of pressure vessel design rules on the basis of fatigue test data. In W. M. Banks & D. H. Nash (Eds.), *Pressure equipment technology-theory and practice* (pp. 237–248). Bury St. Edmunds: Professional Engineering Publications Ltd.

Maddox, S. J. (1994). Fatigue aspects of pressure vessel design. In J. Spence & A. S. Tooth (Eds.), *Pressure vessel design—concepts and principles* (p. 337). London: E&FN Spon.

Maddox, S. J. (2000). Fatigue design rules for welded structures. *Progress in Structural Engineering and Materials,1*(2), 102–109.

Maddox, S. J., & Razmjoo, G. R. (2001). Interim fatigue design recommendations for fillet welded joints under complex loading. *Fatigue and Fracture of Engineering Materials and Structures,24*, 329–337.

Mainardi, G. (1857). Su la teoria generale delle superficie. *Giornale dell'Istituto Lombardo di Scienze, Lettere ed Arti e Biblioteca Italiana, 9*(54), 385–404.

Malvern, L. E. (1969). *Introduction to the mechanics of a continuous medium.* Englewood Cliffs: Prentice-Hall Inc.

Manna, F. (1983). *Corpi cilindrici e contenitori in parete spessa.* Napoli: Pironti.

Manna, F. (1967). Sui recipienti cilindrici in parete spessa parzialmente plasticizzati. *Rassegna Internazionale di Meccanica,3*, 17.

Manna, F. (1973). *Sulla diretta definizione dello stato visco-plastico d'un cilindro di forte spessore.* Quaderni de "La Ricerca Scientifica" (Vol. 78, pp. 71–73). Roma: Consiglio Nazionale delle Ricerche.

Manson, S. S. (1961). Creep under non-steady temperatures and stresses. In J. E. Dorn (Ed.), *Mechanical behaviour of materials at elevated temperatures.* London: McGraw-Hill.

Marin, J. (1962). *Mechanical behaviour of engineering materials.* Englewood Cliffs: Prentice-Hall Inc.

Marin, J., & Pao, Y. H. (1953). An analytical theory of the creep deformation of materials. *Journal of Applied Mechanics, 20.*

Marriott, D. L. (1970). A review of reference stress methods for estimating creep deformations. *2nd IUTAM Symposium on Creep of Structures*, Gothenburg, Berlin: Springer.

Marriott, D. L. (1968). Approximate analysis of transient creep deformation. *Journal of Strain Analysis,3*(4), 288–296.

Martini, D. (1999). *Calcolo Automatico di Recipienti in Pressione.* Bologna: Pitagora Editrice.

Maruyama, K., Tanaka, C., & Oikawa, H. (1990). Long-term creep curve prediction based on the modified θ projection concept. *Journal of Pressure Vessel Technology, 112.*

Mase, G. E. (1970). *Continuum mechanics.* New York: McGraw Hill Book Company.

Massa, E., & Bonfigli, L. (1981). *Costruzione di Macchine* (Vol. I). Milano: Masson Italia Editori.

Massa, E. (1981). *Costruzione di Macchine* (Vol. II). Milano: Masson Italia Editori.

Massonnet, C., & Save, M. (1968). *Calcolo a Rottura delle Strutture: Strutture Spaziali* (Vol. II). Bologna: Zanichelli.

McVetty, P. G. (1943). Creep of Metals at Elevated Temperature – The Hyperbolic Sine Relation Between Stress and Creep Rate. *Transactions of the ASME, 65.*

Megyesy, E. F. (1984). *Pressure vessel handbook.* Tulsa: Publishing Corporation.

Meissner, E. (1915). *Uber elastizitat und festigkeit schalen* Vierteljahrsschr d. Naturf., Gesellsch., Bd. 60, Zürich.

Meissner, E. (1913). Das Elastizitätsproblem für dünne Schalen von. Ringflächen, Kugel-oder Kegelform. *Physikalische Zeitschrift,14*, 343–349.

Meissner, E. (1925). Zur festigkeitsberechnung von hochdruckkesseltrommeln. *Schweizerische Bauzeitung,86*, 1–25.

Melnikov, Y. A. (1995). *Green's functions in applied mechanics.* Southampton and Boston: Computational Mechanics Publications.

Mendelson, A. (1988). *Plasticity: Theory and application.* Melbourne: R. E. Krieger Publishing Co.

Meyer, R. R. (1961). *Discontinuity stresses in shells: General model,* ASTIA (Armed Services Technical Information Agency), Arlington, Virginia. (Report AD-264-376).

Miller, A. K. (1987). *Unified constitutive equations for creep and plasticity.* New York: Elsevier Applied Science Publishers Ltd.

Mirolioubov, I., Engalytchev, S., Serguievski, N., Almamétov, F., Kouritsyne, N., Smirnov-Vassiliev, K., et al. (1973). *Résistance des Matériaux.* Éditions MIR, Moscou: Manuel de Résolution des Problèmes.

von Mises, R. (1913). Mechanik der festen Körper im plastisch-deformablen Zustand. *Göttinger Nachrichten, Akad. Wiss, Math.-Physik., Kl.*, 582–592.

Moaveni, S. (1999). *Finite element analysis. Theory and application with ANSYS.* Upper Saddle River: Prentice-Hall.

Mohr, O. (1906). *Abhandlungen aus dem Gebiete der technischen Mechanik.* Berlin: Wilhelm Ernst and Sohn.

Mohr, O. (1882). *Über die Darstellung des Spannungszustandes und des Deformationszustandes eines Körperelementes*, Zivilingenieur, p. 113.

Moradi, M., Ghorbanpour, A., Khademizadeh, H., & Loghman, A. (2001). Reverse yielding and Bauschinger effect on residual stresses in thick-walled cylinders. *Pakistan Journal of Applied Sciences,1*(1), 44–51.

Moss, D. (2004). *Pressure vessel design manual.* Oxford: Elsevier.

Mushtari, K. M. (1938). Certain generalizations of the theory oh thin shells. *Izv. Kaz. fiz-mat. o-va., pri Kaz Un-te*, 11(8) (in Russian).

Mushtari, K. M., & Galimov, K. Z. (1957). Nonlinear theory of thin elastic shells, Russian translational series, NASA TT-F-62, Washington, DC: US Department of Commerce.

Mushtari, K. M., & Galimov, K. Z. (1961). *Nonlinear theory of thin shells.* Jerusalem: The Israel Program for Scientific Translations.

Muskhelishvili, N. I. (1963). *Some basic problems of the mathematical theory of elasticity* (4th ed.). Groningen: P. Noordhoff Ltd.

Nadai, A. (1950). *Theory of flow and fracture of solids* (Vol. I). New York: McGraw Hill Book Company.

Nadai, A. (1963). *Theory of flow and fracture of solids* (Vol. II). New York: McGraw Hill Book Company.

Naghdi, P. M., & De Silva, C. N. (1954). *Deformations of Elastic Ellipsoidal Shells of Revolution, Proceedings of Second U.S. National Congress of Applied Mechanics*, pp. 333–343.

Naghdi, P. M., & Nordgren, R. P. (1963). On the linear theory of shells under the Kirchhoff hypothesis. *Quarterly of Applied Mathematics,21*, 49–60.

Naghdi, P. M. (1960). Stress-strain relations in plasticity and thermoplasticity. In E. H. Lee & P. S. Symonds (Eds.), *Plasticity* (pp. 121–169). New York: Pergamon Press.

Navier, L. M. H. (1864). *Résumé des Leçons données à l'école des ponts et chaussées sur l'application de la mécanique à l'établissement des constructions et des machines,* (3rd ed.) with notes and appendices by Saint-Venent, Paris.

Norton, F. H. (1929). *The creep of steel at high temperatures.* London: McGraw-Hill.

Novozhilov, V. V. (1964). *the theory of thin elastic shells.* Groningen: P. Noordhoff LTS.

Novozhilov, V. V. (1970). *Thin shell theory* (2nd ed.). Groningen: Wolters-Noordhoff Publishing.

Obodan, N. I., Lebedeyev, O. G., & Gromov, V. A. (2013). *Nonlinear behaviour and stability of thin-walled shells, series: Solid mechanics and its applications* (Vol. 199). Berlin: Springer.

Odqvist, F. K. G. (1974). *Mathematical theory of creep and creep rupture.* Oxford: Clarendon Press.

Ohji, K., & Marin, J. (1964). Creep of metals under non-steady conditions of stress. *Proceedings Conference on Thermal Loading and Creep in Structures and Components*, London: Institution of Mechanical Engineers.

Orowan, E., & Becker, R. (1953). In A. H. Cottrel (Ed.), *Dislocations and plastic flow in crystals.* Oxford: Oxford University Press.

Palumbo, N. (2006). *Fondamenti Teorici delle Formule per il Calcolo dei Fasciami Cilindrici nelle Attrezzature a Pressione*. Convegno SAFAP, Roma: Confronto tra Codici di Progettazione.

Penny, R. K., & Marriott, D. L. (1995). *Design for creep* (1st ed.). UK: McGraw-Hill Book Company (1971) and (2nd ed.). London: Chapman & Hall (1995).

Perzyna, P. (1966). Fundamental problems in viscoplasticity. *Advances in Applied Mechanics,9*, 243–377.

Peterson, K. M. (1853). Über die Biegung der Flächen, Doctoral Thesis, Dorpat University.

Peterson, R. E. (1974). *Stress concentration factor*. New York: Wiley.

Petrov, V. V. (1975). The method of successive loadings in the nonlinear theory of plates and shells. *Izd-vo Saratov University, Saratov* (S.S.S.R).

Piobert, G., Morin, A. J., & Didion, I. (1842). Commission des Principes du Tir. *Mémorial de l'Artillerie,5*, 501–552.

Poincaré, H. (1892). *Leçons sur la Théorie de L'ÉLASTICITÉ*. Paris: Geoges Carré Éditeur.

Poisson, S. D. (1829). Mémoire sur l'équilibre et le mouvement des corps solides. *Mémories de l'Academy Science Paris, 8,* 357.

Ponter, A., & Hayhurst, D. (1981). *Creep in structures*. New York: Springer.

Popov, E. P. (1949). Bending of beams with creep. *Journal of Applied Physics,20*(3), 251–256.

Popov, E. P. (1968). *Introduction to mechanics of solids*. Englewood Cliffs: Prentice-Hall Inc.

Prager, W., & Hodge, P. G. (1951). *Theory of perfectly plastic solids*. New York: Wiley.

Prager, W. (1961). *Introduction to mechanics of continua*. Boston: Ginn & Company.

Prager, W. (1959). *Introduction to plasticity*. Massachusettes: Addison-Wesley Publishing Co.

Prandtl, L. (1924). Spannungsverteilung in plastischen Körpen. In Proceeding 1st International Congress of Applied Mechanics, (pp. 43–54). Delft.

Prosciutto, A. (1952). *Elementi di Costruzione di Macchine*. Bologna: Pàtron.

Przeminiecki, J. S. (1968). *Theory of matrix structural analysis*. New York: MeGraw-Hill Boock Company.

Rabotnov, Y. N. (1969). *Creep problems in structural members*. Amsterdam: North-Holland Publishing Co.

Rabotnov, Y. N. (1953). *Some problems of the theory of creep*. NACA, TM 1353.

Rabotnov, Y. N. (1945). *The boundary conditions in the theory of thin shells* (Vol. XLVII, No. 5). D.A.N., S.S.S.R.

Rabotnov, Y. N. (1945). *The fundamental equations of the theory of shells* (Vol. XLVII, No. 2). D.A.N., S.S.S.R.

Ramberg, W., & Osgood, W. R. (1943). *Description of stress-strain curves by three parameters*. NACA (National Advisory Committee for Aeronautics), Technical Note 902, July 1943.

Reddy, J. N. (1985). *An introduction to the finite element method*. New York: McGraw-Hill Book Company.

Reissner, E. (1941). A new derivation of the equations of the deformation of elastic shells. *American Journal Math,63*(1), 177–184.

Reissner, E. (1952). On axisymmetrical deformation of thin shells of revolution. *Proceedings of Symposium Applied Mathematics,* (Vol. 3, 27–52).

Reissner, E. (1952). Stress-strain relations in the theory of thin elastic shells. *Journal Mathematics Physics,31*, 105–119.

Reissner, H. (1912). *Spannungen in Kugelschalen*. Breslau-Festscher: Müller.

Rekach, V. G. (1978). *Static theory of thin-walled space structures*. Moscow: MIR Publishers.

Reuss, A. (1930). Berücksichtigung der elastischen Formänderung in der Plastizitätstheorie. *Zeitschrift für angewandte Mathematik und Mechanik,10*, 266–274.

Rolfe, S. T., & Barson, J. M. (1977). *Fracture and fatigue control in structure applications of fracture mechanics*. Englewood Cliffs: Prentice-Hall Inc.

Roš, M., & Eichinger, A. (1949). *Die Bruchgefahr fester Korper bei Ruhender*. Beanspruchung, Zurich: Statischer.

Rubnstein, M. F. (1966). *Matrix computer analysis of structures*. Englewood Cliffs: Prentice-Hall Inc.

Rutten, H. S. (1971). *Asymptotic approximation in the three-dimensional theory of thin and thick elastic shells,* Nederl. Boekdruk Ind., 's Hertogenbosch.

Saada, A. S. (1993). *Elasticity: Theory and applications* (2nd ed.). Malabor: Krieger Publishing Company.

de Saint Venant, A. J. C. B. (1870). Mémoire sur l'établissement des équations différentielles des mouvements intérieurs opérés dans les corps solides ductiles au delà des limites où l'élasticité pourrait les ramener à leur premier état. *Comptes Rendus Academy Science,70*, 473–480.

Salvini, P., Amodio, D., Placidi, F., & Vullo, V. (1993). *A machine for crashworthiness test on thin walled structures,* ASME Winter Annual Meeting, November 28–December 3.

Salvini, P., Vullo, V., Mannucci, G., & Amodio, D. (1994). A new theoretical model describig the axisymmetric crashworthiness of tubes and its experimental validation. In *Basic metrology and applications*, Torino: Levrotto & Bella.

Sanders, J. L. (1959). *An improved first-approxmation theory for thin shells,* NASA Report, No. 24.

Sanders, J. L. (1963). Nonlinear theories for thin shells. *Quarterly of Applied Mathematics,21*, 21–36.

Serensen, S. V., Kogaev, V. P., & Shneiderovich, R. M. (1963). *Nesushchaia sposobnost'I raschety detalei mashin na prochnost'* (2nd ed.). Moskva: Mashigiz (in russo).

Shames, I. H., & Cozzarelli, F. A. (1997). *Elastic and inelastic stress analysis*. Philadelphia: Taylor & Francis Ltd.

Shames, I. H. (1980). *Engineering mechanics—statics and dynamics* (3rd ed.). Englewood Cliffs: Prentice-Hall Inc.

Shames, I. H. (1989). *Introduction to solid mechanics* (2nd ed.). Englewood Cliffs: Prentice-Hall Inc.

Shanley, F. R. (1946). The column paradox. *Journal of the Aeronautical Sciences,13*(12), 678.

Shigley, J. E., Meschke, C. R., & Budynas, R. G. (2004). *Mechanical engineering design* (7th ed.). New York: The McGraw-Hill Companies Inc.

Sim, R. G. (1971). Evaluation of reference parameters for structures subject to creep. *Journal of Mechanical Engineering Sciences,13*, 47–50.

Sim, R. G., & Penny, R. K. (1971). Plane strain creep behaviour of thick-walled cylinders. *Journal of Mechanical Sciences,13*, 987–1009.

Sim, R. G. (1973). Reference stresses and temperatures for cylinders and spheres under internal pressure and steady heat flow. *Journal of Mechanical Sciences,15*, 211–220.

Simmond, J. G., & Danielson, D. A. (1972). Nonlinear shell theory with finite rotation and stress-function vectors. *Journal of Applied Mechanics,39*, 1085–1090.

Skrzypek, J. J., & Hetnarski, R. B. (1993). *Plasticity and creep: Theory, examples, and problems*. Boca Raton: CRC Press (English Edition).

Smirnov, V. (1970). *Cours de Mathematiques Superieures*. Tome II, Editions MIR, Moscow.

Smirnov, V. (1972). *Cours de Mathematiques Superieures*. Tome III, deuxième partie, Editions MIR, Moscow.

Smith, J. O., & Sidebottom, O. M. (1965). *Inelastic behavior of load-carrying members*. New York: Wiley.

Soare, M. (1962). *Application des equations aux differences Finis au Calcul des Coques*. Paris: Eyrolles.

Soderberg, C. R. (1936). The interpretation of creep tests for machine design. *Transactions of the ASME,58*, 733–43.

Sokolnikoff, I. S. (1956). *Mathematical theory of elasticity*. New York: McGraw-Hill Book Company.

Soler, A. I., & Caldwell, S. (1990) A proposed ASME Section VIII, Division 1, tubesheet design procedure. In *Proceedings of the 1990 ASME PVP Conference* (Vol. 186) (HOO 605). Nashville.

Sonsino, C. M. (1999). An overview of the state of the art on multiaxial fatigue of welds. In E. Macha, W. Bedkowski, & T. Lagoda (Eds.), *Multiaxal fatigue and fracture* (Vol. 25, pp. 195–217). ESIS Publication.

Steele, M. C. (1952). Partially plastic thick-walled cylinder theory. *Journal of Applied Mechanics,19*(2), 133–140.

Southwell, R. V. (1913). On the general theory of elastic stability. *Philosophical Transactions of the Royal Society of London, Series A,213*, 187–244.

Swift, H. W. (1952). Plastic instability under plane stress. *Journal of the Mechanics and Physics of Solids,1*, 1–18.

Tabakman, H. D., & Lin, Y. J. (1978). Quick way to calculate thermal stresses in cylindrical shells. *Machine Design,50*, 138–143.

Taira, S., Tanaka, K., & Ohji, K. (1960). A mechanism of deformation of metals at high temperatures with special reference to the tension test and creep test. *Bulletin of the JSME, 3*(10), 228–234.

Timoshenko, S. P. (1953). *History of strength of materials*. New York: McGraw-Hill Book Company Inc.

Timoshenko, S. P., & Gere, J. M. (1961). *Theory of elastic stability* (2nd ed.). Tokio: McGraw-Hill Kogakusha Ltd.

Timoshenko, S. P., & Goodier, J. N. (1970). *Theory of elasticity* (3rd ed.). New York: McGraw Hill.

Timoshenko, S. P. (1956) *Strength of materials*. (3rd ed., part I and part II). Princeton (NJ): D. Van Nostrand Co.

Timoshenko, S. P., & Woinowsky-Krieger, S. (1959). *Theory of plates and shells* (2nd ed.). Singapore: McGraw-Hill Book Company.

Tresca, H. E. (1864). Mémoire sur l'écoulement des corps solides. *Comptes Rendus hebdomadalres des Academy of Science,59*, 754–758.

Tricomi, F. G. (1967). *Equazioni differenziali* (4th ed.). Torino: Boringhieri.

Tsui, E. Y. W. (1968). *Stresses in shells of revolution*. Menlo Park: Pacific Cost Publishers.

Ugural, S. C., & Fenster, S. K. (2003). *Advanced strength and applied elasticity* (4th ed.). Upper Saddle River: Prentice Hall.

Van Den Broek, J. A. (1948). *Theory of limit design*. New York: Wiley.

Ventsel, E., & Krauthammer, T. (2001). *Thin plates and shells: Theory, analysis, and applications*. New York: Marcel Dekker Inc.

Verein Deutscher Eisenhüttenleute. (1969). *Stahl im Hochbau, Handbuch für Entwurf, Berechnung und Ausführung von STAHLBAUTEN*. Düsseldorf: Verlag Sthaleisen M.B.H.

Vivio, F., & Vullo, V. (2010). Closed form solutions of axisymmetric bending of circular plates having non-linear variable thickness. *International Journal of Mechanical Sciences,52*(9), 1234–1252.

Vlasov, V. Z. (1944). *The fundamental differential equations of the general theory of elastic shells*. Prikl. Mat. Mekh., Akademiya Nauk. S.S.S.R, (Vol. VIII, No. 2) (also NACA T.M., No. 1241).

Vlasov, V. Z. (1964). General theory of shells and its applications in engineering. NASA Technical Translation TTF-99 (trans from Russian), Washington, DC.

Von Karman, T., & Tsien, H. S. (1941). The buckling of thin cylindrical shells under axial compression. *Journal of the Aeronautical Sciences,8*(8), 302–312.

Vullo, V., & Vivio, F. (2013). *Rotors: Stress analysis and design*. Milan: Springer.

Watts, G. W., & Lang, H. A. (1952). The stresses in a pressure vessel with a flat head. *Transactions of the ASME,74*, 1083.

Webster, G. A., & Piercey, B. J. (1967). An interpretation of the effect of stress and temperature on the creep proprieties of a nickel-base superalloy. *Metal Science Journal,1*, 97–104.

Weertmann, R. (1951). On elastic continua with hereditary characteristics. *Journal of Applied Mechanics,18*, 273–279.

Westergaard, H. M. (1952). *Theory of elasticity and plasticity*. New York: Wiley.

Winkler, E. (1867). Die Lehre von der Elastizition und Festigkeit. *Prague,141*, 182–184.

Zaslavsky, A. (1982). A note on Saint-Venant's principle. *Israel Journal of Technoly,20*, 143–144.

Zhao, J., & Wang, G. (2010). Unloading and reverse yielding of a finite cavity in a bounded cohesive-frictional medium. *Computers and Geotechnics,37*, 239–245.

Zick, L. P. (1951). Stresses in large horizontal cylindrical pressure vessels on two saddle supports. *Welding Journal Research Supplement, 30.*

Zienkiewitcz, D. C. (1969). *The finite element method in engineering science*. New York: McGraw-Hill.

Zudans, Z., Yen, T. C., & Steigelmann, W. H. (1965). *Thermal stress techniques in the nuclear industry*, The Franklin Institute Research Laboratories Philadelphia. New York: American Elsevier Publishing Company, Inc.

Reference Standards and Similar

AD—Merkblatt S 1: *Semplified Analysis for Fatigue Loading, AD* 2000 *Code,* issued by VdTÜV, Carl Heymanns Verlag.

AD—Merkblatt 2000 edition, English translation, Carl Heymans Verlag KG, D-50939 Köln.

ANCC-ISPESL, Raccolta M: *Impiego di materiali nella costruzione e riparazione degli apparecchi a pressione,* Revisione 1995.

ANCC-ISPESL, Raccolta S: *Impiego della saldatura nella costruzione e riparazione di apparecchi a pressione,* Revisione 1995.

ANCC-ISPESL, Raccolta VSG, *Specificazioni Tecniche per la Verifica della Stabilità dei Generatori di Vapore d'Acqua,* Revisione 1995.

ANCC-ISPESL, Raccolta VSR, *Specificazioni Tecniche per la Verifica della Stabilità dei Recipienti in Pressione,* Revisione 1995.

API (American Petroleum Institute) Standards (2002). *Recommended rules for design and construction of large, welded, low-pressure storage tanks,* (10th Ed.). API 620, API: Washington, DC.

ASME (American Society of Mechanical Engineers) (2010). *Boiler and pressure vessel code*, An American National Standard, Section VIII, *Rules for construction of pressure vessels,* Division 1, New York: ASME.

ASME (American Society of Mechanical Engineers) (2010) *Boiler and pressure vessel code*, An American National Standard, Section VIII, *Rules for construction of pressure vessels,* Division 2, *Alternative Rules,* New York: ASME.

ASME (American Society of Mechanical Engineers) (2010). *Boiler and pressure vessel code*, An American National Standard, Section VIII, *Rules for Construction of Pressure Vessels,* Division 3, *Alternativerules for construction of high pressure vessels,* New York: ASME.

ASME (American Society of Mechanical Engineers) (2010). *Boiler and pressure vessel code*, An American National Standard, Section I, *Rules for construction of power boilers,* New York: ASME.

ASME (American Society of Mechanical Engineers) (2010). *Boiler and pressure vessel code*, An American National Standard, Section II: *Materials—Part A—ferrous material specifications,* New York: ASME.

British Standards Institute (2009). PD 5500/2009, Specification for unfired fusion welded pressure vessels. London: BSI Standards.

BS 5500 (1997). *Unfiredfusion welded pressure vessels*. British Standard.

BS PD 5500 (2000). *Specification for unfired fusion welded pressure vessels*. London: BSI Standards.

CNR (1997). Bollettino Ufficiale, Parte IV, *Norme Tecniche: Bollettino Ufficiale N. 182/97,* Roma.

CNR-Consiglio Nazionale delle Ricerche (1997). Bollettino Ufficiale n. 182 *Costruzioni di Acciaio: Istruzioni per il calcolo, l'esecuzione, il collaudo e la manutenzione (*CNR 10011/ 97*)*, Roma.

CODAP French code for the construction of pressure vessels, SNCT, F-92400 Courbevoie.

CODAP 2005 (2009). *French code for construction of unfired pressure vessels*, Ed. 2005, Division 1 and Division 2, with Addedum 09/07, September 2007, and Addedum 03/09, March 2009, SNCT Publications, Paris.

CODAP 95 (1999) Code français de construction des appareils à pression: Section C7. Revised January 1, 1999.

CR ISO/TR 15608/1999(E), *Welding: Guidelines for a metallic grouping system for fabrication purposes.*

CSES, Centre for Strategy and Evaluation Service. (2012). *Evaluation oft he pressure equipment directive, Executive summary,* UK.

CSES, Centre for Strategy&Evaluation Service. (2012). *Evaluation oft he pressure equipment directive, Final Report,* UK.

CTI, Comitato Termotecnico Italiano. (2005). *Raccomandazioni per l'Uso delle Raccolte ISPESL, Revisione 1995, nell'ambito della Direttiva 97/23/CE*, Ed. Maggio.

Directive 2009/105/EC of the European Parliament and of the Council of 16 September 2009 concerning: *Simple pressure vessels* (SPVD).

Directive 75/324/EEC of the European Parliament and of the Council of 20 May 1975 concerning: *Aerosol dispensers* (ADD).

Directive 97/23/EC of the European Parliament and of the Council of 29 May 1997 concerning pressure equipment and gas appliances: *Pressure equipment directive* (PED).

Directive 99/36/EC of the European Parliament and of the Council of 29 April 1999 concerning: *Transportable pressure equipment* (TPED).

EN 13445-1 (2009). *Unfired pressure vessels, Part* 1: *General.*

EN 13445-1 (2009). *Unfired pressure vessels, Part* 2: *materials.*

EN 13445-1 (2009). *Unfired pressure vessels, Part* 3: *Design.*

EN 13445-1 (2009). *Unfired pressure vessels, Part* 4: *Fabrication.*

EN 13445-1 (2009). *Unfired pressure vessels, Part* 5: *Inspection and testing.*

EN 13445-1 (2009). *Unfired pressure vessels, Part* 6: *Requirements for the design and fabrication of pressure vessels and pressure parts constructed from spheroidal graphite cast iron.*

EN 13445-1 (2009). *Unfired pressure vessels, Part* 8: *Additional requirements for pressure vessels of aluminium and aluminium alloys.*

EN 1990 (2002). *Eurocode*: *Basis of structural design* with *appendix A*1: 2005.

EN 1991 (2002/2006). *Eurocode* 1: *Actions on structures.*

EN 1993-1-1 (2005). *Eurocode* 3: *Design of steel structures- Part* 1-1: *General rules and rules for buildings.*

EN 1993-1-3 (2006). *Eurocode* 3: *Design of steel structures- Part* 1-3: *General rules-supplementary rules for cold-formed members and sheeting.*

EN 1993-1-5 (2006). *Eurocode* 3: *Design of steel structures- Part* 1-5: *Plated structural elements.*

EN 764-1 (2004). *Pressure equipment: Part* 1: *Terminology—pressure, temperature, volume, nominal size.*

EN 764-2 (2002). *Pressure equipment: Part* 2: *Quantities, symbols and units.*

EN 764-3 (2002). *Pressure equipment: Part* 3: *Definition of parties involved.*

EN CEN/TR 13445-1 (2009). *Unfired pressure vessels, Part* 9: *Conformance of EN* 13445 *Series to ISO* 16528.

EN CR 13445-1 (2009). *Unfired pressure vessels, Part* 7: *Guidance of the use of conformity assessment procedures.*

EPERC Bulletin Nr 2. (October 1999). European approach to pressure equipment inspection. In J. B. Veyret, & G. Baylac (Ed.), *European Commission JRC*, NL-1755 ZG Petten, S.P.I. 192.

EPERC Bulletin Nr 4. (June 2001). European R&D on fatique strength and hydrotest for pressure equipment, European Commission JRC, NL-1755 ZG Petten, S.P.P. 01.42.

Report 5, *Assessment Procedures for the High Temperature Response of Structures.* UK: Nuclear Electric, 1991.

Report 6, *Assessment of the Integrity of Structures Containing Defects.* UK: Nuclear Electric, 1991.

Report ASME STP—PT—007 (2006). *Comparison of pressure vessels codes ASME Section VIII and EN 13445,* Technical, Commercial, and Usage Comparison, Design Fatigue Life Comparison, ASME Standards Techcnology, LLC.

TEMA (Tubular Exchanger Manufacturers Association, Inc.) Standards. (2007). *Shell and tube heat exchanger codes and standards,* (9th ed.). TEMA Standards.

Name Index

V. Vullo, *Circular Cylinders and Pressure Vessels*, Springer Series in Solid and Structural Mechanics 3, DOI: 10.1007/978-3-319-00690-1,

Subject Index

V. Vullo, *Circular Cylinders and Pressure Vessels*, Springer Series in Solid and Structural Mechanics 3, DOI: 10.1007/978-3-319-00690-1,

D

E

I

T

CPSIA information can be obtained
at www.ICGtesting.com
Printed in the USA
LVHW081516110522
718514LV00003B/12

9 783319 006895